Springer Series in Optical Sciences

Volume 188

For further volumes:
http://www.springer.com/series/624

Springer Series in Optical Sciences

The Springer Series in Optical Sciences, under the leadership of Editor-in-Chief William T. Rhodes, Georgia Institute of Technology, USA, provides an expanding selection of research monographs in all major areas of optics: lasers and quantum optics, ultrafast phenomena, optical spectroscopy techniques, optoelectronics, quantum information, information optics, applied laser technology, industrial applications, and other topics of contemporary interest.

With this broad coverage of topics, the series is of use to all research scientists and engineers who need up-to-date reference books.

The editors encourage prospective authors to correspond with them in advance of submitting a manuscript. Submission of manuscripts should be made to the Editor-in-Chief or one of the Editors. See also www.springer.com/series/624

Axel Donges · Reinhard Noll

Laser Measurement Technology

Fundamentals and Applications

Springer

Axel Donges
nta Hochschule Isny—University of Applied Sciences
Isny
Germany

Reinhard Noll
Fraunhofer-Institut für Lasertechnik
Aachen
Germany

ISSN 0342-4111 ISSN 1556-1534 (electronic)
ISBN 978-3-662-43633-2 ISBN 978-3-662-43634-9 (eBook)
DOI 10.1007/978-3-662-43634-9
Springer Heidelberg New York Dordrecht London

Library of Congress Control Number: 2014942164

Printed on acid-free paper

Springer is part of Springer Science+Business Media (www.springer.com)

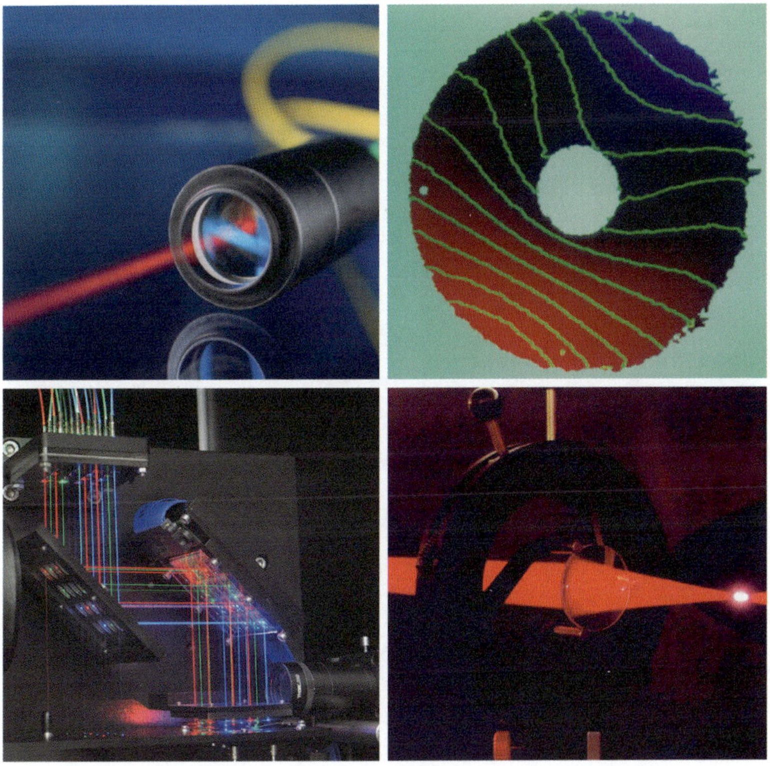

Top left measuring head of a sensor based on optical coherence tomography. *Top right* speckle interferogram showing the out-of-plane deformation field of a measuring object. *Bottom left* multi-wavelength excitation of fluoresceing particles in a microfluidic chip. *Bottom right* focusing of a nanosecond laser pulse inducing a plasma in the atmosphere at irradiances $> 10^7$ W/cm^2

Preface

In the past decade, a rapid development of laser measurement technology took place. Powerful laser measuring methods are available for a variety of measuring and inspection tasks. Their applicability and benefit were demonstrated successfully for offline and inline measurements of physical and chemical quantities. The underlying dominant characteristics of laser measuring methods are simply the fact, that lasers measure without contact and with the velocity of light.

A clear trend originating in the early 1990s of the passed century becomes more and more obvious. The measuring equipment is no longer a delicate instrumentation to be operated in laboratory or measuring rooms only, but it is transferred into production facilities being installed close or even inside a production line. By this, laser measuring methods enable inline acquisition of measurands allowing for a fast insight into the process or product and thus represent the key component for closed loop concepts where the measuring information gained directly in the production line is used for prompt feedback actions to upstream or downstream process stages. In this sense, laser measurement technology creates a novel era of process transparency opening new potentials for process optimization and quality control in industry. This far-reaching potential was not accessible so far.

Scientists and engineers in research, development, and production should therefore study laser measurement technology to understand the physical principles, the set-up of laser measurement equipment and to get an overview of its application potentials.

In this book, the fundamentals of laser measurement technology are described in Chaps. 1–5, laser measuring methods and their applications are subject of Chaps. 6–14. Target audience are physicists and engineers in the job as well as students of applied physics and engineering sciences at universities. The fundamentals, methods, and applications are described to such an extent that the reader is able to track and assess future developments of laser measurement technology. The required previous knowledge is limited to the fundamentals of physics as those teached in the first semesters at universities.

The authors came in touch with laser technology in the early 1980s of the last century as doctoral candidates at the Institute for Applied Physics at the Technische Hochschule Darmstadt. Fascinated by the potentials of laser radiation they took different professional careers. A.D. moved to the educational sector and became a Professor and now dean for physics and optics at the nta Hochschule

Isny. R.N. started at the newly founded Fraunhofer Institute for Laser Technology (ILT) in Aachen and is now head of the competence area laser measurement technology and EUV sources. He has transferred various laser measurement techniques to industrial use and stimulated the foundation of new companies to develop and market new laser measurement devices. Since 2012 he is an Assistant Professor at the RWTH Aachen university. For this book, the authors bring in their comprehensive experience in teaching, research, development, and application in the field of laser measurement technology.

The authors want to express their special thanks to their wives for their patience and encouragement to enable this book project.

Isny, Aachen

Axel Donges
Reinhard Noll

Contents

Chapter 1
Introduction

Abstract In the introductory chapter, we first give an overview of the evolution of optical and laser metrology. Then we outline the wide range of laser measurement technology. Finally, we point to the fact, that laser metrology is a non-contact measurement with high flexibility, high measuring speed and high precision.

1.1 Optical Metrology and Laser Measurement Technology

Today there are plenty of different measurement technologies available that one single scientist or engineer can hardly overlook. Hence measurement technologies are divided into different sub-disciplines. Besides electrical metrology a very important discipline is physical metrology. It was developed in need of special measurement instruments—such as thermometers, clocks, scales or microscopes—for physics and engineering. Nowadays physical metrology is divided into very different autonomous areas such as nuclear physics metrology, acoustic measurement technology or optical metrology.

Optical measurement methods have gained a prominent importance for metrology. Although optical measurement technology is a rapidly growing area, it is not a new discipline. The development of physical sciences has been affected from the very beginning by optical measurement techniques. With the invention of the microscope in the late 16th century and the telescope in the early 17th century, optical measurement technology has contributed significantly to the study of micro- and macrocosm. Other milestones in optical metrology have been the development of photography and spectroscopy in the 19th century. The latter offers important insights into the structure of matter. In the 20th century optical metrology has provided significant contributions to the development of science. This is evidenced inter alia by the awarding of Nobel Prizes for the development of interferometric precision instruments for spectroscopic and metrological studies

A. Donges and R. Noll, *Laser Measurement Technology*, Springer Series in Optical Sciences 188, DOI: 10.1007/978-3-662-43634-9_1,

(A. Michelson), of Raman scattering (C. Raman), of phase contrast microscopy (F. Zernike), of holography (D. Gabor), nonlinear optics with lasers (N. Bloembergen, A. Schawlow), of ultraprecise laser spectroscopy (J. Hall, T. Hänsch) and CCD sensors (charged coupled devices to detect optical radiation; W. Boyle, G. Smith).

In 1960 Theodore Maiman realized the first laser. Lasers are currently used for various applications, e.g. in communication technology, material processing and medicine. With the compact disc, the laser is to be found even in many households. Due to the laser, some conventional methods were substituted by novel technologies. Laser ablation or laser additive manufacturing (LAM) are examples: These processes are capable to produce shapes of workpieces, which cannot be realized with conventional machining. Generally, laser technology is now considered to be a key technology, comparable to microelectronics.

Due to its special characteristics, laser beams are also versatile measurement tools. Immediately after the invention of the laser, a new branch of measurement technology was established: laser measurement technology. Laser measurement technology has developed dynamically in recent years. Its applications range from production engineering via process engineering, quality assurance, recycling technology, environmental protection, biotechnology up to medical technology. The characteristics of laser measurement technology are:

- non-contact measurement,
- high flexibility,
- high measuring speed and
- high precision.

Laser-based measurement methods are increasingly used in manufacturing processes where a routine testing of each produced good is required. The growing importance of process and quality control in manufacturing has accelerated sustainably the development of laser measurement technology in many areas.

The capability of non-contact measurements is often decisive to deploy laser measuring systems in production lines. Lasers are measuring quality-critical features without interrupting the production flow, thus providing in realtime data for direct process control. Utilization of this inline capability enables to automate test procedures straightforwardly, to optimize production processes and to improve product quality significantly in a variety of applications.

The aim of this book is to present the basics of laser measurement technology. In Chaps. 1–5, the physical principles of laser technology are discussed. Chapters 6–14 are dedicated to different laser measurement techniques. From the variety of methods used today, we have made a selection. The selection criteria are technological maturity and commercial availability. For each laser measurement method described, we present typical applications. The Appendix offers supplemental information on the topics interference visibility and spectral radiance as well as some derivations of formulas. A list of abbreviations and symbols completes the Appendix.

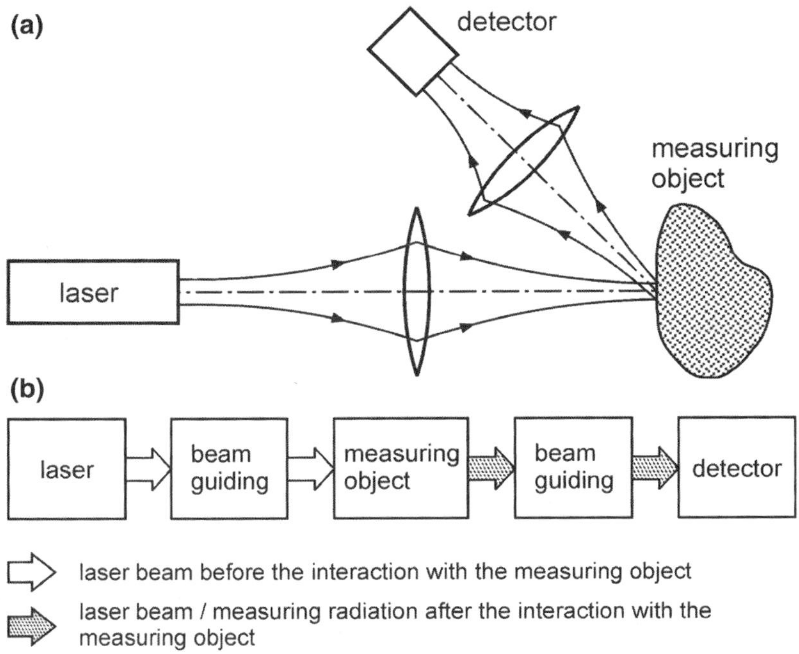

Fig. 1.1 Laser measuring system: **a** principal set-up, **b** block diagram

1.2 Schematic Set-up

Figure 1.1a shows the basic set-up of a typical laser measuring system. The three main components are:

- laser,
- measuring object and
- detector.

The radiation of the laser interacts with the object to be measured. During this interaction, information of the measuring object is impressed to the laser radiation. A part of the radiation reaches the detector, which converts the information stored in the light into a recognizable form for the observer. Between laser and measuring object and between measuring object and detector, elements for beam shaping and guiding are arranged in general. Examples of such elements are focusing optics, beam expanders or optical fibers.

The components shown in Fig. 1.1b are described in Chaps. 2–5. Chapter 2 provides an overview of the properties of laser radiation. The different possibilities of interaction between laser radiation and measuring object are discussed in the third chapter. In Chap. 4, the main elements of beam shaping and guiding are presented. Chapter 5 describes the most common radiation detectors. The knowledge gained in the first part of the book, is necessary to understand the laser measurement methods treated in Chaps. 6–14. In addition, it provides the necessary basic knowledge to follow competently the latest developments in laser measurement technology.

Chapter 2
Properties of Laser Radiation

Abstract At first, we summarize the properties of a transverse electromagnetic wave. This is followed by the discussion of the diffraction of light. We present the concept of coherence—temporal and spatial—and compare laser radiation with thermal light. After that, we treat the Gaussian beam and the higher transverse modes. An overview of typical laser parameters is given. Finally, we deal with the dangers of laser radiation.

2.1 Light as Electromagnetic Wave

The total of all electromagnetic phenomena is governed by Maxwell's equations. They link the temporal and spatial derivatives of the electric and the magnetic field strength, $\vec{E}$ and $\vec{H}$ [1–5]. Starting from Maxwell's equations, J. C. Maxwell predicted the existence of electromagnetic waves that propagate with the speed of light. During the years 1887/88, H. Hertz succeeded in the generation and detection of electromagnetic waves. So, light was recognized as a special case of an electromagnetic wave.

An overview of the investigated and mostly technically used frequency or wavelength spectrum of electromagnetic waves gives Fig. 2.1. The laser radiation—usually $\lambda = 0.1$–$10\ \mu\text{m}$—represents only a small part of the total spectrum.

In the simplest case, a laser beam is represented by a plane harmonic wave

$$\vec{E}(z,t) = \vec{E}_0 \cos(\omega t - kz - \varphi_0), \tag{2.1a}$$

$$\vec{H}(z,t) = \vec{H}_0 \cos(\omega t - kz - \varphi_0) \tag{2.1b}$$

that propagates in z-direction with $\vec{E}$ electric field strength vector, $\vec{H}$ magnetic field strength vector; $\vec{E}_0, \vec{H}_0$ amplitude, ω angular frequency, k angular wave number and φ_0 phase angle. The directions of the field strength vectors and the direction of propagation of that electromagnetic wave are shown in Fig. 2.2. Light is a

A. Donges and R. Noll, *Laser Measurement Technology*, Springer Series in Optical Sciences 188, DOI: 10.1007/978-3-662-43634-9_2,

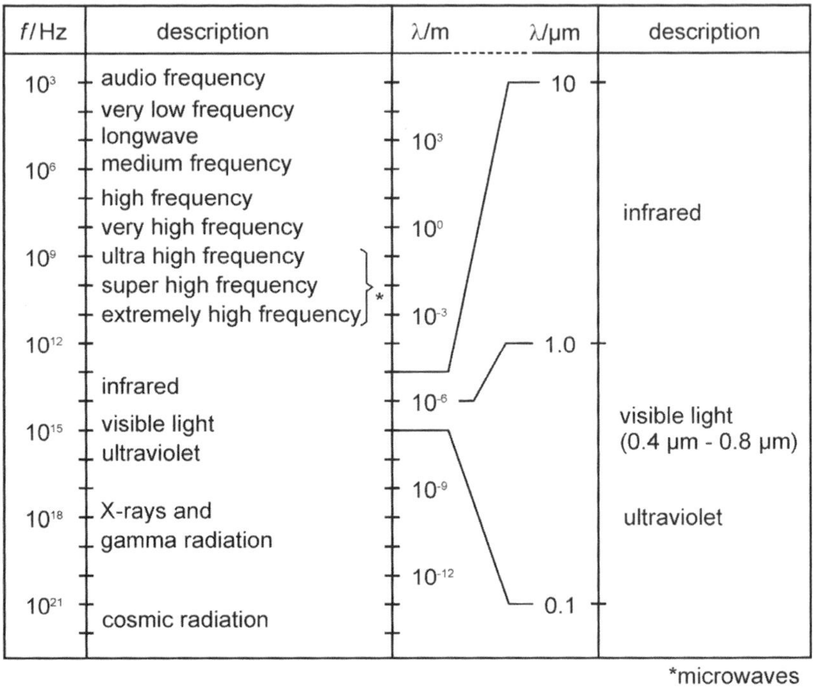

Fig. 2.1 The range of the electromagnetic spectrum

transverse wave, i.e., $\vec{E}$, $\vec{H}$ and the direction of propagation given by the wave vector $\vec{k}$ are oriented perpendicularly to each other. The direction, in which the electric field strength vector oscillates, is called the direction of polarization.[1] The angular frequency ω, the frequency f, the angular wave number k and the wavelength λ are related by the equations

$$\omega = 2\pi f \tag{2.2}$$

and

$$k = \frac{2\pi}{\lambda}. \tag{2.3}$$

The angular frequency ω and the angular wave number k or the frequency f and the wavelength λ follow the dispersion relation

$$\frac{\omega}{k} = f\lambda = c = \frac{c_0}{n}, \tag{2.4}$$

[1] Some authors define the direction of polarization as the direction of the magnetic field strength.

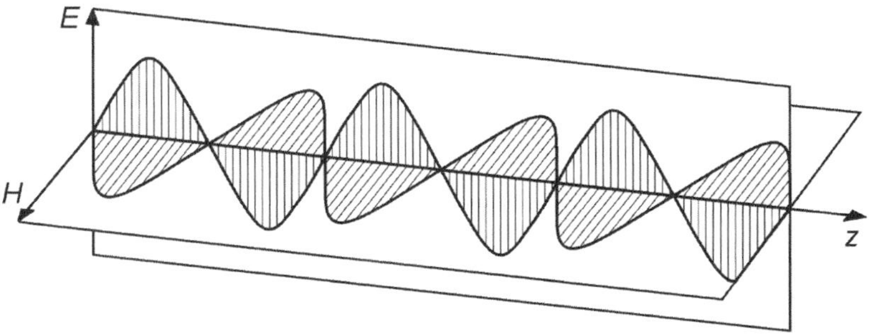

Fig. 2.2 Representation of the field strengths of a plane harmonic electromagnetic wave. The electric field strength, the magnetic field strength and the direction of propagation are oriented perpendicularly to each other. The direction of the electric field strength defines the direction of polarization

respectively. c is the phase velocity of the electromagnetic wave in a homogeneous and lossless medium. It is calculated with the vacuum speed of light c_0 and the refractive index n of the medium. The vacuum speed of light is given by $c_0 \approx 3 \times 10^8$ m/s.

The electric and magnetic field strength of a light wave are proportional to each other:

$$E \propto H. \tag{2.5}$$

So, for describing an electromagnetic wave, it is sufficient to state the electric field strength only.

An electromagnetic wave transports energy. The irradiance I, i.e., the transported power per unit area, is as usually proportional to the square of the amplitude:

$$I \propto E_0^2. \tag{2.6}$$

2.2 Beam Parameters

A plane harmonic wave is uniquely described by indicating five independent parameters. A complete set of independent parameters are, for example,

- irradiance,
- phase angle,
- direction of propagation,
- wavelength and
- direction of polarization.

For many applications in metrology, laser beam parameters have to be controlled actively. The necessary optical components, which can be used to modulate mostly without delay the five beam parameters, are presented in Chap. 4.

By the interaction of the laser beam with the measuring object, the beam parameters, e.g. the wavelength, are changed in a characteristic way. It is the task of laser measurement technology to detect these changes and to deduce the properties or the state of the measuring object. In Chaps. 6–14 it is shown in many examples, how this can be performed. First, we restrict ourselves to the presentation of the measuring methods to detect the laser beam parameters.

2.2.1 *Irradiance*

The frequencies of the visible spectrum are in the range of some 10^{14} Hz. Therefore it is impossible to detect the temporal field strength curve of a laser beam by eye or electro-optical detectors. Metrologically accessible, however, is the irradiance, averaged over many periods of oscillation, which is proportional to the square of the amplitude, cf. (2.6). In the following we will use the terms "intensity" and "irradiance" as synonyms, they have the unit $\mathrm{W/m^2}$. For measuring the irradiance metrology provides a plurality of radiation detectors. All radiation detectors are based on the same principle: the absorption of radiation. The absorbed light energy, depending on the type of detector, can be measured in several ways. Since the irradiance of a laser beam is a very important radiometrical parameter, we dedicated an entire Chapter—Chap. 5—to these detectors.

2.2.2 *Phase Angle*

There are no detectors, which measure the phase angle of a light wave in a direct way. Nevertheless, the phase angle can be detected indirectly by superposition of light waves.

When two or more waves are superposed, then the resulting field strength is obtained by vector addition of the individual field strengths, this is the so called principle of superposition. For two harmonic waves

$$E_1 = E_0 \cos(\omega t - kz) \tag{2.7a}$$

and

$$E_2 = E_0 \cos(\omega t - kz - \varphi_0), \tag{2.7b}$$

which differ only in terms of the phase angle φ_0, the principle of superposition reads

$$E = E_1 + E_2 = 2E_0 \cos\left(\frac{\varphi_0}{2}\right) \cos\left(\omega t - kz - \frac{\varphi_0}{2}\right). \tag{2.8}$$

Since the irradiance of the resultant wave is proportional to the square of the amplitude, we find for the irradiance of the resultant wave

$$I = 4I_0 \cos^2\left(\frac{\varphi_0}{2}\right), \tag{2.9}$$

where I_0 is the irradiance of the wave E_1 or E_2, respectively.

Figure 2.3 illustrates two limiting cases: (a) For $\varphi_0 = 0$, the two waves are in phase. Then the field strength has doubled and the irradiance quadrupled. This case is known as constructive interference. In the case (b) of $\varphi_0 = \pi$, the two field strengths oscillate in opposite phase, and the resulting field strength and irradiance are zero. This case is called destructive interference.

Using relation (2.9) for the resulting wave we can calculate the phase angle φ_0 of wave E_2. The wave E_1 defines the zero phase ($\varphi_0 = 0$) and is referred to as the reference wave. This principle of phase measurement is the basis of holography, which is discussed in Chap. 7.

2.2.3 Direction of Propagation

Let us consider two light beams which differ only in their directions of propagation. In order to measure the angle α between the two propagation directions, we can detect the distances d_1, d_2 and L, illustrated in Fig. 2.4. For the angle α we find

$$\tan\left(\frac{\alpha}{2}\right) = \frac{d_2 - d_1}{2L}. \tag{2.10a}$$

If $|\alpha| \ll 1$, the result simplifies to

$$\alpha = \frac{d_2 - d_1}{L}. \tag{2.10b}$$

For very small angles α, this method is due to measurement errors, too inaccurate. We therefore consider another way of measuring, based on the superposition principle. In Fig. 2.5 the principal set-up is sketched. Two plane waves, which differ only with respect to their propagation direction, are superposed. This creates an equidistant fringe pattern, which can be recorded on a photographic plate. From the spacing Δx of the interference fringes the angle α is deduced. For sufficiently small angles, $|\alpha| \ll 1$, it follows

$$\alpha = \frac{\lambda}{\Delta x}, \tag{2.11}$$

as a simple geometric argument shows. The smaller the angle α between the two waves, the greater the spacing Δx of the interference fringes. By measurement of

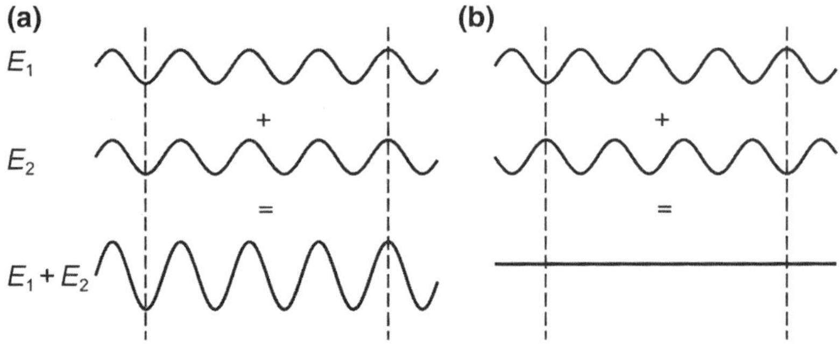

Fig. 2.3 Constructive (**a**) and destructive (**b**) interference of two harmonic waves

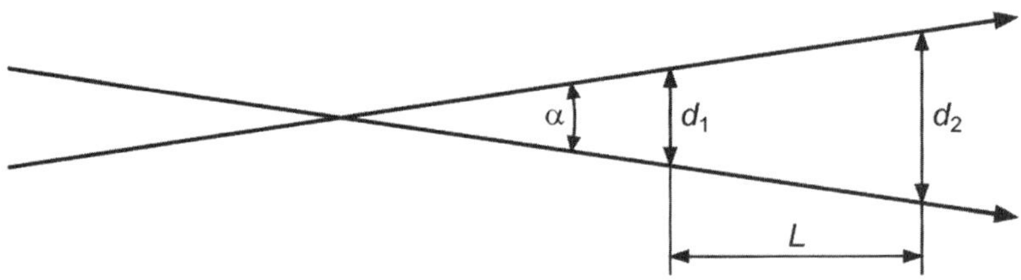

Fig. 2.4 Measuring the angle between two light beams

the fringe spacing Δx, it is possible to determine the angle α between the propagation directions of the two light waves.

There is the question whether an arbitrarily small angle α can be measured. In answering this question, it should be noted that each of the two light waves has a finite diameter $2a$, see Fig. 2.5a. The lower limit for the smallest measurable angle is that the fringe spacing Δx can not be greater than the diameter $2a$. That is why

$$\alpha_{\min} = \frac{\lambda}{2a}. \tag{2.12}$$

Two wave trains, which include an angle α that is not greater than $\alpha_{\min}$, can thus, in principle with respect to the direction of propagation, not be distinguished [6]. The two waves are to be considered as being parallel. In this context, the critical angle $\alpha_{\min} = \lambda/2a$ is referred to as diffraction angle. We will return to this issue in Sect. 2.3.

2.2.4 Wavelength

The wavelength λ of a laser beam can be measured in various ways. One possibility is the diffraction by a grating, see Sect. 2.3. But much more accurate are interferometric measurements, e.g. with a Michelson interferometer, described in Sect. 2.4.2.

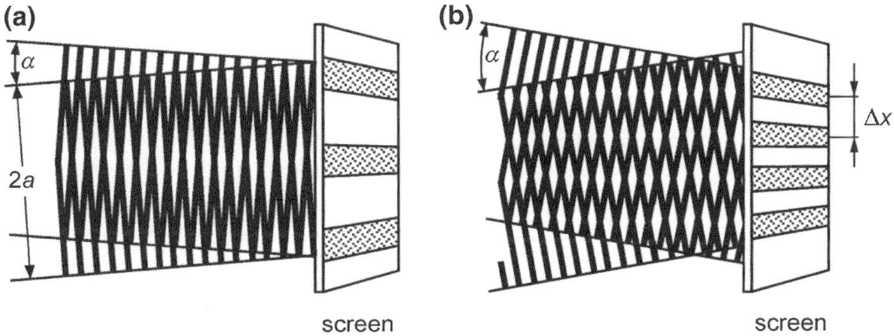

Fig. 2.5 Superposition of two light waves with different propagation directions: **a** $\alpha = 10°$, **b** $\alpha = 20°$

The relative wavelength shifts $\Delta\lambda/\lambda$, occurring in laser measurement technology, are usually very small. A determination of $\Delta\lambda = \lambda_2 - \lambda_1$ by absolute measurements of λ_1 and λ_2 is—due to unavoidable measurement errors—not possible. That is why a direct measurement of the wavelength shift $\Delta\lambda$ is proposed.

A common method for direct measurement of small changes in wavelength is based on the principle of superposition, as shown in Fig. 2.6. Two otherwise identical waves, which differ only slightly with respect to their wavelengths, are superimposed. The wavelengths of the waves are λ and $\lambda + \Delta\lambda$. Then, the irradiance of the resulting wave is modulated in time, cf. (2.4). This phenomenon is known as beating. The time interval between two neighboring maximum values of the irradiance, the so-called beat period T_b is given by:

$$T_b = \frac{\lambda^2}{c\,\Delta\lambda}. \tag{2.13}$$

Measuring T_b allows to determine the absolute wavelength shift $\Delta\lambda$, provided the absolute wavelength λ is known.

The uncertainty of the experimental determination of the beat period T_b is reduced, the more intensity maxima and minima will be involved. The lower limit of the measurement period τ is the beat period T_b itself. That is why for

$$\tau < T_b = \frac{\lambda^2}{c\,\Delta\lambda} \tag{2.14}$$

a determination of the beat period T_b is no longer possible. The two wavelengths λ and $\lambda + \Delta\lambda$ are then, in principle, indistinguishable [6].

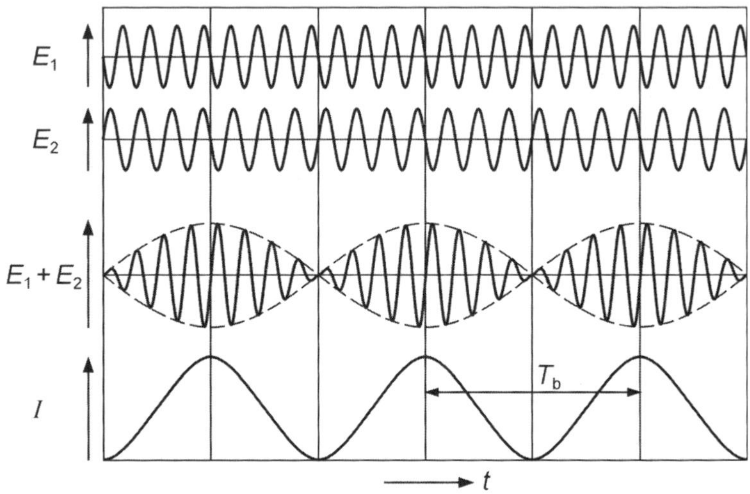

Fig. 2.6 Superposition of two waves with slightly different frequencies

2.2.5 *Polarization*

In a linearly polarized wave, the electric field strength vector oscillates perpendicularly to the propagation direction in a defined direction, see Fig. 2.2. The vibration direction of the electric field strength is referred to as the direction of polarization.

For the experimental determination of the polarization direction of a linearly polarized wave we can use a polarizing filter. Such a filter is an optical element, which is, in best case, transparent for light of a certain polarization direction. If the polarization direction of the light wave and the forward direction of the polarization filter are oriented perpendicularly to each other, the transmittance decreases, ideally, to zero.

If two orthogonal linearly polarized light waves of the same frequency are superimposed, the result is in general elliptically polarized light. An important special case is circularly polarized light. In this case, the propagation direction is also perpendicular to the field vectors, which form, however, a helical structure. We will return to this in Sect. 4.1.2.

In addition to polarized light, there is also non-polarized light. For non-polarized light, the electric field strength vectors oscillate perpendicularly to the propagation direction, too. However, the field strength vectors change quickly their directions of oscillation in a random manner, so that no preferential direction can be observed. This is also partly the case with laser light. However, it is possible to force a defined direction of polarization by installing polarizing elements in the laser cavity. Polarizing elements are often present in a laser, so that the most lasers emit polarized or partly polarized light.

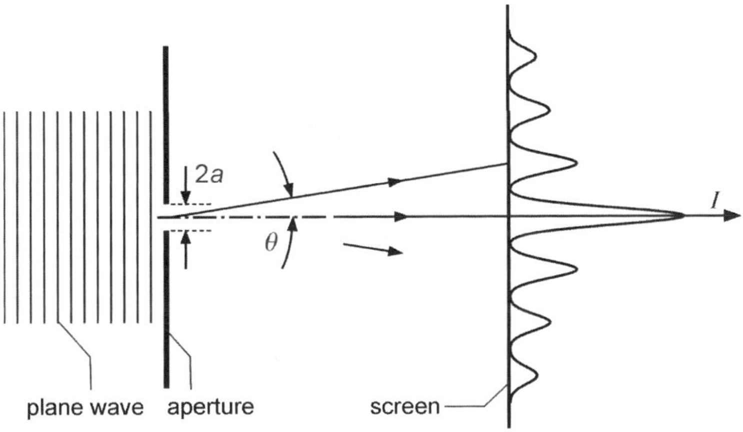

Fig. 2.7 Diffraction of a plane wave by a pinhole

2.3 Diffraction

Light shows, like any other wave, the phenomenon of diffraction. This means that there are deviations of light propagation as described by the laws of geometrical optics. Any transverse limitation of a light beam causes diffraction. Figure 2.7 shows, as an example, a parallel light beam, falling on a pinhole. Behind this aperture, the transmitted light diverges and enters the area of the geometrical shadow (the borders of this geometrical shadow are indicated in Fig. 2.7 as horizontal dashed lines just behind the aperture). The reader can easily convince himself of diffraction by looking through a small hole at a distant light source. The light source is enlarged relative to the direct observation and surrounded by colored rings.

A clear explanation of the diffraction provides the so-called Huygens-Fresnel principle. It says that each point of a wavefront is the starting point of a spherical secondary wave of the same frequency, see Fig. 2.8. These secondary waves are not radiated uniformly in all directions. The directionality of the secondary emissions are described by the inclination factor:

$$K(\theta) = \frac{1}{2}(1 + \cos\theta). \tag{2.15}$$

θ is the angle made with the normal to the primary wavefront. K has its maximum in forward direction ($\theta = 0$) and vanishes in backward direction ($\theta = \pi$).

The wave field at any point behind the aperture results from the superposition of all these secondary waves. The secondary waves travel, in general, different long distances. Thus, as a result of constructive or destructive interferences, a more or less complicated interference pattern, the so-called diffraction pattern, will occur.

In the following, we consider only the so-called Fraunhofer diffraction, which is—compared to the general case of Fresnel diffraction—mathematically simpler to describe. Fraunhofer diffraction occurs, when the incident and diffracted light consist of parallel light, respectively. Fraunhofer diffraction is realized if both, the

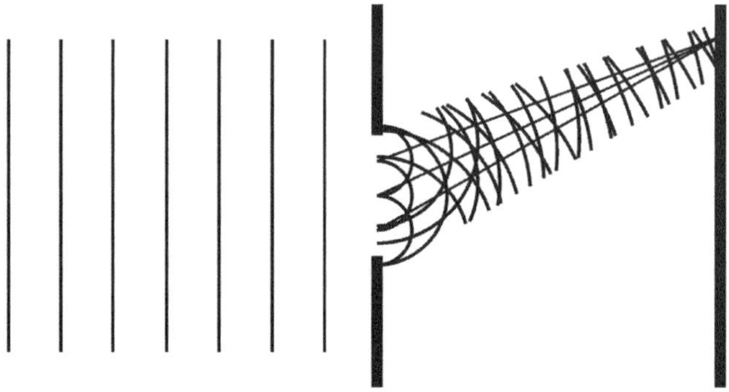

Fig. 2.8 Explanation of diffraction by the Huygens-Fresnel principle

light source and the observation screen, are located very far away from the diffracting aperture. In practice, it is sufficient, if the distance R between the light source and the diffracting aperture and the distance L between the aperture and the observation screen fulfill the condition:

$$R, L \gg \frac{a^2}{\lambda}, \tag{2.16}$$

respectively. Hereby is a a typical dimension of the diffracting aperture, see Fig. 2.7, e.g. in the case of a pinhole the pinhole radius, and λ is the wavelength of the monochromatic light. Another way to realize Fraunhofer diffraction is to place the light source and the viewing screen in the focal planes of collecting lenses, respectively.

2.3.1 Single Slit

Let us consider, as an example, Fraunhofer diffraction of a monochromatic plane wave irradiating a single slit. The application of the Huygens-Fresnel's principle provides, in accordance with the experiments for large observation distance, the normalized intensity curve, see Fig. 2.9:

$$\frac{I(\theta)}{I_0} = \left(\frac{\sin \alpha}{\alpha}\right)^2, \tag{2.17a}$$

with θ observation angle, $2a$ slit width, λ wavelength and the abbreviation:

$$\alpha = \frac{2\pi}{\lambda} a \sin(\theta). \tag{2.17b}$$

The first minimum of the intensity occurs at an angle:

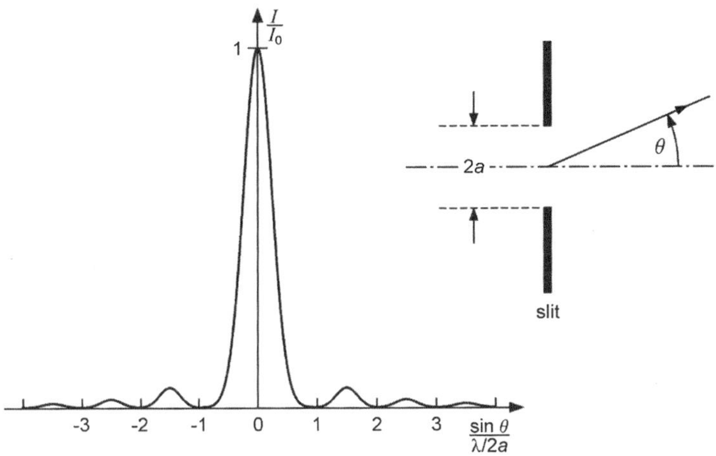

Fig. 2.9 The Fraunhofer diffraction pattern of a single slit

Fig. 2.10 The incident beam is divided into many partial waves

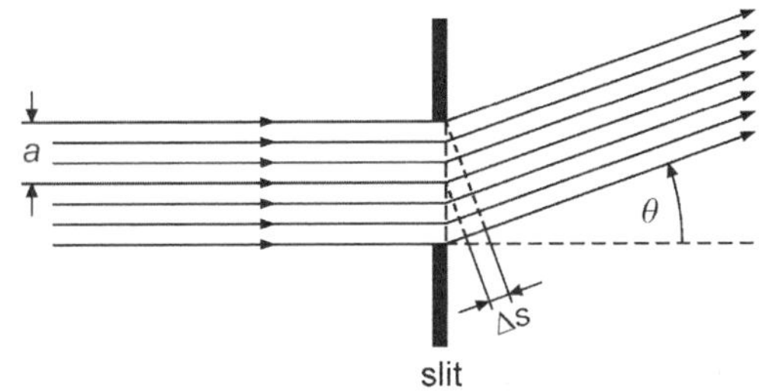

$$\sin(\theta_d) = \frac{\lambda}{2a}. \tag{2.18}$$

θ_d is called diffraction angle. It is usually small, so that we mostly can use the approximation $\sin(\theta_d) \approx \theta_d$. This diffraction angle is identical to the critical angle α_{min} given in relation (2.12) of Sect. 2.2.3.

A clear explanation for the total annihilation of the waves at the diffraction angle provides the following simple argument: We suppose that the incident wave, as shown in Fig. 2.10, is divided into a large number of rays. The path difference Δs between two rays at a distance a is

$$\Delta s = a \sin(\theta). \tag{2.19}$$

If this path difference is just one half wavelength

$$\Delta s = \frac{\lambda}{2}, \tag{2.20}$$

then a complete destructive interference of all partial waves results. The use of (2.20) and (2.19) yields the desired diffraction angle θ_d.

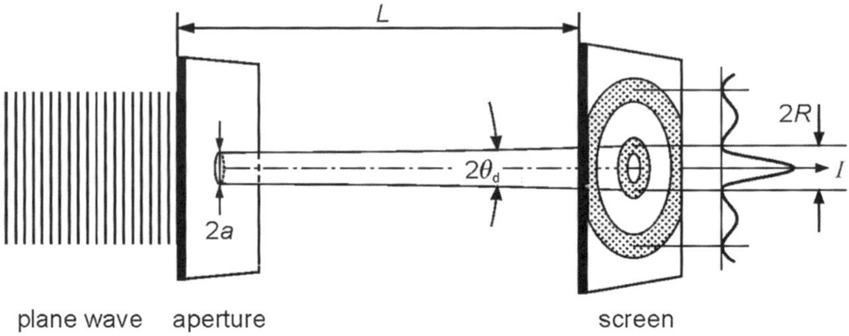

Fig. 2.11 Diffraction by a pinhole

2.3.2 Circular Aperture

For laser measurement technology, the diffraction at a circular aperture is of great interest, see Fig. 2.11. For reasons of symmetry, a rotationally symmetric diffraction pattern is created. The application of the Huygens-Fresnel principle provides in Fraunhofer approximation:

$$\frac{I(\theta)}{I_0} = \left(\frac{2\mathrm{J}_1(\alpha)}{\alpha}\right)^2, \tag{2.21}$$

where J_1 is the Bessel function of 1st order and $\alpha = \frac{2\pi}{\lambda} a \sin(\theta)$ with a radius of the pinhole. The first dark ring caused by destructive interference occurs at the angle:

$$\sin\theta_\mathrm{d} = 1.22\frac{\lambda}{2a}. \tag{2.22}$$

Hence, the first dark ring on the screen has a radius of:

$$R = 0.61\frac{\lambda L}{a}. \tag{2.23}$$

Hereby it was assumed that $a \gg \lambda$ applies. About 84 % of the incident power hits the screen in the so-called diffraction disk, whose border is defined by the first dark ring. This disk is also called the Airy disk.

2.3.3 Double Slit

A double slit consists of two identical slits each having a width of $2a$, respectively, which are arranged at a distance of $2b$, see Fig. 2.12. The mathematical treatment provides for the intensity distribution of the diffracted light:

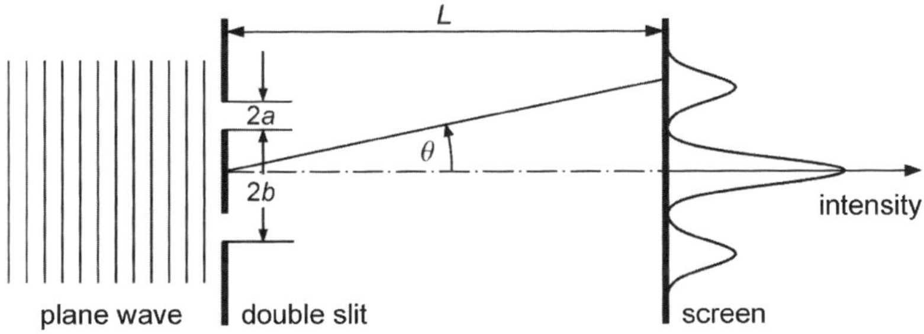

Fig. 2.12 Diffraction at a double slit

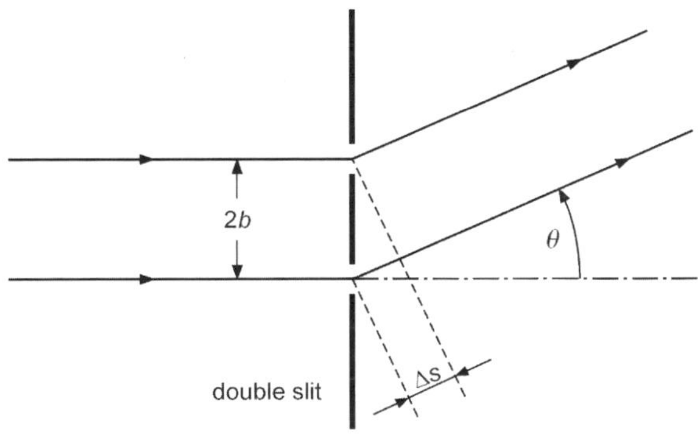

Fig. 2.13 For determination of the diffraction maxima of a double slit

$$\frac{I(\theta)}{I_0} = \left(\frac{\sin\alpha}{\alpha}\right)^2 \cos^2\beta. \tag{2.24a}$$

The quantity α has the same meaning as in the case of the single slit, see relation (2.17b). β is the abbreviation for:

$$\beta = \frac{2\pi}{\lambda} b \sin\theta. \tag{2.24b}$$

The intensity profile is illustrated in Fig. 2.12.

The positions of the maxima are clearly determined by the fact that the diffracted light waves in the two slits must have a path difference Δs, which is an integer multiple of the wavelength, see Fig. 2.13. The positions of the maxima are, therefore, in accordance with (2.24a, 2.24b), given by:

$$\sin(\theta) = n\frac{\lambda}{2b} \qquad n = 0, \pm 1, \pm 2, \ldots, |n| \leq \frac{2b}{\lambda}. \tag{2.25a}$$

On the observation screen, which is placed at a distance of L behind the double slit, a pattern of straight fringes occurs. In the limit of small angles, these interference fringes are equidistant. The distance between neighboring fringes amounts to:

$$\Delta x = \frac{\lambda L}{2b}. \tag{2.26}$$

2.3.4 Grating

Probably the most widely applied diffraction phenomenon, especially in spectroscopy, is the diffraction by a grating. A grating consists of a large number N of individual slits, which are arranged at a distance g. The quantity g is called the grating constant. Diffraction gratings are used both, in transmission or in reflection. The positions of the main maxima of the grating coincide with the maxima of the double slit, i.e.,

$$\sin(\theta) = n\frac{\lambda}{g} \qquad n = 0, \pm 1, \pm 2, \ldots \left(|n| \leq \frac{g}{\lambda}\right). \tag{2.25b}$$

Between the main maxima still so-called secondary maxima occur, see Fig. 2.14.

Provided a sufficiently large number of individual slits, $N \gg 1$, the intensities of the secondary maxima are so small that they may be neglected. The main maxima are more narrow and high, the more slits contribute to the diffracted wave. The width of the main maxima, defined by the distance between adjacent minima, is calculated as:

$$\Delta \theta = \frac{4\pi}{N}. \tag{2.27}$$

In our previous considerations, we have always assumed diffracting objects that transmit a portion of the incident light. However, diffraction phenomena occur also in case of reflective objects or objects causing a phase shift only. As an example, a phase grating is considered.

By a phase grating, not the amplitude, but the phase of an incident wave is influenced. This can, for example, be realized by the fact that the refractive index changes periodically, see Fig. 2.15b. Equation (2.25b) applies also in the case of the phase grating.

2.4 Coherence

In nearly all measurement methods presented later, the coherence of the laser radiation plays a central role. That is why we deal this concept in the following sections.

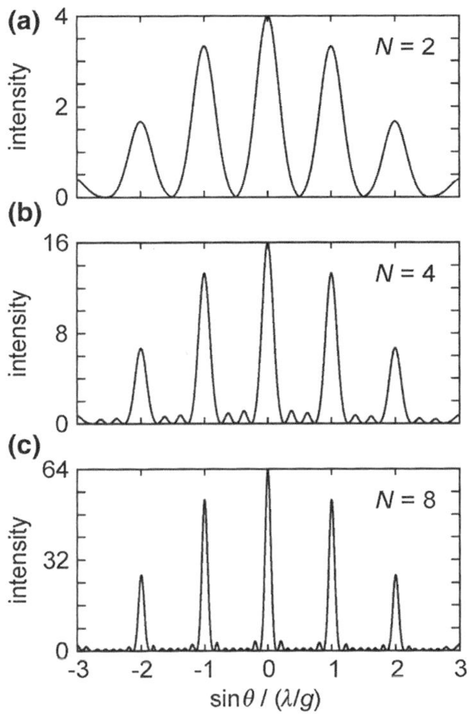

Fig. 2.14 Irradiance distribution for a grating with **a** $N = 2$, **b** 4 and **c** 8 slits. It is assumed $g = 8a$

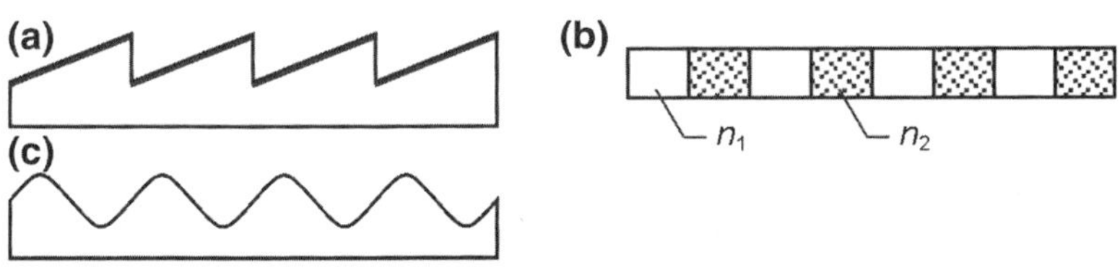

Fig. 2.15 Different phase gratings: **a** reflection grating, **b** transmission grating of constant thickness and periodically varying refractive indices, **c** transmission grating

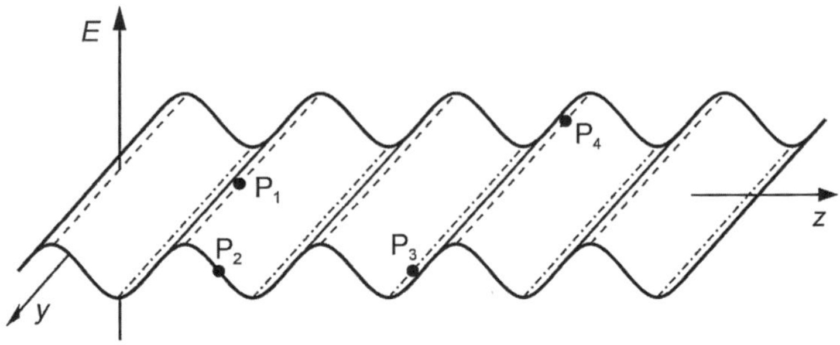

Fig. 2.16 Snapshot of the electric field strength of a plane harmonic wave. Examples: **a** The field oscillations at the two points P_1 and P_4 are in phase ($\varphi_0 = 0$); **b** The field oscillations at the two points P_1 and P_3 or P_3 and P_4 are anti-phased ($\varphi_0 = \pi$); **c** The field oscillations at the two points P_1 and P_2 have a phase shift of $\varphi_0 = \pi/2$

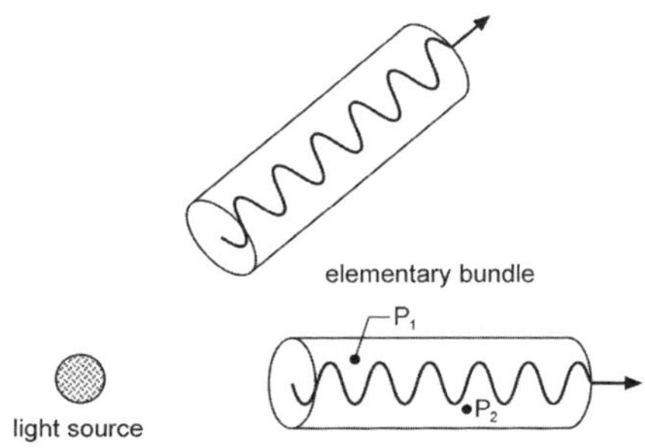

Fig. 2.17 Schematic representation of an elementary bundle. The oscillations of the electric and magnetic field at any two points within an elementary bundle have a fixed phase relationship. The shown cylindrical shape is a simplification. In general an approximately conical shape exists

2.4.1 Definition of Coherence

Coherence literally means related. A coherent light field is, therefore, a field of light, in which the oscillations of the electric and magnetic field strengths are related in space and time. One example of a coherent wave is the plane harmonic wave:

$$E = E_0 \cos(\omega t - kz), \tag{2.1a}$$

see Fig. 2.16.

In contrast to the ideal case of the plane harmonic wave, a defined phase relationship between the field oscillations at two different points P_1 and P_2 exists only, when these points are located within a so-called elementary bundle, see Fig. 2.17.

Often the terms temporal and spatial coherence are considered. The length L_c (in the direction of propagation) or the diameter d_c (perpendicular to the direction of propagation) of an elementary bundle is called its longitudinal or transverse coherence length, respectively. Since the elementary bundle moves with the wave and thus with the speed of light, at a fixed point in space only for the period of time:

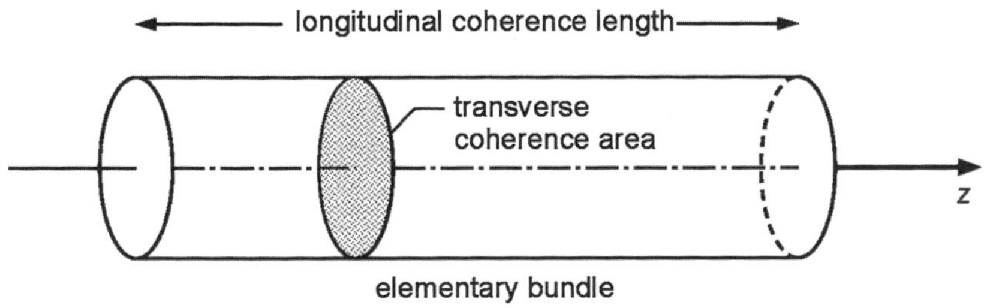

Fig. 2.18 Longitudinal coherence length and transverse coherence area

$$T_c = \frac{L_c}{c}, \tag{2.28}$$

a defined harmonic oscillation occurs, see Fig. 2.18. T_c is referred to as coherence time. The cross-sectional area of an elementary bundle is known as transverse coherence area.

2.4.2 Measurement of Temporal Coherence

The coherence time T_c and the longitudinal coherence length L_c can be measured, for example, with a Michelson interferometer, see Fig. 2.19. A light wave is divided by a beam splitter into two waves of equal intensity. After the two partial waves have traveled different long distances in the interferometer, they are superposed again. In this experiment the intensity of the superimposed partial waves is measured as a function of the path difference $2\Delta L$, with $\Delta L = L_2 - L_1$. As long as the path difference $2\Delta L$ is small compared to the temporal coherence length L_c of the radiation field, the measured intensity varies—due to constructive and destructive interference—as a function of the path difference $2\Delta L$. However, if the path difference $2\Delta L$ is large compared to the coherence length L_c, the measured intensity is independent of $2\Delta L$. The intensity is then the sum of the intensities of the two partial waves, i.e., no interference phenomenon occurs. The two partial waves are apparently losing their interference ability, when they are displaced by more than the longitudinal coherence length L_c, see Appendix A.1 for a more detailed consideration. This is intuitively clear since then wave trains with randomly fluctuating phase relation—i.e. light from different elementary bundles—are superimposed.

Experimental and theoretical studies have proven that the coherence time T_c increases with decreasing angular frequency bandwidth $\Delta\omega$ of the light source. This dependence is described approximately by the following relation [6]:

$$\Delta\omega T_c = \pi. \tag{2.29}$$

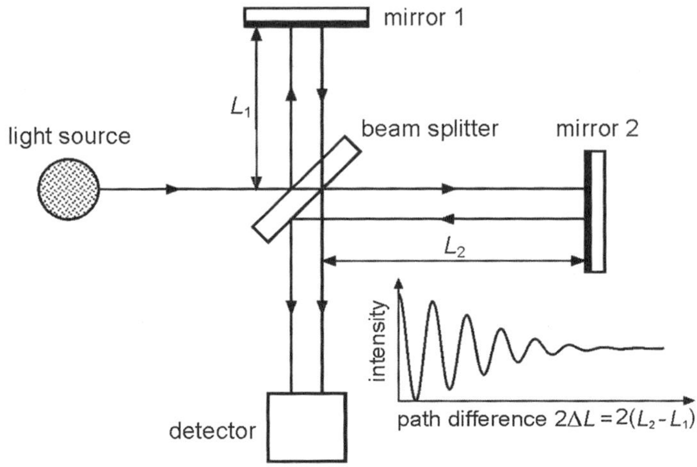

Fig. 2.19 Measurement of the longitudinal coherence length with a Michelson interferometer

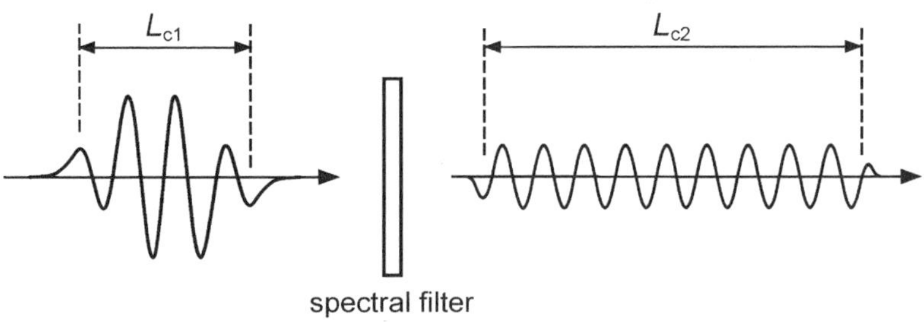

Fig. 2.20 Increasing the temporal coherence length by a spectral filter

This equation is known as temporal coherence condition. The narrower the spectrum of the light, the greater is the coherence time and thus according to (2.28) the longitudinal coherence length. This relationship is deployed often in optical measurement techniques through the use of a spectral filter. With the help of a spectral filter, the bandwidth of the passing light becomes smaller and thus the longitudinal coherence length increases. This is called temporal filtering, see Fig. 2.20, which, however, leads to considerable losses in light intensity.

2.4.3 Measurement of Spatial Coherence

The transverse coherence length d_c can be measured by Young's double-hole experiment. The radiation field to be examined illuminates two pinholes and the diffraction pattern is observed. The two pinholes are arranged at a distance $2b$ and have the same distance from the light source, see Fig. 2.21.

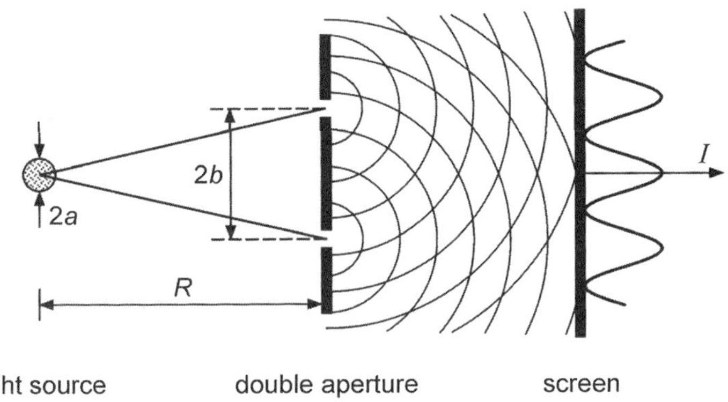

Fig. 2.21 Measurement of the spatial transverse coherence length with Young's double-hole experiment

In this experiment, an observable diffraction pattern only arises, when the distance of the two pinholes $2b$ is chosen to be sufficiently small. If the distance $2b$ increases, the contrast of the interference pattern is more and more fading. Beyond a critical distance $2b_{\max}$ of the pinholes, the interference pattern disappears. The radiation field is obviously spatially coherent, if the critical distance $2b_{\max}$ is not exceeded. That is why $2b_{\max}$ is interpreted as transverse coherence length d_c. The transverse coherence area is determined by measuring the transverse coherence length in different directions perpendicular to the propagation direction of the light.

The condition of spatial coherence is closely related to the diffraction of light. In the following we consider for the sake of simplicity the case of a circular light source. Experiments show that spatial coherence is present only if the condition:

$$1.22\frac{\lambda}{2a} \geq \frac{2b}{R} \tag{2.30}$$

is fullfilled with λ wavelength of light, a radius of the circular light source, $2b$ spacing of the two pinholes, R distance between the light source and the double aperture, cf. Fig. 2.21.

Equation (2.30) implies that light is spatially coherent, when it is observed within the diffraction angle of the light source, i.e. $1.22\lambda/2a$. The maximum distance between the two apertures, wherein the interference pattern can still just be seen, is calculated from (2.30) to:

$$d_c = 2b_{\max} = 1.22\frac{\lambda}{2a}R. \tag{2.31}$$

This is the so-called transverse coherence length. For the transverse coherence area follows in the case of a circular light source:

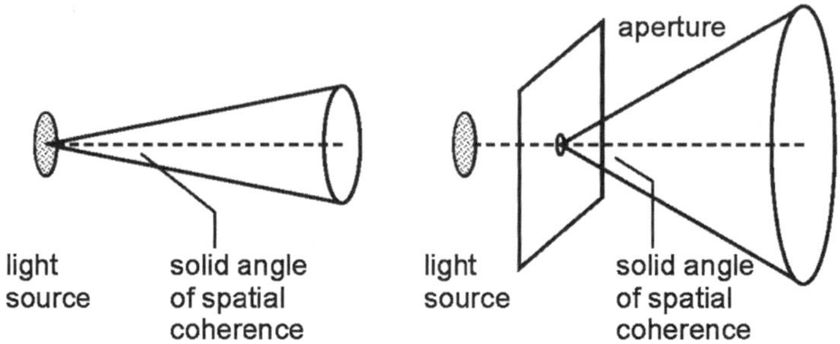

Fig. 2.22 Increasing the spatial coherence angle through an aperture whose diameter is smaller than that of the light source

$$A_c = \pi\, b_{max}^2 = 1.22^2 \pi \frac{\lambda^2}{16a^2} R^2 \approx 0.918 \frac{\lambda^2}{\pi\, a^2} R^2 \approx \frac{\lambda^2}{\pi\, a^2} R^2. \tag{2.32}$$

Rewriting this equation yields the spatial coherence condition:

$$\Delta A\, \Delta\Omega_c \approx \lambda^2. \tag{2.33}$$

Here is $\Delta A = \pi\, a^2$ the effective area of the light source and $\Delta\Omega_c = A_c/R^2$ the maximum solid angle, within which the light is spatially coherent. $\Delta\Omega_c$ is called solid coherence angle. It is determined by the diffraction of light at the finite area ΔA of the light source.

The solid coherence angle and therefore also the transverse coherence area can be increased, if with an aperture only a part of the surface of the light source is utilized, see Fig. 2.22. However, this so-called spatial filtering effect decreases, as in the case of temporal filtering, the intensity of light.

2.5 Comparison of Laser Radiation and Thermal Light

Let us compare a laser beam with the light of an ordinary light bulb. In comparison to the filament lamp, laser radiation has a low beam divergence and a small spectral bandwidth. In other words: the laser emits a nearly monochromatic and parallel light beam. The light from the bulb on the other hand is broadband—white light—and emitted in a spatially isotropic way, see Fig. 2.23.

The advantages associated with the low beam divergence and the small spectral bandwidth of laser radiation, are diverse:

- An important point are the good coherence properties of laser radiation. This is very important for all interferometric measurement methods described below—for example, the holographic interferometry, see Chap. 7.

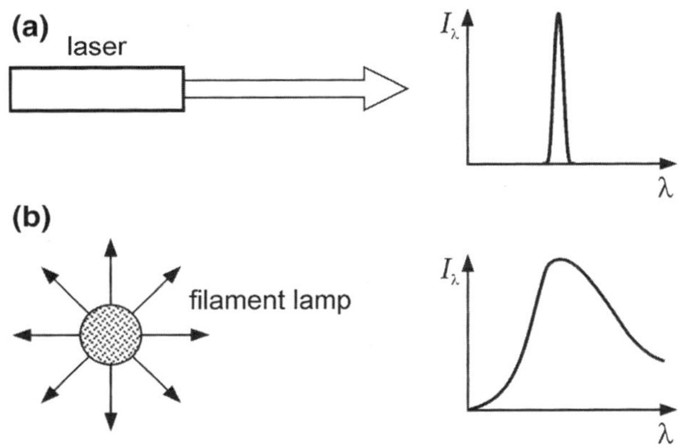

Fig. 2.23 Comparison of the radiation from a laser (**a**) and a filament lamp (**b**), see Appendix A.2 for a more detailed consideration

- Monochromatic laser light passes a narrowband filter almost unattenuated, while the incident broadband light is absorbed or reflected nearly completely. As a consequence, most detectors are equipped with narrow-band filters. Thus the laser light scattered at the measuring object arrives at the detector, while disturbing daylight is blocked, cf. Appendix A.2, Figs. A.4, A.5.
- The low beam divergence allows to guide a laser beam with an almost constant intensity over large distances.
- A further feature of the laser radiation with respect to thermal light is the ability to generate very short pulses of light with high intensity. This property is used, for example, in laser spectroscopy, cf. Chap. 13.

The difference in quality between the light of a laser and a thermal light source is based on the various processes of light generation. For a thermal light source the spontaneous emission plays the dominant role. This is quite different from the light generation by a laser. Here the essential elementary process is stimulated emission. This was expressed already in the naming. The word laser is an acronym, which was formed by **L**ight **A**mplification by **S**timulated **E**mission of **R**adiation.

The schematic design of a laser is sketched in Fig. 2.24. The most important component is the optical amplifier. It is able to amplify an incident light wave in the correct phase, i.e., coherent. This is achieved by a sufficiently large number of excited atoms, which radiate by stimulated emission. Left and right of the optical amplifier mirrors are mounted. These mirrors constitute the so-called optical resonator. A light wave, which has passed through the amplifier, leaves only in part the resonator via the partially transmitting output mirror. The rest of the light remains in the resonator and passes the amplifier again. That is why we call a laser an optical-feedback amplifier.

A necessary condition to start laser oscillation is that the waves propagating back and forth in the resonator superpose with correct phase in one loop. Neglecting the influence of diffraction, this is the case when the wavelengths are given by:

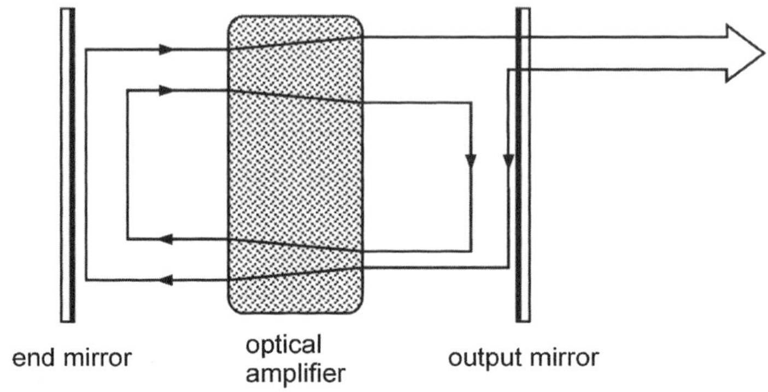

Fig. 2.24 Basic structure and mode of operation of a laser

Fig. 2.25 Examples for longitudinal modes of an optical resonator

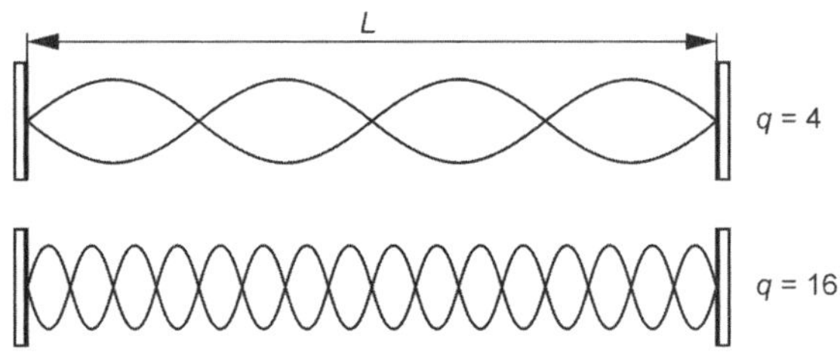

$$\lambda_q = \frac{2L}{q}. \tag{2.34}$$

L is the distance between the mirrors and q is a natural number. Since $L/\lambda_q \gg 1$, q assumes large values. Equation (2.34) also describes precisely the wavelengths, which correspond to standing light waves, which can be formed in a lossless optical resonator. The different waveforms, characterized by different values of q, are designated as longitudinal modes, see Fig. 2.25.

We summarize: A laser is an optical feedback amplifier. With sufficient gain, any feedback amplifier is unstable, i.e., it becomes an oscillator. An optical feedback amplifier in self-excitation is also called a laser oscillator or short laser. For a more detailed presentation of the basics of laser physics, we refer to the literature as e.g. [7–14].

2.6 Gaussian Beam

The propagation of electromagnetic waves is described by the wave equation. It is derived directly from Maxwell's equations of electrodynamics. For a homogeneous and loss-less medium, the wave equation for the electric field strength reads:

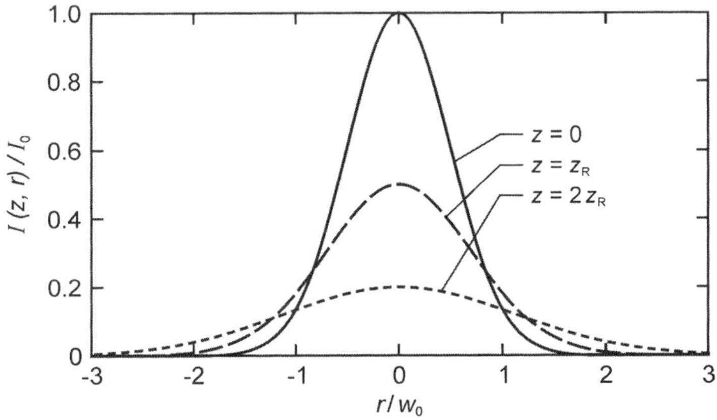

Fig. 2.26 Radial intensity profile of a Gaussian beam as a function of the normalized radius

$$\Delta \vec{E} = \frac{1}{c^2} \frac{\partial^2 \vec{E}}{\partial t^2}, \tag{2.35}$$

with Δ Laplace operator, c speed of light.

A solution of this wave equation, for example, is the planar-harmonic wave (2.1a). But the plane-harmonic wave is only a very rough approximation of a laser beam. In contrast to the fictional plane harmonic wave, the intensity of a real laser beam is not constant over the beam cross section. Due to its finite beam diameter, a laser beam has, in contrast to the plane wave, a diffraction-caused beam divergence. This divergence leads to non-planar wavefronts, i.e., curved surfaces of constant phases.

For the explanation of most measurement methods presented in this book, the model of the plane harmonic wave is sufficient. But in some cases, e.g. focusing of laser beams, this model is too rough. A better description of a laser beam, with respect to the most used transverse fundamental mode, is given by the so-called Gaussian beam. The Gaussian beam is also a solution—in fact, an approximate solution—of the wave equation (2.35).

2.6.1 Description of the Gaussian Beam

A Gaussian beam [7, 13, 14] is a rotationally symmetrical laser beam with respect to the propagation direction, which is assumed to be the z-axis. The irradiance I decreases in the radial direction following a Gaussian function, see Fig. 2.26:

$$I(z, r) = I(0, 0) \left(\frac{w_0}{w(z)} \right)^2 \mathrm{e}^{-2\left(\frac{r}{w(z)} \right)^2}. \tag{2.36}$$

Hereby we use the abbreviations:

$$w(z) = w_0\sqrt{1 + \left(\frac{z}{z_R}\right)^2} \tag{2.37a}$$

and

$$z_R = \frac{kw_0^2}{2} = \frac{\pi w_0^2}{\lambda}. \tag{2.38}$$

The term $e^{-2\left(\frac{r}{w(z)}\right)^2}$ describes the radial decrease in irradiance, according to a Gaussian function. At $r = w(z)$, the irradiance has—in comparison to the beam axis ($r = 0$)—fallen by a factor of $1/e^2$ (= 13.5 %). That is why we call $w(z)$ the beam radius, see Fig. 2.26. The beam radius $w(z)$ attains its minimum w_0 at $z = 0$. This place is called beam waist and w_0 is the waist radius. With increasing or decreasing z, the beam radius $w(z)$ increases according to equation (2.37a).

There are two limiting cases to be distinguished:

- $|z| \ll z_R$: The beam radius $w(z)$ is almost independent of z. It approaches approximately the waist radius w_0.

$$w(z) \approx w_0 \tag{2.37b}$$

- $|z| \gg z_R$: The beam radius $w(z)$ increases linearly with z.

$$w(z) \approx \frac{w_0}{z_R}|z| \tag{2.37c}$$

For $z = \pm z_R$, there is an increase of the beam radius to $\sqrt{2}w_0$. z_R is an important parameter of the Gaussian beam called Rayleigh length. In the far field ($|z| \gg z_R$), the Gaussian beam has a constant opening angle θ_0—also called far-field divergence—which is related to diffraction, see Fig. 2.27. For the opening angle, we find, according to (2.37c):

$$\tan(\theta_0) = \frac{w_0}{z_R} = \frac{\lambda}{\pi\, w_0}. \tag{2.39a}$$

Since the wavelength λ is small relative to the waist radius w_0, we use in the following the approximation $\tan(\theta_0) \approx \theta_0$. By rearranging, we find:

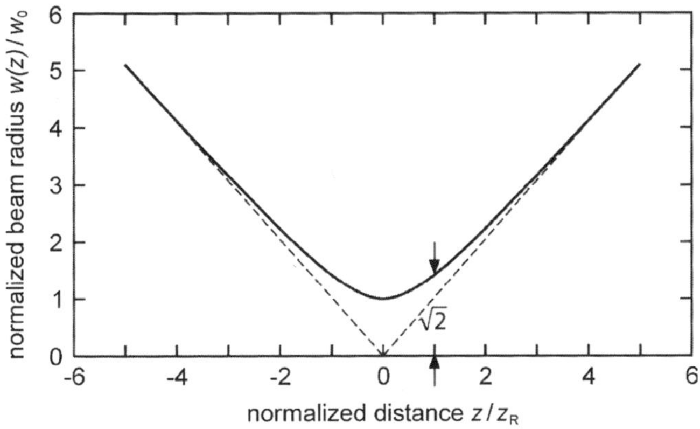

Fig. 2.27 Normalized representation of the beam radius as a function of the normalized distance

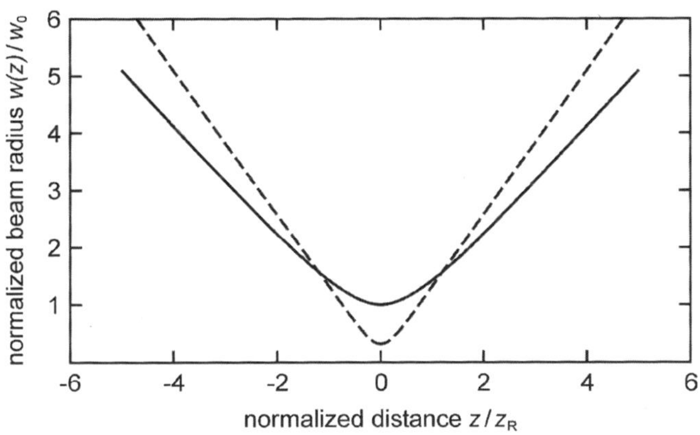

Fig. 2.28 Clarification of the relationship $\theta_0 w_0 = \text{const}$. Normalized beam radius as a function of the normalized distance represented as *solid line*. The *dashed curve* shows a beam with a smaller beam waist and a higher opening angle

$$\frac{\theta_0 w_0}{\lambda} = \frac{1}{\pi}, \tag{2.39b}$$

i.e., the term $\theta_0 w_0/\lambda$ is a conservation quantity, a propagation invariant. With decreasing waist radius w_0, therefore, the divergence of the Gaussian beam increases, see Fig. 2.28.

With increasing beam radius $w(z)$, for reasons of energy conservation, the field strength and irradiance must decrease on the beam axis. This is considered by the factor $(w_0/w(z))^2$ in equation (2.36).

The wavefronts, i.e., the surfaces of the same oscillation state are, as already mentioned, no plane surfaces. Near the beam axis, the wavefront can be

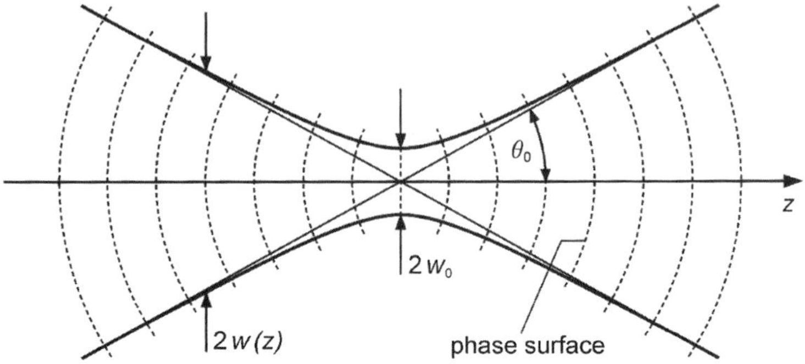

Fig. 2.29 Wavefronts of the Gaussian beam shown as dashed lines

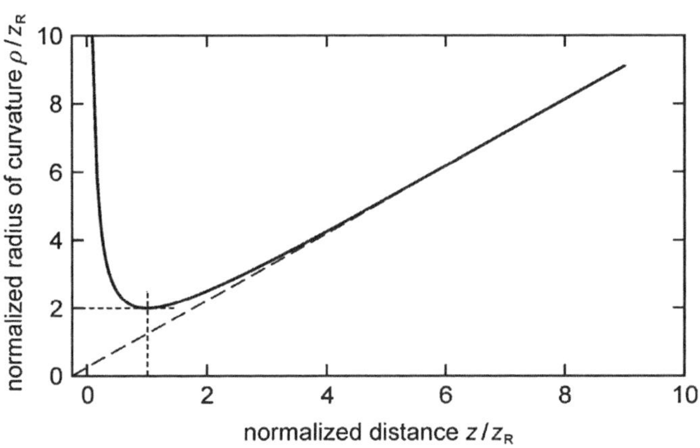

Fig. 2.30 Normalized representation of the radius of curvature of the wavefronts of the Gaussian beam as a function of the normalized distance

approximated by spherical surfaces, see Fig. 2.29. The radii of curvature of the phase surfaces are defined by the equation:

$$\rho(z) = z\left[1 + \left(\frac{z_R}{z}\right)^2\right]. \tag{2.40}$$

At the beam waist ($z = 0$) the wavefront is flat ($\rho(0) \to \infty$). For $|z| \gg z_R$ (2.40) simplifies to

$$\rho(z) \approx z. \tag{2.41}$$

In this case, the wavefronts are identical with those of spherical waves, which have their origin at $z = 0$. The radii of curvature take their minimum at $z = \pm z_R$, see Fig. 2.30.

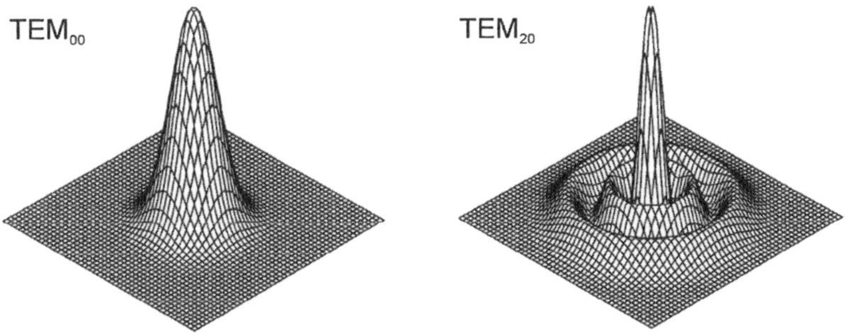

Fig. 2.31 Representations of the intensity distribution of the TEM_{00}-mode and the TEM_{20}-mode

A Gaussian beam is described uniquely, if the position of the beam waist ($z = 0$), the Rayleigh length z_R—alternatively the waist radius and the far-field divergence—and the wavelength λ are known. A user of laser radiation should know these parameters being defined by the optical amplifier and the geometry of the optical resonator.

2.6.2 Higher Transverse Modes

The approach presented in Sect. 2.6.1—the Gaussian beam—is only one approximate solution of the wave equation (2.35). In addition, there exist many other solutions of equation (2.35), which are similar to the harmonics of membranes. These solutions of higher order are called $TEM_{u,v}$-modes. The indices u and v are any non-negative integer values. The abbreviation TEM indicates that the solutions describe transverse electromagnetic waves.

The TEM_{00}-mode coincides with the Gaussian beam (2.36). Figure 2.31 shows, in addition to a Gaussian beam, a TEM_{20}-mode as it is calculated in cylindrical coordinates. In contrast to the Gaussian beam, the likewise rotationally symmetrical TEM_{20}-mode has two circular nodal lines in the finite. In laser measurement technology, the transverse modes of higher order play a minor role. That is why we do not deepen this topic.

Each transverse mode can, in principle, be excited at different wavelengths, see Sect. 2.5, Fig. 2.25. In addition, different states of polarization have to be considered. For a full characterization of a laser mode, it is therefore necessary, to indicate in addition to the transverse mode indices u and v two more indices q and p: $TEM_{u,v,q,p}$. The natural number q describes the longitudinal mode order, analogous to the equation (2.34), which is true only for plane waves. p can have two values and characterizes the state of polarization of a mode.

In a single-mode laser, only one single mode is excited. In a multi-mode laser several modes oscillate in parallel. A real laser beam can always be considered as a superposition of different modes $TEM_{u,v,q,p}$.

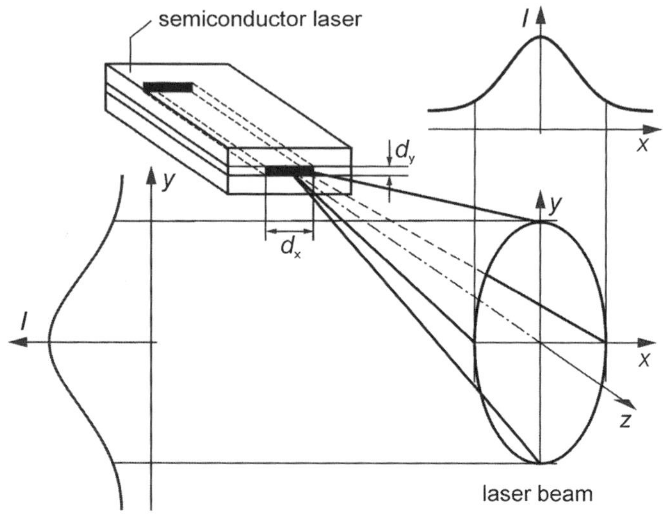

Fig. 2.32 Typical beam divergence of a semiconductor laser

Often, in laser measurement technology fundamental mode lasers (TEM_{00}) are used. These are lasers that have at least approximately, a Gaussian radial intensity distribution. Exceptions are semiconductor lasers that emit no Gaussian beam. The reason for this is the small size and the shape of the emitting surface of the laser, see Fig. 2.32.

Typical dimensions are $d_x = 5$–$100\ \mu m$ and $d_y = 0.1$–$1\ \mu m$ resulting in divergence angles θ_y of up to some 10°. The radiated wave has an elliptical shape, i.e., the beam diameters in two orthogonal directions differ in general. This also applies to the radii of curvature of the wavefronts. The different curvatures in two orthogonal planes are a manifestation of astigmatism.

The elliptical emission characteristic of a semiconductor laser is unsatisfactory for most metrological applications. The corrective is an appropriate optic for beam correction, as depicted in Fig. 2.33. First, the laser beam is expanded with a collimating lens, and thus the beam divergence is reduced. With the following cylindrical lens the astigmatism is corrected. The two prisms expand the laser beam, but only in the plane of projection of the upper diagram in Fig. 2.33. The beam diameter in the orthogonal direction is not affected, see Fig. 2.33, bottom. So we obtain a circular beam with a rotationally symmetrical divergence.

2.6.3 Beam Quality

To characterize the difference between a real laser beam and a theoretical Gaussian beam, we introduce the beam quality number K. In Sect. 2.6.1, we have recognized that the quantity $\theta_0 w_0/\lambda$ of a Gaussian beam is equal to the constant value $1/\pi$,

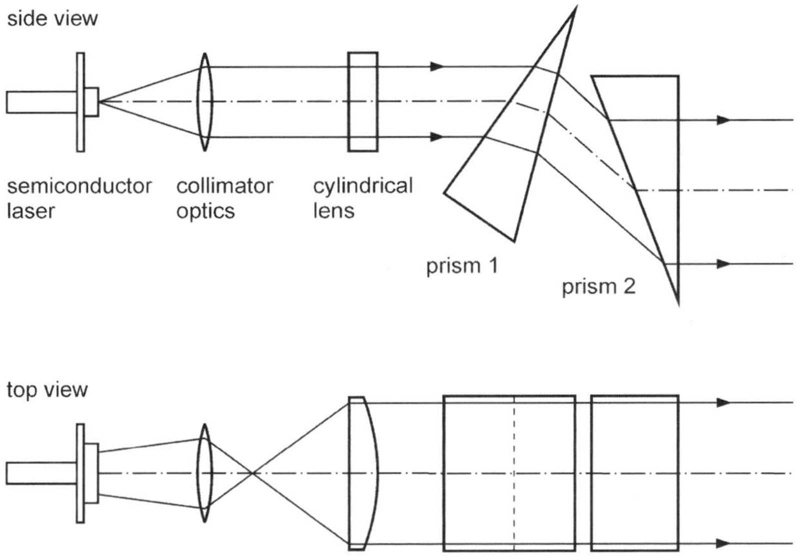

Fig. 2.33 Arrangement for the correction of the elliptical beam distribution of a semiconductor laser

see equation (2.39b). Since a Gaussian beam represents the laser beam with the lowest possible beam divergence [12], for every real and initially assumed rotationally symmetrical laser beam holds:

$$\frac{\Theta_0 W_0}{\lambda} = \frac{1}{K\pi} \quad \text{with} \quad 0 \le K \le 1 \tag{2.39c}$$

or

$$K = \frac{\lambda}{\pi\, W_0 \Theta_0}. \tag{2.39d}$$

K denotes the so-called number of beam quality [15, 16]. W_0 and Θ_0 characterize the waist radius and the opening angle of the real laser beam. The waist radius is the local minimum of the real beam radius $W(z)$ and the opening angle Θ_0 corresponds to the derivative $\mathrm{d}W(z)/\mathrm{d}z$ for large values of $|z|$ ($|z| \gg z_R$). The beam radius $W(z)$ is in this case equivalent to the radius of the smallest circular aperture, which let pass 86.5 % of the beam power.

For a theoretical Gaussian beam, the number of beam quality reaches its maximum value, $K = 1$. For a real laser beam, we find a smaller number ($0 \le K \le 1$). If we divide (2.39b) by (2.39c), we find:

$$K = \frac{w_0 \theta_0}{W_0 \Theta_0} \le 1. \tag{2.39e}$$

Assuming equal waist radii of the Gaussian and the real laser beam, i.e. $W_0 = w_0$, K describes the ratio of the opening angles θ_0/Θ_0 of the ideal Gaussian beam and the real laser beam.

For a non-rotationally symmetrical laser beam, such as for that shown in Fig. 2.32 of a semiconductor laser beam, we find analogously to equation (2.39d):

$$K_i = \frac{\lambda}{\pi\, W_{i,0.954}\Theta_{i,0.954}}, \qquad i = x, y. \tag{2.39f}$$

K_x and K_y describe the beam quality in two mutually orthogonal directions x and y, which are perpendicular to the beam axis. The x- and y-directions are the directions of smallest or greatest beam dimensions, respectively. Under the beam dimension $2W_{i,0.954}$ the smallest width of a slit is understood which transmits 95.4 % of the incident laser power. Compared to the circular aperture, the increased percentage considers that through a slit more power can pass than through a circular aperture of the same diameter. From the variation of $W_i(z)$, $i = x$, y, the waist radius $W_{i,0.954}$ and the far-field divergence angle $\Theta_{i,0.954}$ can be determined, for each direction x and y.

Besides the beam quality number K, also the standardized beam propagation ratio M^2 is used. It corresponds to the reciprocal beam quality number [15]:

$$M^2 = \frac{1}{K}, \tag{2.42a}$$

$$M_i^2 = \frac{1}{K_i} \quad (i = x, y). \tag{2.42b}$$

2.7 Beam Parameters of Specific Laser Systems

The currently available laser systems generate coherent light in a wide parameter range:

- The spectrum of wavelengths ranging almost continuously from the far infrared to the ultraviolet spectral region.
- High power lasers achieve continuous beam powers of over 100 kW.
- In pulsed operation the beam power can be increased to values more than 100 TW.
- With the help of the mode-locked technique it is possible to press the pulse length under one femtosecond.

These examples are intended to keep the reader the wide range of possible beam parameters in mind.

The lasers used in laser measurement systems, depending on the application, have typically an average output power of 1 mW–50 W and pulse power of 0.2 W–250 MW. In pulsed operation, the maximum pulse energy can achieve some joules and pulse lengths range between 5 and 1000 ns.

Each metrological task requires, for its optimal solution, suitable measuring instruments. This applies in particular to laser measurement technology. Here, the user is demanded to select the suitable type of measuring laser. In this both, technical aspects and economic considerations, must be included. As guidance, in Table 2.1 the typical features of these lasers are listed that have importance for the measuring methods described in Chaps. 6–14.

2.8 Dangers of Laser Radiation

Everyone working with lasers, must be aware of the dangers generated by a laser. We therefore briefly review these issues. For a more thorough treatment the reader is referred to the literature, e.g. [17–21].

Generally in laser protection, we distinct between laser-specific and laser-non-specific risks. Examples of non-specific laser hazards, which we will not elaborate, are:

- electrical hazards (high voltage),
- exposure to X-rays being generated by operation of laser systems,
- risk of fire,
- implosion or explosion,
- formation of ozone by laser radiation,
- exposure to cooling medium.

The laser-specific risks are based on the absorption of direct or reflected laser radiation by human tissue. Beam powers and pulse energies deployed in most of the laser measuring methods are so small that we can limit the following considerations to the risks for the eyes.

Figure 2.34 schematically shows the structure of a human eye and the path of a laser beam through the eye. The laser radiation enters through the cornea into the eye, and passes through the front ocular chamber and the circular opening of the iris to the ocular lens. The opening of the iris is called the pupil. Depending on the brightness, the diameter D of the pupil varies between 2 and 7 mm. By the eye lens, the light is concentrated and passes through the so-called vitreous body, a gelatinous mass, to the retina. The retina is the light-sensitive organ that is connected via the optic nerve with the brain. So, on the retina an illuminated spot of diameter d is formed, see Fig. 2.34.

The danger for the eye is caused by the absorption of the laser radiation in the eye, for example, in the cornea, in the eye lens or in the retina. Figure 2.35 shows the degree of transmission T_E of the human eye from the cornea to the retina, as well as the transmittance T_E multiplied by the degree of absorption A_R of the retina.

Table 2.1 Properties of some commercial laser systems

	Excimer laser	Argon ion laser	HeNe laser	CO_2 laser	Dye laser	Ruby laser	Nd:YAG laser	Semiconductor laser
Wavelength [nm]	193 248 308	514	633	9300–10800	570–750	694	1064; 532; 266	780–1500
Mean power [W]	0.01–50	0.5–20	0.001–0.07	1–50	0.1–2	–	0.1–20	0.001–0.5
Pulse power [MW]	2–30	–	–	0.01–20	0.006–5	2–250	1–100	2×10^{-7}
Pulse energy [J]	0.05–1	–	–	0.01–2	0.007–0.2	0.1–5	0.1–2	4×10^{-8}
Pulse length [ns]	10–30	–	–	100–1000	5–15	20–60	5–50	200–350
Application	S	D, H, I, Sp	D, H, I, Sp, T	S	S, LIF	H	H, S, Sp; LIF	D, H, I, Sp, T

Indicated are parameter ranges that are typical for applications in laser measurement technology. D laser-Doppler method, H holographic interferometry, I laser interferometry, S laser spectroscopy, Sp speckle interferometry, T laser triangulation, LIF laser-induced fluorescence

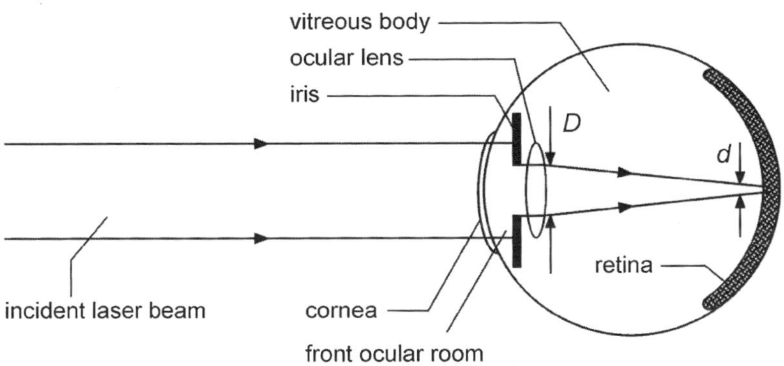

Fig. 2.34 Beam path of a laser beam through an eye

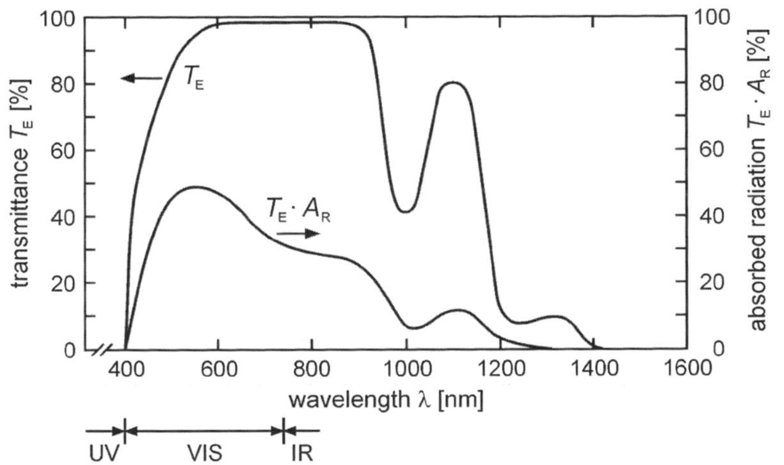

Fig. 2.35 Transmittance of the human eye and degree of absorption of the retina relative to the incident laser beam

This product $T_E\ A_R$ describes the fraction of the incident laser radiation that is absorbed by the retina. We consider as an example firstly the visible radiation of a HeNe laser with a wavelength of $\lambda = 633$ nm. This light gets almost unattenuated to the retina, which absorbs about 40 % of the incident power. Quite different is the situation for light of an UV laser. UV radiation is almost absorbed completely in the front part of the eye that is in the cornea and the eye lens.

A great danger of laser radiation to the retina is its ability to form a very tiny focus, as the following analysis shows. For further considerations, we start from the worst case, which leads to the smallest possible spot size:

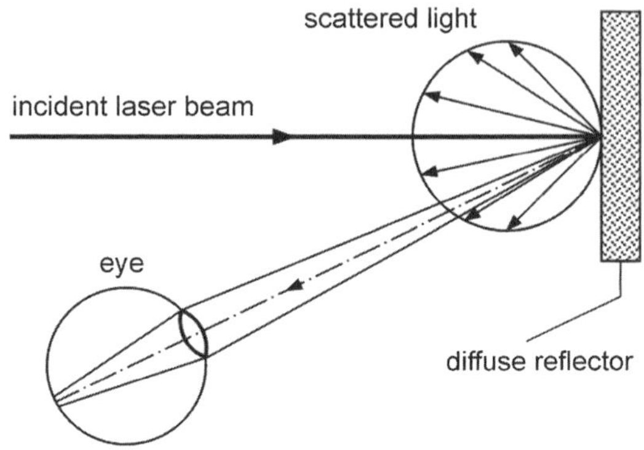

Fig. 2.36 Imaging diffusely reflected light onto the retina of an eye

- The incident laser beam is assumed to be a plane wave.
- The pupil diameter adopted its maximum value of $D_{\text{max}} = 7$ mm. The diffraction related focus diameter is then minimal.
- The eye is not accommodated. In this case, the image-side focal point of the lens is located on the retina.

The diameter of the resulting diffraction disk on the retina is calculated with the help of equation (2.23) to:

$$d_{\text{min}} = 2.44\frac{\lambda f}{D_{\text{max}}} \text{ with } \lambda = \frac{\lambda_0}{n}, \tag{2.43}$$

with f focal length of the eye lens, λ_0 vacuum wavelength of the laser radiation and n refractive index in the eyeball. The ratio of pupil to focus cross-section area is calculated to

$$\left(\frac{D_{\text{max}}}{d_{\text{min}}}\right)^2 = \frac{D^4_{\text{max}} n^2}{(2.44\lambda_0 f)^2}. \tag{2.44}$$

For the light of a HeNe laser, a focal length of the eye lens of 22.8 mm, a refractive index in the interior of the eye of 1.34 and a maximum pupil diameter of 7 mm, the ratio of the areas is calculated as 3.5×10^6. This factor also describes the increase of the irradiance of the laser radiation from the cornea to the retina. This calculation shows clearly the danger of laser radiation to the eye.

In addition to direct or reflected laser radiation at mirrors, diffusely reflected laser radiation is also a risk. By diffuse reflection, the parallelism of the laser beam is destroyed, see Fig. 2.36. If such diffuse light reaches the eye, the risk—compared with direct laser radiation—is lower because only a fraction of the diffusely scattered radiation enters the eye through the pupil. In addition, the diffuse scattering light is distributed, when compared to parallel laser light, over a larger area of the retina. Nevertheless, the danger of diffusely scattered light should not be underestimated.

When operating a laser, certain safety regulations must be noted. These regulations are summarized in different national and international standards. The precautions to be taken depend on the hazard class of the used laser system. According to the potential risk, there are seven categories of risk. The laser classification is, for details see e.g. [17]:

- Class 1: The laser radiation is not dangerous under usual conditions.
- Class 1 M: The wavelength ranges from 302.5 nm to 4 μm. The laser radiation is not harmful to the eye, as long as the cross section of the laser beam is not reduced by optical instruments such as magnifying glasses, lenses, telescopes.
- Class 2: The laser radiation is in the visible spectrum (400–700 nm). For short-term exposure (duration up to 0.25 s), it is eye-safe.
- Class 2 M: It is the same as class 2. But it is dangerous, if the cross section of the laser beam is reduced by optical instruments.
- Class 3R: The radiation is in the wavelength range from 302.5 nm to 1 mm. It is dangerous to the eye. The power or energy is less than five times the limiting values of the allowed radiation of class 2 (in the wavelength range of 400–700 nm) or class 1 (in the remaining wavelength range).
- Class 3B: The laser radiation is hazardous to the eyes and in special cases also for the skin. Diffuse scattered light is safe usually.
- Class 4: The laser radiation is hazardous to the human eye, often also to the skin.

The various precautions to be taken during the operation of a laser, are described, for example, in [17–21]. We therefore do not treat safety precautions in detail here. In practical work with lasers the following rules should be observed always:

- Never look into a direct or not diffusely reflected laser beam!
- Avoid laser beams in the height of the eyes!
- Avoid reflective surfaces, except in case of necessary mirrors for beam guidance!
- Use protective housings and beam stops so the laser light cannot leak out!
- If necessary, suitable laser safety goggles must be worn!
- If possible, work in bright light, so that the eye pupils have small diameters!

References

1. J.D. Jackson, *Classical Electrodynamics* (Wiley, New York, 1998). ISBN 978-0471309321
2. M. Born, E. Wolf, A.B. Bhatia, P.C. Clemmow, D. Gabor, A.R. Stokes, A.M. Taylor, P.A. Wayman, W.L. Wilcock, *Principles of Optics* (Cambridge University Press, Cambridge, 2003). ISBN 0 521 642221
3. L. Bergmann, C. Schaefer, H. Niedrig, *Lehrbuch der Experimentalphysik, Bd. III: Optik* (Walter de Gruyter, Berlin, 2004), ISBN 3-11-017081-7
4. E. Hecht, *Optics* (Pearson, Reading, 2003), ISBN 978-8177583571
5. M.V. Klein, Th.E. Furtak, *Optics* (Wiley, New York, 1986). ISBN 0-471-87297-0

6. A. Donges, *Die Kohärenzbedingungen*. Physik und Didaktik, pp. 109–118 (1991)
7. A. Donges, *Physikalische Grundlagen der Lasertechnik* (Shaker, Aachen, 2007). ISBN 978-3-8322-6392-8
8. W.T. Silfvast, *Laser Fundamentals* (Cambridge University Press, Cambridge, 2004). ISBN 9780521833455
9. P.W. Milonni, J.H. Eberly, *Laser Physics* (Wiley, New York, 2010). ISBN 9780470409701
10. W. Radloff, *Laser in Wissenschaft und Technik* (Springer, Heidelberg, 2011). ISBN 978-3-8274-3
11. F.K. Kneubühl, M.W. Sigrist, *Laser* (Vieweg u. Teubner, Wiesbaden, 2008). ISBN 978-3-8351-0145-6
12. W. Kleen, R. Müller, *Laser* (Springer, Berlin, 1969). ISBN 9783540045977
13. J. Eichler, H.J. Eichler, *Laser—Bauformen, Strahlführung, Anwendungen* (Springer, Berlin, 2003). ISBN 978-3-540-30149-6
14. D. Meschede, *Optik, Licht und Laser* (Teubner, Wiesbaden, 2005). ISBN 3-519-13248-6
15. European Standard DIN EN ISO 11146-1, Lasers and laser-related equipment—Test methods for laser beam widths, divergence angles and beam propagation ratios—Part 1: Stigmatic and simple astigmatic beams, 2005
16. A.E. Siegman, New developments in laser resonators. Proc. SPIE **1224**, 2–14 (1990). doi:10.1117/12.18425
17. DIN EN 60825-1/11.01-IEC 60825-1/01.2012: *Safety of laser products–Part 1: Equipment classification and requirements*
18. J. Eichler, *Laser und Strahlenschutz* (Vieweg, Braunschweig, 1992). ISBN 3-528-06483-8
19. K. Barat, *Laser Safety Management* (CRC Press, Boca Raton, 2008). ISBN 978-1420068542
20. E. Sutter, *Schutz vor optischer Strahlung* (VDE-Verlag, Berlin, 2008). ISBN 978-3800730728
21. Unfallverhütungsvorschrift "*Laserstrahlung*" vom 1. Oktober 1988 i.d.F. vom 1. Januar 1997 mit Durchführungsanweisung vom Oktober 1995 (BGV B 2 bzw. GUV 2.20)

Chapter 3
Interaction of Laser Radiation and Matter

Abstract Initially the particle concept of light is explained. Then we present the different kinds of interaction of light and matter. We start with the reflection and refraction, followed by the discussion of linear and non-linear absorption of light. The most important light scattering processes are described: Rayleigh, Mie and Raman scattering. Finally, the frequency doubling and the optical Doppler effect is treated.

All optical measurement methods are based on the same principle: the scattering of light. The incident laser beam with well defined parameters interacts with the object to be measured. During this interaction, also called scattering, at least one beam parameter of the laser beam is changed. So the scattered laser light carries information about the object to be measured.

The task of laser measurement technology is to draw conclusions from the scattered light of the characteristics of the measuring object. In this chapter, the elementary processes of interaction of light with matter are presented.

3.1 Particle Character of Light

So far, we have always regarded light as an electromagnetic wave. With this idea, a variety of phenomena, e.g. diffraction and interference, can be treated [1–5]. For the description of the interaction of light and matter, the wave concept proves to be inadequate in most cases.

Already at the end of the 19th century, it became clear that not all experimental observations could be explained by the wave model, for example the photoelectric effect. To explain the photoelectric effect, Albert Einstein introduced in 1905 his bold light quantum hypothesis stating that, light consists of discrete particles, called photons, moving at the speed of light. Each photon is assigned to a certain energy W, which is proportional to the light frequency f:

A. Donges and R. Noll, *Laser Measurement Technology*, Springer Series in Optical Sciences 188, DOI: 10.1007/978-3-662-43634-9_3,

$$W = hf. \tag{3.1}$$

The constant of proportionality is Planck's constant $h \approx 6.6 \times 10^{-34}$ Js.

Even in classical electrodynamics, an electromagnetic wave is associated with a momentum. This is also the case in the particle concept of photons. Einstein postulated that a photon having the energy hf, has a momentum

$$p = \frac{hf}{c} \tag{3.2}$$

in the direction of propagation.

In addition to energy and momentum, photons also have an intrinsic angular momentum, called spin. This intrinsic angular momentum can be understood intuitively as a rotation of a photon. Experimental and theoretical studies indicate that the component of the angular momentum in the direction of propagation takes only two values [6]:

$$S = \pm \frac{h}{2\pi} = \pm \hbar. \tag{3.3}$$

This result implies that photons, with respect to their direction of propagation, rotate both, clockwise as well as counterclockwise.

The obvious disagreement between the wave concept and the particle concept could not be resolved by classical physics. But it forced the development of quantum theory. Only in the framework of quantum electrodynamics, a branch of quantum theory, a complete and thorough description of light is possible.

The quantum electrodynamic description of light requires in general an enormous mathematical effort. But in most cases, there is no need for a precise description in terms of quantum electrodynamics. Depending on the problem, mostly one of the two concepts—the wave concept or the particle concept—is sufficient. It is a well-known fact that for the description of propagation phenomena, the wave concept is often suitable. The interaction of light with matter can be treated well with the particle concept.

3.2 Reflection and Refraction

3.2.1 Refractive Index

We consider the propagation of light in a dielectric medium, in which the light propagates with the phase velocity

$$c = \frac{c_0}{n}. \tag{3.4}$$

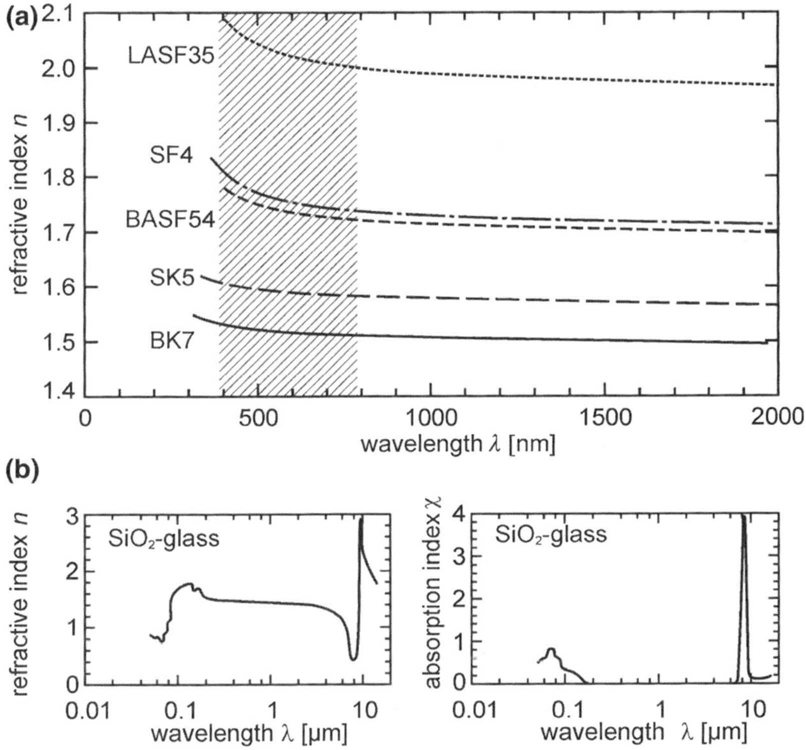

Fig. 3.1 **a** Refractive index for different glasses, **b** refractive index and absorption index of quartz glass as a function of wavelength

c_0 is the speed of light in vacuum and n is the refractive index. The refractive indices of transparent materials have typical values in the range of $1 < n < 3$. The phase velocities in media are, therefore, generally less than the vacuum speed of light, i.e., $c < c_0$. The refractive indices n depend on the wavelength λ or the frequency f of the light. This phenomenon is called dispersion. Figure 3.1 shows dispersion curves for various glasses.

Figure 3.1a shows that the refractive index of glass decreases in the region of the visible spectrum with increasing wavelength. This phenomenon is called normal dispersion. If the refractive index increases with increasing wavelength, then there is anomalous dispersion. In Fig. 3.1b, the refractive index n and, in addition, the absorption index χ of quartz glass are shown for a wide range of wavelengths. The absorption index χ is a measure of the intensity loss of light propagating through a medium. We will come back to this topic in Sect. 3.3.

From Fig. 3.1b we learn that near the wavelengths, at which a large absorption is present, the refractive index changes strongly, too. This statement is valid not only for quartz glass. In general an absorption line or band is accompanied by a

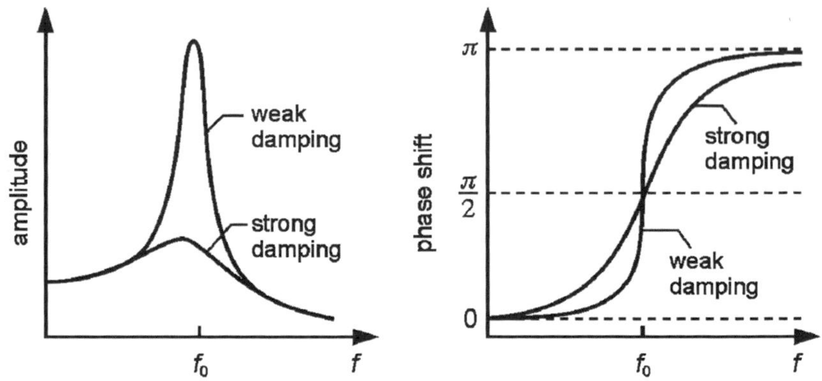

Fig. 3.2 Frequency response of the amplitude and phase shift of the harmonic forced vibration of an atom by a light wave of frequency f. f_0 resonance frequency

change of the refractive index. An absorption band increases the refractive index on the long wavelength side and lowers n on the short wavelength side.

The relation between dispersion and absorption becomes clear when light propagation in the medium is considered more accurately. The medium is composed of individual atoms or molecules. Under the influence of the electric field of the light wave, the atoms are polarized, i.e., the positive and negative charge centers are shifted against each other. This gives rise to electric dipoles which oscillate in time with the light wave. That is why the atoms emit light with the same frequency. The superposition of these secondary waves of the incident wave produces in the medium the ongoing wave. A more detailed analysis reveals that the overall wave in the medium propagates with the phase velocity c_0/n.

The atoms are excited by the incident light wave to forced oscillations. The amplitude of these vibrations and the phase of the oscillation with respect to the incident light wave changes most in the vicinity of the resonance frequency of the oscillator, see Fig. 3.2. So it is not surprising that the largest changes in refractive index are in the vicinity of the resonance frequencies. Resonance frequencies are those frequencies where the medium absorbs the most.

3.2.2 *Reflection and Refraction at a Plane Interface*

When light falls on the boundary between two different dielectrics, see Fig. 3.3, partial reflection occurs. The non-reflected portion passes into the other medium. While the incidence angle and reflection angle are equal ($\theta_1 = \theta_1'$), the transmitted beam is refracted that is, it changes its direction of propagation. The direction of the refracted ray is determined by the law of refraction [7]

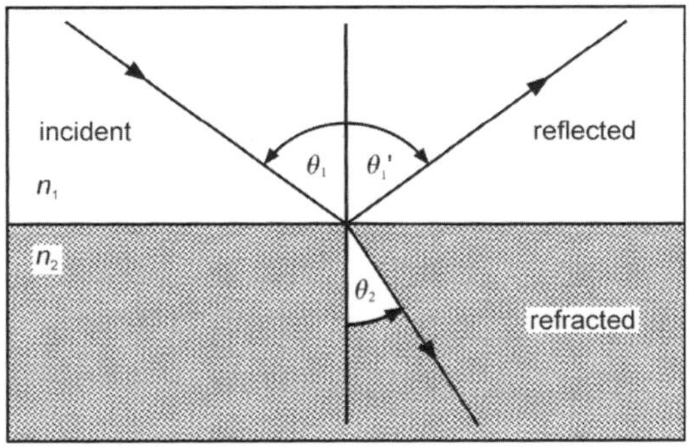

Fig. 3.3 Reflection and refraction of a light beam at an interface between two dielectrics. Incident, reflected and refracted ray and the normal of the boundary surface lie in the plane of projection

$$n_1 \sin\theta_1 = n_2 \sin\theta_2. \tag{3.5}$$

The reflectance R is the ratio of the intensities of reflected and incident beam. The term reflectivity is also used synonymously. The reflectance depends on the polarization direction of the light. Depending on whether the electric field vector oscillates perpendicular ($\perp$, also called s-polarization) or parallel (||, also called p-polarization) to the plane of incidence, the reflectances are given by

$$R_\perp = \left[\frac{\sin(\theta_1 - \theta_2)}{\sin(\theta_1 + \theta_2)}\right]^2, \tag{3.6a}$$

$$R_\parallel = \left[\frac{\tan(\theta_1 - \theta_2)}{\tan(\theta_1 + \theta_2)}\right]^2. \tag{3.6b}$$

These two equations are referred to as Fresnel's formulas. The curves of $R_\perp$ and $R_\parallel$ depend on the angle θ_1 of incidence as illustrated in Fig. 3.4 (for the case $n_1 = 1$ and $n_2 = 1.5$). For normal incidence ($\theta_1 = 0$), there is no difference between $R_\perp$ and $R_\parallel$:

$$R_{\theta_1=0} = \left[\frac{n_1 - n_2}{n_1 + n_2}\right]^2. \tag{3.6c}$$

For the transition from air $n_1 \approx 1$ to glass with $n_2 = 1.5$, $R_{\theta_1=0}$ is calculated to be 4 %.

Since $R_\perp$ and $R_\parallel$ differ, the polarization state of light changes with the reflection, in general. This is especially evident, when the incident light strikes the boundary at the Brewster angle θ_B:

$$\tan(\theta_B) = \frac{n_2}{n_1}. \tag{3.7}$$

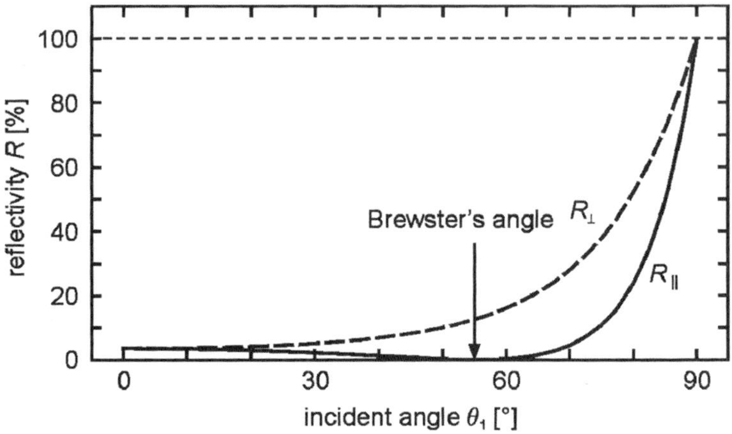

Fig. 3.4 Reflectivity (or reflectance) at the interface between two dielectrics for ⊥- and ||-polarization as a function of angle of incidence. The light beam passes from medium 1 to 2 with $n_1 < n_2$

Then $R_{||} = 0$, i.e., the ||-component is not reflected, see Fig. 3.4. Therefore, a non-polarized light reflected at the Brewster angle θ_B, becomes ⊥-polarized.

The absence of the ||-component in the case of $\theta_1 = \theta_B$ can be explained by the model of the Hertzian dipole. The light field, penetrating into medium 2, excites the atoms of medium 2 to forced oscillations. The vibrating atoms act as Hertzian dipoles and emit light with the same frequency. It should be noted that no light is emitted in the direction of the dipole axis. A reflection is thus impossible, if reflected and refracted ray would be perpendicular to each other. This would be the case, when the light beam would be reflected at Brewster's angle.

Figure 3.5 shows the basic curve of the reflectances $R_\perp$ and $R_{||}$ for the transition from a medium 1 to 2 with $n_1 > n_2$. With respect to Fig. 3.4, there is a qualitatively different behavior. In this case, the law of refraction (3.5) permits only a transition to the optically thinner medium, when the incident angle θ_1 is smaller than the critical angle θ_t of total reflection:

$$\sin(\theta_t) = \frac{n_2}{n_1}. \tag{3.8}$$

If the critical angle of total reflection is exceeded ($\theta_1 > \theta_t$), independent of the polarization direction, a full reflection, so-called total reflection, takes place. The critical angle of total reflection is for the transition from glass ($n_1 = 1.5$) to air ($n_2 \approx 1$) approximately 42°. Total reflection is used in technical applications, such as e.g. optical fibers or deflecting prisms, as depicted in Fig. 3.6.

Often, reflected waves are brought to interference. In those cases, possible phase shifts must be noted. In analogy to the reflection of a mechanical wave at a fixed rope end, during the reflection at the interface between medium 1 and 2 with

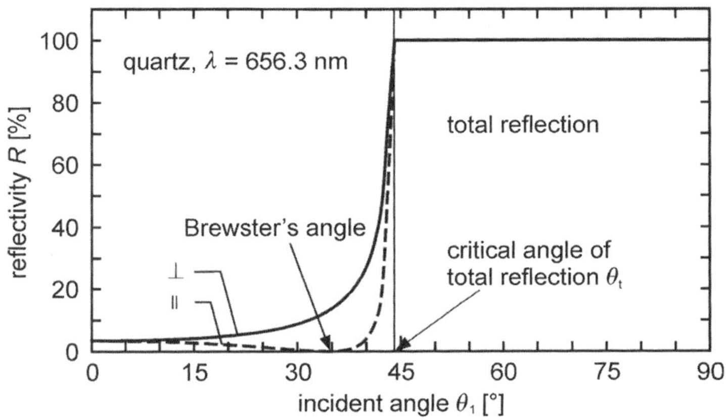

Fig. 3.5 Reflectivity at the interface for a wave propagating from medium 1 to 2 with $n_1 > n_2$ for ⊥- and ||-polarization as a function of angle of incidence. If the angle of incidence exceeds the critical angle, then there is total reflection

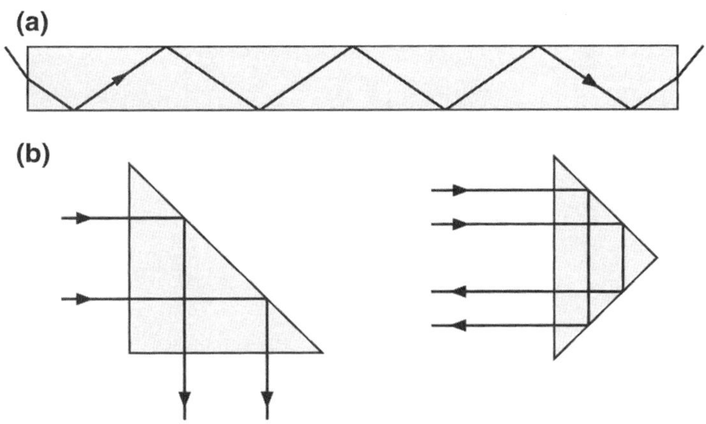

Fig. 3.6 Beam path in **a** an optical fiber, and **b** in deflecting prisms

$n_1 < n_2$, a phase shift of $\Delta\varphi = \pi$ occurs. In the opposite case ($n_1 > n_2$), there is no phase shift.

The reflectivity of metal surfaces is more complicated than the cases discussed so far. For metals the non-reflected portion is absorbed. The reflectance at normal incidence is significantly higher than those observed for transparent media. Figure 3.7 shows the typical curves of the reflectances $R_\perp$ and $R_{||}$ of a polished metal surface. As in the case of the beam reflection by a dielectric, the polarization state of light is also changing with the reflection at a metal surface. This must, if necessary, be taken into account, e.g. at the reflection of a laser beam by a metal mirror.

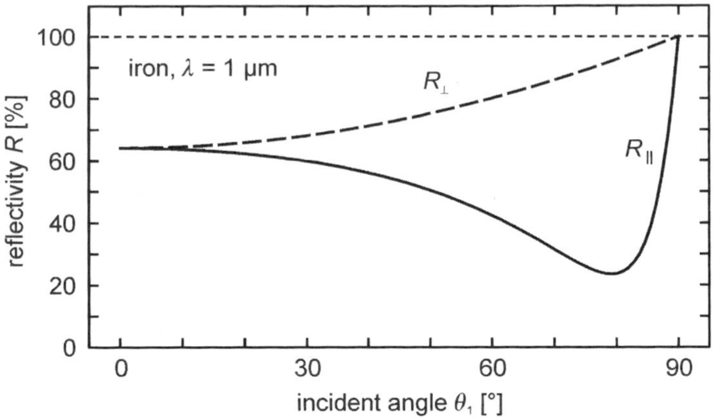

Fig. 3.7 Principle curve of the reflectivity of a polished metal surface for ⊥- and ∥-polarization in dependence on the angle of incidence

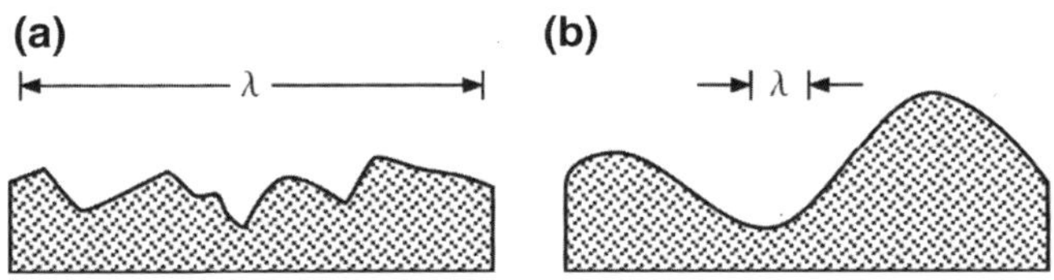

Fig. 3.8 For the definition of **a** a smooth and **b** a rough surface

3.2.3 Reflection at Rough Surfaces

The previous considerations of the reflection were related to smooth or polished boundaries, such as those used for flat or spherically curved mirrors. Now we want to extend our considerations to the case of rough surfaces. A surface is said to be rough, when the dimensions of its spatial texture are large compared to the wavelength of the light, see Fig. 3.8.

If a beam of light meets a rough surface with the incident angle θ_1, the light is diffusely reflected, cf. Fig. 3.9a. The surface then appears, from all angles of view, in the same brightness. Regardless of the angle of incidence, the radiant intensity—that is the reflected power per unit area and unit solid angle—of the diffuse reflected light obeys Lambert's cosine law:

$$I_e = I_{e0} \cos \theta_2. \tag{3.9}$$

Here, θ_2 is the angle of observation, which is measured with respect to the surface normal. The diffuse reflection is also known as remission.

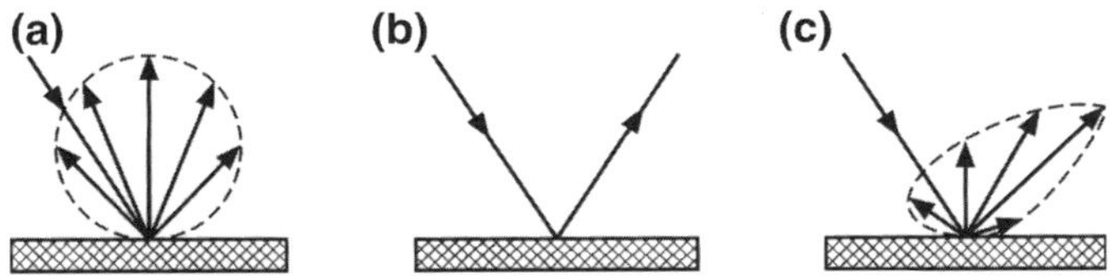

Fig. 3.9 **a** Diffuse reflection, **b** regular reflection, **c** combination of diffuse and regular reflection

Diffuse reflection and the previously discussed regular reflection—see Fig. 3.9b—are idealized limiting cases of common reflection, in which both, a regular and a diffuse fraction, occur, see Fig. 3.9c.

3.2.4 Birefringence

Birefringence is a phenomenon that occurs in some anisotropic crystals more or less pronounced. It is particularly well observed in calcite ($CaCO_3$). In the following we treat the birefringence of calcite, as required to understand the issues described in Chap. 4, where we use polarization-influencing optical components.

If a non-polarized beam impinges at a plate made of calcite, in general the light beam splits into two beams, see Fig. 3.10a. One partial bundle obeys the law of refraction (3.5). Therefore, it is called ordinary ray (o.). The other part of the beam cannot be described by the law of refraction (3.5). This beam is refracted, for example, at normal incidence, see Fig. 3.10b. It is, therefore, called extraordinary beam (e.o.). From investigations of the polarization we know that ordinary and extraordinary beam are polarized perpendicularly to each other. Linearly polarized light can be produced by blocking whether the ordinary or the extraordinary beam.

In general, ordinary and extraordinary rays propagate in the crystal with different phase velocities. Further, it was found that for the ordinary wave, the refractive index is $n_{o.}$, and thus, the phase velocity $c_{o.}$ is independent of the direction of propagation. On the other hand the refractive index $n_{e.o.}$, and hence the phase velocity of the extraordinary wave $c_{e.o.}$, depends on the direction of propagation. In Fig. 3.11, the situation described is summarized. There exists a certain propagation direction, in which the phase velocity is independent of the polarization direction. This direction is termed the optical axis. The existence of this particular direction is due to symmetries in the crystal structure. By the direction of incidence and the optical axis a plane, called the principal plane, is defined. The polarization direction of the ordinary wave is oriented perpendicularly to the principal plane, and is therefore always perpendicular to the optical axis. For the extraordinary wave, the electric field vector oscillates within the principal plane, see Fig. 3.10, where the hatching indicates the direction of the optical axis.

In the direction of the optical axis, there is no birefringence since, in principle, ordinary and extraordinary ray cannot be distinguished. When the light propagates

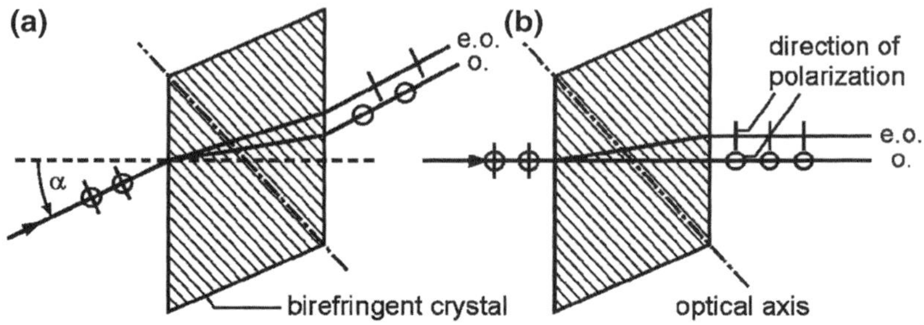

Fig. 3.10 A non-polarized light beam is split by a birefringent crystal in an ordinary (o.) and an extraordinary (e.o.) beam **a** angle of incidence $\alpha \neq 0$, **b** angle of incidence $\alpha = 0$

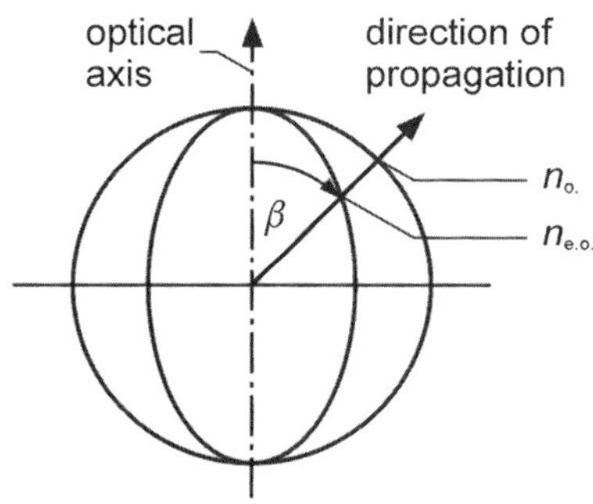

Fig. 3.11 Polar representation of the refractive index ellipsoid. The angle β is measured between the direction of propagation of the light beam and the optical axis. The outer sphere describes the isotropic ordinary index of refraction. The intersection of the propagation direction with the ellipsoid gives the extraordinary index of refraction. The case $n_{o.} > n_{e.o.}$ shown here is called negative birefringence

perpendicularly to the optical axis, the difference in phase velocities attains the greatest value, see Fig. 3.11. Nevertheless, for reasons of symmetry, there is no spatial separation of ordinary and extraordinary ray in that case.

For materials that are not birefringent, external influences can induce birefringence. This applies to many media that are subject to mechanical stress. This phenomenon is called stress induced birefringence. For laser measurement technology it is interesting that electric fields make isotropic materials such as gases, liquids and solids birefringent. This phenomenon is called Kerr effect according to its discoverer John Kerr. It can be well observed in liquids, particularly in nitrobenzene. Liquids are always isotropic, even if they consist of anisotropic molecules. This is due to the random orientations of the individual molecules. By an electric field, the molecules can be aligned, whereby the liquid becomes birefringent. The optical axis is—due to symmetry—given by the direction of the electric field strength. The difference of the refractive indices for light polarized parallel and perpendicular to the applied electric field E increases with the square of the applied field strength:

$$n_{\mathrm{e.o.}} - n_{\mathrm{o.}} = K\lambda_0 E^2. \tag{3.10}$$

λ_0 denotes the vacuum wavelength and K is Kerr's constant, which depends on the material used. For the mostly used nitrobenzene, K is 10^{-12} m/V^2 at room temperature.

Finally, we discuss the so-called Pockels effect. The Pockels effect occurs only in certain crystals. These crystals become birefringent by an electric field or change their birefringent properties. There is a distinction between the transverse and the longitudinal Pockels effect, depending on, whether the applied E-field is oriented perpendicularly or parallelly to the propagation direction of the light beam, respectively.

By the application of an electric field to the crystal, two perpendicular polarization directions occur with different phase velocities and refractive indices. The difference in the refractive indices increases linearly with the applied field strength:

$$\Delta n = rn^3 E. \tag{3.11}$$

n is the refractive index for the observed polarization directions without the applied electric field and r denotes the electro-optical coefficient, which is specific for the crystal and its orientation. For the well-used KDP crystal (potassium dihydrogenphosphate), r has at room temperature a value of approximately 10^{-11} m/V.

3.3 Absorption

3.3.1 Classical Absorption

If a light beam propagates in a homogeneous medium, the light intensity I decreases exponentially along the path. This is Lambert's law of absorption, see Fig. 3.12:

$$I(d) = I_0 \mathrm{e}^{-\alpha d}. \tag{3.12}$$

I_0 is the intensity of the light entering into the absorbent medium, and d is the traveled distance in the absorber. The quantity α denotes the linear absorption coefficient, which depends on the material and the wavelength.[1] Typical values of α range in the visible and near infrared spectrum from 0.05 km^{-1} for low-loss optical fibers up to 0.05 nm^{-1} for metals.

[1] The absorption coefficient α is related to the absorption index χ from Fig. 3.1b: $\chi = \alpha\lambda/4\pi n$.

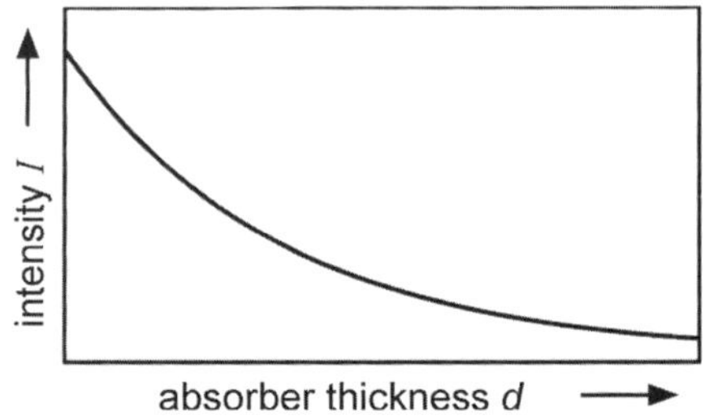

Fig. 3.12 The absorption of light

The phenomenon of absorption can be interpreted atomistically. All matter is composed of atoms (or molecules). These have discrete energy levels, which are denoted by W_i. $i = 1$ describes the minimum energy value. With increasing integer i, the energy values W_i grow, too. In liquids and solids, these energy levels may smeare to continuous energy bands. Provided $kT \ll W_2$, then all atoms are in the lowest energy state, the ground state W_1, where $k = 1.38 \times 10^{-23}$ J/K is Boltzmann's constant and T the temperature. Now let us consider that light with the proper frequency

$$hf_{i1} = W_i - W_1, \; i = 2, 3, 4, \ldots \tag{3.13}$$

is irradiated. Then the atoms may be lifted from the ground state ($i = 1$) to an excited level ($i > 1$). Thus the light field loses a photon with energy hf_{i1}. This quantum process is referred to as absorption.

The atoms, which are excited by the incident light, release the absorbed energy again. There are two possibilities. One of these is the so-called spontaneous emission. Spontaneous emission means that an excited atom without any external influence passes from an excited state W_i to a lower energy level W_K, while a photon with energy

$$hf_{ik} = W_i - W_k, \; i > k \tag{3.14}$$

is emitted. On an individual basis, it is not possible to predict the lifetime of the excited level. However, a mean lifetime τ_{ik} can be indicated that is typical for the respective transition. The spontaneous emission is—by averaging over a lot of emissions—spatially isotropic. That is why the intensity of the directed light beam decreases according to Lambert‘s law (3.12), although no energy of light is actually lost by absorption and subsequent isotropic emission.

In addition to spontaneous emission, atoms can release their energy also in a non-radiative way, e.g. via so-called collisions of the second kind. In a collision of second kind, an excited atom transfers its energy fully or partially to another atom by a collision. Excitation energy is converted into kinetic energy or again to excitation energy of the knocked-on partner. Finally, the absorbed light energy is converted into heat.

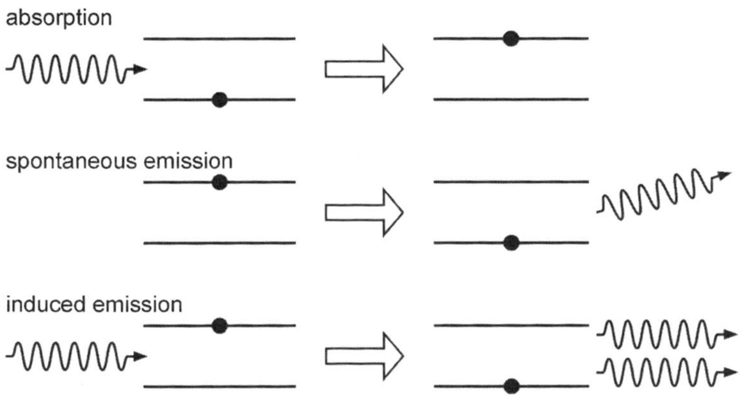

Fig. 3.13 Schematic representation of absorption, spontaneous and induced emission

3.3.2 Non-linear Absorption

In addition to spontaneous emission, there is also so-called induced emission. In this case, the transition from the higher to the lower atomic level is triggered by an external radiation field. Here, as in the spontaneous emission, a photon of energy

$$hf_{ik} = W_i - W_k,\ i > k \tag{3.14'}$$

is released. In contrast to spontaneous emission, the induced emission of the photon is not spatially isotropic. The photon is emitted in the direction of the incident wave. In the wave concept, this means that the induced wave emitted is coherent to the incident wave. The incident and induced wave have the same frequency, the same propagation direction and the same polarization direction and they have a defined phase relationship to each other, see Fig. 3.13.

The number of induced emission processes depends on both, the number of excited atoms and the intensity of the incident light at the resonance frequency of the atoms, i.e., relation (3.14′) is fulfilled. At ordinary light intensities, induced emission processes may be neglected. Then, induced emission processes have no effect on absorption. This case is referred to as linear absorption, see Sect. 3.3.1. The transmittance T defined by the ratio of transmitted to incident intensity, is in the case of linear absorption independent of the incident intensity. However, this statement loses its validity, when intense light—typically focused and pulsed laser light—is used. In this case, the induced emission processes dominate versus spontaneous emissions. For that reason, absorbance decreases and transmittance is growing. This case is called non-linear absorption. In good approximation the transmittance T is given in case of so-called homogeneous line broadening by the function:

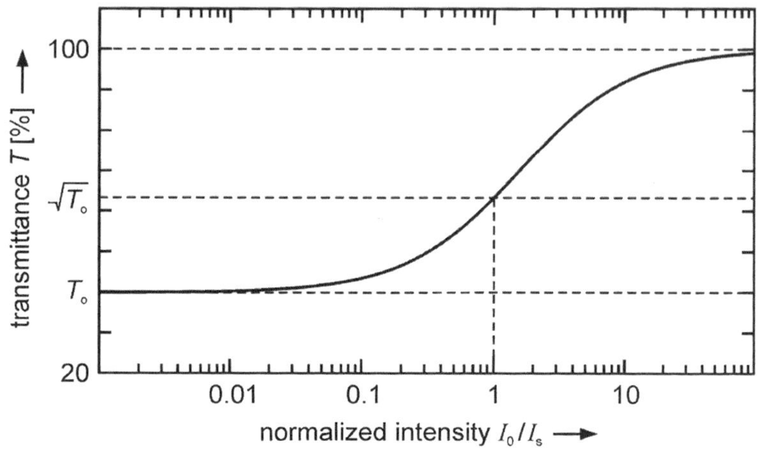

Fig. 3.14 Transmittance T as a function of the incident intensity normalized to the saturation intensity I_s in the case of non-linear absorption

$$T = T_0^{\frac{1}{1+I_0/I_s}}, \tag{3.15}$$

see Fig. 3.14, with T_0 transmittance at low intensities in the range of linear absorption, I_0 incident intensity and I_s the so-called saturation intensity being characteristic for each absorber.

3.3.3 Two-Photon Absorption

Let us consider an atom (or molecule) with the atomic energy levels W_1 and W_2. If the atom is in the level 1, an incident photon can be absorbed, provided the resonance condition

$$W_2 - W_1 = hf_{21} \tag{3.16}$$

is satisfied. Let us assume a relation between two atomic levels and a photon energy of the following form:

$$W_2 - W_1 = 2hf, \tag{3.17}$$

there is usually no absorption. This statement is not valid for large intensities. With sufficient intensity of the incident light, an atom can absorb simultaneously two photons. In this case the atom transits from state W_1—via a prohibited virtual intermediate level W^*—to state W_2. This is called two-photon absorption, see

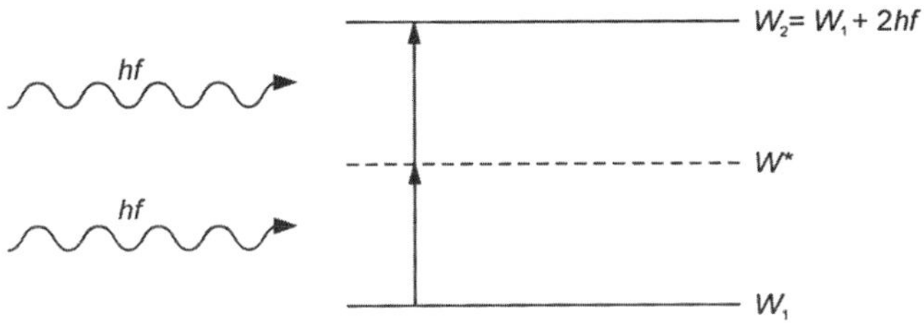

Fig. 3.15 Schematic illustration of two-photon absorption

Fig. 3.15. The probability p for two-photon absorption is proportional to the square of the intensity I of the incident light [6]:

$$p \propto I^2. \tag{3.18}$$

Therefore, two-photon absorption is a non-linear optical effect.

Two-photon absorption also works with two photons of different frequencies. An atom can be ionized by two-photon absorption. At even higher laser intensities also multi-photon absorption and multi-photon ionization processes occur, i.e., more than two photons are absorbed simultaneously by one atom.

3.4 Light Scattering

In general, light scattering comprises any kind of interaction of light with matter, in which the light is not absorbed permanently. This definition includes the previously discussed effects of reflection and refraction. In a narrower sense, the term light scattering deals only with interaction processes with microscopic particles, e.g. atoms or dust particles. So we do in the following discourse.

The scattering of light appears in a great variety. Therefore, we distinguish different types of scattering: elastic scattering, inelastic scattering, coherent scattering and incoherent scattering. In case of elastic scattering, wavelength and frequency of the scattered light remains—unlike inelastic scattering—conserved. Coherent or induced scattering is an elastic scattering, in which between incident and scattered wave a defined phase relationship exists. If not, the scattering is incoherent or spontaneously.

The field of light scattering is very extensive [8, 9]. We therefore restrict our considerations to the cases that are important for laser measurement techniques presented later in this book.

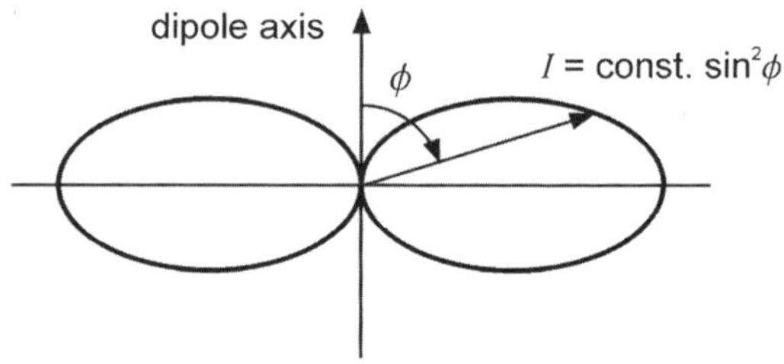

Fig. 3.16 Radiation pattern of a dipole

3.4.1 Rayleigh Scattering

Rayleigh scattering occurs only, when the particle diameter D is small compared to the wavelength λ of the irradiated light. When a particle interacts with the electric field of a linearly polarized light wave, it will see an approximately homogeneous alternating electric field because of $D \ll \lambda$. The scattering particle is polarized dynamically by the alternating electric field of the light wave. It becomes a Hertzian dipole, which oscillates at the frequency of the incident light. The dipole axis is determined by the direction of polarization of the incident wave. The particle emits, as each oscillating dipole, electromagnetic waves. The frequency of the emitted waves coincides with the frequency of the oscillating dipole and therefore with the frequency of the incident light. Hence, Rayleigh scattering is an elastic scattering. The intensity of the scattered light increases with the fourth power of the reciprocal wavelength. Blue light is, therefore, scattered more intense than red. This is the reason, why the sky looks blue on a clear day. The dipole radiation is predominantly emitted perpendicularly to the dipole axis, see Fig. 3.16. No radiation is emitted in the direction of the dipole axis. The emission characteristics are independent of the geometry of the scattering particle. The intensity characteristics reads:

$$I \propto \frac{\alpha^2}{\lambda^4} \frac{\sin^2 \phi}{r^2} I_0, \tag{3.19}$$

where ϕ is the angle between the viewing direction and the dipole axis, λ is the wavelength, α is the polarizability of the scattering particle, I_0 is the intensity of the incident wave and r is the distance from the dipole.

3.4.2 Mie Scattering

Mie scattering is, as Rayleigh scattering, an elastic scattering, i.e., the frequency of light is preserved in the scattering process. For Mie scattering—in contrast to Rayleigh scattering—the dimensions of the scattering particles are at least in the order of the wavelength of light or greater ($D \gtrsim \lambda$). Emanating from different areas

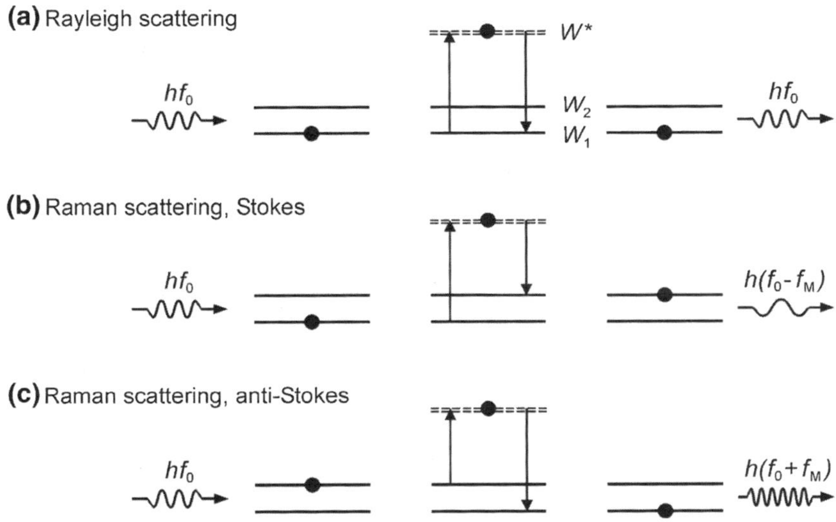

Fig. 3.17 Schematic representation of Rayleigh and Raman scattering

of the scatterer, the scattered waves are, therefore, in general phase shifted. Depending on the direction of observation and the shape and the dimensions of the scattering particle, constructive and destructive interferences will occur. Hence, Mie scattering is a kind of Rayleigh scattering, but the scattered radiation depends on the shape and the size of the scatterer. Using Mie scattering, the radii of spherical scatterers can be determined, experimentally. Further properties of Mie scattering are described in Sect. 11.3.1.

3.4.3 Raman Scattering

Raman scattering is an inelastic scattering since the wavelength emanated by the scatterer becomes both longer and shorter. A simple explanation of the Raman effect is given using the particle concept of light.

When a photon of frequency f_0 bounces a molecule, the photon can transfer a part of its energy to the molecule and excites the molecule to vibrate. During this interaction, the molecule gains the energy hf_M, while the photon loses the same amount of energy. After the interaction, the photon has a reduced frequency of $f_0 - f_M$, which is called Stokes frequency. The case is sketched in Fig. 3.17b. The incident photon excites the molecule from the ground state W_1 to a forbidden, so-called virtual state, W^*. After a short time ($\approx 10^{-12}$ s), the molecule transits from W^* to the excited vibrational state W_2 and emits a photon of frequency $f_0 - f_M$.

Vice versa it is possible that a photon takes energy in a collision from an oscillating molecule. As shown in Fig. 3.17c, a molecule in the vibrational state W_2 absorbs an incident photon of frequency f_0 and transits in a virtual state W^*. Then, a transition from W^* to the ground state W_1 occurs, which is accompanied by the emission of a photon with the increased frequency $f_0 + f_{\mathrm{M}}$. This frequency is called the anti-Stokes frequency.

Normally, more molecules are in the vibrational ground state than in a vibrationally excited state. Stokes-Raman scattering is, therefore, more frequently observed than anti-Stokes Raman scattering.

The Raman effect is applied in Raman spectroscopy. The Raman spectroscopy is used for gases, liquids and solids for the study of vibrational energy levels of molecules, cf. Sect. 13.3. While the classic Raman spectroscopy utilizes spontaneous scattering, the invention of the laser led to the development of new measurement methods based on induced Raman scattering.

3.5 Frequency Doubling

Let us consider the propagation of a light beam through a transparent medium. By the oscillating electric field of the light wave, the atoms are dynamically polarized, as we have already discussed. So the atoms become electrical dipoles oscillating with the frequency of the incident light. These dipoles emit light now themselves. The frequency of the emitted light is determined by the frequency of oscillation of the dipoles and agrees with the frequency of the incident light. This conservation of frequency in the propagation of light in a medium is the basis of linear optics.

The principle of frequency conservation will lose its validity, if the electric field strength of the light wave is too large. This may be the case with intense laser light. We examine this phenomenon more closely. When matter is exposed to an electric field, it is polarized, i.e., the individual atoms become electrical dipoles. The induced dipole moment per unit volume is called polarization P. The following applies:

$$P = \varepsilon_0 \chi E, \tag{3.20a}$$

where ε_0 is the permittivity of free space and χ is the electrical susceptibility. χ is usually a material-specific constant. However, at high field strengths the electric susceptibility depends on the electric field strength. Approximately the following relation prevails:

$$\chi = \chi_0 + \chi_1 E, \tag{3.20b}$$

where χ_0 and χ_1 are material-dependent constants. Thus between polarization P and electric field strength E holds a nonlinear function:

$$P = \varepsilon_0 \chi_0 E + \varepsilon_0 \chi_1 E^2. \tag{3.21}$$

For a harmonically varying field strength

$$E = E_0 \cos \omega t \tag{3.22}$$

the polarization P reads

$$P = \varepsilon_0 \chi_0 E_0 \cos \omega t + \varepsilon_0 \chi_1 E_0^2 \cos^2 \omega t \tag{3.23a}$$

and after trigonometric transformation

$$P = \frac{\varepsilon_0 \chi_1 E_0^2}{2} + \varepsilon_0 \chi_0 E_0 \cos(\omega t) + \frac{\varepsilon_0 \chi_1 E_0^2}{2} \cos(2\omega t). \tag{3.23b}$$

The frequency spectrum of the polarization P contains a DC component, the angular frequency ω and the doubled angular frequency 2ω. The static component of the polarization is of no interest to the emission of radiation of the oscillating dipoles. The angular frequency components ω and 2ω of the polarization P describe the effect of the induced dipoles emitting light with the angular frequencies ω and 2ω. By nonlinear polarization of the dielectric material, a light wave, whose frequency is twice as large as the frequency of the incident wave, is generated. This phenomenon is called frequency doubling.

In Sect. 11.3, the general case of (3.20b) will be considered. As a result, further frequency components will be excited. Finally, we consider frequency doubling in the particle concept. To satisfy the energy conservation principle, two photons of the incident wave with the frequency f_1 are necessary to generate one photon of the frequency-doubled wave of frequency $f_2 = 2f_1$:

$$hf_1 + hf_1 = 2hf_1 = hf_2. \tag{3.24}$$

In addition to the energy conservation law, the conservation of momentum must also be satisfied, i.e.,

$$\frac{hf_1}{c_1} + \frac{hf_1}{c_1} = \frac{hf_2}{c_2} = \frac{2hf_1}{c_2}, \tag{3.25}$$

where c_1 and c_2 are the phase velocities of the light waves in the dielectric with respect to the frequencies f_1 and f_2. It is obvious that the conservation of momentum is satisfied, if

$$c_1 = c_2 \text{ or } n(f_1) = n(f_2), \tag{3.26}$$

with $n(f_i)$ refractive index of the medium for the wave with frequency f_i. The condition that the refractive indices for the two frequencies f_1 and f_2 must be the

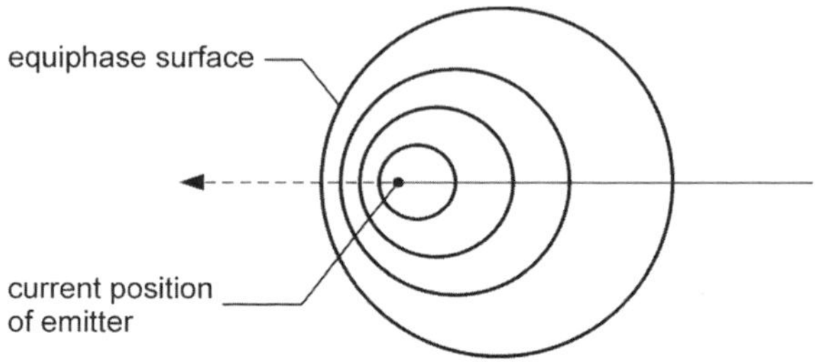

Fig. 3.18 Wave propagation, if the emitter is moving to the left

same, is not easily achievable. All matter shows in fact dispersion, i.e., the refractive index n depends on the frequency f of the light, cf. Fig. 3.1. But the refractive indices can be matched with a birefringent crystal.

3.6 Optical Doppler Effect

Particularly impressive, the Doppler effect can be demonstrated in acoustics. Depending on how a sound emitter and receiver are moving against each other, the receiver will observe an increased or decreased frequency, see Fig. 3.18. The same is observed, when a light source and a light detector are moved relative to each other.

The optical Doppler effect is, in contrast to the acoustic one, a relativistic effect. But for small velocities, there is no difference. That is why we can use the well-known formulas

$$\lambda' = \lambda \mp \Delta\lambda \tag{3.27a}$$

or

$$f' = f \pm \Delta f \tag{3.27b}$$

with

$$\Delta\lambda = \lambda \frac{v_r}{c} \tag{3.28a}$$

or

$$\Delta f = f \frac{v_\mathrm{r}}{c} = \frac{v_\mathrm{r}}{\lambda} \tag{3.28b}$$

in the limit of $|v_\mathrm{r}| \ll c$. v_r is the component of the relative velocity in the direction of the line connecting light source and light detector, λ' or f' is the Doppler shifted wavelength or frequency, $\Delta\lambda$ or Δf is the change in wavelength or frequency and λ

or f is the wavelength or frequency in the case of $v_r = 0$, respectively. If light source and detector are approaching each other, the wavelength is shortened (negative sign in (3.27a)) or the frequency is raised (positive sign in (3.27b)), respectively. In the other case, the signs change. The Doppler effect of a light wave scattered at a moving particle will be discussed in more detail in Chap. 11.

For almost all applications in laser measurement technology, the relative speeds compared with the speed of light, are very small. The non-relativistic approximations (3.27a)–(3.28b) are, therefore, adequate for our purposes.

References

1. M. Born, E. Wolf, A.B. Bhatia, P.C. Clemmow, D. Gabor, A.R. Stokes, A.M. Taylor, P.A. Wayman, W.L. Wilcock, *Principles of Optics* (Cambridge University Press, Cambridge, 2003). ISBN 0 521 642221
2. L. Bergmann, C. Schaefer, H. Niedrig, *Lehrbuch der Experimentalphysik, Bd. III: Optik.* (Walter de Gruyter, Berlin, 2004). ISBN 3-11-017081-7
3. E. Hecht, *Optics* (Reading, Pearson, 2003). ISBN 978-8177583571
4. H. Haken, H.C. Wolf, *The Physics of Atoms and Quanta* (Springer-Verlag, Heidelberg, 2005). ISBN 978-3-540-20807-5
5. D. Meschede, *Optik, Licht und Laser* (Teubner, Wiesbaden, 2005). ISBN 3-519-13248-6
6. A. Donges, *Elementare Quantenoptik* (Hüthig, Heidelberg, 1990). ISBN 3-7785-1901-8
7. H. Haferkorn, *Optik—Physikalisch-technische Anwendungen und Grundlagen* (John Wiley & Sons, Hoboken, 2008). ISBN 9783-527625024
8. *The Mie Theory—Basics and Applications,* ed. by W. Hergert, Th. Wriedt. Springer Series in Optical Sciences, vol 169 (Springer, Heidelberg, 2012), ISBN 978-3-642-28738-1
9. C.F. Bohren, D.R. Huffman, *Absorption and Scattering of Light by Small Particles* (J. Wiley & Sons, Hoboken, 1998). ISBN 978-0471293408

Chapter 4
Beam Shaping and Guiding

Abstract This chapter is divided in three parts. In the first section, various optical components for beam guiding and beam shaping are presented. In particular, we treat beam deflection, beam splitting, the modulation of polarization, intensity, wavelength and phase. This is followed by a section on beam shaping and focusing of a Gaussian beam. In the last section the propagation of light in optical fibers is discussed.

The parameters of a laser beam are determined by the set-up of the laser. In many applications in metrology, it is not possible to guide the laser beam directly to the measuring object or the scattered light to the detector. In most cases, the use of optical elements, such as beam splitters, mirrors or phase shifters, is required [1–12].

A laser beam—for example a Gaussian beam with a known Rayleigh length—is characterized by five beam parameters

- direction of propagation,
- direction of polarization,
- beam power or irradiance or intensity,
- wavelength and
- phase angle,

see Sect. 2.2. If required, these five parameters can be affected actively, i.e., modulated. The most important optical components to modulate a laser beam are described in the following section.

4.1 Optical Elements for Beam Modulation

4.1.1 Beam Deflection and Beam Splitting

To change the direction of propagation of a laser beam, plane mirrors can be deployed. We distinguish between metal and dielectric mirrors.

A. Donges and R. Noll, *Laser Measurement Technology*, Springer Series in Optical Sciences 188, DOI: 10.1007/978-3-662-43634-9_4,

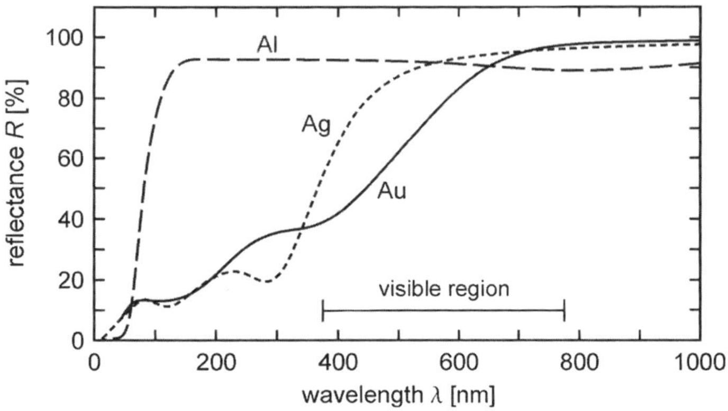

Fig. 4.1 Reflectance R of various, polished metal surfaces as a function of the wavelength λ

With metallic mirrors, the reflection takes place on the polished metal surface. Metals have, especially in the infrared spectral range, a high reflectance. The reflectance (or reflectivity) increases typically while going from the ultraviolet to the infrared region, see Fig. 4.1.

In the simplest case, a dielectric mirror is represented by a glass plate. Usually on one side of the plate an anti-reflection coating is applied, in order to avoid undesired double reflection. The simple glass plate has the disadvantage of a low reflection. The reflectance in the visible range at normal incidence is only about 4 %, see Sect. 3.2.2. That is why up to 30 dielectric layers with alternating refractive indices are applied by vapor deposition on the non-coated side of the plate, see Fig. 4.2. All individual layers have an optical thickness, i.e., the refractive index times the geometrical thickness, equal to a quarter wavelength ($\lambda/4$). The reflected waves add up by constructive interference, so that reflectances R of more than 99.9 % can be achieved, see Fig. 4.3. That is why the dielectric multilayer mirrors are also named interference mirrors. In case of normal incidence of the laser beam on the mirror, the wavelengths for maximum reflection are given by

$$\lambda_q = \frac{4d_{\mathrm{opt}}}{2q-1}, \quad q = 1, 2, 3, \ldots, \tag{4.1}$$

where d_{opt} is the optical thickness that each layer must have.

The reflectance of a dielectric mirror (in a surrounding medium with refactive index $n_s = 1$) at normal incidence is given by

$$R = \left[\frac{n_1^2\left(\frac{n_1}{n_2}\right)^{2l} - n_3}{n_1^2\left(\frac{n_1}{n_2}\right)^{2l} + n_3}\right]^2, \tag{4.2}$$

where $l+1$ layers have the refractive index n_1, l layers have the refractive index n_2 and n_3 is the refractive index of the substrate and $n_1 > n_2$ [13]. Equation (4.2)

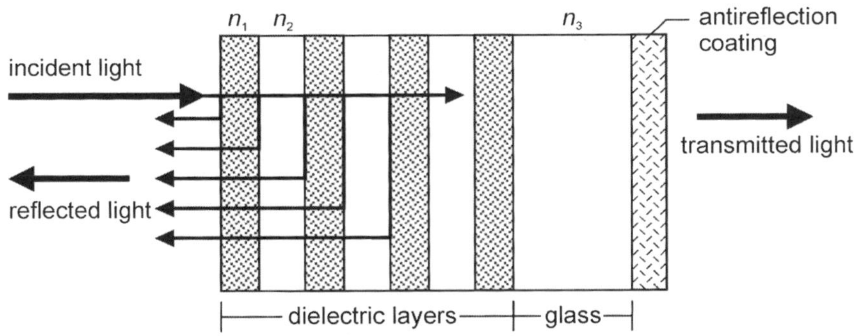

Fig. 4.2 Schematic structure of a dielectric multilayer mirror

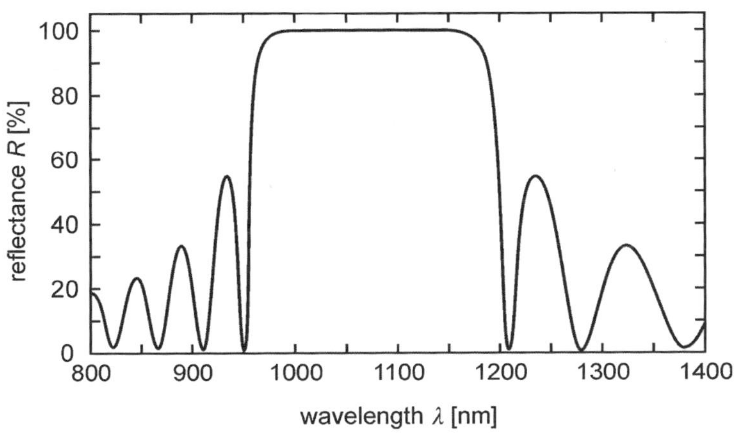

Fig. 4.3 Wavelength dependence of the reflectance of a dielectric multi-layer mirror

refers to Fig. 4.2 which shows that the set of layers starts and ends with a high-index layer (n_1) so that the total number of dielectric layers is an odd number.

As already described in Sect. 3.2, the reflectance of metal mirrors as well as for dielectric multilayer mirrors depends on the angle of incidence, wavelength and polarization state of the light. This disadvantage can be avoided, if for the beam deflection the total internal reflection in deflecting prisms is used, see Fig. 4.4. In order to avoid reflection losses, the entrance and exit surfaces of the prism have an antireflective coating. With deflecting prisms, reflectances R of practically 100 % can be achieved.

Disadvantage of the deflecting prism shown in Fig. 4.4a is that the incident laser beam must necessarily fall perpendicularly to the side surface in order to achieve a rectangular deflection. This disadvantage is avoided with the pentaprism, see Fig. 4.4c. Regardless of its tilting, the pentaprism delivers an output beam, which is oriented perpendicularly to the input beam.

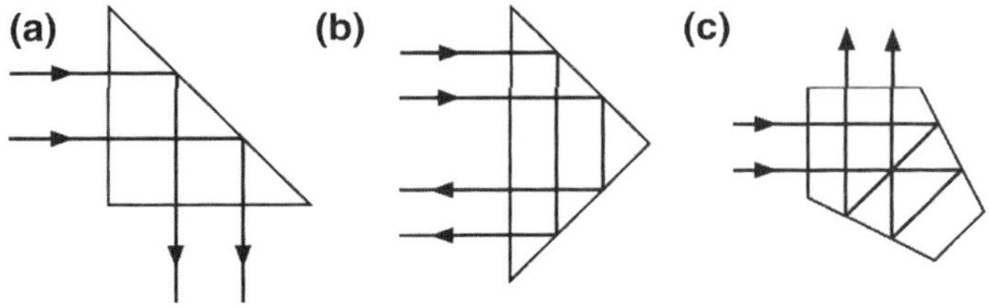

Fig. 4.4 Different kinds of deflecting prisms

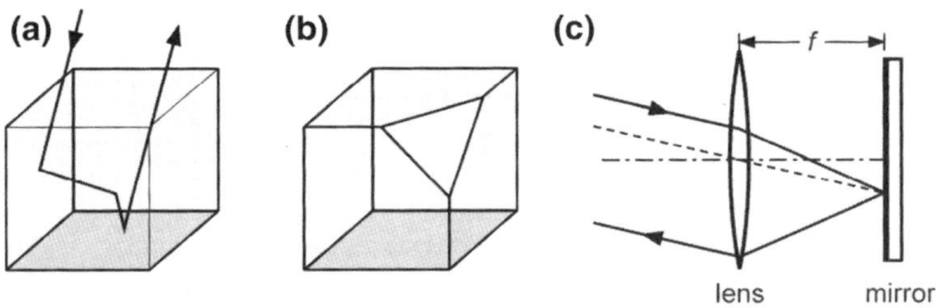

Fig. 4.5 Different kinds of retroreflectors, **a** triple mirror, **b** corner reflector consisting of a cube of transparent material with a truncated corner, **c** cat's eye

For some applications, mirrors are required reflecting the light opposite to the direction of incidence (180° deflection) regardless of their orientation. This is performed by so-called retroreflectors. They consist, for example, of three mirrors—a so-called triple mirror—being arranged in such a way that they represent the corner of a cube. In this triple mirror, each incident beam is reflected once at each mirror, which leads to an overall deflection of 180°, see Fig. 4.5a. Glass cubes with a missing corner are also suited as retroreflectors, see Fig. 4.5b. These elements take advantage of the total internal reflection. Finally, we point to the so-called cat's eye. A cat's eye is in the simplest case a lens with a plane mirror in its back focal plane, see Fig. 4.5c.

As a beam splitter, partially reflecting mirrors—interference or metal mirrors—may be used. Interference mirrors are made of materials, which are characterized by a very low absorption. The non-reflected portion of the radiation propagates, therefore, almost loss-less through the mirror. The absorption losses are typically less than 1 %.

A metallic beam splitter consists of a very thin metal layer since a thickness of some micrometers would absorb the incident light, practically completely. Therefore, such a thin metal layer must have a carrier, for example a glass plate with an antireflective coating at the back.

For the beam splitters discussed so far, a possibly unwanted lateral beam displacement occurs, see Fig. 4.6a. This lateral beam displacement can be reduced if a thin membrane is taken as a carrier of the reflecting layer instead of a relatively thick glass plate, see Fig. 4.6b. These so-called pellicle beam splitters are mechanically very unstable. More suitable are cubic beam splitters. These consist

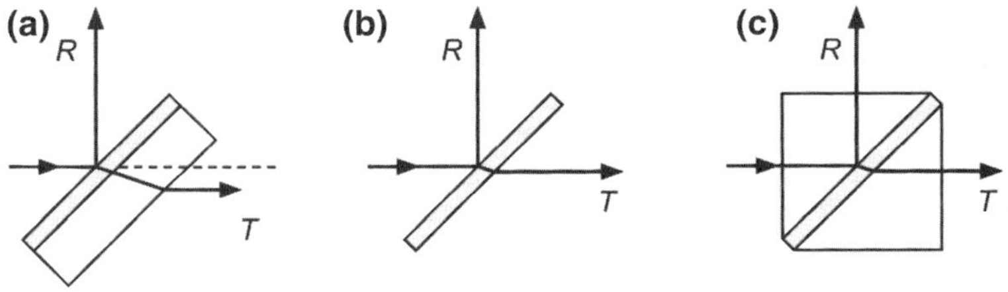

Fig. 4.6 Beam splitters with 90°-deflection

of two rectangular prisms, which are assembled to form a cube, see Fig. 4.6c. Between the opposing base faces of the two prisms, the metallic or dielectric reflective layer is located. Since incident and emergent beams pass the air-glass interfaces perpendicularly, only a negligible lateral beam displacement occurs.

Splitting mirrors can be characterized by specifying the quantities R, T and A. R is the already known reflectance. T and A mean transmittance and absorbance, respectively. They describe the transmitted or absorbed intensity, related to the incident intensity, respectively. Usually, scattering losses are allocated to the absorbing losses. These three quantities are related by:

$$R + T + A = 1 = 100\,\%. \qquad (4.3)$$

It is desirable to keep the absorbance as small as possible. Values of $A < 1\ \%$ are achieved. Deploying beam splitters, the user should take into account that the quantities R, T and A depend on wavelength, angle of incidence and polarization state of the light.

Polarization-sensitive beam splitters, in which the two split beams are polarized perpendicularly to each other, are frequently used in laser measurement technology. They are described in the next Sect. 4.1.2.

In special cases a variable beam deflection is required. Suitable for this purpose are pivoted mirrors and prisms. Two rotatable reflectors arranged in series, allow for a two-dimensional beam deflection (laser scanner), an example will be shown in Sect. 13.1.7.

The previously discussed methods for beam deflection are based on reflection. However, diffraction can also be used for beam deflection. This possibility is realized with acousto-optic modulators. Acousto-optic modulators consist of a transparent liquid or solid medium, usually quartz. In this medium, a standing or moving ultrasonic wave is excited. The spatially periodic compressions and rarefractions of the medium generate a phase grating. The grating constant is equal to the wavelength Λ of the ultrasonic wave. The sound frequencies range typically between 50 and 500 MHz, so that—with the speed of sound of about 5 km/s—grating constants of 10–100 μm can be realized.

In the theory of diffraction at a grating, two limiting cases can be distinguished: the diffraction at thin and the diffraction at a thick diffraction grating. Whether a grating is to be considered as thin or thick, is determined by comparing the

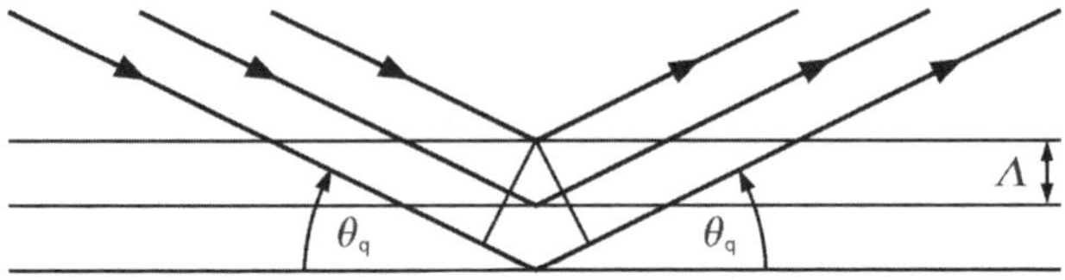

Fig. 4.7 Illustration of Bragg reflection

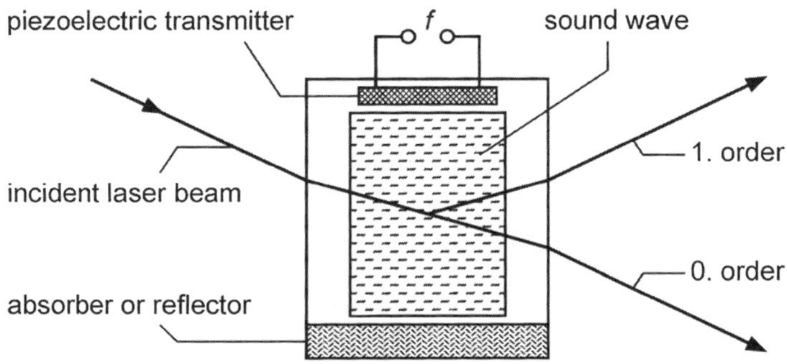

Fig. 4.8 Construction and operation of an acousto-optic modulator. The dashed lines illustrate the wavefronts of the ultrasound wave

thickness D of the grating with the quantity Λ^2/λ. The acousto-optic modulators used in laser technology satisfy in general the condition $D > \Lambda^2/\lambda$. Thus, we consider only the case of diffraction at a thick grating.

The diffraction of light waves at a thick grating is analogous to X-ray diffraction in crystals. W. H. and W. L. Bragg recognized 1913 that the diffraction of X-rays is to be regarded as a reflection on the lattice planes of the crystal. In contrast to normal reflection, the so-called Bragg reflections occur only at very specific angles of incidence—the glancing angles. At these glancing angles the reflected rays at neighboring lattice planes have a path difference, which is an integer multiple of the wavelength of the X-rays, i.e. $\Delta s = q\lambda$ with $q = 1, 2, 3, \ldots$, see Fig. 4.7. The so-called Bragg equation describes the possible glancing angles. Transcribed to acousto-optic modulators, it reads

$$2\Lambda \sin \theta_q = q\frac{\lambda_0}{n} \quad q = 1,\ 2,\ 3,\ \ldots. \tag{4.4}$$

Here Λ and λ_0/n denote the wavelength of the ultrasonic wave and the wavelength of the diffracted light in the diffracting medium with the refractive index n, respectively. An acousto-optic modulator is also referred to as a Bragg cell. In a Bragg cell only a single diffraction order appears. It can be achieved—provided that the laser beam is irradiated under the glancing angle—that almost 100 % of the incident irradiance is diffracted into this single order, see Fig. 4.8.

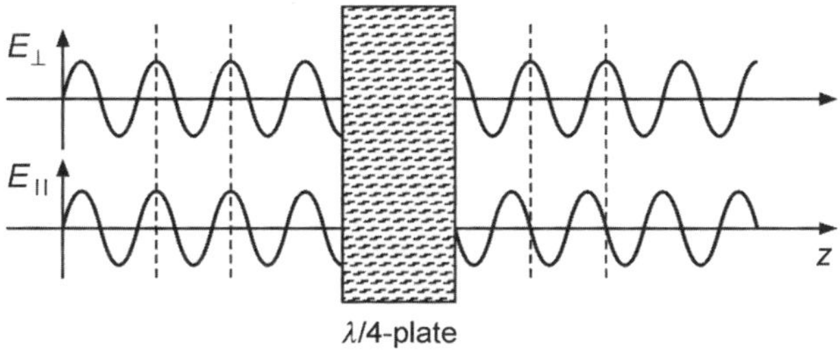

Fig. 4.9 Illustration of the action of a quarter-wave plate

By variation of the ultrasonic frequency or by switching the ultrasonic transmitter on and off, a modulation of the direction of the diffracted laser beam is possible. The achievable bandwidth of some MHz is limited by the running time

$$\tau = \frac{d}{v}, \tag{4.5}$$

in which the ultrasonic wave travels across the laser beam diameter d. v denotes the speed of sound in the Bragg cell.

4.1.2 Modulation of the Polarization

In the following, we discuss optical components allowing to influence actively the polarization direction of the laser beam. They are based mostly on birefringence. The basic principles were explained in Sect. 3.2.4.

The first polarization changing element we consider, is called quarter-wave plate ($\lambda/4$-plate). It is nothing but a thin birefringent crystal of suitable thickness. The incident linearly polarized laser beam can formally be decomposed into two identical waves, whose polarization directions are perpendicular to each other. The polarization directions of these two partial waves are perpendicular or parallel to the optical axis of the crystal, respectively. Therefore, the two partial waves correspond to the ordinary and the extraordinary ray, which propagate with different phase velocities $c_{\text{o.}}$ and $c_{\text{e.o.}}$, respectively. A spatial separation of the two beams does not occur because both waves pass the crystal perpendicularly to the optical axis, cf. Fig. 3.11. Because of the difference of the phase velocities, the two waves are mutually phase-shifted, when they leave the plate. The thickness of a quarter-wave plate corresponds—according to its name—to a phase shift with an absolute value of $|\Delta\varphi| = \pi/2$, see Fig. 4.9. The superposition of the two phase-shifted partial waves leaving the crystal plate forms a circularly polarized wave as illustrated in Fig. 4.10. Contrariwise, a quarter-wave plate converts also circularly polarized light into linearly polarized light.

In a circularly polarized light wave, the tip of the electric field vector is located on a helical line, which moves without rotation with the speed of light in the

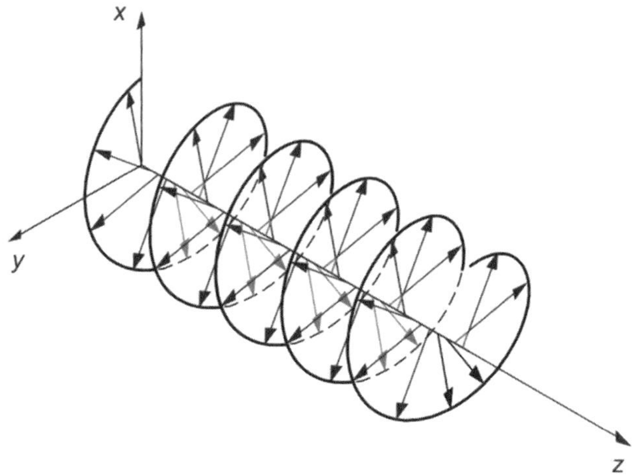

Fig. 4.10 Representation of a circularly polarized wave. The wave propagates in the $+z$-direction. A left circularly polarized wave is shown

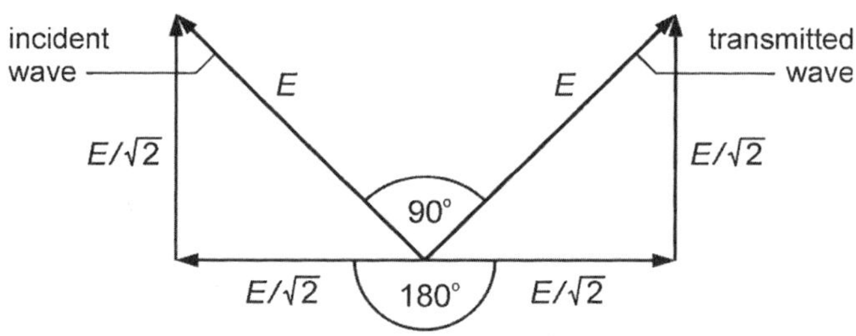

Fig. 4.11 Principle of a half-wave plate

direction of propagation. The pitch of the helix is equal to the wavelength. That is why the field strength vector rotates in a fixed plane (z = const.) with the light frequency. There is a distinction between right and left circularly polarized light, depending on whether the field vector rotates clockwise or anti-clockwise with respect to the direction of propagation.

Circularly polarized light is, just as linearly polarized light, a special case of elliptically polarized light. In elliptically polarized light, the tip of the field strength vector rotates on an elliptical helical line. Elliptically polarized light can be regarded as a superposition of two orthogonally linear polarized waves with any phase shift and different amplitudes.

If the thickness of a quarter-wave plate is doubled, then the phase shift behind the crystal is $|\Delta\varphi| = \pi$. Such a half-wave plate ($\lambda/2$-plate) rotates the polarization direction of a linearly polarized wave by 90°, see Fig. 4.11 or changes the direction of rotation of circularly polarized light from clockwise to anti-clockwise and vice versa.

Quarter-wave or half-wave plates are also made of materials that become birefringent only by applying an electric field, as explained for the Pockels or Kerr

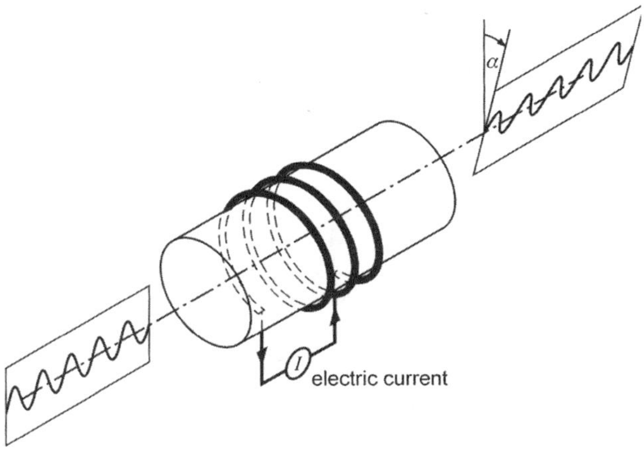

Fig. 4.12 Design and mode of operation of a Faraday rotator

effect, cf. Sect. 3.2.4. Such components allow a simple modulation of the polarization of the laser beam. With components, based on the Kerr effect, the shortest switching times, down to some picoseconds, can be achieved.

Optically active substances, such as a sugar solution or a quartz crystal, are used to rotate the polarization direction of a linearly polarized wave. Experimentally it is found that the rotation angle α is proportional to the thickness d of the optically active medium

$$\alpha = (\alpha)\, d, \tag{4.6}$$

where (α) is the specific rotation. It is a material and concentration-dependent quantity, which increases with decreasing wavelength. For crystalline quartz, the specific rotation (α) ranges in the visible spectrum from 10 to 50°/mm. The strong dependence of the rotation angle on the wavelength is referred to as optical rotary dispersion. By convention, the sign of (α) is chosen to be positive, if the substance is dextrorotatory. This means that for an observer looking towards the beam, the polarization direction rotates clockwise. In the other case, the substance is levo-rotatory and (α) is negative.

Some substances become optically active only under the influence of a magnetic field, which is known as the Faraday effect. This is used in a Faraday rotator, see Fig. 4.12. It consists of a substance, which is located in an externally applied magnetic field H. The angle of rotation α of the polarization direction is proportional to the magnetic field strength H and to the length d of the light path within this substance:

$$\alpha = VHd \tag{4.7a}$$

with

$$V = \frac{e}{m}\frac{\lambda_0}{2c_0}\frac{\mathrm{d}n}{\mathrm{d}\lambda_0}, \tag{4.7b}$$

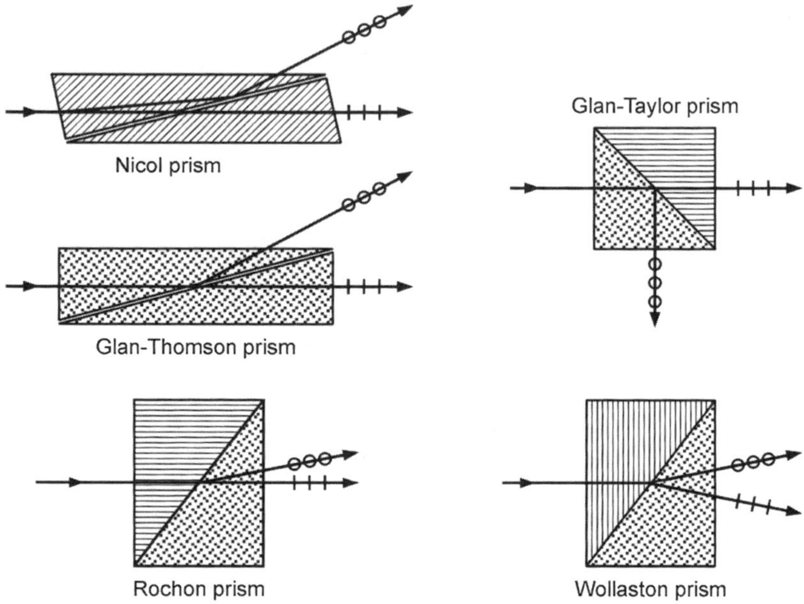

Fig. 4.13 Schematic representation of some polarizing prisms. The incident unpolarized light beam is always divided into two linearly polarized light beams. The symbol I or ○ indicates the direction of polarization. The direction of the optical axis is indicated by the *hatching* (optical axis parallel to the lines) or *dotting* (optical axis perpendicular to the drawing plane)

where *e/m* is the specific charge of the electron, n the refractive index and c_0, λ_0 speed of light, wavelength in vacuum, respectively. V is Verdet's constant. For lead silicate glass in the visible spectrum V amounts to about 20 μrad/A. If a long coil is used to generate the magnetic field, we obtain for the angle of rotation

$$\alpha = VNI, \tag{4.7c}$$

where N is the number of windings and I is the coil current. For 300 windings and a coil current of 10 A, the angle of rotation of lead silicate glass is only 3.5°. When modulating the polarization direction with a Faraday rotator it should be noted that the attainable modulation frequencies are limited by the large inductance of the magnetic field coil.

At the end of this section, we present polarizing prisms. Polarizing prisms allow to divide an incident beam into two beams. These two beams are polarized perpendicularly to each other and they propagate in different directions. This is achieved through a clever use of birefringence. Figure 4.13 gives an overview of the most common polarizing prisms. Exemplary, the principle of polarization prisms is explained for the Glan-Taylor prism.

A Glan-Taylor prism consists of two rectangular birefringent prisms, cemented together to form a cube, see Fig. 4.13. The refractive index of the cement is equal

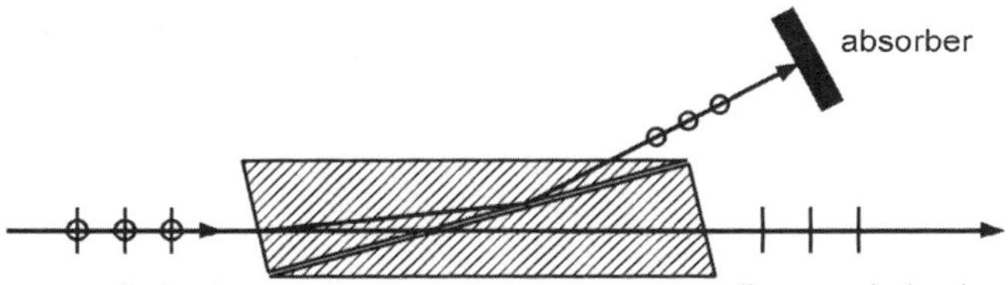

Fig. 4.14 Operating principle of a polarizer

to the ordinary refractive index $n_{o.}$ of the prism. The optical axis of the first prism is oriented perpendicularly to the drawing plane. Thus, the phase velocity and the refractive index in the first prism depends on the polarization direction of the incident light beam. For the ordinary ray, whose polarization direction is oriented perpendicularly to the optical axis, the refractive index is $n_{o.}$. $n_{e.o.}$ is the refractive index for the extraordinary ray polarized perpendicularly thereto. The optical axis of the second prism is oriented parallelly to the beam direction. That is why the refractive index in the second prism is independent of the polarization direction and amounts to $n_{o.}$. For that reasons, the ordinary ray is not reflected nor refracted while passing the interface between the two prisms.

This is in contrast to the extraordinary ray, which is totally reflected, if the condition

$$\frac{n_{o.}}{n_{e.o.}} < \sin(45^\circ) = \frac{\sqrt{2}}{2} \tag{4.8}$$

holds.

Each polarization prism separates a non-polarized light beam into two mutually perpendicularly linearly polarized waves. Therefore, a polarizing prism is referred to as polarizer, where usually one of the two linearly polarized waves will be canceled by a beam stop, see Fig. 4.14. If the polarization direction of an incident wave is analyzed, the polarizing prism is referred to as an analyzer.

With a Glan-Taylor prism and a half-quarter plate, a beam splitter with a variable division ratio can be set up. The design and principle is shown in Fig. 4.15. Let us consider a linearly polarized laser beam, whose direction of polarization coincides with the optical axis of the half-wave plate, cf. Fig. 4.11. Then, the polarization direction of the beam remains preserved and the following Glan-Taylor will not deflect the beam. Next, we turn the half-wave plate so that its optical axis and the polarization direction of the incident light form an angle of $\alpha = 45°$. In this case, the polarization direction of the incident light turns around 90° and the light beam is deflected perpendicularly by the Glan-Taylor prism. In the general case, when the angle α takes any values between 0 and 45°, the light behind the half-wave plate is elliptically polarized.

Accordingly, this light is split by the Glan-Taylor prism into a transmitted and reflected beam. Since practically no absorption occurs, it always reads

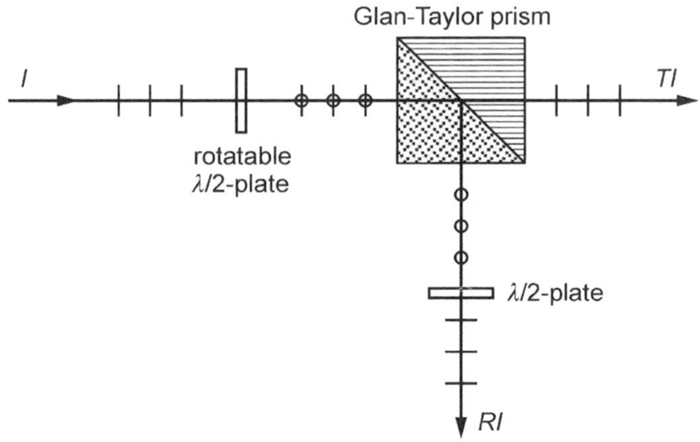

Fig. 4.15 Principle of a variable beam splitter

$$R = 1 - T, \tag{4.9}$$

where the reflectance R and the transmittance T takes continuously—depending on the angular position of the half-wave plate—values between 0 and 1. It should be noted that reflected and transmitted beam are mutually perpendicularly polarized. If this is disturbing, the polarization direction of both beams can be aligned parallel by using another half-wave plate, cf. Fig. 4.15.

4.1.3 Modulation of the Intensity

In many applications, the power or intensity of the laser beam must be modulated. In the simplest case, mechanical components are used. An example is the shutter, as known in photography allowing light to pass for a determined period of time to expose a film in an old fashioned mechanical camera. So-called optical choppers are utilized for periodic modulation. A chopper is a rotating disc with slits or holes along the circumference. Choppers have the disadvantage that only comparatively low modulation frequencies of up to 100 kHz can be achieved.

Widely used are electro-optical modulators. They consist of a combination of an electro-optical element with two polarizing elements. As an electro-optic element, Faraday rotators as well as Kerr cells or Pockels cells are applied, see Sects. 4.1.2 and 3.2.4. The principle of an electro-optic modulator is shown in Fig. 4.16. The incident light is linearly polarized by a polarizer. This polarizer can be omitted, if the laser beam already has a defined direction of polarization. The following electro-optic element changes the state of polarization of the light. Either the direction of polarization is rotated or the incident light is converted into elliptically polarized light. The second polarizer transforms the changes of the polarization into an intensity modulation.

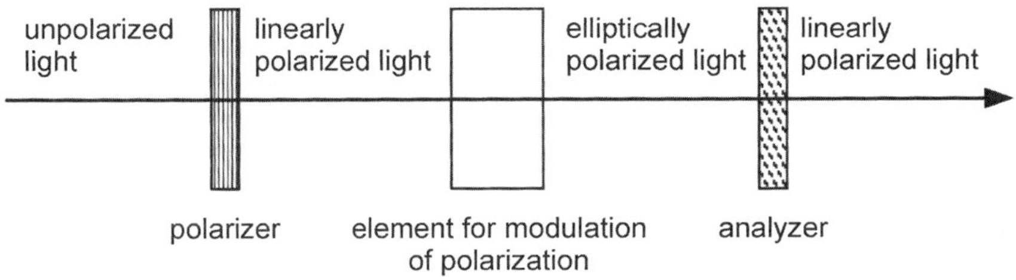

Fig. 4.16 Principle of an electro-optic modulator

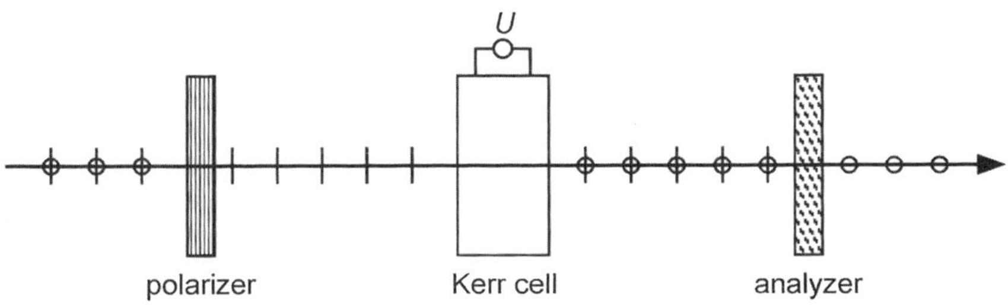

Fig. 4.17 Set-up to modulate the intensity by a Kerr cell

As an example, we consider intensity modulation by a Kerr cell, see Fig. 4.17. When no voltage is applied to the Kerr cell, it has no effect on the polarization of the incident light. Since the forward directions of the polarizer and the analyzer are oriented perpendicularly to each other, practically no light is transmitted by the analyzer ($T = 0$ for $U = 0$). However, if we apply the so-called half-wave voltage $U_{\lambda/2}$, the Kerr cell acts as a half-wave plate. The half-wave voltage varies depending on the respective Kerr cell and wavelength; it amounts 10–20 kV. Therefore, the Kerr cell turns—when it is oriented correctly—the direction of polarization by 90°. Thus, the incident light passes the analyzer practically without absorption ($T = 1$). For operation voltages $0 < U < U_{\lambda/2}$, the transmittance of the modulator ranges between two extreme values. The transmittance as a function of the applied voltage U is described by the following non-linear relationship, see Fig. 4.18:

$$T = \sin^2\left[\frac{\pi}{2}\left(\frac{U}{U_{\lambda/2}}\right)\right]. \tag{4.10}$$

If polarizer and analyzer are aligned in parallel, the sine function in (4.10) is to be replaced by the cosine function.

For an optical shutter, only the limiting cases $T = 0$ and $T = 1$ are of importance. Therefore, the non-linearity of the characteristic curve, see (4.10), is not disturbing. This is quite different, when a linear relation between the transmittance and the applied voltage is needed. In this case, the Kerr cell must be driven near the operating point $U \approx 0.75 \cdot U_{\lambda/2}$, where the relation is quasi-linear, see Fig. 4.18.

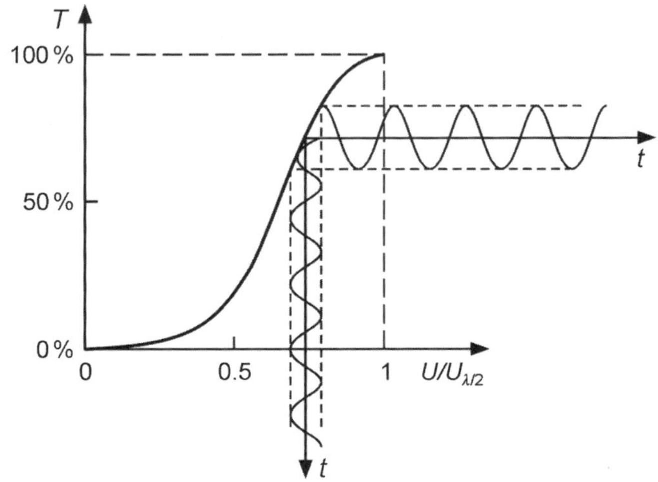

Fig. 4.18 Characteristic curve of a Kerr cell for irradiance modulation

Potentially the Kerr cell has to be biased with a constant voltage U_0. Experimentally it is easier to change the state of polarization with an additional quarter-wave plate making the DC voltage dispensable.

Also optical elements for beam deflection combined with an absorber, are suited as optical shutters. The principle is shown in Fig. 4.19. Absorption and transmission can be selected by switching between the two beam directions. An appropriate component for deflection is an acousto-optic modulator, as already presented in Sect. 4.1.1. Alternatively, birefringence induced by an electric field can be used for beam deflection, see Sect. 3.2.4.

4.1.4 Modulation of the Wavelength

Wavelength or frequency are modulated with the help of the Doppler effect, see Sect. 3.6. A laser beam being reflected by a moving mirror, undergoes a frequency change. The same happens, when light is diffracted by a moving, e.g. rotating, grating. However, of great practical importance is the diffraction of light at a moving ultrasonic wave in an acousto-optical modulator, see Sect. 4.1.1. The frequency change Δf of the diffracted laser beam is given by:

$$\Delta f = q f_a \quad q = 0, \pm 1, \ldots . \tag{4.11}$$

Here f_a is the frequency of the ultrasonic wave and q is the order of diffraction. If the frequency f_a of the ultrasonic wave is modulated, frequency and wavelength of the light change too.

A further optical component, which changes the wavelength or the frequency of a laser beam, is the so-called frequency doubler. As frequency doublers certain

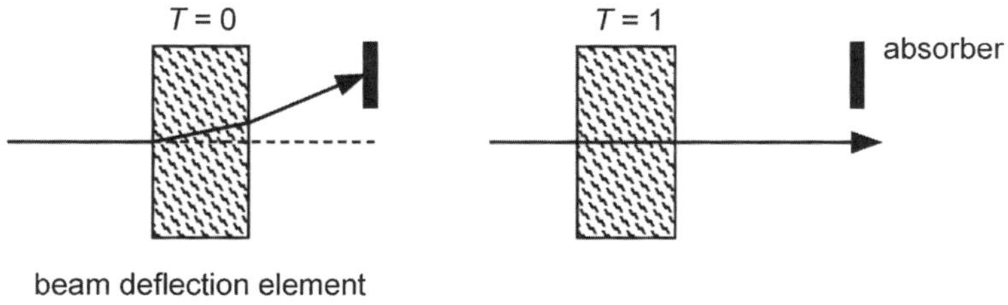

Fig. 4.19 Principle of an optical switch using beam deflection. The beam deflection is caused by birefringence, e.g. Pockels or Kerr effect, or by diffraction in an acousto-optic modulator

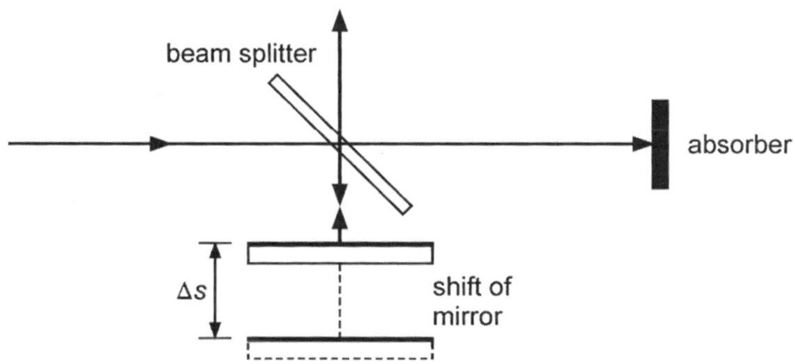

Fig. 4.20 Phase shift by changing the geometric path

crystals, e.g. potassium dihydrogen phosphate (KDP), are applied. Generally, frequency doubling occurs only at high beam irradiances of about 1 $\mathrm{MW/cm^2}$. The physical principle of harmonic generation has been explained in Sect. 3.5.

4.1.5 Phase Shifter

A phase shift of an electromagnetic wave $\Delta\varphi$ is caused by a change of the optical path. The optical path is the product of the geometric path and the refractive index. Therefore, both possibilities are used to shift the phase of a wave: changing of geometric path and changing of refractive index.

Optical components capable for phase modulation, are called phase shifter. We consider some examples.

In the simplest case, a phase shifter varies the geometric path. In the example shown in Fig. 4.20, the phase shift is determined by the displacement Δs of a deflection mirror. The phase shift is given by

$$\Delta\varphi = 4\pi\frac{\Delta s}{\lambda}. \tag{4.12}$$

In order to realize a vibration-free and reproducible shift Δs of the deflection mirror, piezoelectric actuators are used mostly.

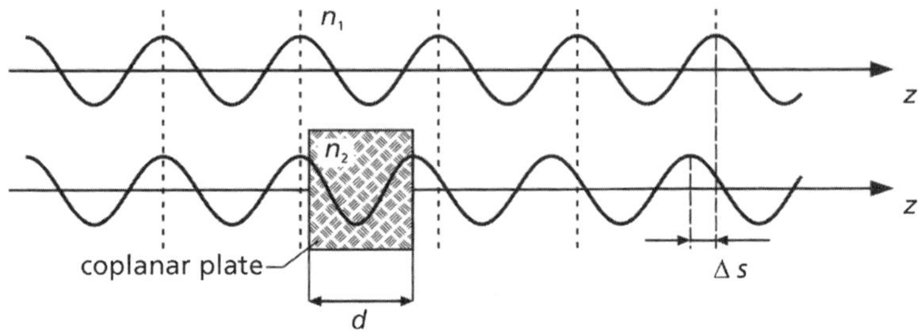

Fig. 4.21 Phase shift of a wave passing a plane parallel plate

Fig. 4.22 Principle of Babinet's phase shifter

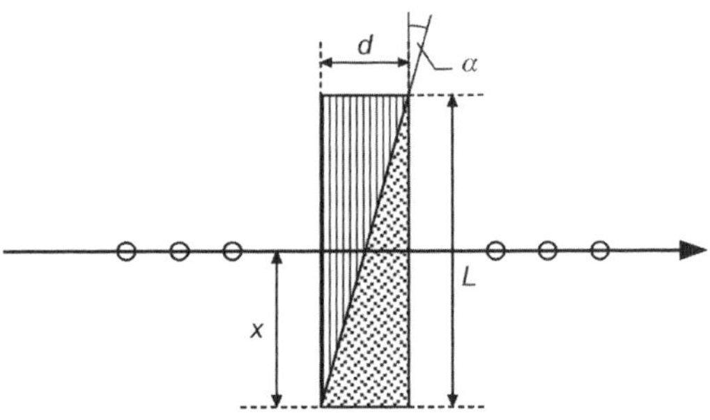

Let us now consider the case, in which the optical path is influenced via a change of the refractive index. A laser beam passes through a plane-parallel transparent plate with thickness d and refractive index n_2, see Fig. 4.21. Since the light within the plate propagates slower than in air with $n_1 \approx 1$, the beam experiences a phase shift corresponding to the difference of the optical paths $\Delta s = (n_2 - n_1)\, d$:

$$\Delta\varphi = 2\pi \frac{(n_2 - n_1)\, d}{\lambda_0}. \tag{4.13}$$

To adjust any desired phase shift, Babinet's phase shifter can be used, which is in principle a Wollaston prism at which the angle $\alpha \approx d/L$ is chosen very small, see Fig. 4.22. Therefore, a passing beam undergoes a negligible deflection only. A light beam with a polarization direction perpendicular to the plane of projection, which enters the prism at a lateral distance x, travels in the first prism the length $d \cdot x/L$ and in the second prism the length $d \cdot (1 - x/L)$. Because of the different orientations of the optical axes of the first and second prism, we have to apply in the first prism the refractive index $n_{\text{o.}}$ and for the second prism the refractive index $n_{\text{e.o.}}$. Therefore, the optical path length is given by:

$$s_{\text{opt}} = n_{\text{o.}} \frac{x}{L} d + n_{\text{e.o.}} \left(1 - \frac{x}{L}\right) d. \tag{4.15}$$

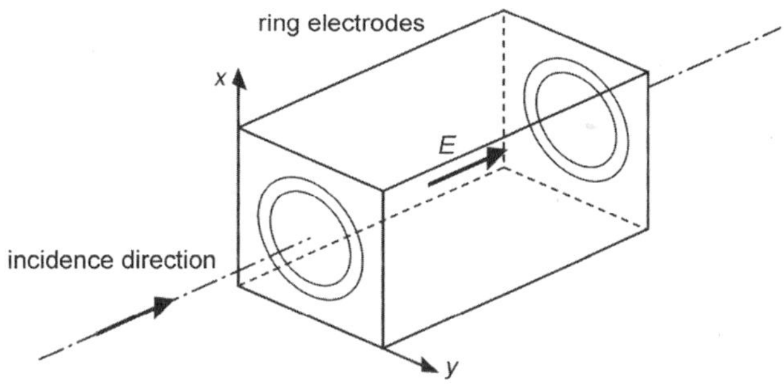

Fig. 4.23 Pockels cell for phase modulation

In comparison to a light beam entering at $x = 0$, the light beam considered has a phase shift of

$$\Delta\varphi(x) = \frac{2\pi\, d(n_{\mathrm{o.}} - n_{\mathrm{e.o.}})}{\lambda_0\, L} x. \tag{4.16}$$

By shifting Babinet's compensator in the lateral x-direction, any arbitrary phase shift in the range of $0 \leq \Delta\varphi \leq \frac{2\pi\, d|n_{\mathrm{o.}} - n_{\mathrm{e.o.}}|}{\lambda_0}$ can be adjusted. The use of Babinet's phase shifter—also called Babinet compensator—requires a sufficiently small beam cross section, so that $\Delta\varphi(x)$ remains practically constant over the cross section of the light beam.

The phase shifters described so far are not appropriate for a high frequency modulation of phase shifts. For dynamic phase modulation, electric field induced changes in the refractive index are used. For that, the Pockels effect is applied because there is a linear relationship between the change in the extraordinary refractive index and the electric field, see Sect. 3.2.4.

Figure 4.23 shows schematically the set-up of a so-called longitudinal Pockels cell for phase modulation. An appropriate birefringent crystal, such as KDP, is equipped with ring electrodes to apply a longitudinal electric field E in the direction of the optical axis. Thus, for a light beam propagating in the direction of the optical axis, the refractive indices n_{x} and n_{y} for states of polarization parallel to the x- or y-axis are changed by the applied longitudinal electric field, see Sect. 3.2.4:

$$n_x = n + \frac{1}{2} r\, n^3\, E, \tag{4.16a}$$

$$n_y = n - \frac{1}{2} r\, n^3\, E. \tag{4.16b}$$

For a light wave being polarized in the x-direction, a change of the optical path length results and, therefore, a phase shift of

$$\Delta\varphi = \frac{\pi\, r\, n^3\, U}{\lambda_0}. \tag{4.17}$$

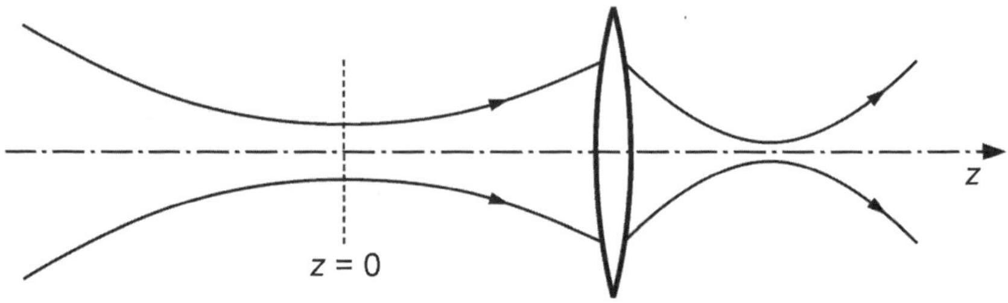

Fig. 4.24 Focusing of a Gaussian beam by a converging lens

Here r is the electro-optic coefficient, n the refractive index in the direction of the optical axis without electric field, U the applied voltage and λ_0 the wavelength in vacuum.

4.2 Propagation of Gaussian Beams

Lasers with stable resonators oscillating in the fundamental mode emit a Gaussian beam. For a given wavelength λ and position of the beam waist ($z = 0$), the geometry of the Gaussian beam can be specified by a single parameter, the Rayleigh length z_R, as described in Sect. 2.6. In many cases of laser measurement technology, the Gaussian laser beam is modified by optical elements. Examples of such elements are lenses or telescopes to focus or to expand the Gaussian beam. In the following, we assume that a Gaussian beam passing through an optical component, remains a Gaussian beam, see Fig. 4.24. This is true, for example, for a lens where the lens boundary causes no additional diffraction. In practice it is sufficient, if the diameter of the optical component is at least six times greater than the radius w of the Gaussian beam at the place of the component.

4.2.1 ABCD *Law*

Beam radius $w(z)$ and radius of curvature $\rho(z)$ of the wavefronts of Gaussian beams were introduced in Sect. 2.6.1. Both are z-dependent functions. The z-axis indicates the direction of propagation, wherein the beam waist is located at $z = 0$. The quantities $w(z)$ and $\rho(z)$ can be obtained from a single quantity, the so-called complex beam parameter $q(z)$. This complex beam parameter is defined by

$$q(z) = z + \mathrm{i}z_R. \tag{4.18}$$

i is the imaginary unit $\mathrm{i} = \sqrt{-1}$. For the quantities $w(z)$, $\rho(z)$ and $q(z)$ the following relationships hold:

$$\frac{1}{w(z)^2} = -\frac{\pi}{\lambda}\mathrm{Im}\left(\frac{1}{q(z)}\right) \tag{4.19}$$

and

$$\frac{1}{\rho(z)} = \mathrm{Re}\left(\frac{1}{q(z)}\right). \tag{4.20}$$

Convince yourself of the validity of the last two relations by inserting (4.18) in (4.19) and (4.20) and comparing the result with (2.37a) and (2.40).

If a Gaussian beam passes an optical element, so in general, its beam properties $w(z)$ and $\rho(z)$ and thus—according to (4.19) and (4.20)—the complex beam parameter $q(z)$ will change. The computation of $q(z)$ behind the optical component using (4.18), $q(z) = z + \mathrm{i}z_\mathrm{R}$, is no longer allowed. Instead, the so-called *ABCD* law has to be applied. It combines the complex beam parameter q_in at the entrance of the optical system with the complex beam parameters q_out at the output:

$$q_\mathrm{out} = \frac{Aq_\mathrm{in} + B}{Cq_\mathrm{in} + D} \tag{4.21a}$$

or

$$\frac{1}{q_\mathrm{out}} = \frac{C + {}^{D}\!/_{q_\mathrm{in}}}{A + {}^{B}\!/_{q_\mathrm{in}}}. \tag{4.21b}$$

A, B, C and *D* are the matrix elements of the so-called ray transfer matrix *M*:

$$M = \begin{pmatrix} A & B \\ C & D \end{pmatrix}. \tag{4.22}$$

M is a 2×2 matrix, which is characteristic for a considered optical element or combination of components. Once you have calculated q_out, beam radius $w(z)$ and radius of curvature $\rho(z)$ of the wavefronts of the outgoing beam can be determined by (4.19) and (4.20).

In many cases, the incident Gaussian beam is focused by an optical component and a second beam waist is formed behind that optical element, see e.g. Fig. 4.24. The beam waist at $z = 0$ is by convention always the waist of the incident beam that is not affected by the optical component.

Figure 4.25 gives an overview of the ray transfer matrices of the most important optical elements. The ray transfer matrices, which we consider in this section, are those used in ray tracing [14, 15]. The ray transfer matrix of an optical component,

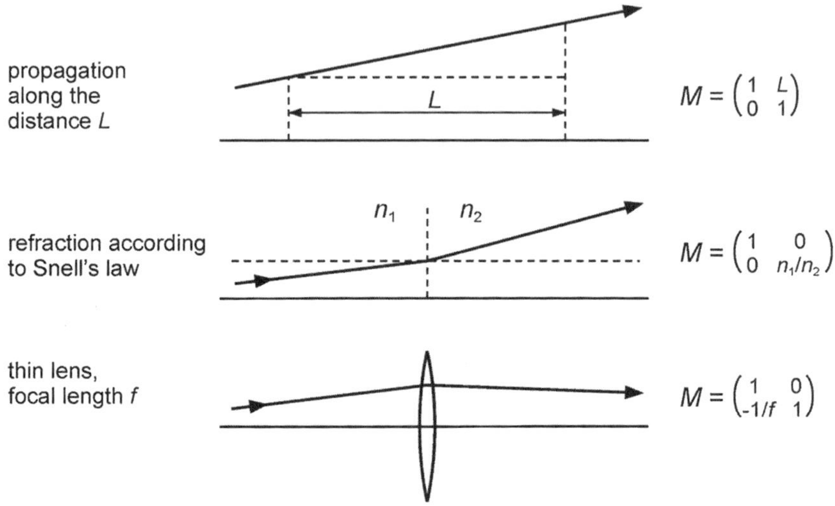

Fig. 4.25 Ray transfer matrices of the most important optical elements

which is composed of several subsystems 1, 2, …, n with the ray transfer matrices $M_1, M_2, \ldots, M_n$ is calculated by multiplying the matrices

$$M = M_n \cdot M_{n-1} \cdot \cdots \cdot M_2 \cdot M_1. \tag{4.23}$$

In the matrix multiplication it is important to ensure that the positions of the sub-matrices M_j are arrayed in reverse order in (4.23), i.e. starting with $j = n$ going to $j = 1$, whereas the Gaussian beam propagates from subsystem $j = 1$ to $j = n$.

Next we consider, as examples, focusing and expansion of a Gaussian beam.

4.2.2 Focusing

For many problems of optical metrology, focusing of Gaussian beams is required. For this purpose, we arrange at a distance g from the beam waist ($z = 0$) a converging lens. Behind the lens, a new Gaussian beam is formed, see Fig. 4.26. Next, we calculate the beam waist of the new beam behind the lens, i.e., the distance b between the new waist and the lens as well as the waist radius w_0' of the focused Gaussian beam.

The ray transfer matrix describing the propagation from the beam waist of the incident beam to the beam waist of the focused beam, is composed of three sub-matrices:

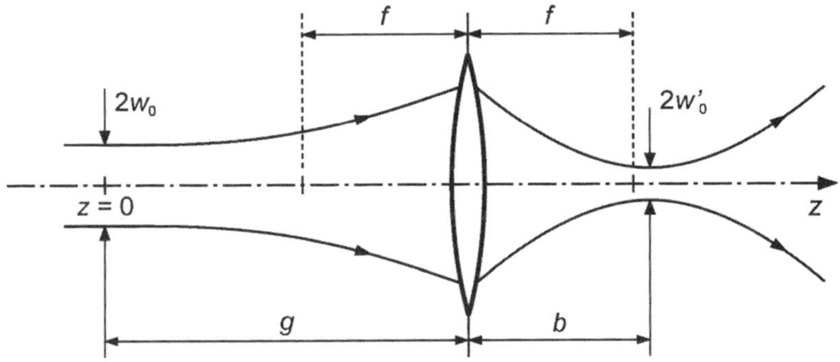

Fig. 4.26 Focusing of a Gaussian beam

- matrix for the beam propagation over the distance b,
- matrix for focusing by a thin lens of focal length f and finally
- matrix for the beam propagation over the distance g.

Observing the order we start with the matrix describing the last section of beam propagation and then backwards:

$$M_{\mathrm{f}} = \begin{pmatrix} 1 & b \\ 0 & 1 \end{pmatrix} \begin{pmatrix} 1 & 0 \\ -1/f & 1 \end{pmatrix} \begin{pmatrix} 1 & g \\ 0 & 1 \end{pmatrix} = \begin{pmatrix} 1 - \frac{b}{f} & g\left(1 - \frac{b}{f}\right) + b \\ -\frac{1}{f} & 1 - \frac{g}{f} \end{pmatrix}. \tag{4.24}$$

The complex beam parameter of the incident beam at the point of the waist ($z = 0$) is

$$q_{\mathrm{in}} = q(z = 0) = \mathrm{i} z_{\mathrm{R}} \quad \text{with} \quad z_{\mathrm{R}} = \frac{\pi\, w_0^2}{\lambda}, \tag{4.25}$$

where z_{R} denotes the Rayleigh length of the incident beam. The complex beam parameter q_{out} of the focused beam at the exit of the system is given by the *ABCD* law

$$q_{\mathrm{out}} = \frac{b + g\left(1 - \frac{b}{f}\right) + \mathrm{i} z_{\mathrm{R}}\left(1 - \frac{b}{f}\right)}{1 - \frac{g}{f} - \mathrm{i}\frac{z_{\mathrm{R}}}{f}}. \tag{4.26}$$

So far, (4.26) for the complex beam parameter is valid for any distance b behind the lens. However, since b should indicate the position of the waist of the focused beam, the real part of q_{out} must vanish, cf. (4.20). From this requirement, it follows

$$b = \frac{f\left[z_{\rm R}^2 + g(g-f)\right]}{z_{\rm R}^2 + (g-f)^2}. \tag{4.27}$$

In the limit of $z_{\rm R} \rightarrow 0$, (4.27) simplifies to the imaging equation of geometrical optics:

$$\frac{1}{f} = \frac{1}{g} + \frac{1}{b}. \tag{4.28}$$

The imaginary part of $1/q_{\rm out}$ delivers, taking (4.27) into account, the Rayleigh length $z_{\rm R}^{'}$ of the focused Gaussian beam:

$$z_{\rm R}^{'} = \frac{z_{\rm R} f^2}{(g-f)^2 + z_{\rm R}^2}. \tag{4.29}$$

So we are able to calculate the waist radius $w_0^{'}$ and the divergence angle $\theta_0^{'}$ of the focused beam:

$$w_0^{'} = w_0 \frac{f}{\sqrt{(f-g)^2 + \left(\frac{\pi w_0^2}{\lambda}\right)^2}}, \tag{4.30}$$

$$\theta_0^{'} = \frac{\lambda \cdot \sqrt{(f-g)^2 + \left(\frac{\pi\, w_0^2}{\lambda}\right)^2}}{\pi\, w_0 f} \tag{4.31a}$$

or

$$\theta_0^{'} = \theta_0 \sqrt{\left(1 - \frac{g}{f}\right)^2 + \left(\frac{\pi\, w_0^2}{\lambda f}\right)^2}. \tag{4.31b}$$

For the quantity $w_0^{'}\theta_0^{'}/\lambda$, we compute the same results as for the incident Gaussian beam, cf. (2.39b):

$$\frac{w_0^{'}\theta_0^{'}}{\lambda} = \frac{w_0\theta_0}{\lambda} = \frac{1}{\pi}. \tag{4.32}$$

This reflects the general statement that in the transformation of a laser beam through a paraxial aberration free optical system, the beam quality factor K (or K_x, K_y, see Sect. 2.6.3) is maintained.

If a Gaussian beam is focused, the irradiance is increased. With a laser output power P, the average irradiance in the focus is:

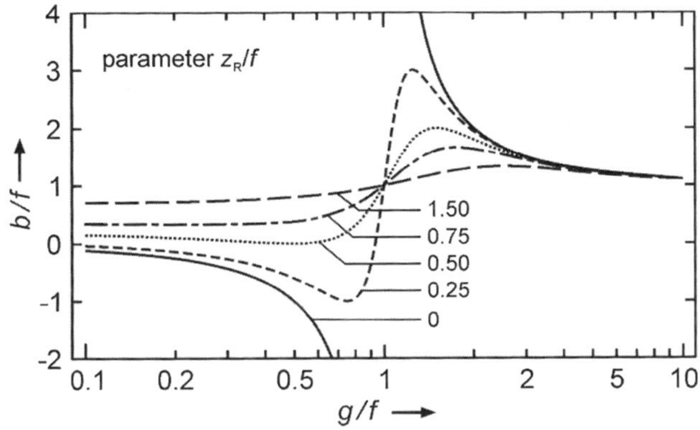

Fig. 4.27 Normalized distance *b/f*—cf. Fig. 4.26—as a function of *g/f*, see (4.27)

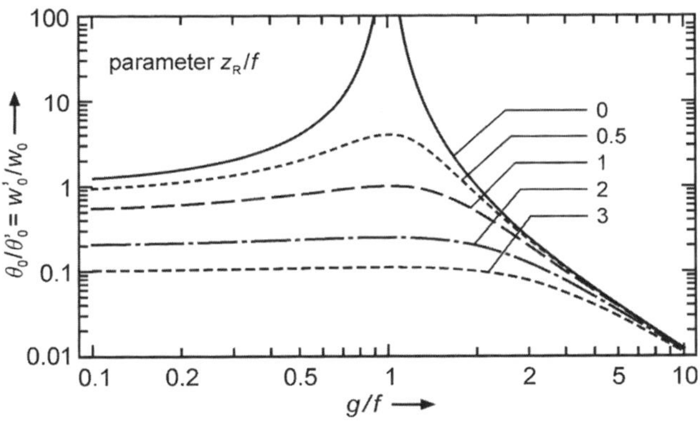

Fig. 4.28 Normalized beam radius w_0'/w_0 in the focus or normalized angle of divergence θ_0'/θ_0 as a function of *g/f*, see (4.30) and (4.31)

$$I_F' = \frac{P}{\pi\, w_0'^2}. \tag{4.33a}$$

With respect to the average irradiance I_L in the beam waist of the unfocused beam

$$I_L = \frac{P}{\pi\, w_0^2}, \tag{4.33b}$$

we find

$$\frac{I_F'}{I_L} = \left(1 - \frac{g}{f}\right)^2 + \left(\frac{\pi\, w_0^2}{\lambda f}\right)^2. \tag{4.34}$$

The normalized profiles of the quantities b, w_0', θ_0' and I_F' are shown in Figs. 4.27, 4.28 and 4.29.

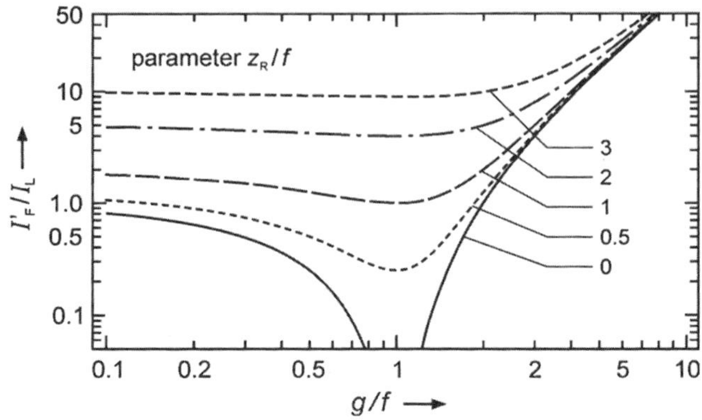

Fig. 4.29 Normalized focus irradiance I'_F/I_L as a function of *g/f*, see (4.34)

An interesting special case occurs, when the beam waist of the incident beam lies in the front focal plane of the lens:

$$g = f. \tag{4.35}$$

Then the focus is exactly in the back focal plane:

$$b = f. \tag{4.36}$$

Furthermore follows

$$\frac{w'_0}{w_0} = \frac{f}{z_R}, \tag{4.37}$$

$$\frac{\theta'_0}{\theta_0} = \frac{z_R}{f} \tag{4.38}$$

and

$$\frac{I'_F}{I_L} = \left(\frac{z_R}{f}\right)^2. \tag{4.39}$$

In the discussion of (4.27) and (4.29)–(4.34), it has to be considered that the beam diameter $2w(z = g)$ at the lens increases with increasing values of g and w_0. This means that diffraction no longer may be ignored and (4.27) and (4.29)–(4.33a, 4.33b) lose their validity. We consider as an example again the focus radius in the case $g = f$:

$$w'_0 = \frac{f\lambda}{\pi w_0}. \tag{4.40}$$

If we choose as maximum possible value of $w(z = g)$ half of the diameter D of the lens, we can estimate the smallest possible value of the focal radius as

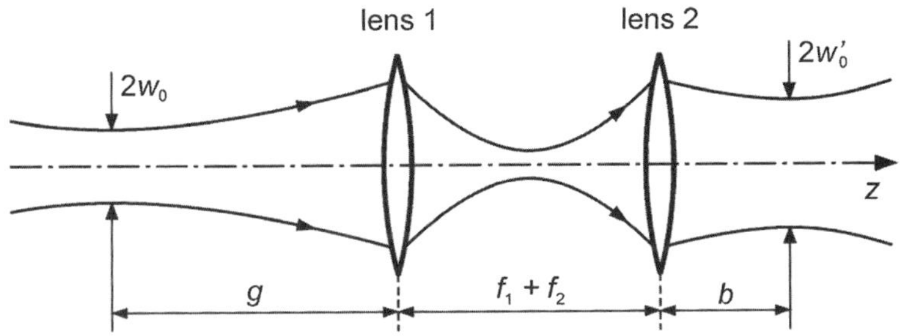

Fig. 4.30 Expansion of a Gaussian beam with a telescope

$$w'_{0,\min} = \frac{2f\lambda}{\pi D}. \tag{4.41}$$

Because of physical and technical reasons, the diameter D of a lens cannot be significantly larger than its focal length f. That is why the smallest achievable focus diameter following (4.41) is in the range of the wavelength λ used:

$$w'_{0,\min} \approx \lambda. \tag{4.42}$$

This result is generally applicable to any distance g since (4.41) represents the lower limit of $w'_{0,\min}$ also for $g \neq f$.

4.2.3 Beam Expansion

Many problems of laser measurement technology require an expansion of the laser beam. Therefore, a telescope—called Kepler's telescope—is used. A telescope consists of two convergent lenses with focal lengths f_1 and f_2, which are arranged at a distance $f_1 + f_2$, see Fig. 4.30. If the distance between the two lenses is $f_1 + f_2$, the telescope is adjusted to infinity. We restrict our considerations to this case.

The ray transfer matrix M_T of the telescope is composed of four sub-matrices:

$$M_T = \begin{pmatrix} 1 & b \\ 0 & 1 \end{pmatrix} \begin{pmatrix} 1 & 0 \\ -1/f_2 & 1 \end{pmatrix} \begin{pmatrix} 1 & f_1 + f_2 \\ 0 & 1 \end{pmatrix} \begin{pmatrix} 1 & 0 \\ -1/f_1 & 1 \end{pmatrix}. \tag{4.43a}$$

The first and third matrix describes the beam propagation corresponding to the distances b and $f_1 + f_2$, respectively. The second and the last matrix applies to both thin lenses of the telescope. The multiplication of the four matrices yields:

$$M_T = \begin{pmatrix} -f_2/f_1 & f_1 + f_2 - bf_1/f_2 \\ 0 & -f_1/f_2 \end{pmatrix}. \tag{4.43b}$$

For an incident Gaussian beam, the complex beam parameter—just in front of the first lens—reads

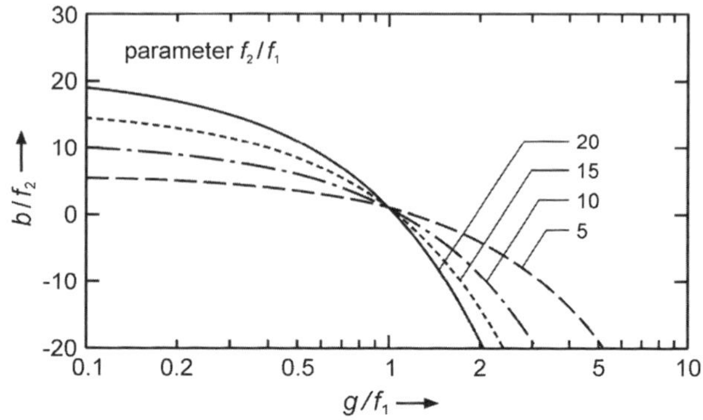

Fig. 4.31 Normalized distance b/f_2 as a function of g/f_1 for the telescope, see (4.46)

$$q_{\text{in}} = g + \mathrm{i}z_{\text{R}}. \tag{4.44}$$

With the help of the *ABCD* law, we calculate the complex beam parameter q_{out} at an arbitrary distance b behind the second lens as

$$\frac{1}{q_{\text{out}}} = \frac{f_1/f_2}{\frac{(g+\mathrm{i}z_{\text{R}})f_2}{f_1} - f_1 - f_2 + \frac{bf_1}{f_2}}. \tag{4.45}$$

Since b denotes the distance from lens 2 to the waist of the expanded beam, the real part $1/q_{\text{out}}$ must disappear. That is why, according to (4.20), the wavefront in the waist is flat. Thus

$$b = f_2^2\left(\frac{1}{f_1} + \frac{1}{f_2} - \frac{g}{f_1^2}\right). \tag{4.46}$$

Equation (4.46) describes the position of the beam waist of the expanded beam. Figure 4.31 shows the normalized distance b/f_2 as a function of g/f_1. b can have both, positive and negative values. If the beam waist of the expanded beam lies to the right of lens 2, then we have $b > 0$. In the other case holds $b < 0$.

For the case that the beam waist of the incident beam is at the front focal plane of lens 1 (i.e. $g = f_1$), the beam waist of the expanded beam is then at the back focal plane of lens 2 (i.e. $b = f_2$).

The reciprocal complex beam parameter at the waist of the expanded beam is calculated with the equations (4.45) and (4.46) as

$$\frac{1}{q_{\text{out}}(b)} = -\mathrm{i}\frac{f_1^2}{f_2^2 z_{\text{R}}}. \tag{4.47}$$

Using (4.19) and (2.38) we calculate the waist radius of the expanded beam as

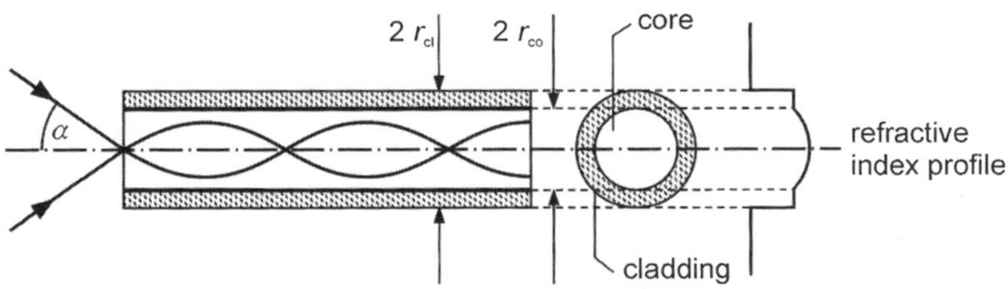

Fig. 4.32 Set-up and working principle of a gradient-index fiber with a parabolic refractive index profile

$$w_0' = w_0 \frac{f_2}{f_1}. \tag{4.48}$$

Beam expansion can be realized only in the case of $f_2 > f_1$. Otherwise, there is a beam constriction. Associated with the beam expansion is a reduction of the divergence angle, see (2.39):

$$\theta_0' = \theta_0 \frac{f_1}{f_2}. \tag{4.49}$$

Laser beams that should propagate over long distances, should be expanded with a telescope, in order to reduce the beam divergence.

4.3 Optical Fibers

For the beam guidance from the laser to the measuring object as well as from the measuring object to the detector, see Chap. 1, optical fibers are deployed depending on the measuring task and the chosen method. Hence, we will discuss this topic in the next two sections presenting gradient-index fibers and step-index fibers [16–19].

4.3.1 Gradient-Index Fiber

The basic design of a gradient-index fiber is sketched in Fig. 4.32. The gradient-index fiber consists of a core, which is surrounded by a cladding. In the core region, the rotationally symmetrical refractive index decreases in radial direction. Mostly a parabolic refractive index profile is present, see Fig. 4.32, given by:

$$n(r) = n_0 \left[1 - \frac{1}{2} \left(\frac{r}{a} \right)^2 \right] \quad \text{for} \quad 0 < r < r_{co}, \tag{4.50}$$

with n_0 maximum refractive index on the axis, $r = 0$; a parameter, r_{co} radius of core, cf. Figure 4.32.

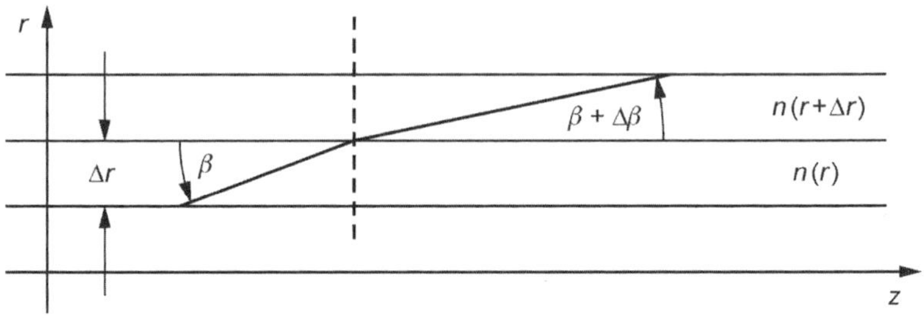

Fig. 4.33 Refraction of light in a gradient fiber

The quantity a describes the radial distance from the optical axis, where the refractive index drops to the value of $n_0/2$ ($r_{co} < a$). Because of the spatially dependent refractive index profile in the core, this kind of optical fiber is referred to as gradient-index fiber. Typical values of a gradient-index fiber are: $n_0 = 1.46$, $r_{co} = 25$ μm and $a = 175$ μm. The core of an optical fiber is surrounded by a cladding with constant refractive index

$$n_{cl} = n_0\left[1 - \frac{1}{2}\left(\frac{r_{co}}{a}\right)^2\right] \text{ for } r_{co} < r < r_{cl}, \tag{4.51}$$

with r_{cl} radius of cladding, see Fig. 4.32. The difference between n_0 and n_{cl} amounts to usually about 1 % only. The cladding diameter $2r_{cl}$ is typically 125 μm.

The light in the fiber cannot propagate, if the angle of the light path relative to the optical axis is too big. We consider, in geometric optical approximation, the refraction of the light beam as illustrated in Fig. 4.33. The application of the law of refraction (3.5) yields in this case:

$$n(r)\cos(\beta) = n(r + \Delta r)\cos(\beta + \Delta\beta). \tag{4.52}$$

We derive from this equation, in the limit of Δn, $\Delta\beta \to 0$, the differential equation

$$\frac{d^2 r}{dz^2} = \frac{1}{n}\frac{dn}{dr}. \tag{4.53}$$

With the parabolic refractive index profile (4.50), it follows

$$\frac{d^2 r}{dz^2} = -\frac{2r}{2a^2 - r^2}. \tag{4.54}$$

In the limit of $2a^2 \gg r^2$ results, with the initial conditions $r(z = 0) = 0$ and $dr/dz\ (z = 0) = \tan\beta_0 \approx \beta_0$, the solution:

$$r(z) = \beta_0\, a\, \sin\left(\frac{z}{a}\right). \tag{4.55}$$

Figure 4.32 shows, how two beams meander through a gradient-index fiber. In addition, there are other propagation modes such as a spiral-shaped beam path. Again and again, the beams are broken by the transitions between layers with different refractive indices. Therefore, gradient-index fibers are also called self-focusing fibers. The spatial period of the meandering light is

$$\Lambda = 2\pi\, a. \tag{4.56}$$

With $a = 175\ \mu m$, Λ is calculated to be 1.1 mm.

The maximum distance of the light beam to the optical axis is, according to (4.55), $\beta_0 a$. This value must be less than the core radius r_{co}

$$\beta_0\, a < r_{co}, \tag{4.57a}$$

otherwise the beam will leave the core of the fiber. The angle β_0 is linked with the incident angle α, see Fig. 4.32, at $r = 0$ by the law of refraction:

$$\sin\alpha = n_0 \sin\beta_0. \tag{4.58}$$

Here we assumed that the refractive index outside the fiber is 1. The maximum angle of incidence,

$$\beta_{0,\max} = \frac{r_{co}}{a}, \tag{4.57b}$$

in which propagation is still possible, is given by:

$$\sin\alpha_{\max} = n_0 \sin\left(\frac{r_{co}}{a}\right). \tag{4.59}$$

With the (4.51) and the approximation $r_c \ll a$, it follows finally

$$\sin\alpha_{\max} = \sqrt{n_0^2 - n_{cl}^2}. \tag{4.60a}$$

$\alpha_{\max}$ is called the angle of acceptance. Only if $|\alpha| \leq \alpha_{\max}$ is fulfilled, then beam propagation in the gradient-index fiber is possible. The sine of the angle of acceptance is referred to as numerical aperture. A typical value for the numerical aperture is 0.2, which corresponds to an acceptance angle of about 12°.

The numerical aperture and the angle of acceptance take their maximum value, when the incident beam is coupled into the fiber entrance face at $r = 0$. Only in this case, (4.60a) is correct. If the beam is injected at a distance r from the fiber axis, the numerical aperture becomes smaller:

$$\sin \alpha_{\max} = \sqrt{n(r)^2 - n_{\mathrm{cl}}^2}. \tag{4.60b}$$

So far, light propagation in gradient-index fibers was treated within the frame of geometric optics. In the further, we briefly discuss the mode structure in the wave concept of physical optics. Due to the radially decreasing refractive index inside the core, the propagating light undergoes a focusing effect. Therefore, we can regard formally the gradient-index fiber as a series of convex lenses. Just as a Gaussian beam can propagate between lenses—see Sect. 4.2—this is also possible in a gradient-index fiber. The beam radius w_0 of the Gaussian beam takes such a value that its diffraction-induced beam divergence is just compensated by the focusing effect of the gradient-index fiber. In the following, we estimate the beam radius of the Gaussian beam.

The Gaussian beam has a diffraction-induced beam divergence. The divergence angle is according to (2.39):

$$\theta_0 = \frac{\lambda}{\pi w_0} \ll 1.$$

This diffraction-induced divergence is compensated by the focusing effect of the gradient-index fiber. Since a parallel light beam reaches the fiber axis after a distance of $\Lambda/4$, we assign formally a focal length to the gradient-index fiber of

$$f = \frac{\Lambda}{4} = \frac{1}{2}\pi\, a. \tag{4.61}$$

Therefore, we calculate for a paraxial ray at a distance r a convergence angle of

$$\theta_{\mathrm{c}} = \frac{r}{f} = \frac{2r}{\pi\, a} \ll 1. \tag{4.62}$$

Because of the assumed compensation of the two angles, we calculate for the radius w_0 of the propagating Gaussian beam, assuming $r = w_0$:

$$w_0 = \sqrt{\frac{\lambda a}{2}}. \tag{4.63a}$$

Replacing the wavelength by the vacuum wavelength and taking the refractive index n_0 $(n(r < r_{\mathrm{co}}) \approx n_0)$, leads to

$$w_0 \approx \sqrt{\frac{\lambda_0 a}{2 n_0}}. \tag{4.63b}$$

A more precise calculation yields, instead of the number 2, the number π in (4.63a), (4.63b). However, (4.63a), (4.63b) is only valid, if the core radius r_{co} is

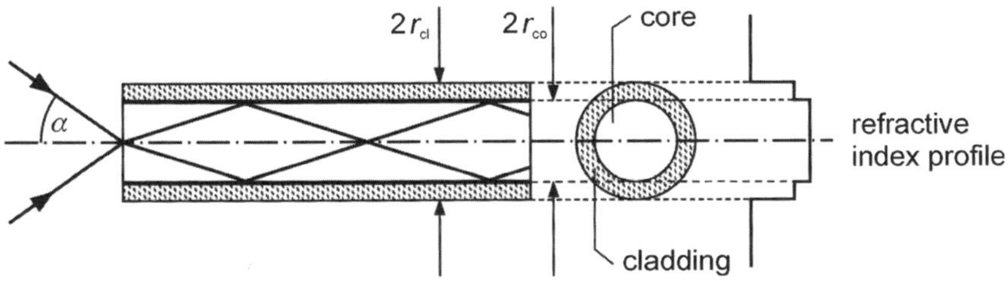

Fig. 4.34 Set-up and working principle of a step-index fiber

sufficiently large compared to the beam radius w_0. This assumption is in general fulfilled.

Besides the considered Gaussian beam, which is also referred to as basic mode, there are, as in a laser cavity, additional modes with more complicated field distributions that are capable to propagate in a gradient-index fiber. Their number is limited. The reason for this is that, with an increasing mode order, the diffraction-caused divergence increases, while the focusing effect of the gradient fiber is constant. The number M of modes that can propagate in a gradient-index fiber is given approximately by [17, 18]:

$$M \approx \pi^2 \left(n_0^2 - n_{\rm cl}^2\right) \left(\frac{r_{\rm co}}{\lambda_0}\right)^2. \tag{4.64}$$

Using numerical values of a typical gradient-index fiber ($n_0 = 1.460$, $n_{\rm cl} = 1.445$, $r_{\rm co} = 25$ μm and $\lambda_0 = 1$ μm), we estimate the number of propagable modes as approximately 269. Such a fiber is, therefore, also referred to as a multimode fiber.

4.3.2 Step-Index Fiber

In addition to gradient-index fibers with a parabolic refractive index profile, there are fibers with a step-index profile available. These fibers are called step-index fibers. In comparison to gradient-index fibers, they can be produced more easily and cost-efficiently.

A step-index fiber is composed of a cylindrical core with a high refractive index $n_{\rm co}$ and radius $r_{\rm co}$, which is surrounded by a cladding with a lower refractive index $n_{\rm cl}$ and a diameter of $2r_{\rm cl}$, see Fig. 4.34. Typical values are: $n_{\rm co} = 1.48$, $n_{\rm cl} = 1.46$, $2r_{\rm co} = 100$ μm and $2r_{\rm cl} = 140$ μm.

Beam propagation in these step-index fibers is based on total internal reflection at the interface between core and cladding, see Sect. 3.2.2. The total reflection in the step-index fiber replaces the continuous light refraction in the gradient-index

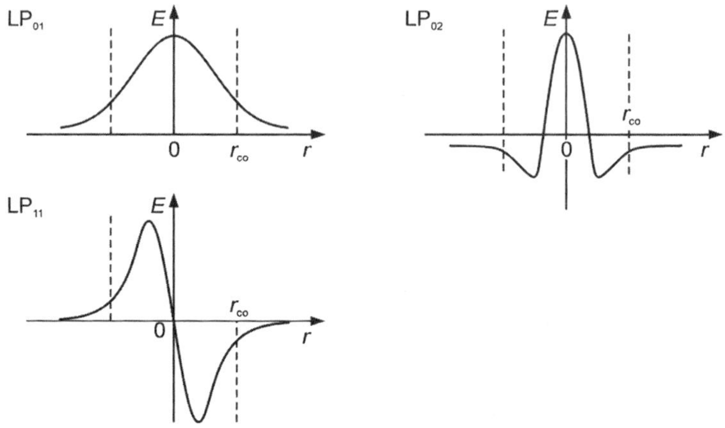

Fig. 4.35 Electric field distributions of some modes of step-index fibers

fiber. In order to fulfill the condition for total internal reflection, the incident beam at the entrance face of the fiber may not exceed a maximum angle of incidence α_{max}. Applied to the numerical aperture of a step-index fiber, this means:

$$\sin(\alpha_{max}) = n_{co}\sin(90^o - \theta_t) = n_{co}\cos(\theta_t). \tag{4.65a}$$

Here, θ_t is the angle of total reflection for the transition from the core to the cladding. With (3.8), we finally find:

$$\sin(\alpha_{max}) = n_{co}\cos\left(\arcsin\left(\frac{n_{cl}}{n_{co}}\right)\right) = \sqrt{n_{co}^2 - n_{cl}^2}. \tag{4.65b}$$

Example: Using the values $n_{co} = 1.48$, $n_{cl} = 1.46$, the numerical aperture is calculated as approximately 0.24, which corresponds to an acceptance angle of about 14°.

As for gradient-index fibers, in general several modes can propagate in a step-index fiber. Some of the possible modes are shown schematically in Fig. 4.35. The number M of modes being able to propagate, can be calculated approximately as [17–19]:

$$M \approx 2\pi^2\left(n_{co}^2 - n_{cl}^2\right)\left(\frac{r_{co}}{\lambda_0}\right)^2. \tag{4.66}$$

With the above stated values of a typical step-index fiber and a vacuum wavelength of 1 μm, the number of modes that can propagate, is over 2,900. Hence, the considered step-index fiber is a multi-mode fiber.

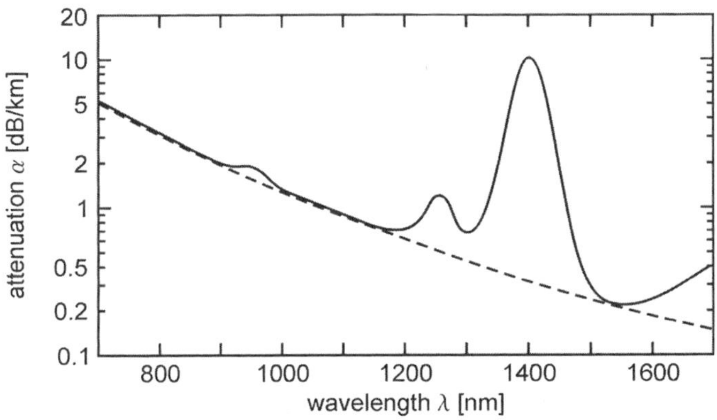

Fig. 4.36 Attenuation of an optical fiber in the near infrared

If the core radius is decreased until it is approximately equal to the wavelength of light,

$$\lambda_0 \approx r_{co} \tag{4.67}$$

we calculate, with $n_{co} = 1.48$, $n_{cl} = 1.46$, the number M of modes to 1. Such a fiber where only one single mode is capable to propagate, is called single-mode fiber. A closer look shows that in a single-mode fiber two perpendicularly polarized modes can propagate.

4.3.3 Absorption and Dispersion of Optical Fibers

Core and cladding of optical fibers are typically made of ultra pure quartz glass. The purity of the glass ensures a low absorption. For fused silica the minimum of the optical attenuation is attained at a wavelength of about 1.6 μm, see Fig. 4.36. So, an attenuation of 0.2 dB/km can be realized, i.e., more than 95 % of the irradiance coupled into a fiber arrives at the end of a 1 km long fiber. For comparison: At a frequency of 100 MHz, coaxial cables have an attenuation of typically 38 dB/km, at 1 GHz the attenuation is even higher: 134 dB/km.[1]

In addition to attenuation, the dispersion behavior of an optical fiber is important for technical applications. We distinguish two different types of dispersion:

- material dispersion,
- mode dispersion.

[1] These values belong to a 50-Ohm-cable, Typ Aircom-Plus from SSB-Electronic [20].

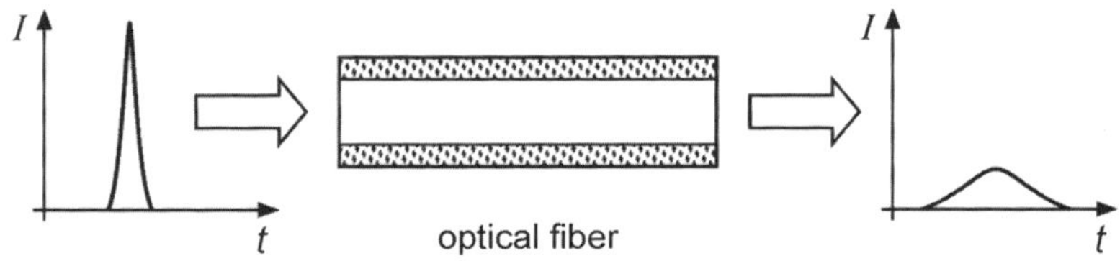

Fig. 4.37 Temporal broadening of a light pulse caused by dispersion

We first consider material dispersion. This refers to the fact that the refractive index of the fiber material is wavelength-dependent, see Sect. 3.2.1. Light waves with different wavelengths, therefore propagate inside the fiber, in general, with different speeds of light.

Also the different modes of an optical fiber propagate, in general, with different speeds inside a fiber. This phenomenon is called mode dispersion. Mode dispersion occurs, especially, in multi-mode step-index fibers. For multi-mode gradient-index fibers, it plays only a minor role since the speeds of the various modes differ only slightly.

Material dispersion and mode dispersion lead to a temporal broadening of propagating light pulses, as shown schematically in Fig. 4.37.

References

1. M. Born, E. Wolf, A.B. Bhatia, P.C. Clemmow, D. Gabor, A.R. Stokes, A.M. Taylor, P.A. Wayman, W.L. Wilcock, *Principles of Optics* (Cambridge University Press, Cambridge, 2003). ISBN 0 521 642221
2. L. Bergmann, C. Schaefer, H. Niedrig, *Lehrbuch der Experimentalphysik, Bd. III: Optik* (Walter de Gruyter, Berlin, 2004). ISBN 3-11-017081-7
3. E. Hecht, *Optics* (Pearson, Reading, 2003) ISBN 978-8177583571
4. F.K. Kneubühl, M.W. Sigrist, *Laser* (Vieweg + Teubner, Wiesbaden, 2008). ISBN 978-3-8351-0145-6
5. D.H. Goldstein, *Polarized Light* (CRC Press, Boca Raton, 2010). ISBN 978-1439830406
6. R. Waynant, M. Ediger, *Electro-Optics Handbook* (McGraw-Hill, New York, 2000). ISBN 978-0070687165
7. M. Bass, C. DeCusatis, J.M. Enoch, V. van Lakshminarayanan, G. Li, C. MacDonald, V.N. Mahajan, E. Van Stryland, *Handbook of Optics (Volume V)—Atmospheric Optics, Modulators, Fiber Optics, X-Ray and Neutron Optics* (McGraw-Hill, New York, 2009). ISBN 978-0071633130
8. M. Mansuripur, *Classical Optics and Its Applications* (Cambridge Universitiy Press, Cambrigde, 2009). ISBN 978-0521881692
9. A. Gerrard, J.M. Burch, *Introduction to Matrix Methods in Optics* (Dover Publications, Mineola, 2012). ISBN 9780486680446
10. J.N. Damask, *Polarization Optics in Telecommunications* (Springer, Heidelberg, 2004). ISBN 0-387-22493-9
11. D. Meschede, *Optics, Light and Lasers: the Practical Approach to Modern Aspects of Photonics and Laser Physics* (Wiley-VCH, Weinheim, 2007). ISBN 3-527-40628-X
12. A. Nussbaum: *Geometric Optics, An Introduction* (Addison-Wesley, Reading, 1968)

13. W. Koechner, *Solid State Laser Engineering* (Springer, Heidelberg, 2006). ISBN 978-0-387-29338-7
14. W. Brouwer, *Matrix Methods in Optical Instrument Design* (Benjamin, New York, 1964)
15. A. Donges, Behandlung optischer Abbildungen mit der Matrix-Methode. MNU **8**(46), 454–459 (1993)
16. V. Alwayn, *Fiber-Optic Technologies* (Cisco Press, San Jose, 2004)
17. G. Schiffner, *Optische Nachrichtentechnik—Physikalische Grundlagen, Entwicklung, Moderne Elemente und Systeme* (Vieweg + Teubner, Wiesbaden, 2005). ISBN 3519004461
18. G. Mahlke, P. Gössing, *Fiber Optic Cables—Fundamentals, Cable Design, System Planning* (Wiley-VCH, Weinheim, 2001) ISBN 3895781622
19. C.C. Timmermann, *Lichtwellenleiter* (Vieweg, Braunschweig, 1981). ISBN 3-528-03341-X
20. SSB-Electronic: Data sheet 50-Ohm-cable, Typ Aircom-Plus. http://ssb.de/pdfs/ssb_090721_datenblatt_aircomplus.pdf, Accessed 28 July 2013

Chapter 5
Detection of Electromagnetic Radiation

Abstract Different kinds of light detectors and their working principles are explained covering the following detector types: thermal detectors, photoelectric detectors, semiconductor detectors, space-resolving detectors. This is followed by the discussion of some general principles of electronic metrology, which must be considered when applying these detectors. Finally, we compare and summarize the properties of the discussed detectors.

The complete information on the measuring object, which we can obtain by optical methods, is included in the scattered light. Therefore, an essential task of laser measurement technology is to capture, store and interpret the scattered light.

Today, a large number of detectors are available, to transform the scattered light in an analyzable signal [1–36]. In this chapter, the most important detectors are discussed. In our representation, we emphasize the physical basics of the detectors. Only a deep understanding of the basic principles allows an optimal usage of the different detectors and helps to avoid errors.

The large number of detectors can be divided into several classes according to the respective underlying physical principles of measurement. We begin with thermal detectors.

5.1 Thermal Detectors

Thermal detectors exploit the fact that each body, which absorbs electromagnetic radiation, is heated. The increase in temperature is a measure of the absorbed radiation power. Thermal detectors are used mainly in the visible and infrared spectrum.

A thermal detector absorbs a portion of the incident radiation according to its degree of absorbance A. In addition to this energy gain, losses by heat conduction and heat convection must be taken into account, see Fig. 5.1. Heat losses, due to thermal radiation, can be neglected in most cases. Therefore, the change in temperature T is given by the differential equation:

A. Donges and R. Noll, *Laser Measurement Technology*, Springer Series in Optical Sciences 188, DOI: 10.1007/978-3-662-43634-9_5,

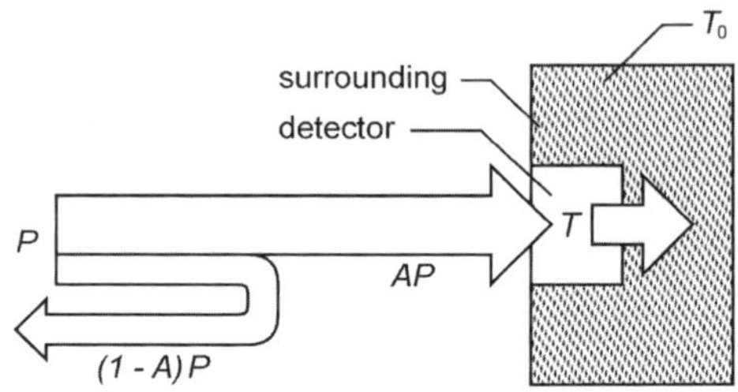

Fig. 5.1 Functional principle of a thermal detector

$$C_Q \frac{dT}{dt} = AP - \frac{T - T_0}{R_Q}. \quad (5.1)$$

Here, C_Q is the heat capacity of the detector, P the incident radiant power also called radiant flux, R_Q the effective thermal resistance of the detector and T_0 the ambient temperature. In the stationary case (P = const.), the solution is:

$$T = T_0 + R_Q AP. \quad (5.2a)$$

To study the time response, we solve (5.1) for the periodically oscillating power:

$$P(t) = P_0(1 - \cos(\omega t)), \quad P_0 = \text{const.} \quad (5.3)$$

The solution is:

$$T(t) = T_0 + R_Q AP_0 \left[1 - \frac{1}{\sqrt{1 + (\omega R_Q C_Q)^2}} \cos(\omega t + \Phi) \right] \quad (5.4a)$$

with

$$\Phi = -\arctan(\omega R_Q C_Q). \quad (5.4b)$$

Φ is the phase shift between $P(t)$ and $T(t)$. The thermal detector shows the typical behaviour of a low-pass filter with an upper angular cutoff frequency of, see Fig. 5.2:

$$\omega_c = \frac{1}{R_Q C_Q}. \quad (5.5)$$

The evaluation of the measurement signals is substantially simplified, if all frequencies occurring are small compared to the cutoff frequency. In the following,

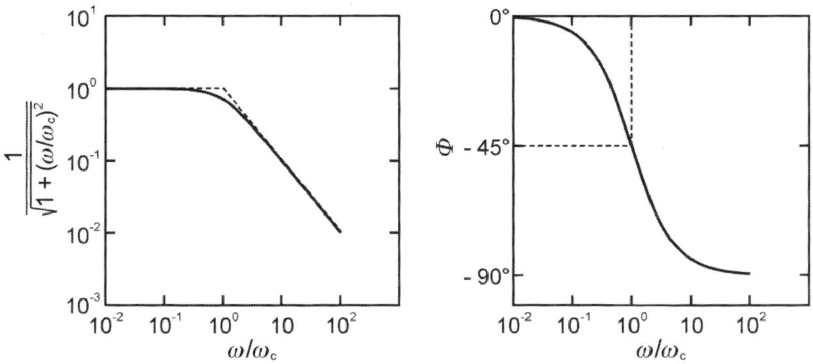

Fig. 5.2 Frequency response of a thermal detector

this will be assumed mostly, so that we can use the stationary solution (5.2a) also in the non-stationary case:

$$T(t) = T_0 + R_Q A P(t), \; \omega \ll \omega_c. \tag{5.2b}$$

So, the incident radiant flux $P(t)$ is directly proportional to the temperature difference $T(t) - T_0$:

$$P(t) = \frac{T(t) - T_0}{A R_Q}, \; \omega \ll \omega_c. \tag{5.6}$$

The sensitivity Σ of a detector is generally defined by the ratio of the measured signal to the incident radiant flux. In case of a thermal detector, this means:

$$\Sigma = \frac{T(t) - T_0}{P(t)} = A R_Q. \tag{5.7}$$

To ensure a high sensitivity,

- the absorber surface of the detector should be black ($A \approx 1$) for the relevant spectral range and
- the thermal resistance should be large, i.e., the detector is well insulated from the environment.

The rise time τ_r, see Fig. 5.3, of any detector is linked with its cutoff angular frequency by the relationship [1, 2]:

$$\tau_r \approx \frac{2.2}{\omega_c}. \tag{5.8a}$$

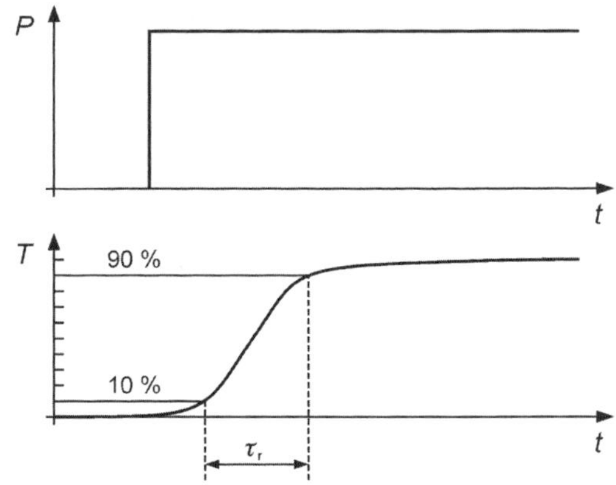

Fig. 5.3 Definition of the rise time τ_r as the time interval, in which the response signal $T(t)$ rises from 10 to 90 % of its final value after irradiation with a step function excitation $P(t)$

For a thermal detector, we have:

$$\tau_r \approx 2.2\, R_Q C_Q. \tag{5.8b}$$

High cutoff frequencies and short rise times require both, a low heat capacity—and hence low mass—and a low thermal resistance. However, the last condition leads to a low sensitivity of the detector. So, high sensitivity and a short rise time exclude each other. Therefore, the user has to compromise depending on the application.

The resolving power of all thermal detectors is limited by omnipresent temperature fluctuations causing temperature noise. The mean square fluctuation of temperature is calculated, using statistical thermodynamics, as [1]:

$$< \Delta T^2 > \; = \frac{kT^2}{C_Q}, \tag{5.9}$$

where $k = 1.38 \times 10^{-23}$ J/K is the Boltzmann constant. With this, we can calculate the signal-to-noise ratio *S/N* of a thermal detector. With the definition:

$$\frac{S}{N} = \frac{T(t) - T_0}{\sqrt{< \Delta T^2 >}}, \tag{5.10}$$

it follows, that:

$$\frac{S}{N} = \sqrt{\frac{C_Q}{k}} \frac{R_Q A P}{T}. \tag{5.11}$$

The minimum radiant flux, that can be detected barely by a thermal detector is calculated from the requirement $S/N \approx 1$ as

$$P_{\min} \approx \sqrt{\frac{k}{C_Q}\frac{T}{AR_Q}} = \sqrt{kC_Q}\frac{2.2T}{A\tau_r}. \tag{5.12}$$

For the calculation of S/N and $P_{\min}$, we have only considered the thermal noise of the thermal detector. Further possible sources of thermal noise, such as those related to electrical resistance measurements, were ignored.

The detector can be cooled to improve the signal-to-noise ratio S/N or to reduce the absolute limit of detection. In addition, it is recommended to increase the absorbance A, heat capacity C_Q and thermal resistance R_Q. However, the two last requirements have a negative effect on the rise time of the detector, cf. (5.8b). This is intuitively clear: The more inert a detector reacts, the less it is disturbed by thermal fluctuations.

Finally, it should be pointed out that a thermal detector can be used to measure the energy of short laser pulses directly. If the laser pulse duration τ is much smaller than the rise time τ_r, $\tau \ll \tau_r \approx 2.2R_QC_Q$, the heat loss $(T - T_0)/R_Q$ in (5.1) can be neglected. Solving the differential equation yields in this case a linear relationship between the pulse energy ΔW, i.e., the time integral of the radiant flux, and the increase in temperature ΔT of the detector.

$$\Delta W = \int_{\tau} P(t)dt = \frac{C_Q\Delta T}{A}, \ \tau \ll \tau_r. \tag{5.13}$$

In this case it makes sense, to define the sensitivity Σ as the ratio of the temperature increase ΔT to the irradiated energy ΔW:

$$\Sigma = \frac{\Delta T}{\Delta W} = \frac{A}{C_Q}. \tag{5.14}$$

A high sensitivity for pulse energy measurement requires a small heat capacity and thus a small detector mass. Due to unavoidable temperature noise, the measured pulse energy has an uncertainty corresponding to the minimum detectable pulse energy:

$$\Delta W_{\min} \approx \frac{\sqrt{kC_Q}T}{A}. \tag{5.15}$$

The increase in temperature in a thermal detector is measured in various ways. In the simplest case, it is carried out with a thermometer. With an appropriate calibration, the thermometer indicates directly the laser power or the laser pulse energy. Such detectors are used, for example, for the simple determination and monitoring of the beam power of carbon-dioxide lasers. In general, especially when a broadband measurement is required, electrical measurement methods are preferred.

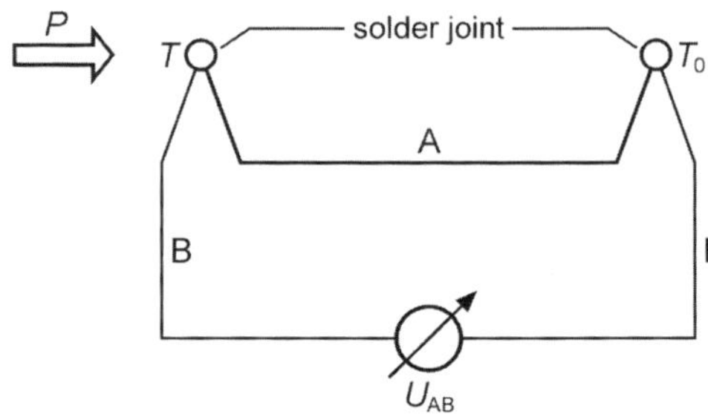

Fig. 5.4 Set-up of a thermocouple

5.1.1 Thermocouple and Thermopile

In a thermocouple, the thermoelectric effect is utilized. Figure 5.4 shows the basic structure. A thermocouple is a conductive loop, which consists of two different metals or semiconductors. For different temperatures at the two solder joints, where the two materials are connected, a so-called thermoelectric voltage is observed. Over a wide range this voltage is proportional to the temperature difference of the solder joints, see Fig. 5.5:

$$U_{\mathrm{AB}} = \varepsilon_{\mathrm{AB}}(T - T_0), \tag{5.16}$$

where $\varepsilon_{\mathrm{AB}}$ is the Seebeck coefficient or the thermoelectric power of the respective conductor combination.

Graphically, $\varepsilon_{\mathrm{AB}}$ represents the slope of a characteristic curve in Fig. 5.5. For the commonly used combination of iron and constantan, the Seebeck coefficient is 53.7 μV/K in the temperature range of 0–100 °C. Particularly large Seebeck co-efficients provide semiconductor thermocouples. For a lead-telluride/zinc-antimonide element, for example, $\varepsilon_{\mathrm{AB}}$ is 180 μV/K in the temperature range of 0–600 °C.

Only one of the two solder joints of the thermocouple is heated by the incident radiation. The other solder joint is maintained at a constant temperature, e.g. by ice water or a thermostat with a heating element. The resultant thermoelectric voltage is a measure of the incident radiant flux.

The sensitivity Σ of a thermocouple is given by the ratio of the thermoelectric voltage U_{AB} and the incident radiant flux P:

$$\Sigma = \frac{U_{\mathrm{AB}}}{P} = \varepsilon_{\mathrm{AB}} A R_{\mathrm{Q}}. \tag{5.17}$$

Typical sensitivities of metallic combinations are about 10 V/W. About five times larger values can be achieved with semiconductor elements.

If n thermocouples are connected in series, the n-fold thermoelectric voltage is measured. Accordingly, the sensitivity will increase by a factor n. Such arrangements are referred to as thermopiles, see Fig. 5.6. Using modern methods of semiconductor technology, compact thermopiles with sensitivities greater than 100 V/W can be produced.

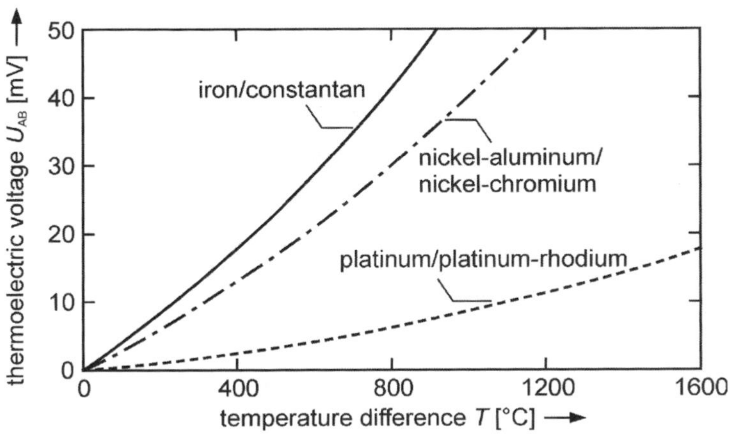

Fig. 5.5 Characteristic curves of common thermocouples. The denominations at the curves refer to the material combinations used for the solder joint

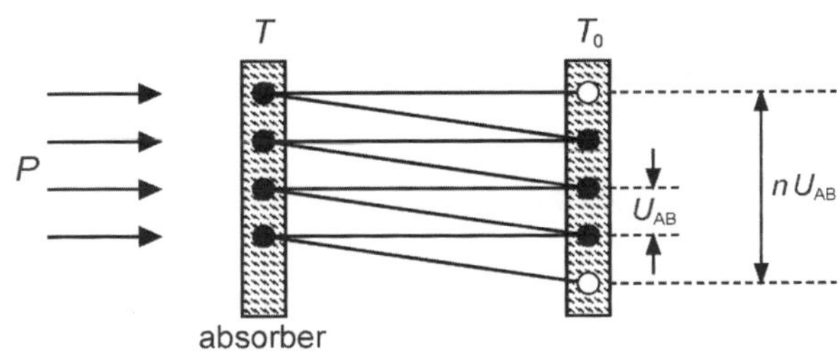

Fig. 5.6 Set-up of a thermopile

Thermocouples or thermopiles are used in the visible and infrared spectral range (up to about 30 μm). Typical rise times are, depending on design and sensitivity, in the range of 10 μs–10 s.

5.1.2 Bolometer

If an electrical conductor or semiconductor is heated, its electrical resistance R changes with temperature T. Figure 5.7 summarizes the typical behaviour of the electrical conductivity of metals and semiconductors. The change in resistance R with temperature T is given by

$$\frac{\mathrm{d}R}{\mathrm{d}T} = \alpha(T)R. \tag{5.18}$$

For metals, the so-called temperature coefficient $\alpha(T)$ is, over a not too wide temperature range, constant and positive, i.e., the resistance increases with increasing temperature. Most pure metals have a temperature coefficient of about

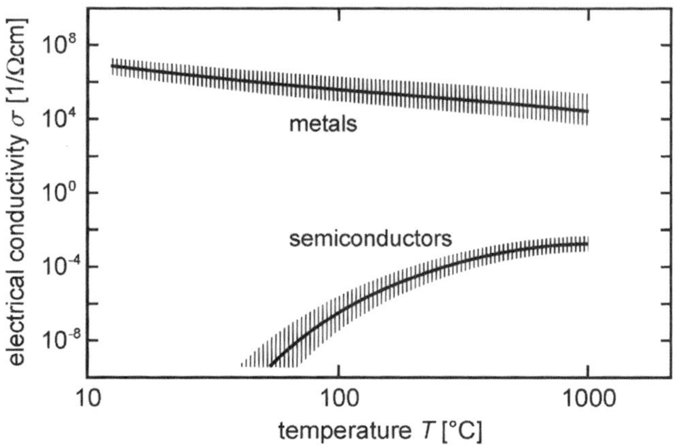

Fig. 5.7 Basic curve progressions of the electrical conductivity of metals and semiconductors as a function of temperature

1/250 K^{-1}. Contaminated metals or alloys have smaller values of α. Unlike metals, the electrical resistance of semiconductors decreases with increasing temperature. That is why the number of free charge carriers increases with increasing temperature. For ordinary semiconductors the following relation holds approximately:

$$\alpha(T) = -\text{const.}\frac{1}{T^2}. \tag{5.19}$$

Semiconductors have a negative temperature coefficient. At room temperature the temperature coefficients of semiconductors are approximately one order of magnitude higher than the values of pure metals.

A bolometer is a thermal detector where the temperature of the detector is determined by means of an electrical resistance measurement. A simple circuit is sketched in Fig. 5.8. The voltage drop U_m across the measuring resistor R_m depends on the temperature of the bolometer:

$$U_m = \frac{R_m U_0}{R(T) + R_m}. \tag{5.20}$$

The change of the measured voltage U_m with temperature T is obtained by derivation:

$$\frac{dU_m}{dT} = -\frac{\alpha(T)R(T)R_m U_0}{[R(T) + R_m]^2}. \tag{5.21}$$

If the resistance change is small, $R(T)$ can be replaced by R_0, which is the resistance of the bolometer without irradiation. In case of resistance matching, i.e., $R_m = R_0$, the measured voltage becomes maximum. With (5.6) and (5.21) it follows

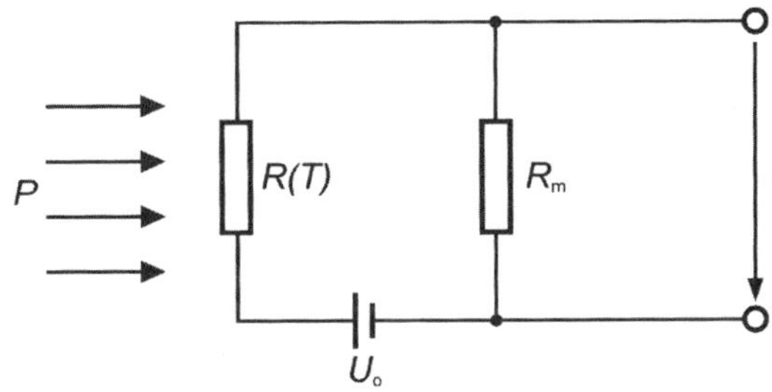

Fig. 5.8 Circuit for resistance measurement

$$\Delta U_{\mathrm{m}} = -\frac{1}{4}\alpha U_0 \Delta T = -\frac{1}{4}\alpha A R_{\mathrm{Q}} U_0 P. \tag{5.22}$$

Hence, the sensitivity of the bolometer is determined by

$$\Sigma = \frac{-\Delta U_{\mathrm{m}}}{P} = \frac{1}{4}\alpha A R_{\mathrm{Q}} U_0. \tag{5.23}$$

Therefore, bolometers should have preferably large temperature coefficients $\alpha(T)$.

Two variants of bolometers are to be distinguished: metal bolometers and semiconductor bolometers. Semiconductor bolometers make use of thermistors whose resistance changes significantly with temperature, see Fig. 5.7. For metal bolometers, mostly thin blackened platinum or nickel foils are used. In the range of −200 to 600 °C both elements have an almost constant temperature coefficient α. The sensitivities reach values of up to 50 V/W. The rise times are less than 5 ms, provided an appropriate electrical design is chosen.

The sensitivities of semiconductor bolometers are much greater. Even at room temperature their temperature coefficients α are much larger. Especially with cooling, e.g. taking liquid helium, semiconductor bolometers achieve sensitivities of up to 10^7 V/W. Superconducting bolometers have even greater sensitivities. They take advantage of the large changes in resistance in the transition region to super-conductivity. The rise times of semiconductor bolometers are slightly smaller than those of metal bolometers. Superconducting bolometers have yet shorter rise times of $\tau_r < 100$ µs. Bolometers, depending on the model, are used in the visible spectrum up to the microwave region.

A disadvantage of the circuit shown in Fig. 5.8—particularly for small values of ΔU_{m}—is the DC component of the measuring voltage U_{m}, which is independent of the incident light power, for $R_{\mathrm{m}} \approx R(T)$:

$$U_{\mathrm{m}} = \frac{1}{2}U_0 + \Delta U_{\mathrm{m}}. \tag{5.24}$$

Therefore in practice, often a compensation circuit is chosen, such as a simple Wheatstone bridge. In this way errors resulting from heating of the wires of the circuit can be compensated also [1, 36].

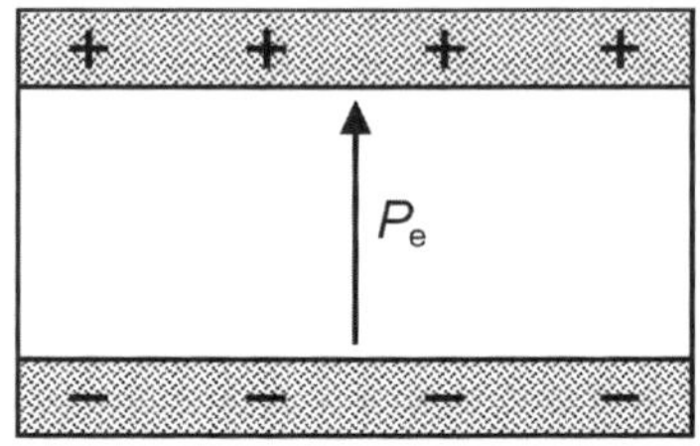

Fig. 5.9 Dielectric polarization of a pyroelectric crystal

5.1.3 Pyroelectric Detectors

Pyroelectric detectors consist of materials having a permanent electric polarization P_e due to their crystal structure. Hence, on two opposite sides of the crystal with the surfaces F, surface charges

$$Q_p = \pm P_e F \tag{5.25}$$

occur, see Fig. 5.9. These surface charges are always compensated by free charges Q_f from the surroundings, so that the resulting surface charges disappear:

$$Q = Q_p - Q_f = 0. \tag{5.26}$$

Therefore, in the static case a measurement of Q_p is not possible.

If a temperature change dT occurs in the time interval dt, the heated crystal will expand. This results in a change of the polarization dP_e and hence a change of the surface charges dQ_p. If the change of the polarization charge dQ_p happens in a sufficiently short time, the number of free compensating charges remains constant. Thus dQ_p appears as resulting surface charge and can be measured. Typical times to compensate polarization charges by free charges are in the range of >1 s.

The change in the polarization charge dQ_p is measured capacitively. To this aim, the pyroelectric crystal is used as dielectric of a capacitor, which is connected with a measuring resistor R_m, see Fig. 5.10. While the surface charge dQ_p is generated, a displacement current caused by influence flows:

$$I = \frac{\mathrm{d}Q_p}{\mathrm{d}t} = F\frac{\mathrm{d}P_e}{\mathrm{d}t} = F\frac{\mathrm{d}P_e}{\mathrm{d}T}\frac{\mathrm{d}T}{\mathrm{d}t}. \tag{5.27}$$

The change of polarization with temperature is called pyroelectric coefficient:

$$p(T) = \frac{\mathrm{d}P_e}{\mathrm{d}T}. \tag{5.28}$$

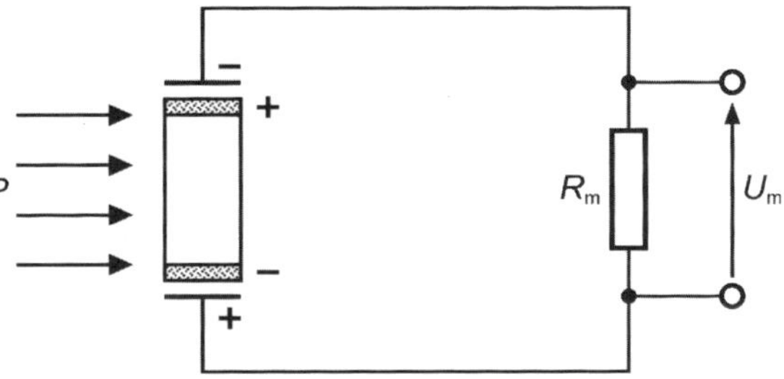

Fig. 5.10 Set-up of a pyroelectric detector

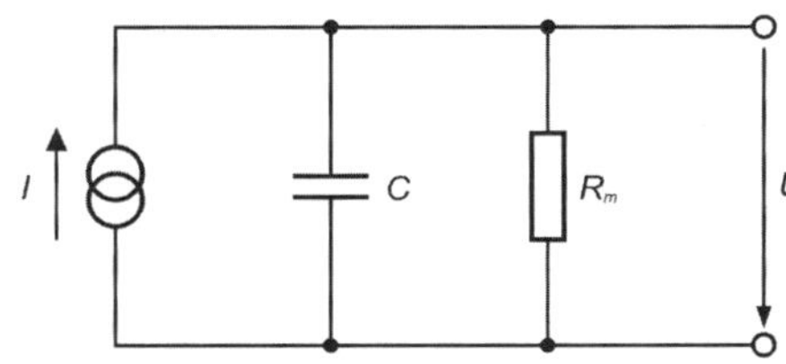

Fig. 5.11 Equivalent circuit diagram of a pyroelectric detector

For a not too wide temperature range this coefficient is constant.

In terms of circuit design, the pyroelectric detector represents a current source with a capacity C and a measuring resistor R_m connected in parallel. Figure 5.11 shows the equivalent circuit diagram. For the calculation of the voltage drop U_m at the resistor R_m, we assume in the following the limiting case $\omega \ll 1/(R_m C)$. Then, the current I flows practically only through the measuring resistor and causes the voltage:

$$U_m = R_m I = F R_m p(T) \frac{dT}{dt}. \tag{5.29}$$

Thus, the measured signal is proportional to the time derivative of the detector temperature. This implies, that the detector temperature changes sufficiently rapidly, so that the generated polarization charges are not compensated by free charges from the surroundings. If this is not the case, the radiation flux has to be modulated with a chopper, see Sect. 4.1.3. A chopper is, for example, a rotating perforated disk to interrupt a light beam periodically. Required chopping frequencies must be higher than 1 Hz.

For measurements with a pyroelectric detector two limiting cases can be distinguished:

(a) Quasi-static mode ($\omega \ll \omega_c$): If all occurring angular frequencies in $P(t)$ are small compared to the angular thermal cutoff frequency, as usually assumed, see (5.6), then the irradiated power and the detector temperature are proportional to each other, cf. relation (5.2b):

$$U_m(t) = F R_m p(T) A R_Q \frac{dP}{dt}. \tag{5.30a}$$

The measured signal is proportional to the derivative with respect to time of the incident radiant flux. Assuming a temporal variation of the power according to:

$$P(t) = P_0(1 - \cos(\omega t)), \quad P_0 = \text{const.}, \tag{5.3}$$

then the measured signal is:

$$U_m(t) = FR_m p(T) A R_Q \omega P_0 \sin \omega t. \tag{5.31}$$

In this case, the sensitivity is defined by the ratio of the amplitude of the measured signal and the amplitude of the incident power:

$$\Sigma_s = FR_m p(T) A R_Q \omega. \tag{5.32a}$$

In the quasi-static operation mode the sensitivity of a pyroelectric detector grows, linearly with the angular frequency ω.

(b) Pulse mode ($\omega \gg \omega_c$): If the incident radiation has sufficiently short pulses or is modulated strongly, i.e., all frequency components of $P(t)$ are satisfying the condition $\omega \gg \omega_c$, then each thermal detector has an integrating effect. Since the measured voltage U_m is proportional to the time derivative of the detector temperature, see (5.29), both effects compensate each other. That is why $U_m(t)$ and $P(t)$ are proportional:

$$U_m(t) = FR_m p(T) A P(t)/C_Q = FR_m p(T) A R_Q \omega_c P(t). \tag{5.30b}$$

Division by $P(t)$ gives the pulse mode sensitivity, which is independent of the frequency:

$$\Sigma_p = FR_m p(T) A R_Q \omega_c = \Sigma_s \frac{\omega_c}{\omega} \ll \Sigma_s. \tag{5.32b}$$

Often pyroelectric detectors are operated in the pulse mode, in spite of the reduced sensitivity compared to the quasi-static mode. The advantage is obvious: the proportionality between $U_m(t)$ and $P(t)$. In addition, the rise time of the pyroelectric detector is no longer limited by the thermal rise time, see (5.8b). It is determined by the electrical time constant $R_m C$, which is in general much smaller. With careful design, rise times less than 100 ps can be realized. Such short rise times require a small measuring resistance, which in turn has a negative effect on the sensitivity.

The quasi-static mode can be transferred to the pulse mode. To this end, the incident radiant flux has to be fractionized by a chopper into individual pulses, see Fig. 5.12. It is important to ensure that the chopping frequency is well above the thermal cutoff frequency.

For the construction of pyroelectric detectors, materials with a large pyroelectric coefficient, such as TGS (triglycine sulfate) and sintered ceramics of lead zirconate titanate or barium strontium niobate are used.

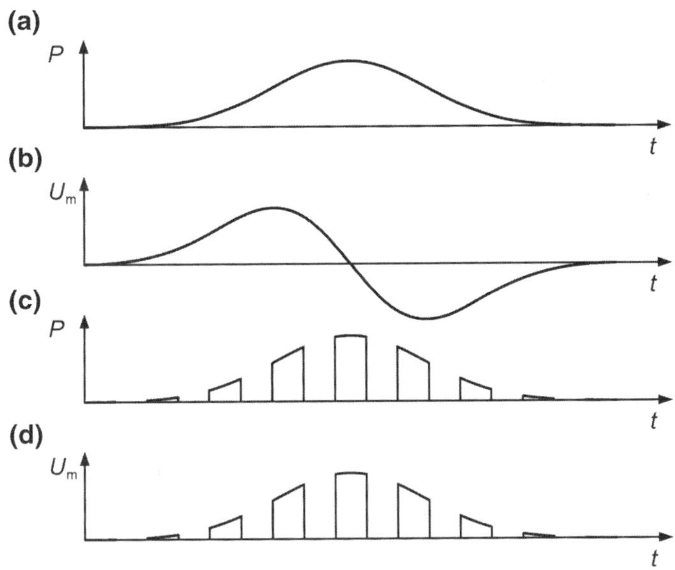

Fig. 5.12 Time dependence of **a** quasi-static flux of the incident light, **b** measured voltage for the quasi-static mode, **c** modulated incident light flux with chopping frequency greater than thermal cutoff frequency, **d** measured voltage for pulse mode operation

Pyroelectic detectors are operated in a wide spectral range of 1 nm–1 mm. In the infrared spectrum, pyroelectric detectors have, among the non-cooled detectors, the largest sensitivities, e.g. 900 V/W with a rise time of about 10 μs. Compared with other thermal detectors, pyroelectric detectors have the shortest rise times. Among the thermal detectors, the pyroelectric detector is of most importance for laser measurement technology.

5.2 Photoelectric Detectors

A short time after the discovery of the electron at the end of the 19th century, it was observed that metal surfaces illuminated by light can emit electrons, see Fig. 5.13. This effect is called external photoelectric effect, the emitted electrons are called photoelectrons. To explain the photoelectric effect, A. Einstein introduced in 1905 his light quantum hypothesis, see Sect. 3.1. Therefore, light consists of photons moving with the speed of light c. If a photon interacts with a metal electron, it transfers its whole energy hf to the electron, i.e., the photon is absorbed. If the absorbed energy hf exceeds the work function W_W of the metal:

$$hf > W_W, \tag{5.33}$$

the electron can leave the metal. Therefore, the photoelectric effect occurs only, if the wavelength of light falls below the cutoff wavelength:

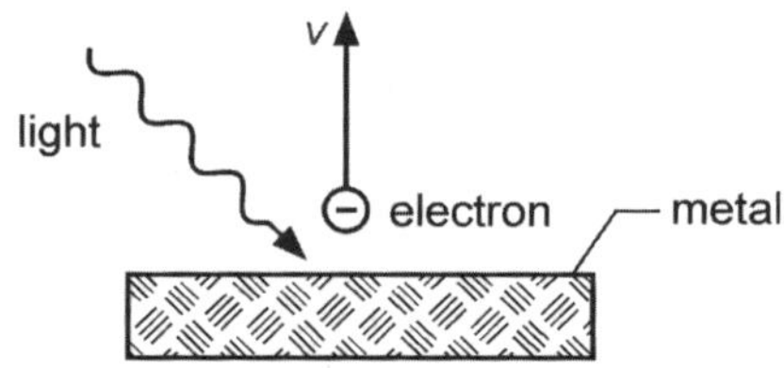

Fig. 5.13 Illustration of the external photoelectric effect

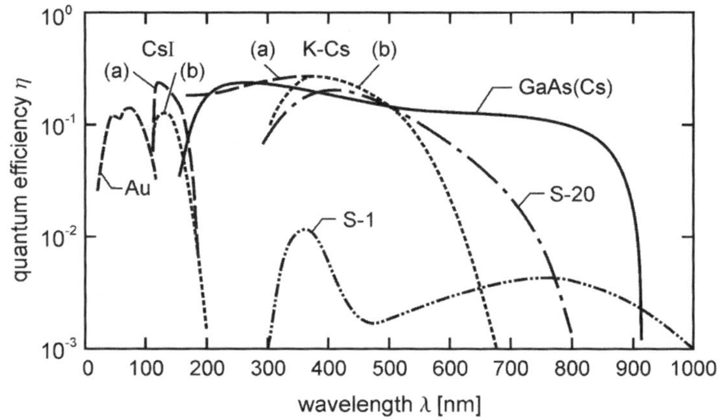

Fig. 5.14 Quantum efficiency η of various cathode materials as a function of the wavelength. Composed materials are often characterized by their *S-number*, e.g. *S-20* denotes a mixture of the elements Na, K, Sb and Cs

$$\lambda_c = \frac{hc}{W_W}. \tag{5.34}$$

Since there are no materials having work functions of less than 1 eV, the external photoelectric effect arises at light wavelengths $\lambda < 1.2\ \mu m$ only.

By the external photoelectric effect light is converted to free electrons. It is self-evident to use this effect to measure electromagnetic radiation. Hereby it is helpful, that the number of induced photoelectrons is proportional to the number of incident photons. The ratio of these numbers is referred to as quantum efficiency, see Fig. 5.14.

5.2.1 Photoelectric Cell

In metrology the external photoelectric effect is applied in the so-called photoelectric cell. A photoelectric cell consists of a cathode and an anode, which are in general placed inside a vacuum glass or quartz tube, see Fig. 5.15. If photons of sufficient energy irradiate the cathode, photoelectrons are released. Under the influence of the applied voltage U_0, the photoelectrons are accelerated towards the anode. In the external circuit, the so-called photocurrent I flows.

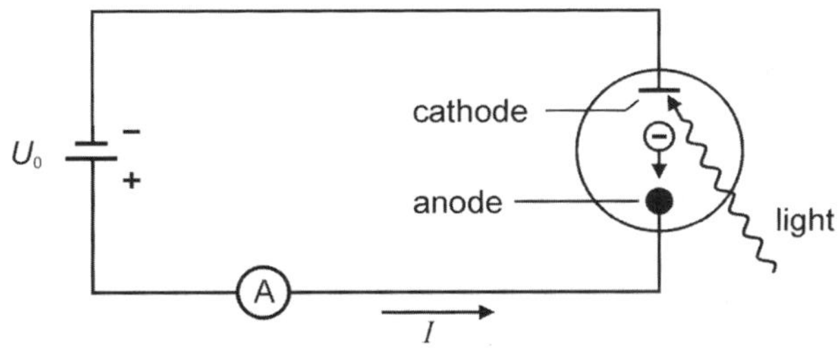

Fig. 5.15 Schematic set-up of a photoelectric cell

If the applied voltage is high enough, all photoelectrons reach the anode, i.e., the photocurrent becomes independent of the applied voltage U_0. Figure 5.16 illustrates this saturation of the photocurrent.

At a fixed acceleration voltage U_0 the saturated photocurrent I_s is directly proportional to the radiant flux of the incident light over a wide range. This linearity is lost, when the cathode is exposed, to high radiant fluxes for a too long period of time. As a result the quantum efficiency decreases. This fatigue effect or aging effect depends on whether the cathode material recovers or not after an idle-period. Cathodes, particularly if they contain caesium, should therefore be protected from unnecessary radiation, e.g. from direct sunlight, even if they are not in use. Besides the fatigue and aging effect, the influence of the space charge of the released photoelectrons must be taken into account. These space charges lead to field distortions and at high radiant fluxes to a space-charge limited, i.e., nonlinear photocurrent.

If ΔN photons with the wavelength λ per time interval Δt irradiate the cathode, this corresponds to a radiant flux of the incident radiation of

$$P = \frac{\Delta N}{\Delta t} \frac{hc}{\lambda}. \tag{5.35}$$

Within the time interval Δt depending on absorbance A and quantum efficiency η, the charge

$$\Delta Q = \eta \, A \, \Delta N \, e \tag{5.36}$$

is set free as photoelectrons. Here e is the elementary charge, $e = 1.6 \times 10^{-19}$ C.

This corresponds to a saturation current of

$$I_s = \frac{\Delta Q}{\Delta t} = \eta \, A \, e \frac{\Delta N}{\Delta t}. \tag{5.37}$$

The ratio of the saturation current I_s and the incident radiant flux P is the wavelength-dependent sensitivity of the photoelectric cell, see Fig. 5.17:

$$\Sigma_I = \frac{I_s}{P} = \frac{\eta(\lambda) \, A(\lambda) \, e \, \lambda}{hc}. \tag{5.38a}$$

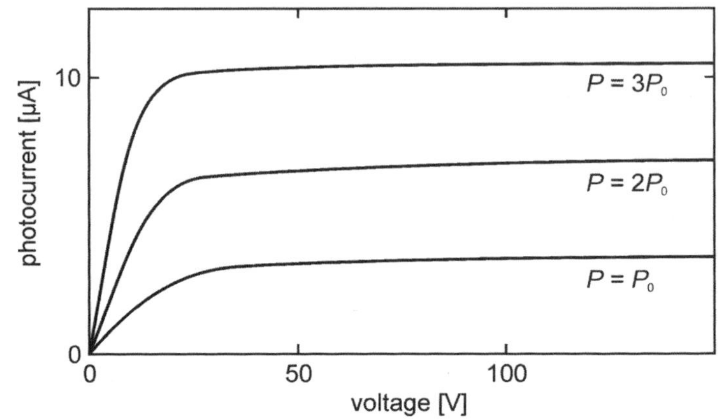

Fig. 5.16 Characteristic current–voltage curves of a high vacuum photoelectric cell at three different light fluxes

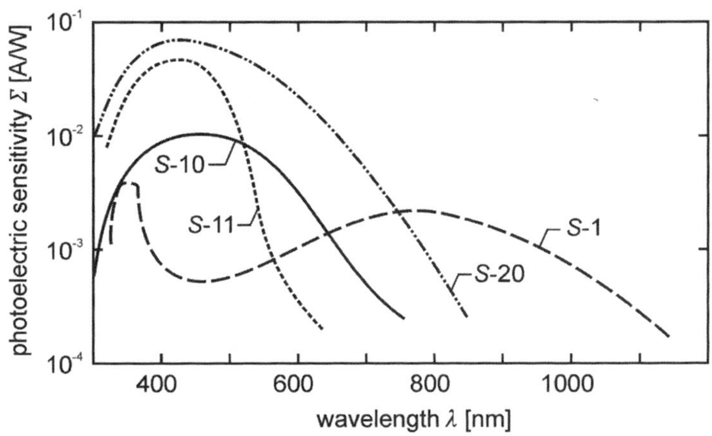

Fig. 5.17 Sensitivity of typical photoelectric cells as a function of the wavelength

In comparison to thermal detectors—with the exception of pyroelectric detectors—photoelectric cells have a fast rise time linked with a high bandwidth. Minimum rise times of 0.1 ns can be achieved, which are determined by the transit time spread of electrons moving from cathode to anode.

In addition to high vacuum photocells, gas-filled photocells are used. In comparison to high vacuum photocells, a gas-filled cell provides a photocurrent that is up to a factor of 100 greater. This is caused by photoelectrons ionizing gas particles by impacts on their way to the anode thus generating additional charge carriers. The advantage of higher sensitivity is accompanied by the disadvantage of a lower bandwidth. The cause is the significantly longer run-time of the heavy ions between anode and cathode. This permits only rise times τ_r greater than 100 μs.

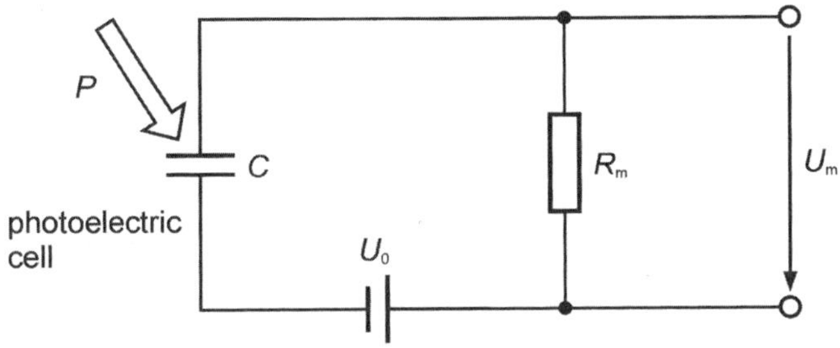

Fig. 5.18 Photoelectric cell with voltage source and measuring resistor. C capacitance of photoelectric cell

Particularly for broadband photoelectric cells the photocurrent is determined by a voltage measurement at a measuring resistor R_m, as shown in Fig. 5.18. R_m should be as large as possible since the measured voltage U_m increases with R_m:

$$U_m = R_m I_s. \tag{5.39}$$

Therefore, the voltage sensitivity of the photoelectric cell rises with increasing measuring resistor:

$$\Sigma_U = \frac{U_m}{P} = \frac{\eta(\lambda) A(\lambda) e \lambda R_m}{hc}. \tag{5.38b}$$

A measuring resistor being too large is disadvantageous. Firstly, the full source voltage U_0 is no longer available at the cell. This leads to a decrease in the photocurrent. Secondly, the $R_m C$ time constant of the circuit increases. Here, $R_m C$ is the product of measuring resistor R_m and capacitance C of the cell, including all stray capacitances. This has a negative effect on the rise time τ_r, if the $R_m C$ time constant is greater than the transit time spread τ of the electrons between cathode and anode.

5.2.2 Photomultiplier

In principle the photomultiplier is a photocell, wherein the photocurrent is amplified. Figure 5.19 shows schematically the set-up. The photocathode is connected with an electron multiplier, where electrons are multiplied by secondary emission.

Secondary emission denotes the process where accelerated electrons hitting the surface of a solid, transfer their energy and release one or more secondary electrons. The material-specific secondary emission yield δ, i.e., the average number of secondary electrons per incident electron, depends on the energy of the electrons, on the surface condition and on the angle of incidence. Typical values of the secondary emission yield δ in an electron multiplier are 4–14.

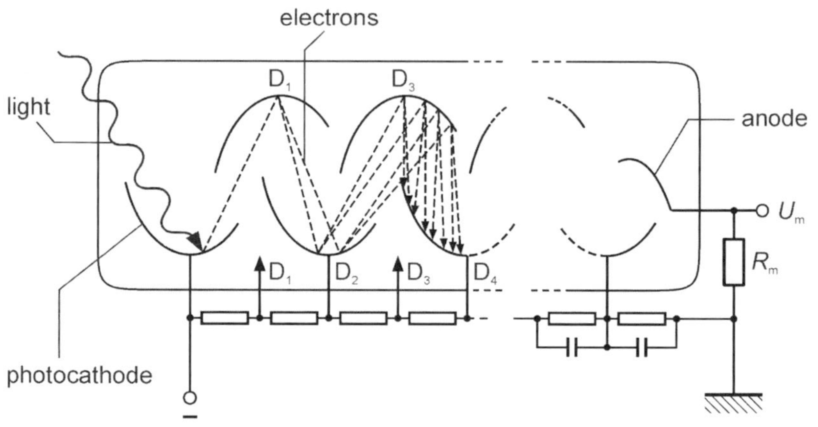

Fig. 5.19 Schematic set-up of a photomultiplier

In a photomultiplier several electrodes, which are called dynodes, are arranged in series. The potential difference between two dynodes is typically in the range of 100–300 V. If a single photoelectron strikes the first dynode, δ_1 secondary electrons are emitted on average and move to the second dynode. When these δ_1 electrons impinge onto the second dynode, the number of electrons increases on average once again by a factor δ_2, and so on. With n dynodes, this results in an overall gain of

$$G = \delta_1 \cdot \delta_2 \cdot \delta_3 \cdots \delta_n. \tag{5.40}$$

The number n of the dynodes is usually 6–14. Therefore, a total gain G of 10^6–10^9 can be obtained. Thus, the sensitivity of a photomultiplier tube is by a factor of G greater than a simple photoelectric cell. The amplified photocurrent, related to the incident radiant flux, is given by:

$$\Sigma_I = G\frac{\eta(\lambda)\,A(\lambda)\,e\,\lambda}{hc}. \tag{5.41}$$

The sensitivity is so high, that single photoelectrons emitted at the cathode can be detected.

The different potential values of the dynodes are provided by a voltage divider, see Fig. 5.19. The voltage supply of the voltage divider has to be highly stable to avoid variations in the gain. In order to prevent collapses of the voltages between the dynodes, the current in the voltage divider circuit has to be significantly greater than the current flowing between the dynodes in the tube. While measuring short light pulses high peak currents may appear. To keep the dynode potentials stable the last voltage divider resistors are bypassed with capacitors in this case, see Fig. 5.19. These decoupling capacitors supply the photomultiplier tube with electric charges during the forming of the signal pulse and reduce a change of the voltage between the last dynode and the anode thus improving the linearity for pulse measurements [4].

The rise times τ_r of photomultipliers are usually in the range of 2–15 ns. However, values of less than 1 ns can be achieved. The rise time is determined mainly by two effects: the run-time and the variation of the run-time of the electrons through the photomultiplier. In order to keep the run-time and the dispersion of the run-time as small as possible, the acceleration voltage should be large and the distances between the dynodes should be small. Also, a small number of dynodes is positive, but at the cost of the total gain. The dispersion of the run-time is essentially caused by different possible paths, which the electrons have taken. To avoid this, specially shaped dynodes and, eventually, an electro-optical entrance system, are deployed. The electro-optical entrance system has the task to compensate different starting conditions, i.e., different starting points, different starting speeds and different starting angles of the emitted photoelectrons at the photocathode. By external magnetic fields, e.g., the magnetic field of the earth, the electrons are deflected from their ideal trajectory. Therefore, a photomultiplier is shielded from magnetic fields.

The photomultiplier is a radiation detector with high sensitivity, low rise time and low noise. The combination of these three properties cannot be exceeded by any other detector system.

5.2.3 Channel Electron Multiplier

The channel electron multiplier is basically nothing else but a secondary electron multiplier with continuously distributed dynodes. It consists of a thin tube with a typical length of 10 cm and an inner diameter of 1 mm. The inner wall of the tube, usually made of lead glass, is specially treated, to achieve a large secondary emission yield and an electrical resistivity of about 1 GΩ. At both ends of the pipe, a potential difference of some kilovolt is applied. To prevent destructive electric arc discharges, the channel electron multiplier is operated under vacuum ($p < 10$ mPa).

Figure 5.20 shows schematically the principle of channel electron multipliers. An incident photon striking the inner wall, generates an electron by the external photoelectric effect. This photoelectron is accelerated in the direction of the higher potential. When it hits the inner wall again, δ secondary electrons are released. These electrons are accelerated and the process of secondary emission is repeated again and again along the inner pipe wall, i.e., an electron avalanche evolves. Per incident photoelectron 10^3–10^5 electrons reach the anode finally. Larger gains cannot be realized because of the so-called ion feedback where the electron avalanche ionizes, particularly near the end of the tube, residual gas molecules. These ions are accelerated towards the entrance. When striking the inner wall, the ions also release secondary electrons, which lead to new electron avalanches. This is disturbing the measurement, especially, because a permanent glow discharge can occur. With curved tubes, ion feedback can be avoided. Ions strike the inner wall, before they have received sufficient energy for secondary electron emission. With curved tubes gain factors of up to 10^8 can be achieved for a single electron.

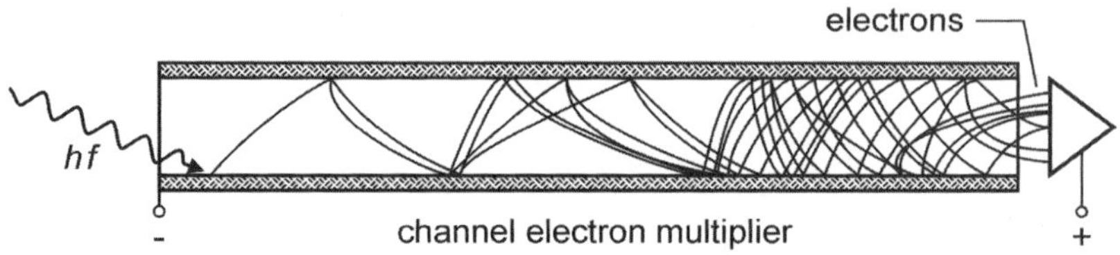

Fig. 5.20 Principle of a channel electron multiplier

In continuous operation, the wall current—also called bias current corresponding to the current in the voltage divider considered previously—should be great compared to the anode current to avoid non-linearities in the gain. Channel electron multipliers will provide a linear output up to a value equal to approximately 10 % of the bias current [5]. Typical wall currents, for operating voltages of 2.5 kV and wall resistances of 60 MΩ are about 40 μA.

In a conventional photomultiplier, some voltage divider resistors are bypassed by capacitors as described above. For pulsed operation and at high gains these capacitors keep the dynode potentials constant. This approach cannot be transferred to channel electron multipliers. If the potential difference along the tube collapses, it must first be refreshed by the low wall currents. This gives rise to dead times of typically 0.5–15 μs.

Channel electron multipliers are commonly used to count individual primary electrons. They can also detect incident photons that strike the beginning of the tube and release photoelectrons from the inner wall, see Fig. 5.20. The width of each current pulse amounts to approximately 20 ns.

5.3 Semiconductor Detectors

Semiconductors can absorb light. In this process, free charge carriers—electrons and holes—are generated. This is called internal photoelectric effect.

5.3.1 Photoresistor

If light is absorbed by a photoresistor, its electrical conductivity increases. Let us consider that light with a radiant flux P falls on a detector with the absorbance A. Then, in the time interval Δt, the energy

$$\Delta W = AP\Delta t \tag{5.42}$$

is absorbed. Thus, in the semiconductor

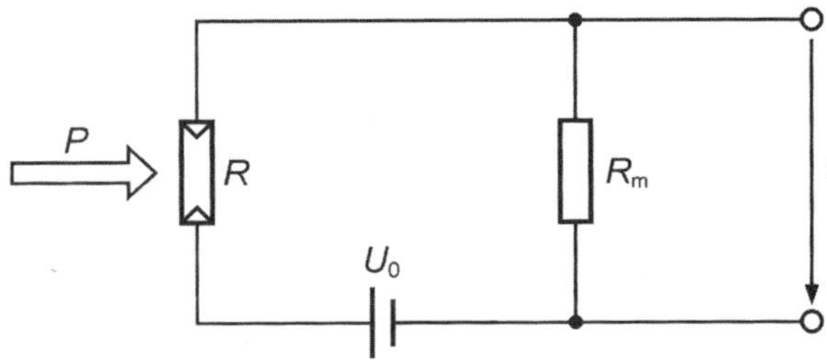

Fig. 5.21 Electrical circuit to measure the change of the resistance of a photoresistor

$$\Delta N_{\text{gain}} = +\frac{\eta A P \Delta t}{hf} \tag{5.43}$$

pairs of charge carriers are produced. Here, f, η and h are the frequency of the incident light, the quantum efficiency, i.e., the number of pairs of charge carriers, generated per absorbed photon, and the Planck constant, respectively. The number N of pairs of charge carriers also decreases, due to recombination. For a small period of time Δt, the following relation holds

$$\Delta N_{\text{loss}} = -\frac{N \Delta t}{\tau_L}. \tag{5.44}$$

τ_L is the mean lifetime of an electron-hole pair. In the stationary case, gain and loss compensate each other. That is why the number of pairs stabilizes at

$$N = \frac{\eta A \tau_L}{hf} P. \tag{5.45}$$

Nevertheless, the number N fluctuates as a function of time. This is referred to as generation-recombination noise.

Even without incident light, charge carriers are present in the photoresistor. Their number is N_0. The change of the resistance of the photoresistor is given by

$$\frac{\Delta R}{R_0} = -\frac{N}{N_0} = -\frac{A \eta \tau_L}{hf N_0} P, \tag{5.46}$$

where R_0 is the resistance without irradiation. It can be determined with the electric circuit shown in Fig. 5.21. We know this circuit already from the discussion of the bolometer, see Sect. 5.1.2. The change of the measured voltage caused by incident light, is calculated with the approximation $DeltaR \ll R_0$ as:

$$\Delta U_m = \frac{R_0 R_m}{(R_0 + R_m)^2} \frac{A \eta \tau_L}{hf N_0} U_0 P. \tag{5.47a}$$

When the measuring resistor matches the photoresistor, i.e., $R_m = R_0$, the signal attains a maximum:

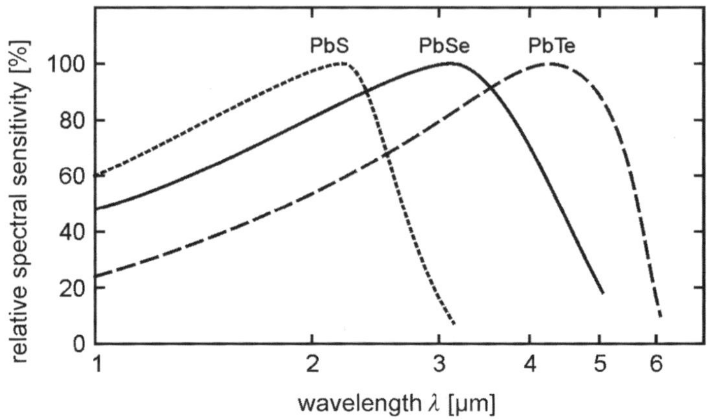

Fig. 5.22 Relative spectral sensitivity of various photoresistors versus wavelength

$$\Delta U_{\mathrm{m}} = \frac{1}{4}\frac{A\eta\tau_{\mathrm{L}}}{hfN_0}U_0P. \tag{5.47b}$$

If the measuring signal is related to the irradiated radiant flux P, we obtain the wavelength-dependent sensitivity of the photoresistor. For impedance matching, we find, see Fig. 5.22:

$$\Sigma(\lambda) = \frac{\Delta U_{\mathrm{m}}}{P} = \frac{A(\lambda)\,\eta(\lambda)\,\lambda\,\tau_{\mathrm{L}}}{4hcN_0}U_0. \tag{5.48}$$

In order to obtain high sensitivity, voltage U_0 and lifetime τ_{L} of the charge carriers should be as large as possible. However, the last requirement leads to an increase of the rise time τ_{r}, which is limited by the mean lifetime τ_{L}:

$$\tau_{\mathrm{r}} \approx 2.2\,\tau_{\mathrm{L}}. \tag{5.49}$$

The charge carriers existing without irradiation are generated by thermal excitation. Therefore, the number N_0 can be reduced by cooling, which affects positively the sensitivity. Noise is also reduced by cooling. It consists of thermal noises of measuring resistor and photoresistor as well as of generation-recombination noise of the photoresistor.

Photoresistors are available for the visible and infrared spectral range, see Table 5.1. With extrinsic photoconductors having doped impurities rise times in the nanosecond range can be achieved provided an appropriate carefully designed circuit is taken.

Table 5.1 Spectral ranges and operating temperatures of some photoresistors

Intrinsic photoconductor			Impurity photoconductor		
Material	Wavelength range [μm]	Temperature [K]	Material	Wavelength range [μm]	Temperature [K]
Cd S	0.5–0.9	295	Ge : Au	1–9	77
Pb S	0.6–3.0	295	Ge : Cu	6–29	4.2
Pb S	0.7–3.8	77	Ge : Zn	7–40	4.2
Pb Se	0.9–4.6	295	Ge : Sb	40–140	4.2
In Sb	0.5–6.5	195			
Hg Cd Te	2–24	77			

5.3.2 Photovoltaic Element and Photodiode

A photovoltaic element is composed of a semiconductor with a p- and a n-doped region, see Fig. 5.23, forming a p-n junction. The conducting electrons of the n-type side diffuse into the p-type region. Conversely, positive holes migrate from the p-type to the n-type side. The remaining stationary charged acceptor and donor atoms form a contact potential between the two regions, the so-called built-in voltage. It prevents a further diffusion of free charge carriers. Due to the recombination of electrons and holes between the n- and p-type material a zone free of mobile charge carriers—i.e. a non-conductive layer—with a typical thickness of about 1 μm is formed.

If photons are absorbed in the non-conductive layer of a photovoltaic element, free electron-hole pairs are generated. Due to the built-in voltage, electrons migrate to the n-doped side and, reversely, holes to the p-type region, respectively. If the semiconductor is short-circuited by an outer conductor, see Fig. 5.24a, a short circuit photocurrent flows being proportional over many orders of magnitude to the number of generated electron-hole pairs per unit time and thus to the incident radiant flux. At open circuit conditions, the open-circuit voltage increases logarithmically with the radiant flux. The general relationship between photocurrent I and potential drop U at the photovoltaic element is given in Fig. 5.25. Circuitry-wise, the photovoltaic element is represented by a current source, wherein both, the short-circuit current and the internal resistance, depend on the radiant flux of the irradiated light.

A photovoltaic element may also be operated as a photodiode. To this end, the photovoltaic element is reverse-biased, see Fig. 5.24b. By this reversed voltage, the electrons are pulled out of the n-layer and holes from the p-layer, i.e., the non-conducting depletion layer is increased. So, only a very small reverse current will flow. This reverse current is also called dark current. It is fast increasing with the reverse voltage to saturation. The dark current I_{d} can be largely suppressed by cooling. If light of a suitable wavelength is absorbed in the depletion layer, electron-hole pairs are created and the reverse current increases. In addition to the dark current, a so-called photocurrent flows, see Fig. 5.25, induced by the

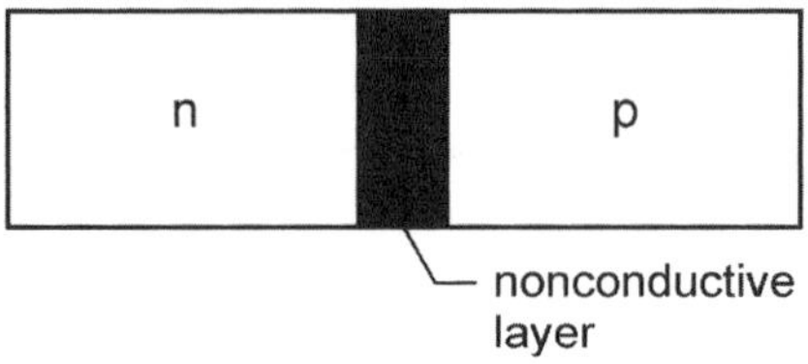

Fig. 5.23 Schematic diagram of a p-n junction

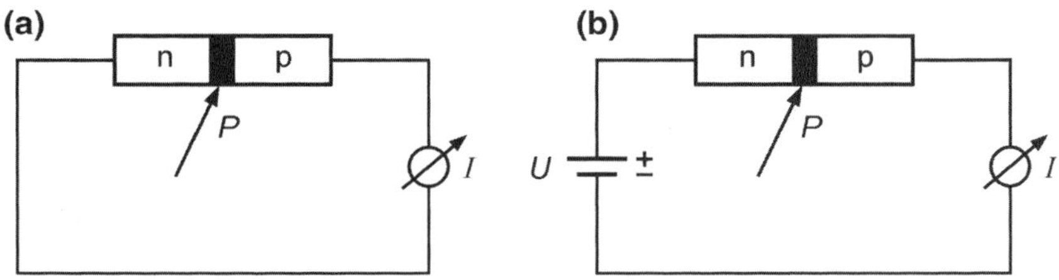

Fig. 5.24 Schematic representation of **a** photovoltaic element and **b** photodiode, in both cases short-circuited

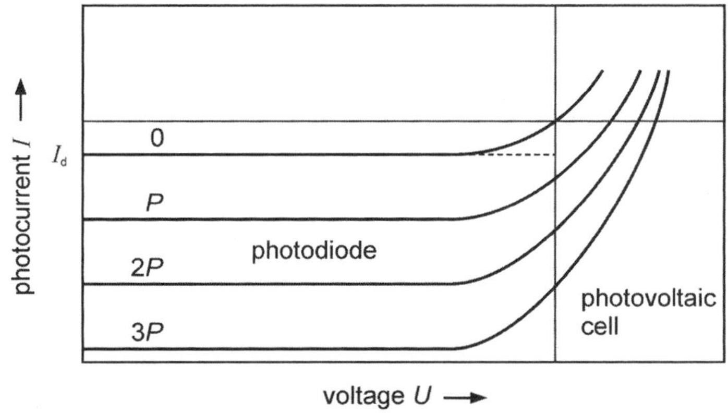

Fig. 5.25 Current to voltage characteristics of a photovoltaic element ($U > 0$) and a photodiode ($U < 0$) at different radiant fluxes P of the irradiated light

irradiated light. For the measurement technique it is important that the saturated photocurrent I_{ph}, i.e., the reverse current minus the dark current in case of a sufficiently high reverse voltage, is proportional to the absorbed radiant flux P:

$$I_{ph} = -\frac{A\eta e}{hf} P. \tag{5.50}$$

Thus the wavelength-dependent sensitivity of the photodiode, see Fig. 5.26, is:

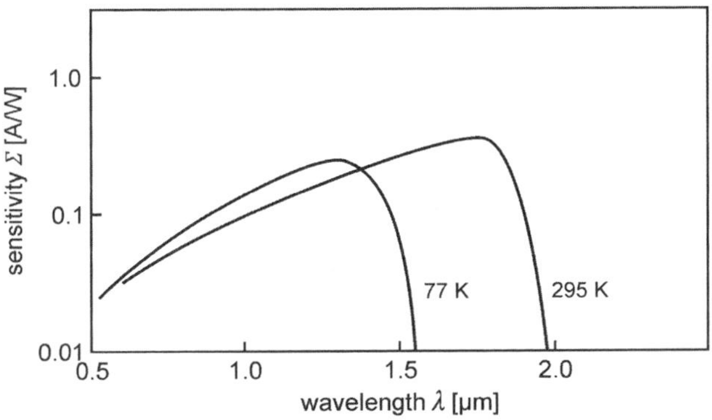

Fig. 5.26 Sensitivity of a germanium photodiode at 77 and 295 K as a function of the wavelength

$$\Sigma = \frac{-I_{\mathrm{ph}}}{P} = \frac{A(\lambda)\,\eta(\lambda)\,\lambda\,e}{hc}. \tag{5.51}$$

Especially for broadband measurements the photocurrent is measured with the help of a measuring resistor R_{m}, see Fig. 5.27. In order to obtain a large measuring voltage U_{m}, the measuring resistor R_{m} should be chosen as large as possible. This has, in the case of the photovoltaic element, the disadvantage that the measuring voltage U_{m} is no longer proportional to the absorbed optical flux P. The measuring resistor R_{m} must therefore be relatively small, so that practically the short circuit current is measured. The situation is more favorable for the photodiode. In this case the photocurrent I_{ph} is independent of the applied voltage U_0, as far as the voltage is sufficiently high to reach saturation. The resistance R_{m} can be chosen relatively large. But it is important to ensure that the voltage drop across the photodiode must be sufficiently high to saturate the photocurrent.

The minimum time resolution of a photovoltaic element and a photodiode is determined by the capacitance C_{dl} of the depletion layer and the measuring resistor R_{m}. The shortest possible rise time is:

$$\tau_{\mathrm{r}} = 2.2\,R_{\mathrm{m}}C_{\mathrm{dl}}. \tag{5.52}$$

For this reason, both, measuring resistor R_{m} and depletion layer capacitance C_{dl}, must be kept as small as possible. The greater the reverse voltage in the photodiode, the greater is the thickness of the depletion layer. Therefore, the capacitance C_{dl} of the depletion layer decreases with increasing reverse voltage. This is the reason why photodiodes have smaller rise times than photovoltaic elements. In order to realize high frequency measurements, the maximum reverse voltage should be applied to the photodiode. Thus it is possible to achieve bandwidths in the GHz range.

The noise of photovoltaic elements or photodiodes is determined by the quantum noise of the photocurrent, the dark current noise, the generation-

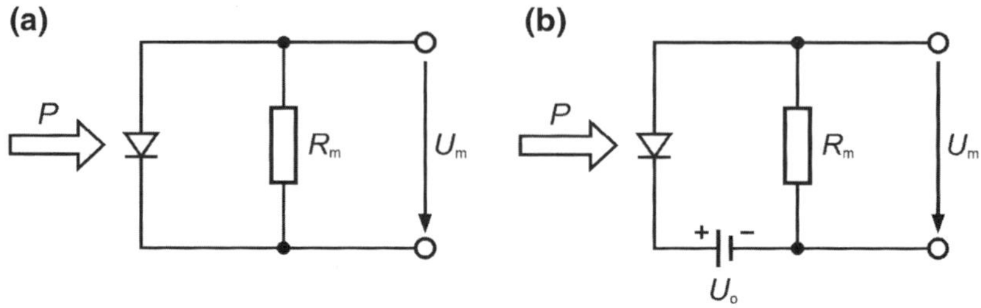

Fig. 5.27 Circuit to measure the photocurrent of **a** a photovoltaic element, **b** a photodiode

Table 5.2 Spectral sensitivity ranges and working temperatures of some photodiodes

Material	Wavelength range [μm]	Temperature [K]	Material	Wavelength range [μm]	Temperature [K]
Si	0.4–1.1	295	InAs	0.6–3.2	77
Ge	0.5–1.8	295	InSb	0.6–5.6	77
InGaAs	0.9–1.7	295	HgCdTe	2–15	77
InAs	1.0–3.7	295	PbSnTe	8–12	77

recombination noise and the thermal noise of the measuring resistor. The two latter noise components can be reduced by cooling.

Commercially offered photodiodes cover the ultraviolet, visible and infrared spectrum from 0.13 to 26 μm. Partly, they have to be operated cooled, because of both, the signal-to-noise ratio or the high dark current. Some examples are given in Table 5.2.

For special applications, a series of modified photodiodes is available. Firstly, we mention the PIN photodiode. Typical for a PIN photodiode is the relatively wide undoped zone, which is arranged between the p- and n-layer. In this zone, the incident radiation generates electron-hole pairs, see Fig. 5.28. The name of the PIN diode reflects the doping of the three layers (p-i-n). Here i stands for the non-doped, so-called intrinsic layer.

Compared to the normal photodiode the broad, high-ohmic zone has two advantages. On the one hand, the dark current is quite low. The other advantage is the very small junction capacitance. PIN photodiodes can therefore reach bandwidths of up to 20 GHz. These large bandwidths, however, are no longer limited by the $R_m C_{dl}$ time constant, but by the drift time of the generated electron-hole pairs in the intermediate layer.

Next, we discuss the avalanche photodiode. An avalanche photodiode is operated with a very large reverse voltage of several hundred volts. Therefore, the charge carriers, which are generated by the incident light in the depletion layer, gain much energy in the electric field of the depletion layer. By impact ionization the number of charge carriers increases strongly, like an avalanche. That is why the photocurrent is amplified with low noise. Typical gains are between 50 and 200. However, with increasing gain, the bandwidth decreases. For an avalanche photodiode, the product of gain G and upper cutoff frequency f_c is a characteristic quantity:

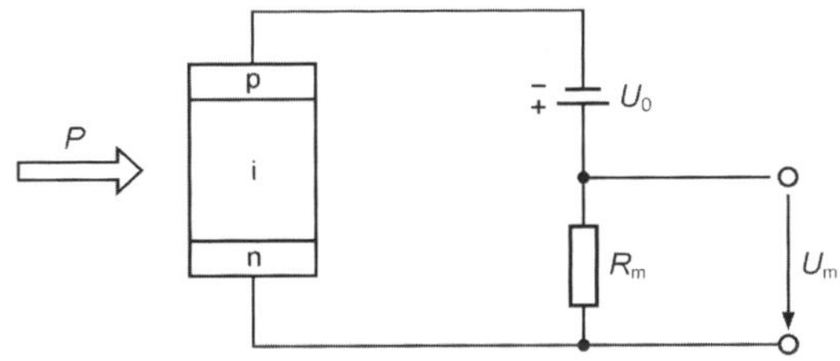

Fig. 5.28 Set-up and circuit of a PIN photodiode. p p-doped region, i intrinsic region, n n-doped region

$$Gf_c = \text{const.} \tag{5.53}$$

The product reaches values of up to 200 GHz.

Finally, it should be pointed to Schottky diodes. For a Schottky diode a semiconductor is brought into contact with a metal. An intermediate layer is created having properties similar to those of a p-n junction but with a lower forward voltage. Schottky diodes can achieve rise times of less than 1 ns.

5.4 Space-Resolving Detectors

For many tasks of laser measurement technology, spatially resolving detectors are required. For this aim, a great variety of detectors are available. The most important and mostly used detectors are summarized below.

5.4.1 Photographic Films

We start with the oldest position-sensitive detector, the photographic film. Photographic films have the advantage that they do not only serve as a detector, but also as a storage medium for the detected radiation.

Photographic films or plates are light-sensitive layers, which are deposited on a carrier, mostly made of plastic, paper or glass. The light-sensitive layer consists of gelatin, in which the photosensitive substance is suspended as microscopic particulates. Usually, a salt of a silver halide, e.g. AgBr, serves as photosensitive substance.

When light falls on the photographic film, the grains of silver halide are more or less photo-chemically activated depending on the exposure. So a latent image is created. In a subsequent chemical development process, the activated silver halide grains are reduced to metallic silver. By this means the latent image is made visible, since the finely divided metallic silver appears black. The silver halide being not activated by incident light, is removed in a subsequent fixing process, see Fig. 5.29. The final washing with water removes all of the soluble chemicals. The individual steps must be performed according to the advice of the manufacturer.

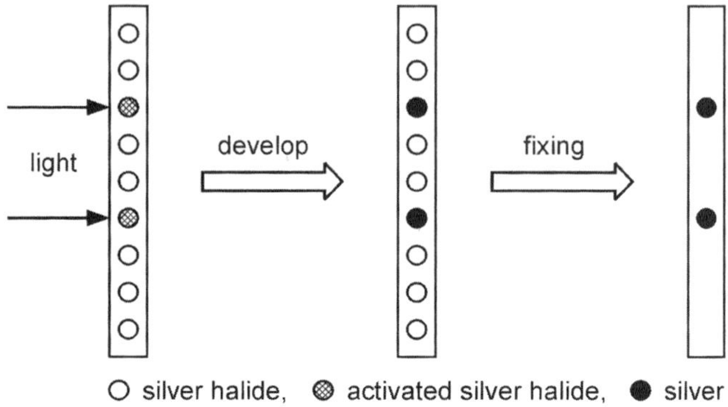

Fig. 5.29 Development of a photographic film

The number of metallic silver grains per unit area is a measure for the applied radiant exposure. Their number can be counted with an optical microscope. However, it is customary, to measure the local transmittance T of the developed film. Therefore, a small area is illuminated with white light. T is given by the ratio of the transmitted intensity I and the incident intensity I_0:

$$T = \frac{I}{I_0}. \tag{5.54}$$

The decadic logarithm of $1/T$ is called optical density:

$$D = -\log_{10} T = \log_{10} \frac{I_0}{I}. \tag{5.55}$$

An optical density of $D = 2$ means, for example, that 1 % of the incident light penetrates the developed film. The optical density depends on the radiant exposure H, which is the product of incident irradiance I and exposure time τ:

$$H = I\,\tau. \tag{5.56}$$

In addition to the *radiant exposure*, there is also the *luminous exposure*. Instead of irradiance (unit: W/m^2) in (5.56) the *illuminance* (unit: lux, 1 lux = 1 lumen/m^2) is used. In this book we denote both quantities with H.

A plot of the optical density D as a function of the decadic logarithm of the exposure H is called characteristic curve, see Fig. 5.30. At low exposure, there is a constant optical density known as optical fog density. It is not caused by the light exposure. This optical density is related to the absorption in the carrier material or is caused by the development process itself. At a certain exposure, the so-called threshold, a small increase of the optical density begins. It is followed by a region with a constant slope. In this region, with a range of about two decades, the photographic layer is most commonly used. Finally, the slope decreases and the

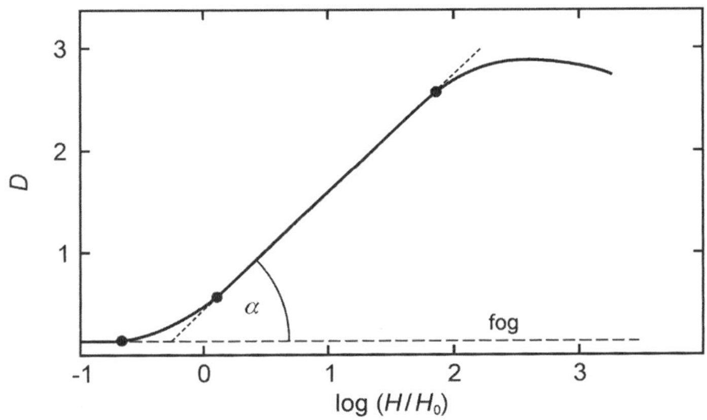

Fig. 5.30 Optical density of a photographic film as a function of the logarithm of the radiant exposure

optical density levels off to saturation. For even higher exposures the optical density decreases again. This effect was first observed in photographs of the sun and is, therefore, called solarization. This is caused by unwanted photochemical reactions, which occur at high exposures.

The optical density of a photographic emulsion is produced by the exposure $H = I\tau$. Usually, the law of reciprocity is valid, i.e., it is irrelevant, whether the exposure is caused by a short-term and high irradiance or by a small irradiance over a longer period. Deviations appear particularly at very short exposure times using bright light, for example, in high-speed photography, and at slow shutter speeds with low irradiances, e.g., in astronomy known as Schwarzschild effect. For the same exposure, the optical density depends on, whether the light energy is provided continuously or at several time intervals. This phenomenon is called intermittence effect.

The sensitivity of a photographic layer is characterized by the exposure $H_{\min}$, which leads to an optical density that exceeds the optical fog density by a certain value, usually $\Delta D = 0.1$. Pure silver halide films are much more sensitive to blue and UV light than to green and red wavelengths. That is why sensitizing dyes are added to the film emulsion in so-called *ortho chromatic* and *pan chromatic* films to expand the spectral range, see Fig. 5.31.

According to the standard ISO 5800, the sensitivity of a photographic emulsion is given by two redundant ISO numbers, which summarize both, the ASA number and the DIN number corresponding to an arithmetic and logarithmic scale, respectively [6]. Example: If a film is specified as ISO 100/21°, it means that the ASA number is 100 and the DIN number is 21. If the sensitivity of a film is doubled or halved, the ASA number is doubled or halved, too, while the DIN number rises or falls by 3 DIN, respectively. The relation between ASA and DIN numbers is provided by Table 5.3. For digital still cameras the standard ISO 12232:2006 describes different techniques to determine the exposure index rating at each sensitivity setting [7].

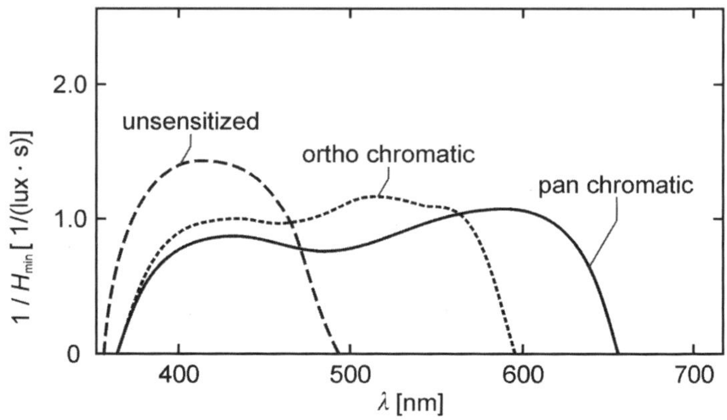

Fig. 5.31 Reciprocal minimum exposure ($1/H_{min}$) for a typical unsensitized, ortho chromatic and pan chromatic film as a function of the wavelength

Table 5.3 Comparison of the sensitivity specifications according to ASA and DIN

ASA	2.5	8	25	80	250	800	2500
DIN	5	10	15	20	25	30	35

The sensitivity of a photographic film can be increased by a pre-exposure, e.g., with the light of an electrical spark. By this, the layer is sensitized to such an extent that a further minor exposure leads to an observable blackening. In addition, there is the possibility to increase the sensitivity of photographic film by special chemical pre-treatments or by special additives used in the development process.

Another quantity, which is defined by the optical density curve, is the γ-value. It describes the slope of the straight part of the characteristic curve, see Fig. 5.30:

$$\gamma = \tan \alpha. \tag{5.57}$$

α is the angle of inclination of the straight part of the characteristic curve. The γ-value is a measure of the contrast.

The spatial resolution of a photographic layer describes its ability to record smallest details. The resolution is defined as the number of line pairs per mm, which can be resolved by the photographic film. A line pair is a light and a dark line. Special high-resolution layers, e.g. for applications in holography, see Chap. 7, have a resolution of about 5,000 line pairs per mm. This corresponds to a maximum storage capacity of 2.5 Gbit/cm^2. The limit of resolution is mainly due to two phenomena: the finite size (granularity) of the silver halide and the scattering of light in the substrate. High-resolution layers require small, homogeneously distributed grains.

In the previous, it was tacitly assumed that the exposure took place with white light, i.e., daylight. If the optical density curve is recorded with monochromatic

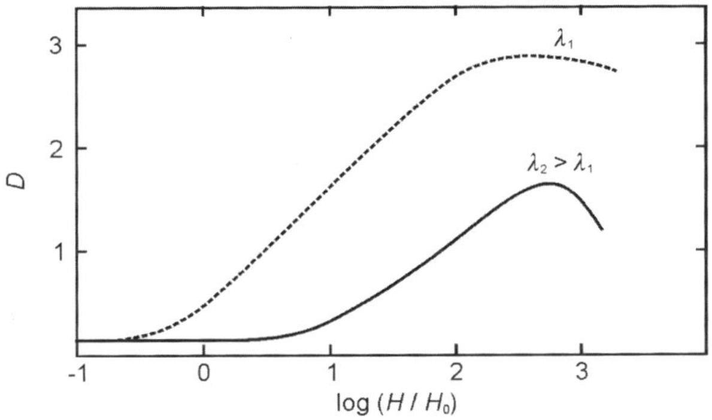

Fig. 5.32 Optical density curves for two different wavelengths

light, there are clear differences, see Fig. 5.32. All quantities that are defined by the characteristic optical density curve—threshold, sensitivity, γ-value, etc.—are wavelength-dependent.

Photographic films are not available for the spectral range of $\lambda > 1.3$ μm. Therefore, the radiation of a CO_2 laser ($\lambda = 10.6$ μm) is not able to blacken a photographic layer. However, under certain circumstances infrared radiation may sensibilize photographic film. By a subsequent irradiation with visible light, the previously sensitized areas become visible.

5.4.2 Phase-Modulating Recording Materials

The photosensitive films can be divided into two classes:

- amplitude-modulating and
- phase-modulating materials.

The films discussed in the section before, belong to the amplitude-modulating ones. When light falls on an exposed and developed silver halide layer, intensity and hence amplitude of the transmitted light are modulated according to the blackening pattern of the developed film. The situation is quite different, when using a phase-modulating material. In this case, the amplitude or intensity remains constant while the phase of the transmitted wave is modulated. Phase-modulating recording materials are almost exclusively used in holography, see Chap. 7. They are produced, for example, by bleaching from an ordinary exposed and developed film layer. By suitable wet chemical processes, the silver grains, which are responsible for the blackening, are transformed into a transparent silver salt. According to the original existing blackening pattern, a space-dependent salt concentration is formed causing a spatial modulation of the refractive index.

Other phase-modulating recording materials, e.g. thermoplastic photographic plates, convert the incident light into a surface relief pattern, see Fig. 5.33. In the following, we will discuss the principles of thermoplastic recording materials.

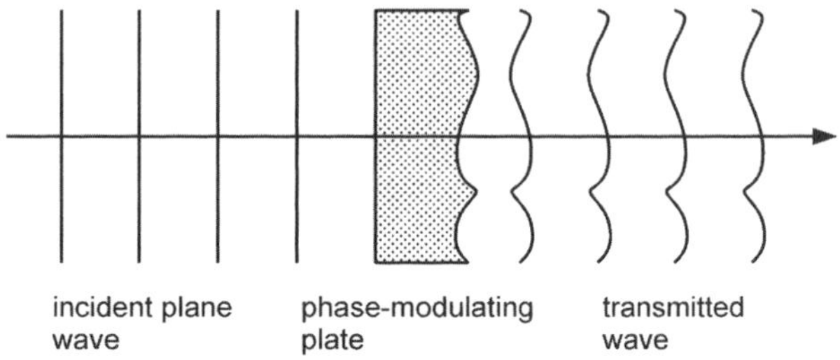

Fig. 5.33 Change of the phase front of a plane light wave passing through a transparent layer of varying thickness. The refractive index of the layer is greater than that of the surrounding atmosphere

In general, photothermoplastic recording materials utilize a glass plate as a carrier being coated with an optically transparent and electrically conducting layer. Above this layer, there is a photoconducting layer. This layer becomes electrically conductive, when it is illuminated. On the photoconducting layer, the actual thermoplastic material is disposed, see Fig. 5.34.

In the following, the different steps of exposure and development of a thermoplastic plate are described:

(a) Before recording, the plate must be sensitized in darkness. This is done by spraying positively charged ions to the surface of the thermoplastic plate. The electrically conductive layer is connected to ground and thus loads negatively, see Fig. 5.34a.
(b) After this initial charging, the exposure can take place. As a result the photoconducting layer becomes partially conductive and—depending on the local exposure—the applied potential difference decreases. Thereby, the negative charges from the conductive layer approach the thermoplastic layer, see Fig. 5.34b.
(c) By a second charging process, the different potential differences, which were induced by the exposure, are compensated. By this, different strong electric fields are created in the thermoplastic material, see Fig. 5.34c.
(d) In the last step, a current is allowed to flow through the conducting layer. Because of Joule heating, the thermoplastic material is heated above its softening point. The electric fields of different sizes, corresponding to the applied spatial exposure pattern, deform the surface of the thermoplastic material. This structure is frozen by rapid cooling and is stable at room temperature, see Fig. 5.34d.

The maximum spatial resolution is about 2,000 line pairs per mm. The process described, from exposure to freezing, takes only a few seconds.

Already exposed plates can be deleted in about a minute and be reused. To recover the plate, it must be re-heated. If heating temperature and heating time are matched with the surface tension of the softened thermoplastic material, the surface relief will be smoothed again.

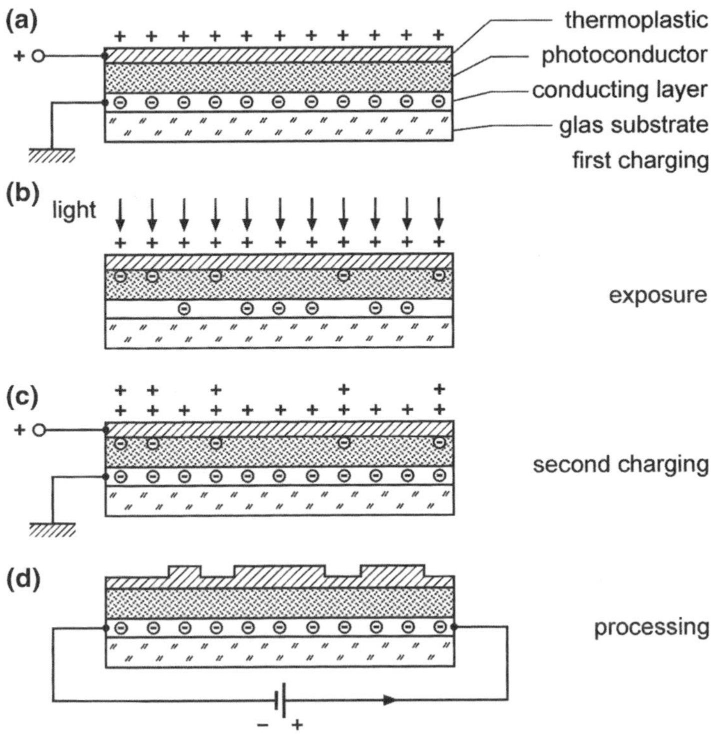

Fig. 5.34 Principle of a photothermoplastic plate. **a** sensibilization by a first charging, **b** exposure with subsequent charge transport through the photoconductor, **c** second charging and generation of the deforming electric field, **d** heating and subsequent deformation of the thermoplastic material

5.4.3 Radiation Converter

Radiation converter transform infrared (IR) or ultraviolet (UV) light into visible (VIS) light. Correspondingly, we distinguish between IR-VIS- and UV-VIS converters. Both types of converters are based on the same principle, the phosphorescence. According to this phenomenon, some materials, so-called phosphors, emit visible light, after they were exposed, for example, to UV light. Phosphorescent material does not immediately re-emit the radiation it absorbs. It lasts, depending on the phosphor, fractions of a second to several months, before re-emission occurs.

IR-VIS and UV-VIS converters have the same structure. They consist of a carrier plate, on which a phosphorescent material is deposited. Firstly, we consider UV-VIS converters. If UV light of a suitable wavelength strikes on the phosphor, then visible luminescent radiation is released. The system operates as a UV-VIS converter. In order to get an intense afterglow, the phosphor should re-emit the visible light in the shortest possible period of time.

For the construction of IR-VIS converters, phosphors having a long afterglow are chosen. Before using, IR-VIS converters must be activated, e.g. by UV light or daylight. Due to the long storage time, luminescent light is emitted over hours or days. Hence, virtually no afterglow is observed. If infrared light of a suitable wavelength irradiates an activated converter, then immediately the emission of visible luminescent light is triggered. With commercially available IR-VIS converters, the spectral range between 0.7 and 2 μm is covered. They are also called IR sensor cards.

The phosphorescence is a temperature-dependent effect, i.e., the storage time of the phosphors decreases with increasing temperature. This effect is used in the so-called thermal converters. If IR radiation falls on an activated thermal converter, visible light is emitted under the influence of the heat generated by the absorbed IR radiation. Thermal converters can also be used in the range $\lambda > 2$ μm. Therefore, they are suited to detect radiation of the CO_2 laser, which has a wavelength of 10.6 μm.

IR-VIS and UV-VIS converters are not suitable for quantitative and time-resolved measurements since they only provide a visual impression of the spatial irradiance distribution. This visual impression is often sufficient, for example, to adjust optical components. Therefore, IR-VIS and UV-VIS converters are indispensable tools while working with IR and UV lasers. Continuous minimum laser irradiances of about 2 μW/cm^2 to 1 W/cm^2—the last value is typical for thermal converters—are required to observe the luminescence of a converter in a darkened room with the naked eye.

5.4.4 *Image Converter*

The structure of an image converter is sketched in Fig. 5.35. The object is imaged onto the photocathode via an optical imaging system. There, the incident light induces the emission of electrons from the cathode, according to the photoelectric effect. Thus, the optical image is converted to a corresponding negative charge pattern on the cathode. This charge pattern is then mapped onto a phosphor screen acting as anode. Finally, a visible image of the object arises. In order to store the image, it can either be photographed or recorded with a video camera.

The imaging of photoelectrons from the photocathode to the phosphor screen can be realized in several ways. In the electron-optical image converter, an electron lens is used. Such an electron lens maps the electrons to the anode by suitably selected electrical and magnetic fields.

In the so-called image converter diode, the electron lens is omitted. Photocathode and anode face each other at a small distance of a few mm. Applying a high voltage of several kilovolts, the photoelectrons are accelerated along the shortest path to the phosphor screen and anode. This is referred to as a 1:1 imaging. Such an image converter is called proximity focus image intensifier diode.

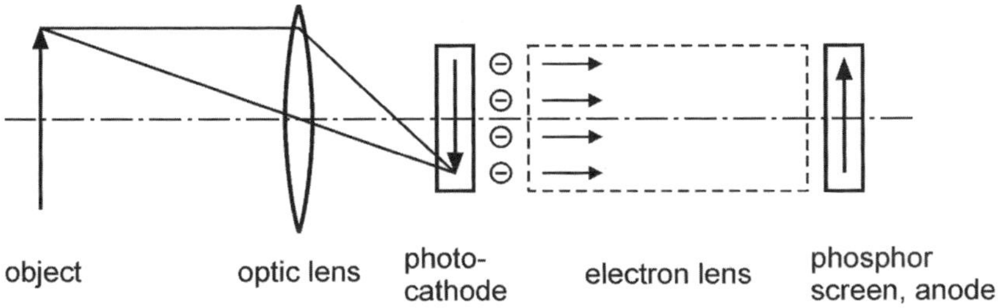

Fig. 5.35 Set-up and working principle of an image converter

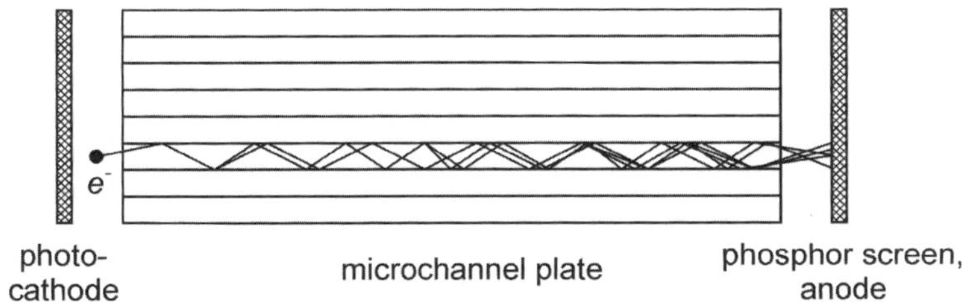

Fig. 5.36 Image converter with a microchannel plate

In recent years, microchannel plates (MCP) are used increasingly in image converters. A microchannel plate consists of a large number of typical 10^7 miniaturized channel electron multipliers, see Sect. 5.2.3. The single channels have typically only a length of 1 mm and a diameter of 10 μm. The charge pattern of the photocathode is transferred to the input of the microchannel plate with the help of an electron lens or, as sketched in Fig. 5.36, achieved by a 1:1 mapping using the proximity approach as described above. The electrons then pass through the channels of the microchannel plate, where they are multiplied by secondary emission. At the exit side of the microchannel plate, the amplified charge pattern of the photocathode occurs. Finally, the electrons are accelerated towards the phosphor anode, where they form the visible image. Image converters cannot reach the high resolution of photographic films. On the phosphor screen up to 80 line pairs per mm can be distinguished.

Depending on the spectral sensitivity of the photocathode, an image converter allows to record images in the invisible spectrum, e.g. the UV and X-Ray region. Image converters have found widespread use as infrared cameras. Another application focus of image converters is in high-speed measurement techniques. In this case, either the voltage applied to the electron-optical imaging system is pulsed, or the flow of the photoelectrons from the photocathode to the phosphor screen is controlled by an electrical grid. By these ways it is possible to make recordings with exposure times of less than 1 ns. Finally, it should be pointed to the fact that an image converter can also be used as an image intensifier. Because

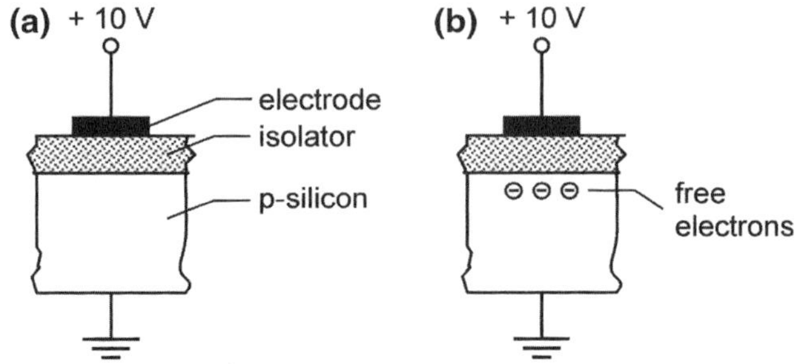

Fig. 5.37 Schematic representation of a MOS capacitor. **a** Prior to exposure, **b** after exposure

of the acceleration of electrons between cathode and anode or because of electron multiplication in case of the microchannel plate, bright images of faint objects can be recorded. Image converters operate as image intensifiers, for example, in night vision equipment and in astronomical telescopes.

5.4.5 Solid-State Image Sensors

Solid-state image sensors have in general two parts:

- They consist of one or more rows of closely spaced light-sensitive detectors (pixels). Commercially available are lines with partly more than 10^3 pixels and two-dimensional arrangements with partly over 10^8 pixels.
  The light-sensitive elements of solid-state sensors are either reversed biased *photodiodes*, as explained in Sect. 5.3.2, or so-called *MOS capacitors*.
  MOS stands for metal-oxide-semiconductor. A MOS capacitor is depicted schematically in Fig. 5.37a. It is composed of three layers. On a, e.g., p-doped silicon substrate, a thin insulation SiO_2-layer is applied. On top of that an electrode follows, which is initially held at a positive potential of about 10 V. Incident light generates electron-hole pairs in the p-doped substrate. The positively charged holes flow to the ground, while the electrons migrate towards the positive electrode. However, due to the insulating layer the electrons do not reach the electrode. They accumulate underneath the electrode, see Fig. 5.37b. The number of stored electrons in a MOS capacitor is a direct measure of the local exposure.
- Solid-state image sensors have a readout circuit, which relays and multiplexes the signals from each pixel to the output. There are different designs for readout electronics. Most common are *MOSFET readout*, *CCD readout* and *CMOS readout*.

Today, three types of solid-state image sensors are established. We briefly look at these three most used types of sensors.

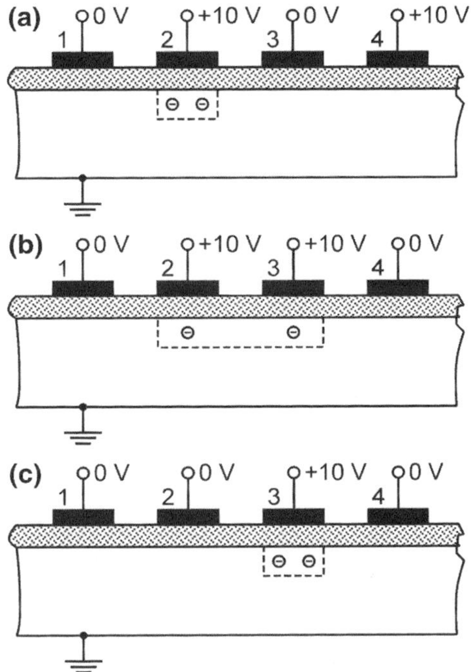

Fig. 5.38 Illustration of the principle of charge-coupled transport. 1–4 electrodes cf. Fig. 5.37

(a) Photodiodes with MOSFET readout—photodiode array

Photodiode arrays (PDAs) consist of a silicon substrate on which up to some thousands of photodiodes are monolithically integrated as e.g. lines with 128, 256, 512 or 1024 pixels or 2-dimensional arrays with 256 × 256 pixels. At the beginning of a recording cycle, the PDA, which are reversed biased, are charged, see Sect. 5.3.2. Subsequently, the individual p-n junctions are—more or less—discharged depending on the exposure by the incident light. Via a MOSFET switch, each p-n junction is connected to the video output. During a recording cycle, each p-n junction is connected sequentially to the video output. This is realized with a shift register that works with a clock frequency of about 1–10 MHz. Thereby, the partially discharged p-n capacities are recharged. The charging current is used as video signal. If a photodiode array consists of N p-n junctions, the video signal is composed of a series of N pulses per image. The pulse height of a single pulse is a measure of the incident radiant energy at the corresponding pixel.

(b) MOS capacitors with CCD readout—CCD image sensor

Let us now consider construction and operation of a *CCD image sensor*. The abbreviation CCD stands for Charge-Coupled Device. For the invention of CCD detectors, W. Boyle and G. Smith received in 2009 the Nobel prize in physics.

A CCD chip consists of a large number of MOS capacitors. In order to read out the stored charges in the various MOS capacitors, the charge-coupled device transport principle is applied. This principle is shown schematically in Fig. 5.38.

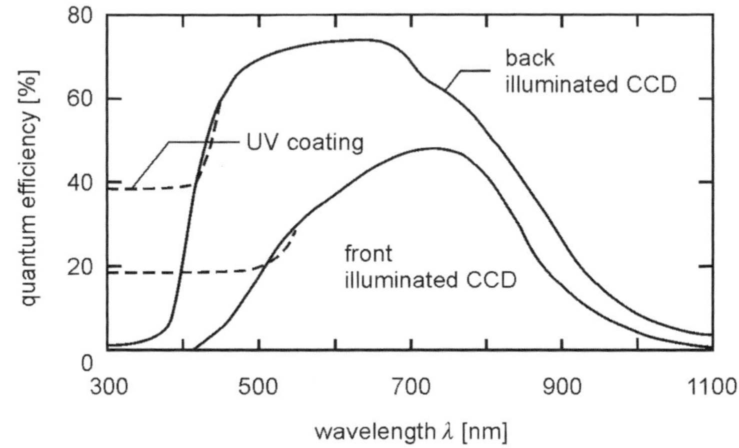

Fig. 5.39 Quantum efficiency of back illuminated and front illuminated CCD image sensors based on silicon as a function of the wavelength, data from [33]. By an UV coating the sensitivity range can be extended significantly into the UV spectral range

Let us consider four juxtaposed MOS capacitors. At the beginning, only capacitor number 2 is energized and charged, see Fig. 5.38a. Subsequently, electrode 3 is also set to a positive potential. Then, the stored charges are redistributed between the two electrodes 2 and 3, see Fig. 5.38b. Then, electrode 2 is grounded. Therefore, all charges are gathered under electrode 3, see Fig. 5.38c. Hence, the goal is achieved: charges stored in a certain MOS capacitor have been pushed into the adjacent capacitor. The charge transfer from one cell to another is virtually lossless, i.e., the charge transfer efficiency is about 100 %.

The principle of charge-coupled device transport is repeated in all MOS capacitors again and again. The charge packets generated by the exposure and stored in the MOS capacitors travel in the direction of the output stage along the interface between silicon substrate and insulator. The output stage converts the individual charge packets into a voltage signal, i.e. the actual video signal. Silicon-based photodiode arrays and CCD detectors are sensitive in the visible and near infrared spectral range from 0.4 to 1.1 μm, see Fig. 5.39. CCD sensors are available with up to some 10^7 pixels. The spatial resolution of an image sensor depends on the distance between two neighboring pixels. A resolution of more than 100 line pairs per mm is achievable. With modern CCD detectors, the displacement of charges from one MOS capacitor to the adjacent MOS capacitor lasts a period of a few 10 ns.

Electron-hole pairs are not only induced by incident light, but also due to the temperature of the semiconductor material. This leads to a dark current. Hence, for special applications where low radiant fluxes have to be detected, the sensors are cooled to reduce the noise level.

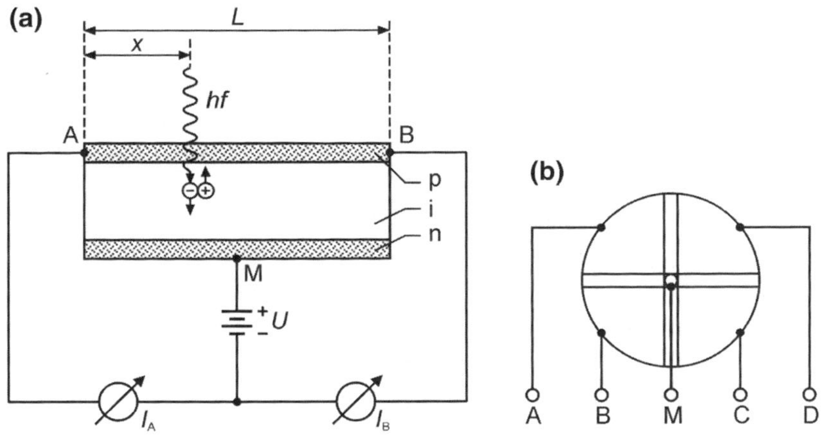

Fig. 5.40 Set-up of **a** a lateral effect diode, **b** a four-quadrant diode

(c) MOS capacitors or photodiodes with CMOS readout—CMOS image sensor

A *CMOS image sensor* is based on CMOS technology. CMOS is the abbreviation of **C**omplementary **M**etal **O**xide **S**emiconductor. A CMOS sensor is composed of a greater number of MOS capacitors or photodiodes of up to some 10^7. In contrast to a CCD sensor, each pixel of a CMOS image sensor contains some transistors to read out the signal. Partly, components for exposure control and A/D-conversion are integrated too. Therefore, each pixel can be processed individually.

CMOS image sensors can be read out faster and reach higher frame rates. Moreover, CMOS sensors have no unwanted charge transfers between adjacent pixels called blooming effect or smear effect. They have a lower power consumption and are cheaper to produce. On the other hand, CCD image sensors are more sensitive with lower noise. In the 70 and 80th of the last century, CCD image sensors had clearly a better performance than CMOS image sensors. Nowadays, with an improved CMOS technology, both are equivalent alternatives in measuring techniques.

5.4.6 Lateral Effect Diode

A simple realization of one-dimensional position measurement is provided by the lateral effect diode. The principle is shown in Fig. 5.40a. An elongated PIN photodiode, see Sect. 5.3.2, of length L and three connections, A, B and M, is reverse biased. A light beam incident at location x generates electron-hole pairs and a photocurrent I_{ph} starts to flow. Like a potentiometer, the photocurrent I_{ph} corresponding to position x, divides into two currents I_A and I_B, which drain off via the connections A and B, respectively, see Fig. 5.40a. The partition ratio is:

$$x = \frac{I_B}{I_A + I_B} L = \frac{I_B}{I_{ph}} L. \tag{5.58}$$

Under favorable conditions, a spatial resolution of about 1 μm can be achieved. In a non-spot-light illumination, (5.58) gives the "center of mass" of the irradiance distribution.

In the quadrant-photodiode, the photoelectric layer is mechanically divided into four quadrants. When light falls on the reverse biased diode, the photocurrent is distributed to four electrodes A, B, C and D, see Fig. 5.40b. The current distribution yields the "center of mass" of the incident radiation distribution. Provided all four quadrants deliver the same photocurrent, then the center of the irradiance distribution is in the middle of the diode.

5.5 Measuring of Detector Signals

Most detector signals are available both, as electric current or voltage signals. In this section we want to deal three tasks of electrical metrology, which are important for measurements with detectors.

5.5.1 Broadband Signals

All previously discussed detectors convert absorbed light energy into an electrical signal. Usually, the signal is available in the form of a voltage, which is picked off at a measuring resistor R_m. Thereby, we speak of a broadband signal, if the frequency spectrum spans a wide range. The width of the frequency spectrum increases for shorter rise times and shorter pulse widths of the signal, cf. relation (5.8a).

When measuring the voltage drop across R_m, it must be ensured that the transmission of the signal from the measuring resistor to the measuring electronics occurs without significant distortions. As an example, we consider the voltage measurement with an oscilloscope, see Fig. 5.41.

Commercially available coaxial cables have wave impedances in the range of 50–93 Ω. For most measurement purposes, 50 Ω cables are used. In order to avoid a signal reflection at the end of the cable, which would lead to a distortion of the measured signal, the input resistance R_i of the oscilloscope must be equal to the characteristic impedance Z of the coaxial cable. This is for some oscilloscopes the case. However, for many oscilloscopes typical input resistors are about 1 MΩ. That is why the input resistor R_i of the oscilloscope has to be bridged with a matching resistor $R_{ma} = Z$, see Fig. 5.41. A coaxial cable being bridged at the end with its characteristic wave impedance operates like a parallel circuit of an ohmic impedance Z with the measuring resistance R_m, see Fig. 5.42. The effective resistance of this equivalent circuit is:

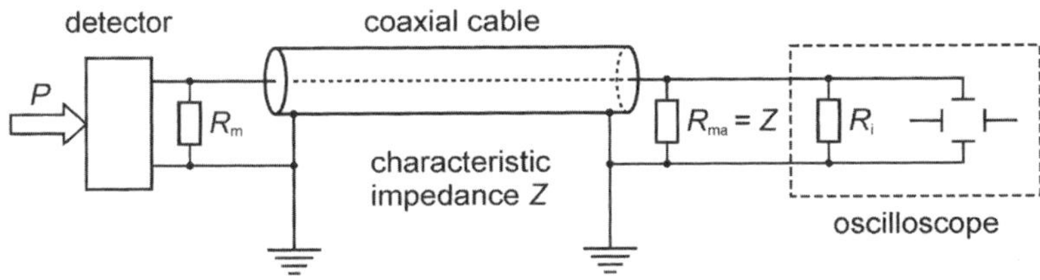

Fig. 5.41 Schematic diagram for the measurement of broadband signals using an oscilloscope with a high input resistance of, e.g., $R_i = 1\ \text{M}\Omega$

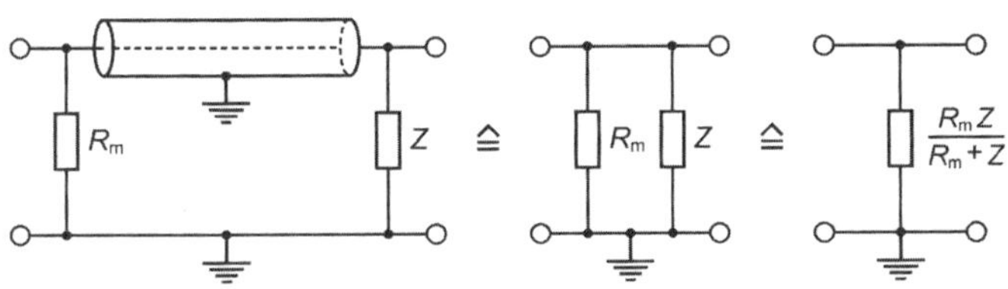

Fig. 5.42 Equivalent circuit of a coaxial cable, which is matched with a resistor ($R_{ma} = Z$)

$$\tilde{R}_m = \frac{R_m Z}{R_m + Z}. \tag{5.59a}$$

Since mostly $R_m \gg Z$, (5.59a) simplifies to:

$$\tilde{R}_m \approx Z. \tag{5.59b}$$

In this case, the measuring resistor can be omitted since the bridged coaxial cable acts as a measuring resistor.

The sensitivity of the system is changed by the matched coaxial cable. In the common case that the detector is—for a given radiant flux—a constant current source and $R_m \gg Z$ holds, the measured voltage is reduced by a factor of $Z/R_m \ll 1$.

While measuring static signals, no matching resistor R_{ma} is necessary since no disturbing reflections occur at the end of the cable. This also applies to 'slowly' varying signals. In this context, 'slowly' means that the largest frequency components of the signal are small compared to the reciprocal roundtrip transit time of the cable. For a typical cable length of 2 m and a signal speed 2×10^8 m/s, signals up to a bandwidth of about 1 MHz, which corresponds to rise times of about 350 ns, can be considered as 'slow'.

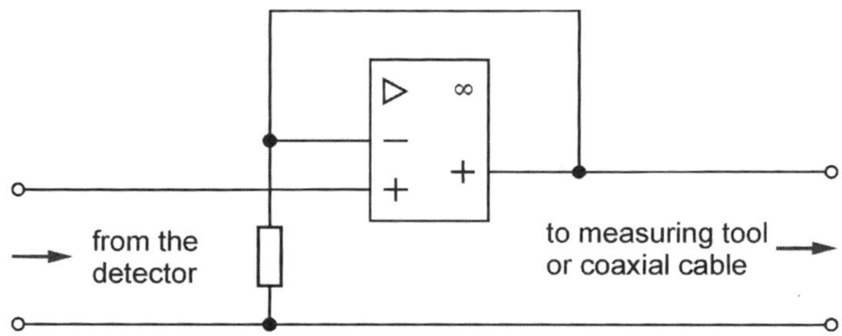

Fig. 5.43 Non-inverting operational amplifier with an amplification factor $G = 1$ for impedance conversion

5.5.2 Impedance Conversion

In many measurement problems a big measuring resistor, e.g., 10 MΩ is needed to achieve a high sensitivity. In this case, voltage measurements can be problematic if the input resistance of the voltmeter is no longer small compared to the high-impedance measuring resistor. In case of an oscilloscope, for example, the input resistor is usually 1 MΩ or, when using a matched 50 Ω coaxial cable, only 50 Ω. Therefore, in general, the voltage drop across R_m changes, when a measuring instrument is connected. This must be considered for the evaluation of the measured signal.

This problem can be avoided, if a suitably chosen preamplifier is arranged directly behind the measuring resistor. The preamplifier must have an input resistance R_{in}, which is large compared to the measuring resistor R_m. On the other hand, the output resistance of the preamplifier must be small compared to the input resistance of the subsequent measuring instrument or the characteristic impedance of the coaxial cable to be connected. In other words, the preamplifier operates as an impedance converter. Operational amplifiers as well as FET amplifiers are used as preamplifiers. Figure 5.43 shows a typical circuit of an operational amplifier with amplification factor $G = 1$. Thus, impedance conversion allows to connect a measuring instrument, which displays the unaltered measuring voltage.

Many manufacturers offer radiation detectors with an integrated preamplifier. The preamplifier serves not only as an impedance converter, but also as a multiplier of the sensitivity ($G > 1$). However, an increase of the amplification factor is always accompanied by a decrease of the bandwidth of the amplifier.

5.5.3 Current–Voltage Conversion

In the matters discussed in Sect. 5.3.2, we learned that the short circuit current I_{sc} of a photovoltaic cell is proportional to the radiant flux of the incident light. The measured voltage drop U_m across the measuring resistor R_m is proportional to the

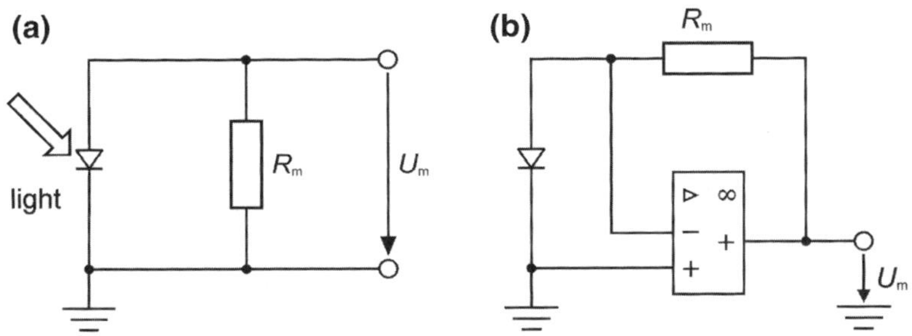

Fig. 5.44 Measurement of the short-circuit current for, e.g., a photovoltaic element. **a** without current–voltage conversion, **b** with current–voltage conversion

short circuit current I_{sc}, if the resistor R_m is chosen to be sufficiently small, see Fig. 5.44a. This inevitably leads to a very small voltage U_m to be measured. Hence in practice a current-voltage converter is used mostly. For the circuit sketched in Fig. 5.44b, even with larger measuring resistors R_m, the equation

$$U_m = R_m I_{sc} \tag{5.60}$$

is valid, as long as the current flowing through R_m is large against the input current of the operational amplifier. With a resistor of $R_m \approx 1$ MΩ, this should apply generally.

5.6 Summary and Comparison

In Chap. 5 an overview of the most used light detectors in laser measurement technology was given. The most important characteristics of the different detectors are summarized in the following tables.

In Table 5.4 the maximum sensitivity given refers to the minimum rise time. If the rise time increases, in general, the sensitivity increases, too. For detectors, which provide a current signal, the sensitivity was always based on a 50 Ω resistor. The specified spectral range will not be covered by a single detector type, in general. Some special detectors are being used outside the specified spectral range. Partly, a cooling (to 4.2 K) is necessary.

In Table 5.5 the most important spatially resolving detectors are compiled for the use in laser measurement technology.

Table 5.4 Summary of typical characteristics of non-position-sensitive light detectors

Detector version	Min. rise time τ_{min} [ns]	Max. sensitivity at τ_{min} [V/W]	Typ. spectral application area [μm]
Thermal element	10^4	10	0.2–30
Thermopile	10^4	150	0.2–30
Metallic bolometer	5×10^6	40	0.4–1000
Thermistor	5×10^6	10^7 ($T = 4.2$ K)	0.4–1000
Pyroelectric detector	0.1	10^{-5}	0.1–1000
Phototube	0.1	15	0.1–1.2
Photomultiplier	0.1–1	10^4–10^7	0.1–1.2
Photoresistor	1	5×10^4	0.4–20
Photovoltaic cell	1	./.	0.2–12
Photodiode	0.2	30	0.3–12
PIN photodiode	0.02	90	0.4–1.8
Avalanche photodiode	0.1	1,500	0.4–12
Schottky diode	10	5	0.2–1

Table 5.5 Maximum achievable spatial resolution and spectral range of position sensitive light detectors

Detector type	Max. spatial resolution [line pairs /mm]	Typ. spectral application area [μm]
Photographic layer	5×10^3	<1.2
Thermoplastic layer	2×10^3	0.3–0.6
IR-VIS—converter	15	0.7–1.7; 10.6
UV-VIS—converter	15	0.3–0.5
Image converter	80	0.1–1.2
Photodiode arrays	70	0.4–1.1
CCD	70	0.4–1.1
Position sensitive detector	10^3	0.4–1.1

References

1. H.-J. Kunze, *Physikalische Meßmethoden. Eine Einführung in Prinzipien klassischer und moderner Verfahren* (Teubner, Stuttgart, 1986). ISBN 978-3519030645
2. W. Pfeiffer, *Impulstechnik* (Hanser, München, 1985). ISBN 978-3446122161
3. J. Ohta, *Smart CMOS Image Sensors and Applications*. (CRC Press, Boca Raton, 2007), ISBN 9780849336812
4. *Photomultiplier Tubes, Basics and Applications*. 3rd Ed. (Hamamatsu, 2007), p. 323
5. *Channeltron—Electronmultiplier Handbook for Mass Spectroscopy Applications*. Burle (2013). https://www.tokyoinst.co.jp/products/hikari/cat/hikari12/cem_handbook.pdf, downloaded 2nd Aug 2013
6. G.R. Mitchell, Silver halide film as a linear detector for 10.6 μm radiation. Rev. Sci. Instrum. **53**, 111–112 (1982)
7. ISO 5800:1987/Cor 1:2001: Photography—Colour negative films for still photography—Determination of ISO speed
8. U. Tietze, C. Schenck, E. Gamm, *Halbleiter-Schaltungstechnik* (Springer, Heidelberg, 2012). ISBN 978-3642310256

9. L. Bergmann, C. Schaefer, H. Niedrig, *Lehrbuch der Experimentalphysik* (Bd. III: Optik. Walter de Gruyter, Berlin, 2004). ISBN 3-11-017081-7
10. R. Waynant, M. Ediger, *Electro-Optics Handbook* (McGraw-Hill, New York, 2000). ISBN 978-0070687165
11. F. Kohlrausch, *Praktische Physik, Bd. I,* ed. by V. Kose, S. Wagner (Teubner Verlag, Stuttgart, 1996), ISBN: 978-3519230014
12. R.J. Keyes, P.W. Kruse, D. Long, A.F. Milton, E.H. Putley, M.C. Teich, H.R. Zwicker, *Optical and Infrared Detectors (Topics in Applied Physics)* (Springer, Heidelberg, 1980), ISBN 978-3540101765
13. E.L. Dereniak, D.G. Crowe, *Optical Radiation Detectors* (Wiley, New York, 1984)
14. M.J. Cohen, E.L. Dereniak, *Semiconductor Photodetectors* (SPIE, Bellingham, 2006)
15. R.H. Kingston, *Detection of Optical and Infrared Radiation* (Springer, Heidelberg, 1978). ISBN 3 540 086127 X
16. S.M. Sze, K.K. Ng, *Physics of Semiconductor Devices* (Wiley, Hoboken, 2007). ISBN 978-0-471-14329-3
17. L. Bergmann, C. Schaefer, W. Raith, *Lehrbuch der Experimentalphysik, Bd. II: Elektromagnetismus* (Walter de Gruyter, Berlin, 1999), ISBN 3-11-016097-8
18. G.F. Knoll, *Radiation Detection and Measurement* (Wiley, Hoboken, 2010). ISBN 978-0-470-13148-0
19. G. Rieke, *Detection of Light—From the Ultraviolet to the Submillimeter* (Cambridge University Press, Cambridge, 2002). ISBN 9780521816366
20. J.F. Cox, *Fundamentals of Linear Electronics: Integrated and Discrete Circuitry* (Delmar, Albany, 2001), ISBN 0-7668-3018-7
21. J. Fraden, *Handbook of Modern Sensors: Physics, Designs, and Applications* (Springer, Heidelberg, 2010). ISBN 9781441964656
22. H. Budzier, G. Gerlach, *Thermal Infrared Sensors: Theory, Optimisation and Practice* (Wiley, Chichester, 2011). ISBN 978-0-470-87192-8
23. J.G. Webster, *The Measurement, Instrumentation, and Sensors Handbook* (Springer, Heidelberg, 1999). ISBN 3-540-64830-5
24. F. Unterseher, *Holography Handbook* (Ross Books, Berkley, 1987). ISBN 0-89496-018
25. ISO 12232: *Photography—Digital still cameras—Determination of exposure index, ISO speed ratings, standard output sensitivity, and recommended exposure index*. International Organization for Standardization (2006)
26. S.F. Ray, *Scientific Photography and Applied Imaging* (Focal Press, London, 1999). ISBN 978-0240513232
27. W. Krug, H.-J. Weide, *Wissenschaftliche Photographie in der Anwendung - Wege zur Informationsausschöpfung photographischer Schwarzweiss-Negative* (Geest & Potig, Leipzig, 1976)
28. L. Rossi, P. Fischer, T. Rohe, N. Wermes, *Pixel Detectors—From Fundamentals to Applications* (Springer, Heidelberg, 2006), ISBN 3540283323
29. A.J. Theuwissen, *Solid-State Imaging with Charge-Coupled Devices* (Springer, Heidelberg, 1995). ISBN 9780792334569
30. J. Nakamura (ed.), *Image Sensors and Signal Processing For Digital Still Camera (Optical Science and Engineering Series)* (CRC Press, Boca Raton, 2006). ISBN 9780849335457
31. J.R. Janesick, *Scientific Charge-Coupled Devices* (SPIE Press, Bellingham, 2001). ISBN 978-0-8194-3698-6
32. N. Blanc, CCD versus CMOS—has CCD imaging come to an end, in *Photogrammetric Week '01*, ed. by D. Fritsch, R. Spiller (Wichmann, Offenbach, 2001), ISBN 9783879073597
33. Hamamatsu learning center (2012). http://learn.hamamatsu.com/articles/quantumefficiency.html downloaded 11-24- 2012
34. O. Zinke, H. Brunswig, *Hochfrequenztechnik 1: Hochfrequenzfilter, Leitungen, Antennen,* eds. by A. Vlcek, H.L. Hartnagel, K. Mayer (Springer, Heidelberg, 1999), ISBN 978-3540664055
35. M.G. Kangl, *Selected papers on CCD and CMOS imagers (Bd. 177 SPIE milestone series)* (SPIE Press, Bellingham, 2003), ISBN 9780819451149
36. H. Bertele, *Impulstechniken*, vol. I (Oldenbourgh, München, 1974). ISBN 3-486-43271-1

Chapter 6
Laser Interferometry

Abstract First, we explain the basics of interferometry. Then we deal with distance measurement using a polarization interferometer and a dual-frequency interferometer. This discussion also includes angular and straightness measurements. Finally, the Twyman-Green interferometer is presented, which can be used for testing the quality of optical imaging components.

In the preceding chapters, the basics of laser measurement technology were presented. In this chapter and following Chaps. 7–14, some of the most significant methods of laser measurement technology are explained. We start with laser interferometry [1–5].

In interferometry light waves are superimposed that have traveled, in general, differently long distances. The intensity—or more precisely the irradiance—of the resulting wave is measured and from this, the path difference is determined. By using interferometry, path differences can be determined with an accuracy of less than a half wavelength. Interferometers have been known for a long time. In 1882, A. A. Michelson performed interferometric experiments with the arrangement described in Sect. 2.4.2. Because of the small coherence lengths and the low spectral radiance (unit: $\mathrm{Wsr^{-1}m^{-2}nm^{-1}}$) of thermal radiation, measurable path differences were initially limited to less than 1 mm.

With the invention of the laser, i.e., a strong and coherent radiation source, interferometry has gained new impetus. Through the use of lasers, the measuring range was extended by orders of magnitude. Further, the accuracy was improved and novel measuring methods could be developed.

A. Donges and R. Noll, *Laser Measurement Technology*, Springer Series in Optical Sciences 188, DOI: 10.1007/978-3-662-43634-9_6,

6.1 Fundamentals of Interferometry

6.1.1 Principle of Superposition and Complex Notation

The basis of interferometry is the principle of superposition, which was already explained in Sect. 2.2.2. According to this principle different light waves do not influence each other and can be brought to undisturbed superposition. If n light waves are superimposed in a point in space, then the resultant field can be calculated by vector addition of the individual sub-field strengths:

$$\vec{E} = \sum_{i=1}^{n} \vec{E}_i. \tag{6.1}$$

When applying the principle of superposition (6.1), one should have in mind that, in general, the summands describing the sub-waves have different amplitudes, wavelengths and polarization directions and are shifted in phase with respect to each other.

The determination of the sum (6.1) is simplified, if the real notation of a light wave

$$\vec{E} = \vec{E}_0 \cos(\omega t - kz - \varphi_0) \tag{6.2}$$

is substituted by the complex notation

$$\underline{\vec{E}} = \underline{\vec{E}}_0 \mathrm{e}^{\mathrm{i}(\omega t - kz - \varphi_0)} \tag{6.3a}$$

or

$$\underline{\vec{E}} = \underline{\vec{E}}_0' \mathrm{e}^{\mathrm{i}(\omega t - kz)} \tag{6.3b}$$

with the complex amplitude

$$\underline{\vec{E}}_0' = \vec{E}_0 \mathrm{e}^{-\mathrm{i}\varphi_0}. \tag{6.3c}$$

In the following, we will always use the complex notation (6.3a), (6.3b) and (6.3c). However, we should observe, that only the real part of (6.3a)–(6.3c) has a physically demonstrative meaning:

$$\vec{E} = \mathrm{Re}\left[\vec{E}_0 \mathrm{e}^{\mathrm{i}(\omega t - kz - \varphi_0)}\right] = \vec{E}_0 \cos(\omega t - kz - \varphi_0). \tag{6.4}$$

The advantage of the complex notation is that additions can be executed easily, without any knowledge of trigonometric addition theorems. After application of the superposition principle using complex notation, the real part of the resultant wave gives the same result as a calculation with real notation.

The field strength of a light wave cannot be measured by direct observation. However it is possible to measure the irradiance I, which is proportional to the square of the amplitude, $I \propto E_0^2$, see (2.6). A look at (6.3a–6.3c) shows that

$$I \propto E_0^2 = \underline{E}\underline{E}^* = \underline{E}_0\underline{E}_0^*. \tag{6.5a}$$

Here, the asterisk * denotes the complex conjugate quantity. Therefore, the irradiance I of a light wave is given—except for a constant factor—by the absolute square of the complex field strength or the absolute square of the complex amplitude.

More precisely, the irradiance I corresponds to the time-averaged absolute value of the so-called Poynting vector $\vec{S} = \vec{E} \times \vec{H}$, i.e.,

$$I = \left\langle \left|\vec{S}\right| \right\rangle = \frac{\langle |\vec{E}^2| \rangle}{Z} = \frac{E_0^2}{2Z} = \frac{E_{\mathrm{eff}}^2}{Z} \tag{6.5b}$$

with

$$E_{\mathrm{eff}}^2 = \frac{E_0^2}{2} = \frac{1}{T}\int_0^T E_0^2 \cos^2(\omega t)\,\mathrm{d}t. \tag{6.5c}$$

Z is the so-called *wave impedance*. For a plane wave, Z is given by:

$$Z = \sqrt{\frac{\mu_r \mu_0}{\varepsilon_r \varepsilon_0}} = \sqrt{\frac{\mu_r}{\varepsilon_r}} Z_0, \tag{6.5d}$$

where

$$Z_0 = \sqrt{\frac{\mu_0}{\varepsilon_0}} \approx 120\pi\Omega \tag{6.5e}$$

is the *wave impedance of the vacuum*.

In this section the complex field strength was underlined in contrast to the real field strength to provide a better distinction. In the following we will use the complex notation only. Hence, for the sake of simplicity, we do not underline the symbols for the complex field strengths as done in (6.3a–6.3c), (6.5a).

6.1.2 Principle of a Laser Interferometer

In the widest sense an interferometer is any optical arrangement, in which two or more light waves are caused to interfere. In case of a laser interferometer, a laser is used as light source. All laser interferometers are based on the same principle: an

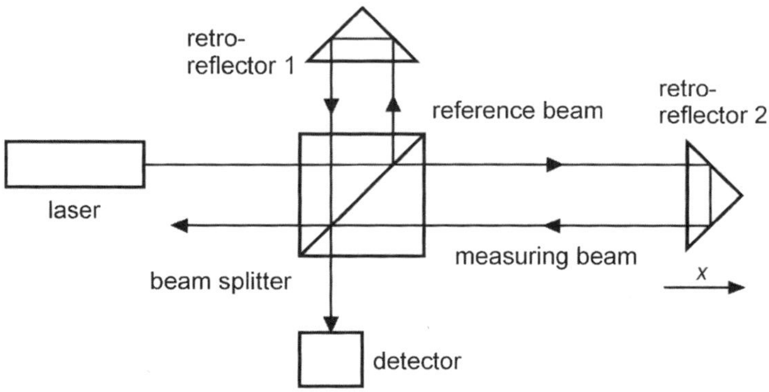

Fig. 6.1 Set-up of a Michelson interferometer with retroreflectors

incident laser beam is split into, mostly two, coherent sub-beams. Subsequently, the sub-beams, after they have traveled, in general, different long optical paths, are superimposed again. Finally, the occurring interference pattern is analyzed.

If the two beams are shifted exactly an integer multiple of a wavelength against each other, the interference pattern is the same as before the shift. Therefore an interferometer determines distances in multiples of the laser wavelength.

Interferometers are used to solve a large number of metrological tasks. Interferometric distance and angle measurements have attained a particular industrial importance, for example in mechanical engineering. The principle of interferometric distance and angle measurement is explained in Sect. 6.2. Section 6.3 treats the Twyman-Green interferometer which allows to test the optical quality of optical components, such as e.g. lenses and mirrors. Other interferometric measurement methods are described in Chap. 7, holography, and Chap. 8, speckle interferometry.

In laser interferometry, mostly optical arrangements are used, which are similar in their basic structure to the interferometer proposed by A. A. Michelson, see Sect. 2.4.2. Their set-up is sketched in Fig. 6.1. Compared to Fig. 2.19, retroreflectors are often used in laser interferometry. Regardless of their orientation, these mirrors reflect incident radiation back into the direction from which the light came, see Sect. 4.1.1.

If mirror 2 in Fig. 6.1 is moved along the x-direction, then the interference pattern, i.e., a system of bright and dark fringes, moves in the detector plane, see Fig. 6.2. (In Fig. 6.2 the fringes are illustrated in a simplified way with only two intensity levels. In reality the fringe intensity varies continuously between maxima and minima).

The active detector surface F should be minimized in order to gain a detector signal with a large degree of modulation in terms of a high difference between high and low signal levels. On the other hand, in the limiting case $F \rightarrow 0$, the absorbed light power and thus the detection signal approaches zero. As a compromise, the diameter of a circular active detector surface should be selected equal to the width of a bright or dark interference fringe.

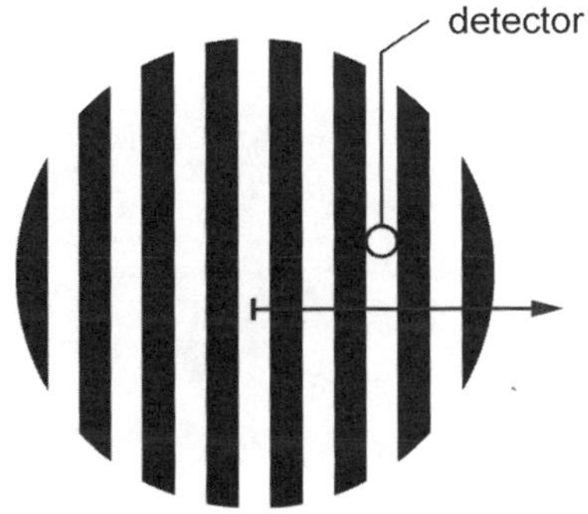

Fig. 6.2 Simplified pattern of interference fringes with a detector to count *bright-dark* transitions. The *arrow* indicates the direction of movement of the interference fringes relative to the fixed detector position

From the number of bright-dark transitions of the detector signal, the displacement of mirror 2 can be determined. However, the arrangement sketched in Fig. 6.1 represents not a powerful interferometer since the direction of displacement is not discriminable. If mirror 2 vibrates, for example, the caused bright-dark transitions are incorrectly added leading to nonsensical results.

6.2 Distance Measurement Using Laser Interferometers

In practice polarization interferometers and dual frequency interferometers are established [3–28]. The first commercial interferometer, which was used for length measurement, was a polarization interferometer. In 1970, the first dual frequency interferometers were put on the market. These two types of interferometers are presented in the following.

6.2.1 Polarization Interferometer

Figure 6.3 shows the basic structure of a polarization interferometer. A helium-neon laser (HeNe laser) generates a linearly polarized light beam, whose direction of polarization encloses an angle of 45° against the plane of projection of Fig. 6.3. With a telescope the beam diameter is expanded to a few millimeters thus reducing the beam divergence, see Sect. 4.2.3. The subsequent beam splitter 1 reflects a small part of the laser beam, which is required for frequency stabilization of the laser. By a further beam splitter 2 the not deflected beam is split into a reference beam and a measuring beam of equal irradiances. Measuring and reference beams are reflected by retroreflectors. Both returning beams are split again and superposed. With the detectors A and B constructive and destructive interferences are measured.

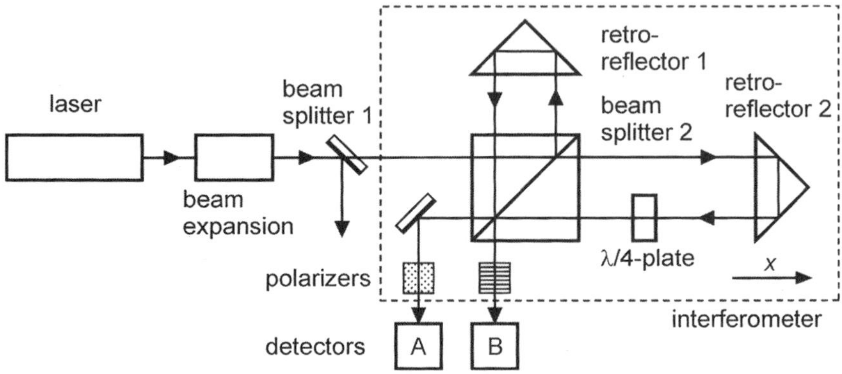

Fig. 6.3 Set-up of a polarization interferometer

Let us ignore for a moment the quarter waveplate in Fig. 6.3. If we denote the phase shift between the two waves at detector A with φ, then the phase shift of the two waves at detector B is $\varphi + \pi$. This phase difference of π is a consequence of the different numbers of reflections and transmissions that the waves have experienced at the beam splitter. If, e.g., the intensity at detector A is zero then—due to the phase difference of π—the signal is maximal at detector B. However, this is not the case here. The quarter wave plate converts the linearly polarized measuring beam into a circularly polarized wave, i.e., in two linearly polarized beams oriented perpendicularly to each other. The phase shift between these two beams amounts to $+\pi/2$ or $-\pi/2$. The actual sign depends on the quarter wave plate. That is why the phase shift of the two waves is $\varphi + \pi \pm \pi/2$ at detector B whereas it is φ at detector A. In order to enable the waves to interfere at detectors A and B, polarizers are required in front of these. The forward directions of the polarizers are chosen to be parallel or perpendicular to the plane of projection, respectively, cf. Fig. 6.3. The field strengths at the two detectors are then given by:

$$E_\mathrm{A} \propto E_0 \mathrm{e}^{\mathrm{i}(\omega t)} + E_0 \mathrm{e}^{\mathrm{i}(\omega t + \varphi)} \tag{6.6a}$$

and

$$E_\mathrm{B} \propto E_0 \mathrm{e}^{\mathrm{i}(\omega t)} + E_0 \mathrm{e}^{\mathrm{i}(\omega t + \varphi \pm 0.5\pi)}. \tag{6.6b}$$

The respective irradiances are calculated as

$$I_\mathrm{A} \propto 1 + \cos\varphi, \tag{6.7a}$$

$$I_\mathrm{B} \propto 1 \pm \sin\varphi. \tag{6.7b}$$

When the retroreflector 2 is displaced in the x-direction by the distance Δx, this changes the phase φ of the measuring beam at the location of both detectors according to:

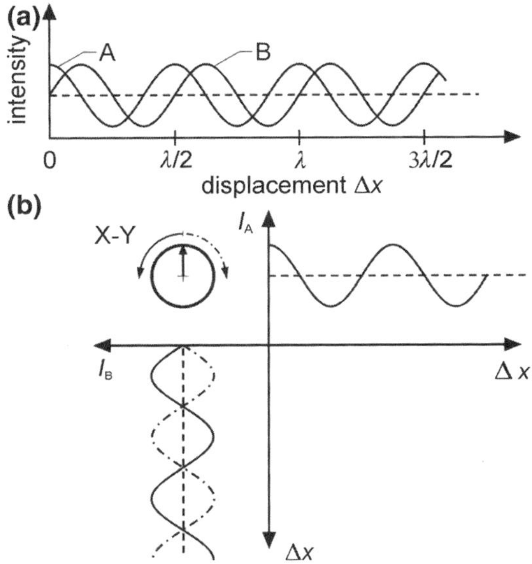

Fig. 6.4 **a** Signals of detectors A and B having a phase shift of π/2, **b** Representation of the two detector signals in the X-Y mode of an oscilloscope (see second quadrant) and input signals I_A at the Y-input, I_B at the X-input of the oscilloscope, respectively, as a function of the displacement Δx

$$\Delta\varphi = 4\pi\frac{\Delta x}{\lambda}. \tag{6.8}$$

For that reason, we can write:

$$I_A \propto 1 + \cos\left(\varphi_0 + 4\pi\frac{\Delta x}{\lambda}\right), \tag{6.9a}$$

$$I_B \propto 1 \pm \sin\left(\varphi_0 + 4\pi\frac{\Delta x}{\lambda}\right). \tag{6.9b}$$

Here, φ_0 is a system-specific constant. Its value is not required for the evaluation.

The displacement of retroreflector 2 is measured in relation to the position of beam splitter 2. The positions of beam splitter 2 and retroreflector 1 must therefore be kept absolutely stable. However, a movement of the laser source is without influence on the measurement. Therefore, it is advantageous to separate mechanically the laser from beam splitter 2 and retroreflector 1.

Previously, it was tacitly assumed that the polarization state of the laser beam is not changed at the beam splitters and retroreflectors. This is valid only in special cases. Generally, compensators must be used to correct these undesirable changes in polarization.

The curves of the intensities I_A and I_B as a function of the displacement Δx, according to (6.9a, 6.9b), are shown in Fig. 6.4a. The phase shift between the two signals is $\pm\,\pi/2$. Their spatial period is $\lambda/2$. The absolute value of the displacement can be determined by counting the minima or maxima. But we can also gain

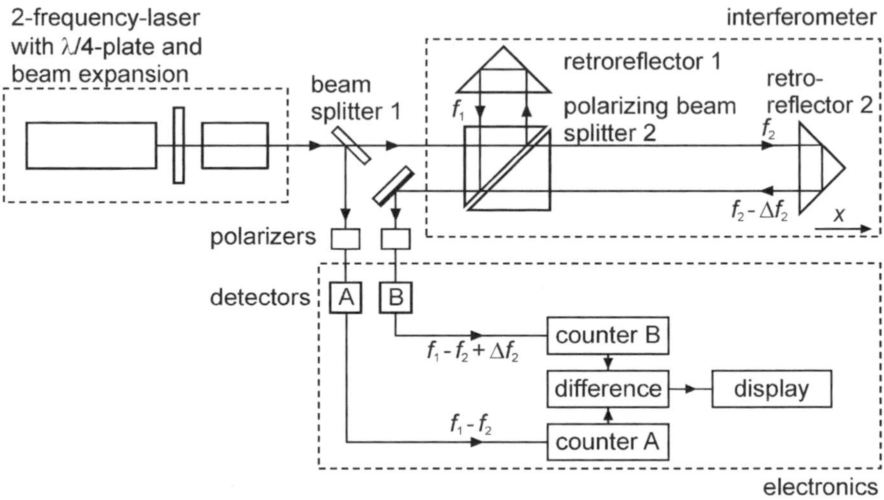

Fig. 6.5 Set-up of a dual-frequency interferometer

information about the direction of displacement. To see this, we suppose that the two detector signals are applied to the two inputs of an oscilloscope, operating in the X-Y-mode. Then, a displacement Δx of retroreflector 2 causes a circle on the oscilloscope screen. Each round corresponds to a displacement of $\Delta x = \lambda/2$. Shifts in the $+x$- or $-x$-direction differ in the sense of rotation, see Fig. 6.4b. For the signals shown in Fig. 6.4a the resulting sense of rotation is anti-clockwise, see leftward curved arrow in the second quadrant in Fig. 6.4b. If the phase of signal B is shifted by π, see dash-dotted line of signal B in the third quadrant of Fig. 6.4b, then the sense of rotation changes to clockwise, see rightward curved dash-dotted arrow in second quadrant.

6.2.2 Dual-Frequency Interferometer

Figure 6.5 illustrates the basic structure of a dual-frequency interferometer. Usually, a wavelength-stabilized HeNe laser is used where the discharge tube is located in a longitudinal magnetic field. The magnetic field causes the laser to emit two oppositely circularly polarized waves with slightly different frequencies f_1 and f_2. This is due to the so-called Zeeman effect. Without loss of generality, it is further assumed that $f_1 > f_2$.

The subsequent quarter wave plate converts the two circular waves into two linearly polarized waves with orthogonal polarization directions. Their frequencies differ typically by some MHz.

The subsequent telescope expands the laser beam. Beam splitter 1 extracts a portion of the laser beam and guides the beam via a polarizer to detector A. The forward direction of the polarizer is oriented in such a way that approximately equal fractions of the mutually perpendicularly polarized waves reach the detector. Detector A yields a signal being modulated with the difference frequency $f_1 - f_2$ and is deployed, inter alia, for wavelength stabilization of the laser source.

The non-deflected laser beam, which is not reflected by beam splitter 1, propagates towards the interferometer. With the polarization-dependent beam splitter cube 2, the incident wave is split into a reference beam with frequency f_1 and a measuring beam with frequency f_2. Both beams are linearly polarized with mutually perpendicular directions of polarization. The retroreflectors deflect reference and measurement beam by 180°, respectively. After reunion in the beam splitter cube, they pass through a polarizer to detector B. Detector B delivers in the static case (i.e. retroreflector 2 is not moving)—as detector A—a beat signal oscillating with the difference frequency $f_1 - f_2$. However, if retroreflector 2, which is fixed at the object to be measured, moves in the positive x- or negative $-x$-direction, the frequency of the reflected measuring beam decreases or increases due to the Doppler effect:

$$\Delta f_2 = \frac{2v_x}{\lambda_2}. \tag{6.10}$$

Here, v_x is the velocity component in the x-direction of retroreflector 2, which is mounted at the measuring object, and λ_2 is the wavelength of the incident measuring beam without Doppler effect. Therefore, the signal measured by detector B is modulated with the frequency $f_1 - f_2 + \Delta f_2$.

The beat signals yielded by the two detectors are fed to two separate counters. For the further discussion we assume that one counter step corresponds to the half of a beat period. Therefore, the counters A and B measure the following numbers of steps:

$$N_A = 2(f_1 - f_2)\Delta t \tag{6.11a}$$

and

$$N_B = 2(f_1 - f_2 + \Delta f_2)\Delta t \tag{6.11b}$$

in the time interval Δt, respectively. Only the difference between the two counter readings is displayed:

$$\Delta N = N_B - N_A = 2\Delta f_2 \Delta t = \frac{4v_x\Delta t}{\lambda_2} = \frac{4\Delta x}{\lambda_2}. \tag{6.12}$$

Here, Δx is the displacement of retroreflector 2—being mounted at the measuring object—in the x-direction relative to the position of beam splitter cube 2. Hence, the difference ΔN between the counter readings gives directly the shift Δx in multiples of a quarter wavelength ($\lambda_2/4$). The sign of ΔN indicates the direction—$+x$- or $-x$-direction—of the movement.

The term $f_1 - f_2 + \Delta f_2$ in (6.11b) becomes negative, if $\Delta f_2 < -(f_1 - f_2)$. Since a counter measures only the absolute value $2|f_1 - f_2 + \Delta f_2|\Delta t$, (6.12) can therefore be used only if $\Delta f_2 > -(f_1 - f_2)$ or

$$v_x > v_{x,\min} = -\frac{\lambda_2 (f_1 - f_2)}{2} \tag{6.13}$$

holds. With the wavelength of the HeNe laser ($\lambda_2 = 633$ nm) and a difference frequency $f_1 - f_2$ of 2 MHz, the minimum allowed velocity is $v_x = -63$ cm/s, i.e., the absolute value of the speed $|v_x|$ of retroreflector 2 towards beam splitter 2 must be less than 63 cm/s.

In practice, the maximum displacement speed is usually limited by the cutoff frequency of the counting electronics to about 30 cm/s.

6.2.3 Wavelength as Length Scale

An interferometer measures distances in multiples of the laser wavelength. So, the measuring accuracy depends on the exact knowledge and the constancy of the laser wavelength. Thus only laser radiation with low bandwidth should be used for laser interferometry.

In practice, usually frequency stabilized HeNe laser with a vacuum wavelength of $\lambda_0 = 632.991399$ nm [13] are deployed, which have a relative wavelength stability of typically 10^{-7}–10^{-8}. The advantages of the HeNe laser are its good beam quality, long service life and low price. As disadvantages we mention its relatively large size, its sensitivity to acoustic and mechanical disturbances, as well as its relatively low power output. Alternatives are semiconductor lasers and diode-pumped solid-state lasers.

For an accurate distance measurement a precise knowledge of the laser wavelength is required. The wavelength λ depends on the refractive index n of the propagation medium. It is:

$$\lambda = \frac{\lambda_0}{n}. \tag{6.14}$$

Here, λ_0 is the laser wavelength in vacuum.

Mostly, laser beams are traveling through atmospheric air. The refractive index n of the surrounding air depends slightly on temperature, pressure and humidity. In addition, other gases, e.g. solvent vapors, can influence the refractive index. At normal conditions, i.e., temperature $T = 20$ °C, air pressure $p = 1013.25$ hPa and humidity 59 % (corresponding to a partial pressure of water vapor of $p_{\mathrm{water}} = 13.3$ hPa), the refractive index n of air is 1.000271274. In case of deviating ambient conditions, the change in the refractive index can be calculated with the approximate formula [13]:

$$\Delta n \approx 10^{-8}\left(\frac{27\Delta p}{\text{hPa}} - \frac{92\Delta T}{\text{K}} - \frac{4.2\Delta p_{\text{water}}}{\text{hPa}}\right). \tag{6.15}$$

In the so-called parameter method, the air parameters along the measuring path are measured at several positions. With this data the laser wavelength is calculated by means of (6.14) and (6.15).

Another option is to measure interferometrically the refractive index of the surrounding air with a refractometer. Absolute measuring refractometers compare the refractive indices of air and vacuum. Commercial devices achieve a measurement accuracy of $\Delta n = 1.4 \times 10^{-7}$ [17].

6.2.4 Measuring Range and Measuring Uncertainty

The measuring range of interferometric measurements is principally limited by the coherence length of the utilized laser light. Frequency-stabilized HeNe laser being applied for interferometric measurements, have at most a relative bandwidth of typically 10^{-7}–10^{-8}. This corresponds, according to (2.28) and (2.29), to coherence lengths between 3 and 30 m. If the length difference between the two interferometer arms of a Michelson interferometer is approximately equal or becomes greater than half the coherence length, the contrast of the interference fringes decreases, see Sect. 2.4.2 and Appendix A.1. Therefore, the measuring range of an interferometric system is limited by the order of magnitude of the coherence length of the laser light.

With a more elaborate frequency control it is possible to achieve relative bandwidths of less than 10^{-8}. Experimentally, under ideal laboratory conditions, a relative bandwidth of 2×10^{-13} was reached for several seconds corresponding to a measuring range of over 1,000 km [29]. Nevertheless, such large measuring ranges cannot be realized practically.

Due to air turbulences, e.g., the measuring range of commercial systems is limited to about 50–60 m.

Interferometric length measurements are among the most accurate methods at all. The uncertainty of measurement will, inter alia, be determined by:

- the wavelength stability of the laser,
- the mechanical and thermal stability of the whole mechanical structure,
- the resolution of the electronics, which evaluates the detector signals and by
- the knowledge and constancy of the refractive index of the ambient air.

In most practical cases, the effects of the ambient air contribute dominantly to the measurement errors [20]. Uncertainties and fluctuations in the air temperature of 0.3 K (0.1 %), the air pressure of 1 hPa (0.1 %) and the relative humidity of 30% points, for example, lead together to a relative error of about 8×10^{-7}. For a measured distance of 1 m this corresponds to an absolute error of 0.8 μm, which is more than one wavelength of the HeNe laser. In fact, the measurement uncertainty is even greater. With the parameter method, see Sect. 6.2.3, relative accuracies of typically 3×10^{-6} can be reached. More accurate results can be achieved, if the refractive index of the air is monitored by a refractometer. For commercial devices the relative measurement error amounts to about 3×10^{-7}.

An interferometer measures a distance in steps of $\lambda/2$ or $\lambda/4$. For a HeNe laser this is equivalent to a resolution of about 300 or 150 nm. The resolution can be further increased by electronic interpolation. The resolving power can reach values of about 10 to even 1 nm [17].

The absolute measurement error ΔL is given by:

$$\Delta L = A + FL. \tag{6.16a}$$

Here, A is the resolution, F the relative measurement error and L the length of the path to be measured. In case of short measurement paths ($L \ll L_c = A/F$), the absolute measurement error is determined essentially by the resolution A:

$$\Delta L \approx A. \tag{6.16b}$$

In the other limiting case ($L \gg L_c = A/F$), we have:

$$\Delta L \approx FL. \tag{6.16c}$$

Example: If the resolution A and the relative measurement error F are 6 nm and 3×10^{-7}, respectively, $L_c = A/F$ is 2 cm.

6.2.5 Angle Measurements

A modified arrangement of a laser interferometer outlined above allows to measure rotations of components. For this purpose the reference beam is deflected in the direction of the measurement object and the previously fixed retroreflector 1 is mounted on the measuring object. Thereby the original reference beam is converted into a second measuring beam, see Fig. 6.6. If the component is rotating, both retroreflectors move at different speeds. This results in two different Doppler shifts. In case of a dual-frequency interferometer, the displayed difference of counts is:

$$\Delta N = 4\left(\frac{\Delta x_2}{\lambda_2} - \frac{\Delta x_1}{\lambda_1}\right), \tag{6.17}$$

where Δx_1 and Δx_2 are the displacements of the two retroreflectors in the x-direction. If we neglect the slight difference between λ_1 and λ_2 the result is in a good approximation:

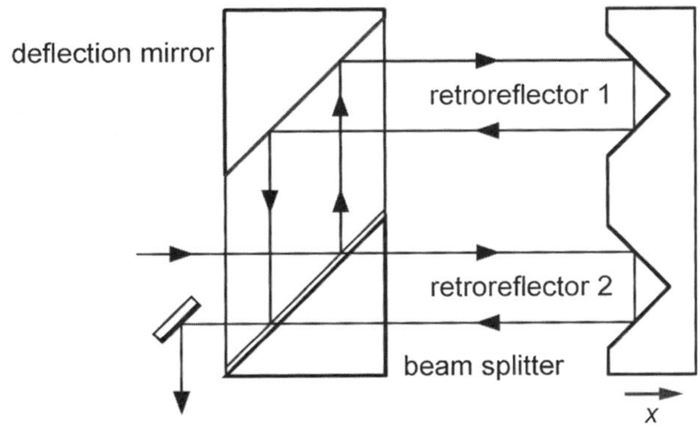

Fig. 6.6 Set-up for interferometric angle measurements

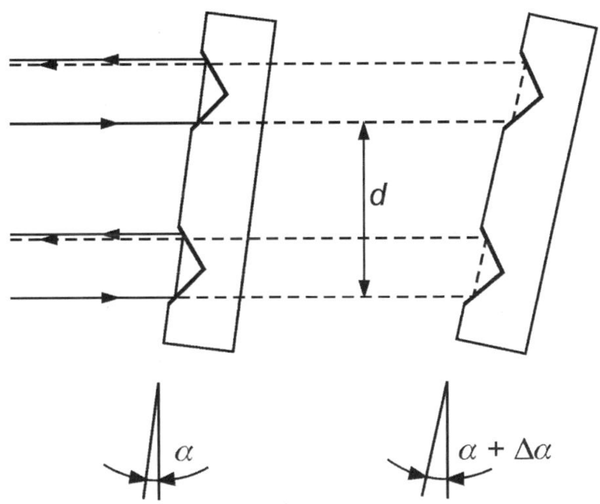

Fig. 6.7 Determination of the angle of rotation Δα

$$\Delta N = 4\frac{\Delta x}{\lambda}. \tag{6.18}$$

Here, $\Delta x = \Delta x_2 - \Delta x_1$ denotes the relative displacement of the two retroreflectors in x-direction and λ is the averaged wavelength. The sign of ΔN indicates the sense of rotation.

If the distance d between the two retroreflectors is known, the angle of rotation $\Delta\alpha$ can be calculated, see Fig. 6.7. In case of small angles, i.e. $\alpha \ll 1$, the following relation holds:

$$\Delta\alpha = \frac{\Delta x}{d\cos\alpha} = \frac{\lambda}{4d\cos\alpha}\Delta N. \tag{6.19}$$

The angle α describes the position of the two retroreflectors at the start of measurement ($\Delta\alpha = 0$). Usually $\alpha = 0°$ is chosen, i.e., $\cos\alpha = 1$.

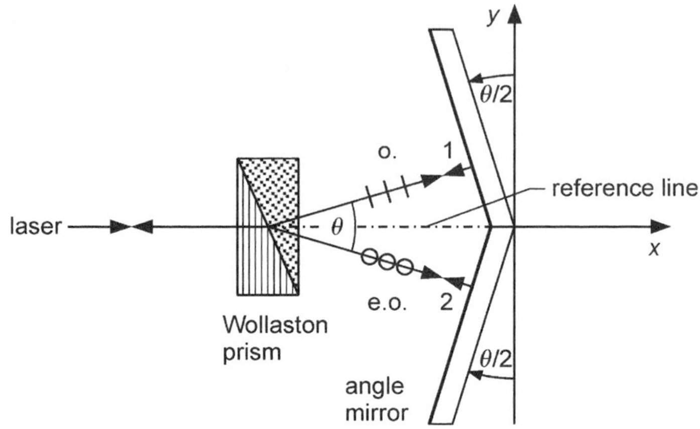

Fig. 6.8 Interferometer set-up for straightness measurements

The uncertainty of interferometric angle measurements is mainly determined by the measurement error of the relative displacement Δx of the two retroreflectors in x-direction, see Sect. 6.2.4. The measurement range is limited to about $\pm 10°$ [17].

6.2.6 Straightness Measurements

The previously discussed methods of measurement were performed with a Michelson interferometer. For straightness measurements slightly different light paths are chosen [23].

A modified dual-frequency interferometer—cf. Figure 6.5—is deployed to measure straightness. The orthogonal arms of the Michelson interferometer are replaced by the set-up shown in Fig. 6.8. A Wollaston prism, see Sect. 4.1.2, splits the incident laser beam into two linearly polarized light waves with orthogonal polarization directions having the frequencies f_1, f_2, respectively. The two beams diverge at an angle θ and travel to the two planar surfaces of an angle mirror, respectively, see Fig. 6.8. The reflected beams are reunited by the Wollaston prism and, finally, directed via a polarizer to detector B, see Fig. 6.5.

For the measurement the Wollaston prism is moved, while the angle mirror remains stationary. Alternatively, the angle mirror can be moved, while the Wollaston prism is fixed. Tilting movements of the Wollaston prism have only a small influence on the measurement in contrast to tilting of the angle mirror [24]. Therefore, in practice the first option, i.e., a moving prism and a fixed angle mirror is selected usually.

For the sake of simplicity, we explain the measurement principle in the case of a moving angle mirror. If the angle mirror is shifted precisely in the x-direction with speed v_x then both laser beams reflected on the moving mirrors experience the same Doppler shift of:

$$\frac{\Delta f_i}{f_i} = -2\frac{v_x}{c}\cos\left(\frac{\theta}{2}\right), \quad i = 1,\ 2. \tag{6.20}$$

Therefore, the measured beat frequencies of both detectors are equal and the indicated counter difference is zero.

The situation is quite different, when the angle mirror is moved with the velocity v_y perpendicularly to the x-direction, see Fig. 6.8. In this case, the upper plane mirror 1 departs with speed $v_y \sin(\theta/2)$ from beam 1, while the lower plane mirror 2 runs with the same speed towards beam 2. Both reflected beams are Doppler shifted with opposite signs:

$$f_1' = f_1 - 2\frac{v_y}{\lambda_1}\sin\left(\frac{\theta}{2}\right), \tag{6.21a}$$

$$f_2' = f_2 + 2\frac{v_y}{\lambda_2}\sin\left(\frac{\theta}{2}\right). \tag{6.21b}$$

The detectors A and B yield signals, which are modulated with different beat frequencies. Analogous to (6.18), we obtain for the difference between the two counts:

$$\Delta N = \frac{8\Delta y \sin\left(\frac{\theta}{2}\right)}{\lambda}. \tag{6.22}$$

The counter difference ΔN is a measure of the deviation Δy from the reference line x, which is defined by the laser beam. If the angle mirror or the Wollaston prism is moved exactly in the x-direction, the counter difference remains always zero. With this method the straightness of the movement can be controlled.

In order to detect deviations perpendicular to the xy-plane, see Fig. 6.8, the measurement must be repeated with Wollaston prism and angle mirror twisted by 90° about the x-axis.

The maximum range of straightness measurements is limited to 30 m. At a measuring length of 1.5 m, for example, a minimum measurement uncertainty of about 1.5 μm can be achieved.

Finally, we consider the case that the laser beam defining reference line x is deflected by 90° with a pentaprism, see Sect. 4.1.1. This creates a second reference line y oriented perpendicularly to x. If Wollaston prism and angle mirror are arranged in the laser beam deflected by 90°—as shown in Fig. 6.9—the deviation Δx of the angle mirror or the Wollaston prism from reference line y can be determined. Thus, the perpendicularity of the movement of the angle mirror or the Wollaston prism with respect to reference line x can be measured.

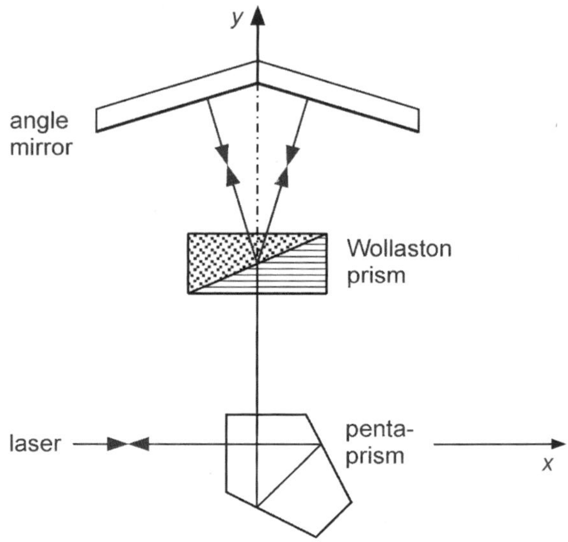

Fig. 6.9 Set-up to inspect perpendicularity

6.2.7 Examples of Applications

There is a broad range of applications of laser interferometers. They are widely used in mechanical engineering, especially in machine tool building and precision engineering [24].

The more and more precisely working machine tools—especially the numerically controlled devices—enabling production tolerances of less than 1 μm, require accurate length measurement systems. Laser interferometers with their large measuring ranges and high resolutions represent a nearly perfect scale of length with sufficient accuracy. Therefore, laser measuring systems are used to calibrate secondary length standards, such as gauge blocks, line scales, etc.

With a laser interferometer, not only length, but also other mechanical parameters, such as tilt and rotation angle, ripple, straightness, parallelism and perpendicularity, can be measured. Hence, laser interferometry is an essential tool for calibration and uncertainty testing of machine tools, robots, and coordinate measuring machines, see Fig. 6.10.

6.3 Twyman-Green Interferometer

With a so-called Twyman-Green interferometer [1, 2, 25–31], the quality of optical imaging components can be tested.

A Twyman-Green interferometer is a modified Michelson interferometer, see Sect. 6.1. Figure 6.11 shows e.g. a Twyman-Green interferometer to test a convex lens. A plane mirror of the conventional Michelson interferometer, see Fig. 2.19, is

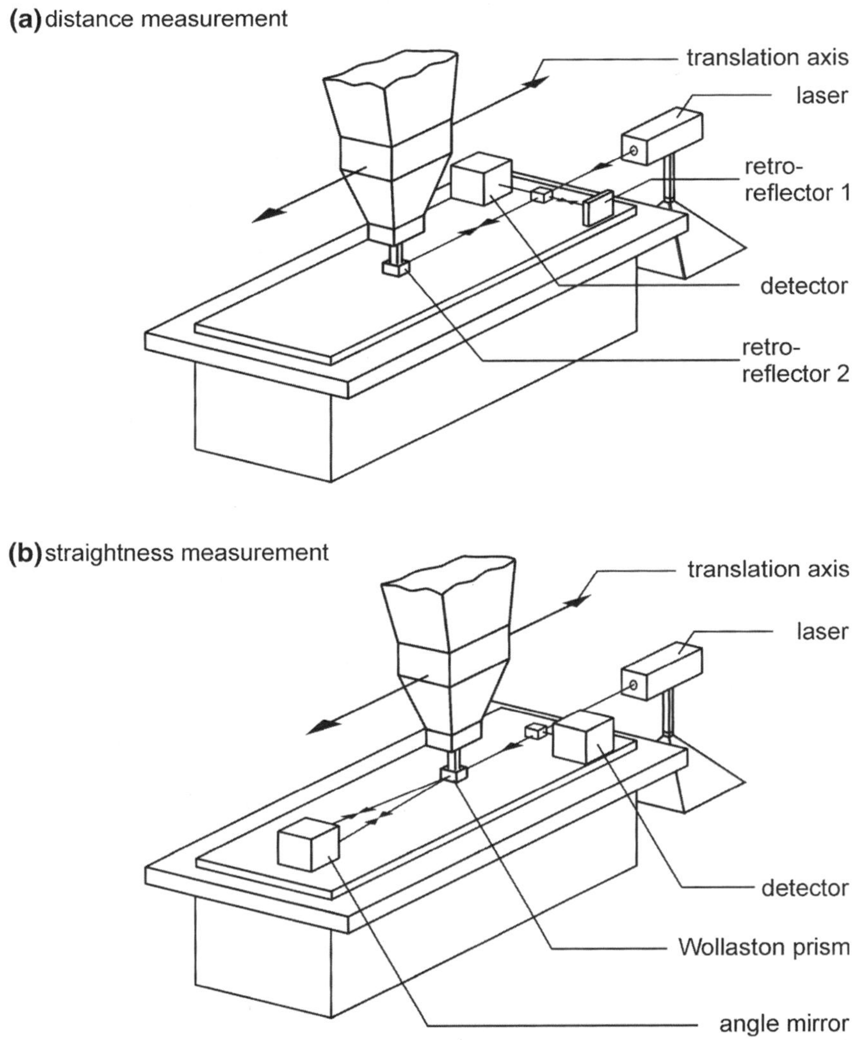

Fig. 6.10 Controlling the feed rate of a moving part of a machine tool. **a** Interferometric length measurement, **b** interferometric straightness measurement

replaced by the combination of a convex mirror and the convex lens to be tested. The focus of the lens and the center of the convex mirror, which is assumed to be perfect, fall together.

For an understanding of the Twyman-Green interferometer, we consider the beam path through the combination of lens and convex mirror. We assume that the incident expanded laser beam is a plane wave being converted by the convex lens into a spherical wave. This spherical wave is reflected by the convex mirror and propagates, after passing through the convex lens again, back as a plane wave.

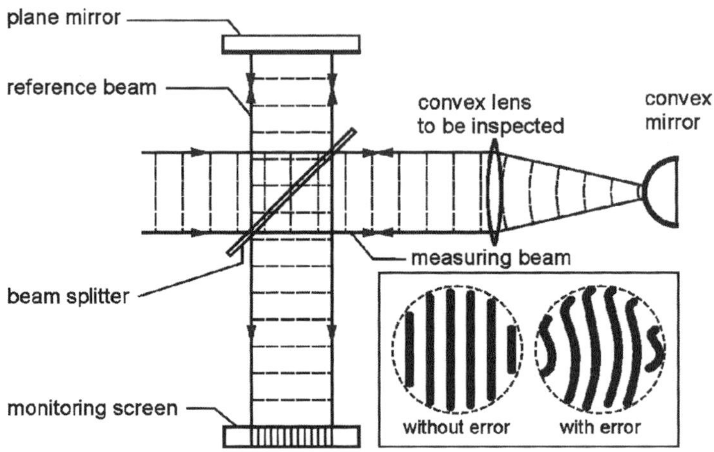

Fig. 6.11 Basic design of a Twyman-Green interferometer to test a convex lens. *Bottom right* Typical interferograms for a perfect and a non-perfect optical component

That is because the focus of the lens and the center of the convex mirror coincide. Therefore, two plane waves are superimposed on the observation screen. If the plane mirror in the reference arm is tilted slightly by an angle α, not indicated in Fig. 6.11, then the propagation directions of the superimposed waves enclose the angle 2α. On the screen results a system of parallel interference fringes. In case of $\alpha \ll 1$ the fringe spacing is given by, see Sect. 2.2.3:

$$\Delta x \approx \frac{\lambda}{2\alpha}. \tag{6.23}$$

The straight fringe pattern on the observation screen arises only under ideal conditions. If the lens is not perfect or misaligned, then the reflected measuring wave is no longer plane. Depending on the aberrations, the fringe pattern is more or less distorted, see Fig. 6.11. From the distortion of the fringes the so-called fitting error can be determined [30]. The fitting error describes the deviation between the surface of the test object—in the considered case the two surfaces of the convex lens—and the surface of the reference object. Here the latter is the surface of the convex mirror assumed to be ideal. In practice, the interferogram is scanned by a video camera and evaluated by a computer.

In our previous discussion we assumed that the reference surface of the convex mirror is perfect. However, in reality ideal convex mirrors transforming an ideal plane wave into an ideal spherical wave do not exist. For that reason, the accuracy of the measurement is limited by the quality of the reference object. Commercial Twyman-Green interferometers are able to detect deviations down to $\lambda/80$ from the desired shape of the optical surface.

The set-up outlined in Fig. 6.11 allows also to test convex mirrors. In this case a very well manufactured lens is used as reference object. In addition, there is a great

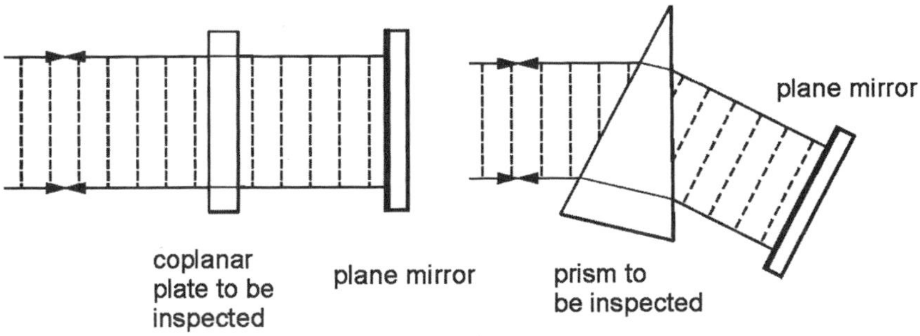

Fig. 6.12 Configurations to test coplanar plates and prisms

variety of optical elements that can be tested with Twyman-Green interferometers. Figure 6.12 shows e.g. appropriate configurations to test coplanar plates and prisms.

With Twyman-Green interferometers errors of optical components are not only determined at the final inspection. Partly, optical components are monitored continuously during the process of production. In this way, optical components of highest quality can be manufactured.

References

1. M. Born, E. Wolf, A.B. Bhatia, P.C. Clemmow, D. Gabor, A.R. Stokes, A.M. Taylor, P.A. Wayman, W.L. Wilcock, *Principles of Optics* (Cambridge University Press, Cambridge, 2003). ISBN 0 521 642221
2. E. Hecht, *Optics* (Pearson, Reading, 2003). ISBN 978-8177583571
3. P. Hariharan, *Basics of Interferometry* (Academic Press, New York, 2007). ISBN 9780123735898
4. P. Hariharan, *Optical Interferometry* (Academic Press, San Diego, 2003). ISBN 978-0123116307
5. A. YaKarasik, V.U. Zusov, *Laser Interferometry Principles* (Mir Publishers, Moscow, 1995). ISBN 978-0849375422
6. Swapan Kumar Saha, *Principles of Interference* (Springer, New York, 2011). ISBN 978-1-4419-5709-2
7. H. Gross (ed.), B. Dörband, H. Müller, H. Gross, *Handbook of Optical Systems. Vol. 5: Metrology of Optical Components and Systems* (Wiley-VCH, Weinheim, 2012) ISBN 978-3-527-40381-3
8. M. Bass (ed.), *Handbook of Optics: Volume II – Design, Fabrication, and Testing; Sources and Detectors; Radiometry and Photometry* (McGraw Hill, New York, 2009). ISBN 9780071498906
9. D. Rosenberger, *Technische Anwendungen des Lasers* (Springer, Berlin, 2012). ISBN 978-3-642-93031-7
10. H.-H. Schüssler, Industrielle Lasermeßtechnik und ihr Einsatz in Produktion und Entwicklung, Teil 1: Grundlagen und Geräteausführungen. Automobil-Industrie **4**, 475–488 (1983)
11. J.N. Dukes, G.B. Gordon, A two-hundred-foot yardstick with graduations every microinch. Hewlett Packard J. **21**(12), 2–8 (1979)

12. H. Bauer, *Lasertechnik: Grundlagen und Anwendungen* (Vogel, Würzburg, 1991). ISBN 3-8023-0437-3
13. F. Lebowsky, Refraktometer mit Zweifrequenzlaser. PTB-Mitt. **4**, 273–276 (1991)
14. T. Pfeifer, J. Waltar, Frequenzstabilisierte Halbleiterlaser für die interferometrische Meßtechnik. FhG-Berichte **1**, 49–54 (1990)
15. W. Luhs, H.-P. Meiser, *Neue Strahlungsquellen für die Laserinterferometrie*. VDI Berichte 749 - Laserinterferometrie in der industriellen Meßtechnik, Tagung Braunschweig, 26. und 27. April 1989
16. VDI-Berichte 749, *Laserinterferometrie in der Längenmeßtechnik* (VDI-Verlag, Düsseldorf, 1989)
17. Agilent Technologies, *Laser and Optics User's Manual* (order number: p/n 05517-90045). Santa Clara, CA, 2002
18. H. Walcher, *Winkel- und Wegmessungen im Maschinenbau* (VDI-Verlag, Düsseldorf, 1985). ISBN 3-18-400708-1
19. Hewlett Packard: *Laser-Meßsystem HP5528A*, Santa Clara, Agilent Technologies, 1992, in: http://cp.literature.agilent.com/litweb/pdf/05528-90022.pdf. Accessed 18 Aug 2013
20. H. Höfler, J. Molnar, C. Schröder, Interferometrische Wegmessung mit automatischer Kompensation von Brechzahlschwankungen. FhG-Berichte **1**, 55–58 (1990)
21. *Dual-Frequency Laser Interferometer ZLM 700 ZLM 800*. Jena, JENAer Meßtechnik GmbH, 2003
22. Teachspin brochure: *Modern Interferometry*. Buffalo, Techspin, 2007, in: http://www.teachspin.com/instruments/moderni/ModInter_full.pdf. Accessed 17 Aug 2013
23. W. Dutschke, P. Grossmann, Fluchtungsmessung mit einem Laser-Interferometer. Zeitschrift für industr. Fertigung **68**(4), 209–212 (1978)
24. H.-H. Schüssler, Industrielle Lasermeßtechnik und ihr Einsatz in Produktion und Entwicklung, Teil 2: Anwendungen bei Werkzeugmaschinen, Koordinatenmeßgeräten und Industrierobotern. Automobil-Industrie **1**, 73–83 (1984)
25. F. Twyman, A. Green, *Improvements in finishing prisms or lenses or combinations of the same and in apparatus therefor*, British Patent No. 103832 (prisms and microscopes), 1916
26. F. Twyman, Interferometers for the experimental study of optical systems from the point of view of the wave theory. Phil. Mag. **35**(6), 49–58 (1918)
27. F. Twyman, An interferometer for testing camera lenses. Phil. Mag. **42**(6), 777–793 (1921)
28. D. Bimberg et al., *Laser in der Industrie* (expert-Verlag, Sindelfingen, 1985). ISBN 3-88508-879-7
29. G. Herziger, H. Lindner, Dynamische Eigenschaften von Laser-Oszillatoren. I. Z. f. Physik **206**, 446–456 (1976)
30. DIN ISO 10110-8:2012-02: *Optics and photonics - Preparation of drawings for optical elements and systems - Part 8: Surface texture; roughness and waviness* (Beuth, Berlin, 2010)
31. Data sheet "*Möller-Wedel Interferometer V-100/P*". http://www.measureshop.biz/files/catalogue/systems/measuring_systems_catalogue_114.pdf. Accessed 27 Nov 2012

Chapter 7
Holographic Interferometry

Abstract The principle of holography is introduced comprising the formation of virtual and real images of the reconstructed object. Holographic interferometry allows to make visible slight changes of the object shape as those induced by deformations or vibrations. The quantitative determination of the displacement vector with the help of the phase-shifting method is explained. We present the fundamentals of digital holography and digital holographic interferometry. Measurement set-ups of holographic interferometry are shown and various examples of applications are given as e.g. the visualization of vibrational modes of mechanical components.

Holographic interferometry is a measuring technique that is used for non-destructive testing and for the analysis of vibrations. With this method it is for the first time possible to measure technical objects with rough surfaces with interferometric exactness, i.e. to the fraction of a wavelength. The basis of this is holography by which a three-dimensional picture of the object to be inspected is produced. The wavefronts of the holographically recorded images of an object can interfere with one another, similar to the interference of light waves which were described in Chap. 6. The following chapter puts forward the holographic principle.

7.1 Holographic Principle

The holographic principle was proposed by Dennis Gabor as early as 1948, but the process has only obtained practical significance through the development of lasers in 1960 [1, 2]. In 1971 Gabor received the Nobel Prize for his pioneering invention. Holography made possible the storage and reconstruction of a wave scattered by an object, preserving amplitude as well as phase. This process goes far beyond that of conventional photography, where light intensities—without any phase information—are recorded.

A. Donges and R. Noll, *Laser Measurement Technology*, Springer Series in Optical Sciences 188, DOI: 10.1007/978-3-662-43634-9_7,

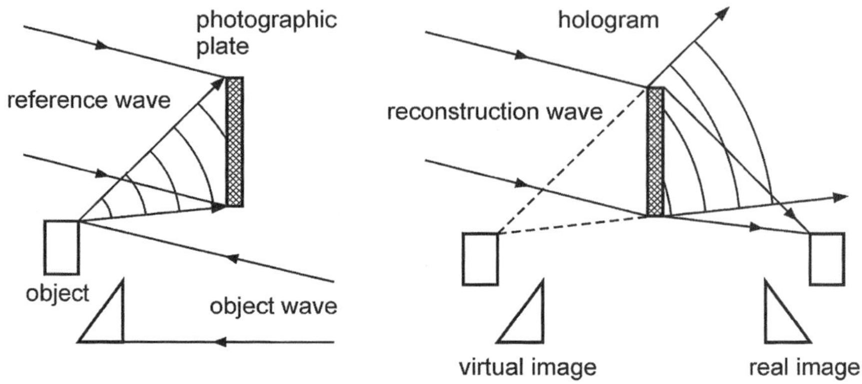

Fig. 7.1 Holographic principle, *left* recording of the hologram, *right* reconstruction of the hologram. The object portrayed consists of two object pieces—*rectangle* and *triangle*—which are placed at different distances to the photographic plate. In the reconstruction the original spatial position of the object is reconstructed in the so-called virtual image, while in the real image both foreground and background as well as near side and far side of the object are transposed, see text

Figure 7.1 shows the principle of holographic recording and reconstruction. The wave scattered by an object is superimposed with a coherent reference wave. The interference pattern created is registered on a photographic plate. The developed plate is called a hologram. To reconstruct the image of the recorded object the hologram is exposed with a reconstruction wave. This wave is diffracted by the recorded interference pattern in the hologram. A virtual and a real image of the object are therefore created. If the reconstruction wave has the same wavelength, wavefronts and direction as the reference wave, the virtual image is visible to an observer at the position where previously the object was. The reconstructed image is three-dimensional, i.e. the object is reproduced in its spatial dimensions.

The superposition of reference and object wave causes destructive and constructive interference in the plane of the photographic plate according to the phase difference between the waves, cf. Sect. 2.2.2. Hence the interference pattern produced, and with it the exposure of the photographic plate, depend not only on the amplitude of the incoming waves but also on their respective phases. In conventional photography the phase of a lightwave has no influence on the exposure of the photographic film.

Since with holography additional phase information of the light wave is stored on the photographic plate it is possible for a recorded object to be reconstructed in its spatial structure. Similar to this is stereophony in acoustics [3]. The single sound waves of a sound source are by phase precisely recorded through several microphones. On playback over several loudspeakers the sound waves reach the human ear with definite phase differences. An impression of a sound pattern is created.

In the following, hologram recording and reconstruction are described with the aid of complex amplitudes of the waves concerned, cf. Sect. 6.1.1. The complex amplitudes of the object wave and of the reference wave in the plane of the photographic plate are indicated by E_o and E_r. For simplification the vector character of these quantities is not considered in the following. The interference pattern is described by the irradiance distribution as a function of the coordinates (x, y) in the recording plane:

$$I(x,y) = \frac{1}{Z}\left(E_o + E_r\right)\left(E_o^* + E_r^*\right),$$
$$I(x,y) = \frac{1}{Z}\left(|E_o|^2 + |E_r|^2 + E_r E_o^* + E_o E_r^*\right), \tag{7.1}$$

where Z is the wave impedance, cf. Sect. 6.1.1, (6.5d), E_o the complex amplitude of the object wave, E_r the complex amplitude of the reference wave and E^* the conjugated complex amplitude. The quantities E_o and E_r are functions of the coordinates (x, y) in the recording plane.

Through this intensity distribution the photographic plate is exposed corresponding to its sensitivity. The transmission of the developed hologram plate is reduced with greater exposure time and greater intensity. This behaviour of a light sensitive material is displayed by the density curve, cf. Sect. 5.4.1. In the following the amplitude transmission t as a function of intensity and exposure time, instead of the density curve, is examined. In parts this function can be approximated by a linear course described as follows:

$$t = 1 - \varepsilon\tau I, \tag{7.2}$$

where ε is a factor, proportional to the slope of the straight line, with which the amplitude transmission curve is approximated in the linear regime; τ is the exposure time and I the irradiance.

If the developed photographic plate—the hologram—on reconstruction is illuminated with the same reference wave E_r, the transmitted wave E_t immediately behind the hologram is given with (7.1) and (7.2):

$$E_t = (1 - \varepsilon\tau I)E_r,$$
$$E_t = \left(1 - \varepsilon\tau\frac{1}{Z}\left(|E_o|^2 + |E_r|^2\right)\right)E_r - \varepsilon\tau\frac{1}{Z}|E_r|^2 E_o - \varepsilon\tau\frac{1}{Z}E_r^2 E_o^*. \tag{7.3}$$

The first term is proportional to the reference wave E_r. The factor in the brackets describes the transmission degree of the hologram for this wave. The second term is directly proportional to the original object wave E_o. That means, through the illumination of the hologram with the reference wave E_r, the original object wave was again reconstructed except for a factor. This object wave appears to the observer to be coming from the object, cf. Fig. 7.1, right hand side, dotted beam paths. In this case a virtual image is spoken of. For this image the objects are reproduced in their spatial positions as they were at the recording of the hologram.

The third term in (7.3) is proportional to the conjugated object wave with a complex coefficient. The wave generated is the so-called real image of the object as shown in the right hand portion of Fig. 7.1.

What is, then, the physical reason for the emergence of a virtual and a real image? Figure 7.2 illustrates a simplified case of a plane reference wave and a scattering point-like object P_1. The spherical wave starting from the object interferes with the plane wave and produces a ring shaped distribution recorded on the photographic plate. The hologram displays in this case the so-called Fresnel zone plate [4]. This consists of a system of concentric light and dark zones, whose widths decrease with increasing diameter of the rings (not shown in Fig. 7.2).

On reconstruction the reference wave is diffracted by the zones. The Fresnel zone plate acts like a grating bent into a ring with a variable grating constant. With the diffraction on a grating with a sinusoidal transmission the ± 1. diffraction order occurs as well as the 0. order, cf. Sect. 2.3.4. Analogous to it, the reference beam is diffracted on the ring shaped grating, namely the zone plate, both in the direction of and away from the axis of symmetry. Thus the diffraction angle is enlarged as the distance between the adjacent zones is decreased. That means the diffraction angles are larger on the edge of the zone plate than in the center.

Hence the Fresnel zone plate in a single diffraction order therefore works like a convex lens, where the beams with increasing distance from the axis of symmetry are likewise more strongly deflected. The other diffraction order corresponds to a concave lens. In this way both orders each produce a virtual and a real image of the object point.

Each point of an extended object produces a corresponding zone plate on the hologram. On reconstruction the virtual and the real image of the recorded object are produced point for point in the manner described above.

Through the overlapping of a number of zone plates in the hologram, which correspond to the single object points, undesirable intermodulation effects also occur. Assuming that the wave scattered by the object possesses an appreciably smaller amplitude than the reference wave, these phenomena can be neglected.

The virtual image, P_{1v}, appears undistorted at the location where the object P_1 was situated as would be expected from (7.3). The real image, P_{1r}, is produced between the observer and the hologram. It shows, however, a so-called pseudoscopic distortion. For clarification a second point, P_2, is shown in Fig. 7.2, above. The distance to the hologram plane is smaller and correspondingly the widths of the Fresnel zones are also smaller. On reconstruction the light is more strongly diffracted, the real image thus likewise occurs a smaller distance away from the hologram. To the observer 'back' and 'front' are transposed. Convex forms, therefore, appear concave and vice versa. In this sense the distorted image is not the same as an object observed from the rear. This characteristic conceals itself in the conjugated phase of the object wave in the third term of the equation (7.3).

Until now it has been assumed that the reconstruction process is carried out in the same configuration and with the same wavelengths as the hologram recording. In addition it was tacitly taken for granted that on developing and processing no change in size or shape of the hologram occurred. The above mentioned assumptions are in general not rigorously fulfilled and the reconstructed picture shows a row of aberrations. A detailed examination of this effect is referred to in the literature [5].

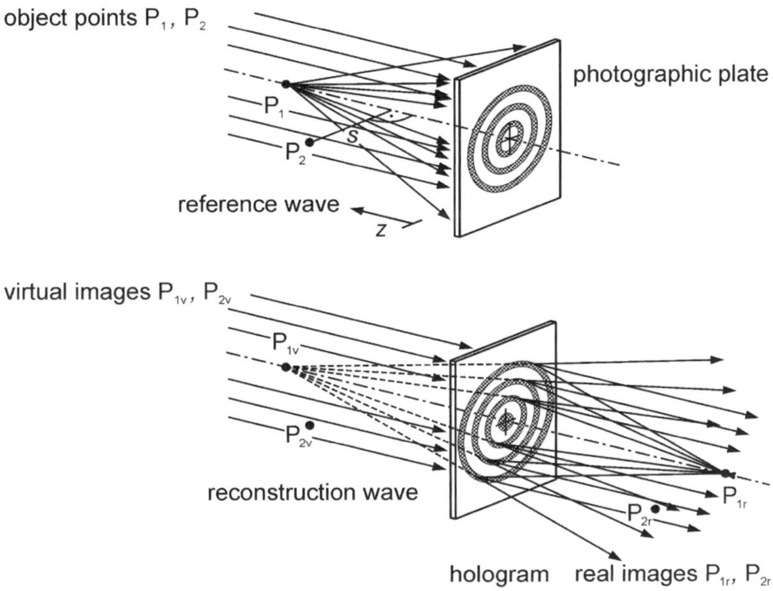

Fig. 7.2 *Top*: recording of the interference pattern generated by the superposition of a plane wave and a spherical wave emanating from the scattering point P_1; for P_2 and s see text. *Bottom*: reconstruction of the hologram by illumination with a reconstruction wave, formation of virtual image P_{1v} and real image P_{1r}

In the following we want only to discuss the effect of different wavelengths on the recording and reconstruction of the hologram. In Fig. 7.2 a plane reference wave is examined. The wave used for the reconstruction is likewise a plane wave with the same propagation direction as the reference wave. The perpendicular distance of the virtual or real image point from the hologram plane is then equal to the focal length of the zone plate which is working as a lens. This focal length is a function of the wavelength as illustrated in Fig. 7.3. If Δr is the width of a transmitting ring of the zone plate the light would be diffracted at the angles $\pm\lambda/\Delta r$ in the ± 1. order, cf. Sect. 2.3, assuming small angles.

For short wavelengths the angle of diffraction becomes smaller, i.e. the focal length becomes greater. It scales there with the wavelength as $f \propto 1/\lambda$. On reconstruction with another wavelength the perpendicular distance to the hologram plane would change correspondingly for the points P_1, P_2 of Fig. 7.2. The mutual distances in the lateral direction remain, on the other hand, unchanged; cf. e.g. the distance s in Fig. 7.2, top. The angle under which an observer sees the reconstructed virtual image P_{2v} is proportional to $1/f$ and thus proportional to λ. If the reconstruction wavelength λ_2 is different to the recording wavelength λ_1 the so-called angle enlargement amounts to: $M_\phi = \lambda_2/\lambda_1$.

The distance of a reconstructed object point P from the hologram plate is z. As we have seen above, $z_1 \propto 1/\lambda_1$ is valid, for another wavelength correspondingly is $z_2 \propto 1/\lambda_2$. From this follows the equations:

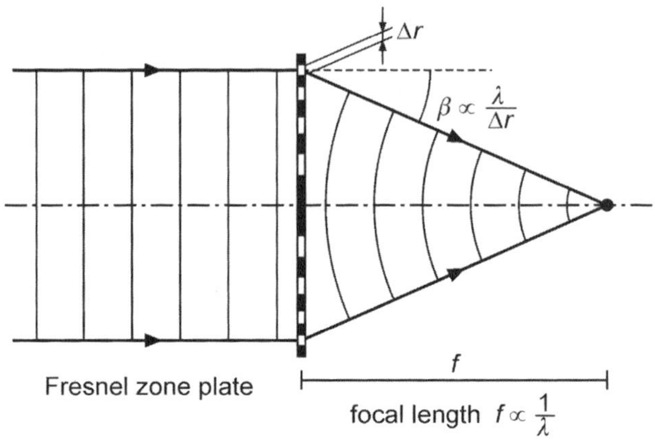

Fig. 7.3 Focal length of a Fresnel zone plate

$$z_1\lambda_1 = z_2\lambda_2,$$
$$\delta z_1\lambda_1 = \delta z_2\lambda_2.$$

The enlargement in the z-direction is defined as $M_z = \delta z_2/\delta z_1$, i.e. the following is valid:

$$M_z = \frac{\lambda_1}{\lambda_2}. \tag{7.4}$$

A more detailed examination of scale relationships is referred to in the literature [5]. There are several types of holograms described in the following.

(a) *Plane hologram and volume hologram*

Until now we have assumed that the hologram behaves like an ordinary diffraction grating, which consists of the superposition of all Fresnel zone plates of the object points. This assumption is, however, only valid if the thickness of the recording medium is small in comparison to the average distance of the interference fringes or rings of the hologram. This hologram type is therefore—in contrast to the volume hologram described below—denoted as a plane hologram. The thermoplastic film used in holography, cf. Sect. 5.4.2, behaves approximately like a plane hologram.

If the thickness of the recording medium is larger than the average distance between the interference fringes a volume hologram is produced [6]. Figure 7.4 shows a simplified structure of a volume hologram for the case of a plane reference and object wave propagating at an angle of $\pm\theta$ against the normal vector of the hologram. The resulting interference fringes have a mutual distance of:

$$d = \lambda/(2\sin\theta). \tag{7.5}$$

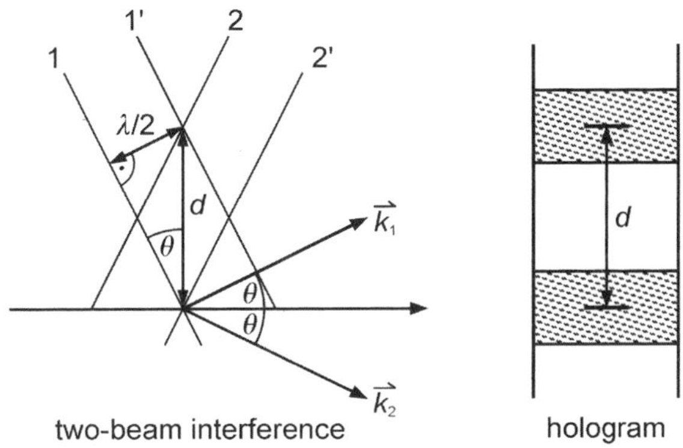

Fig. 7.4 Structure of a volume hologram. A simplified situation is shown where two plane coherent waves interfere leading to interference fringes with a distance d oriented perpendicularly to the plane of the hologram

The particular characteristics of a volume hologram are the angle and wavelength selectivity. On reconstruction the amplitude of the object wave is at a maximum, if the angle θ_r and the wavelength λ_r of the wave used on reconstruction fulfil the condition:

$$\sin\theta_r = \lambda_r/(2d). \tag{7.6}$$

In the case $\lambda_r = \lambda$ relation (7.6) is identical with (7.5). If the reconstruction wave with the same wavelength strikes the volume hologram at an angle of $\theta_r = \theta$ (in Fig. 7.4 in the direction of $\vec{k}_2$) then an object wave of maximum amplitude emerges (in Fig. 7.4 in the direction of $\vec{k}_1$).

Equation (7.6) is completely analogous with the Bragg condition, which describes the diffraction of X-rays in crystals. In this sense the volume hologram behaves like a three dimensional crystal lattice [7]. According to (7.6) each angle has a corresponding definite wavelength λ_r. This characteristic makes it possible to store several holograms with different reference beam angles in a single volume hologram. The terms "plane hologram" and "volume hologram" are used as generic terms for different hologram types. The differentiation of holograms described below refers to the way of recording the interference pattern on the photographic plate.

(b) *Amplitude hologram and phase hologram*

In general on reconstruction of a hologram, the amplitude as well as the phase of the transmitted light of the reconstruction wave is altered. This characteristic is described by the complex amplitude transmission t:

$$t(x,y) = \mathrm{e}^{-\alpha D}\mathrm{e}^{-iknD}, \tag{7.7}$$

with (x,y) coordinates in the hologram plane, $\alpha = \alpha(x,y)$ absorption coefficient at (x,y), $D = D(x,y)$ thickness of the hologram at (x,y), $n = n(x,y)$ refractive index at (x,y) and k angular wavenumber.

Table 7.1 Comparison of the experimentally achieved diffraction efficiencies for different hologram types

Hologram type	Diffraction grating	Diffraction efficiency [%]
Plane/amplitude	$\alpha(x, y)$	6
Plane/phase	$D(x, y)$	33
Volume/amplitude	$\alpha(x, y)$	3
Volume/phase	$n(x, y)$	90

With a normal photographic emulsion the absorption coefficient α changes on exposure, the hologram produced is an amplitude hologram. In contrast to this if no absorption takes place, i.e. $\alpha = 0$, and only D or n change on exposure, a phase hologram is produced. With thermoplastic film a surface relief is generated, i.e. the thickness D of the hologram varies, cf. Sect. 5.4.2.

(c) *Transmission hologram and reflection hologram*

The holograms up to now shown in Figs. 7.1 and 7.2 are called transmission holograms. Reference and object waves from the same side meet on the recording medium. On reconstruction of the hologram with the reference wave the virtual image is observed from the opposite side in the direction of transmission. In contrast to that with a reflection hologram the reference and object waves are directed onto the photographic plate from different sides. On reconstruction the virtual image can be observed from the side of the recording medium where the reference wave was irradiated.

The light intensity of the reconstructed virtual image varies greatly for different hologram types. Here the diffraction ratio is taken as the parameter, which is given through the relationship between the radiant flux in the reconstructed image to the incoming flux of the reconstruction wave. Table 7.1 shows the maximum efficiencies for transmission holograms achieved in experiments [5].

7.2 Principle of Holographic Interferometry

With the aid of holographic interferometry static and dynamic deformations of objects can be measured. The geometric variations of test objects, which are exposed to mechanical forces, temperature changes or oscillations, can be registered with high accuracy. Holographic interferometry simply presupposes that the test objects possess a diffuse scattering surface. With conventional interferometry on the other hand only specular reflective surfaces can be measured, cf. Chap. 6.

Of the different holographic interferometry methods, the double exposure and the real-time method will be described below. A description of the time-averaged method for the observation of stationary harmonic oscillations of an object will not be given in this chapter. The reader should refer to Sect. 8.3.4 where the analogous time-averaged method of speckle interferometry is discussed.

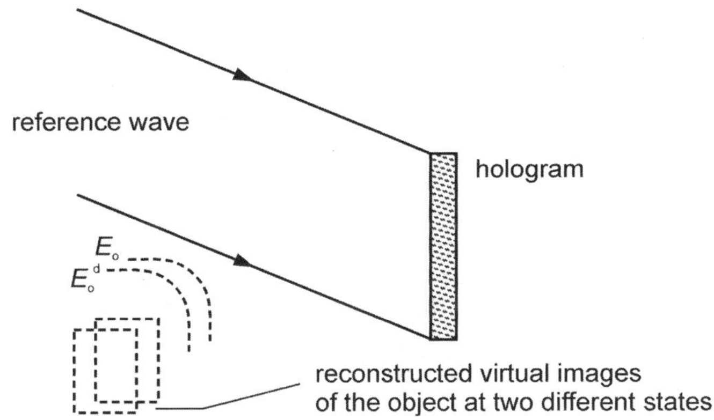

Fig. 7.5 Principle of double-exposure holographic interferometry

7.2.1 Double-Exposure Method

Figure 7.5 shows the principle of holographic interferometry by the double exposure method [8, 9]. The object wave of a temporally changing object is recorded at two different times on a hologram plate.

The test object is holographically recorded once at the time t_1. Between the times t_1 and t_2 ($t_2 > t_1$) the object undergoes a geometric change, which, for example, comes about through an oscillation or through forces acting on the test object. The test object is again recorded on the same hologram plate at t_2. On reconstruction the waves scattered by the object at t_1 and t_2 are simultaneously reproduced. These waves interfere with one another and the emerging interference pattern reproduces the geometric alterations of the object between the two recording times. The double-exposure method is suitable for the analysis of momentary object changes, e.g. due to oscillations, as well for the observation of object deformations through a static load.

In Fig. 7.5 the object waves resulting from the reconstruction of the double exposed hologram are denoted by E_o and E_o^d. For simplification, we assume that when the object geometrically changes only the phase ϕ of the object wave has altered:

$$E_o^d = E_o e^{i\phi}. \tag{7.8}$$

For the wave resulting from the superposition of the reconstructed object waves, the following is valid with (7.8):

$$E = E_o + E_o^d = E_o\left(1 + e^{i\phi}\right), \tag{7.9}$$

where E_o is the amplitude of the object wave at first recording time t_1 and E_o^d the amplitude of the object wave at second recording time t_2.

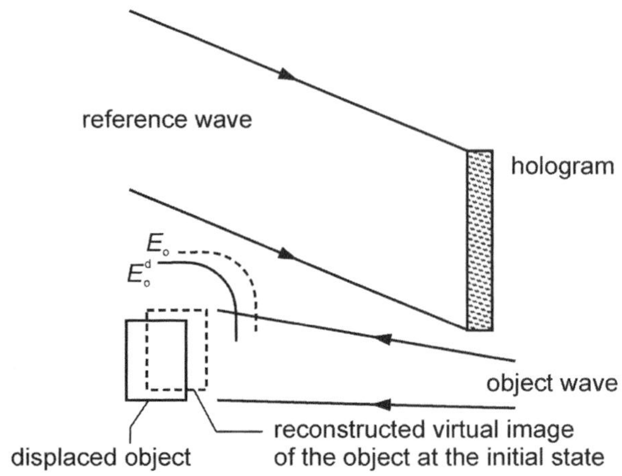

Fig. 7.6 Principle of real-time holographic interferometry

This interference is described by the following irradiance distribution using relation (7.9):

$$I = |E|^2/Z = 2|E_o|^2(1 + \cos(\phi))/Z. \tag{7.10}$$

Bright interference fringes are observed if the following condition is fulfilled:

$$\phi = n\ 2\pi,\ n \text{ integer}. \tag{7.11}$$

If the object did not change between both exposures then $\phi = 0$ holds for all object points and no interference fringes occur.

7.2.2 Real-Time Method

Another variant of holographic interferometry is the real-time method, which is suitable for the on-line control of the changes of an object. The principle of the real-time method is displayed in Fig. 7.6. The wave scattered by an object is recorded in a hologram, as in Sect. 7.1. After development the hologram is again brought into the original recording position. The object is continuously exposed with the laser beam which was also employed for the recording. At the same time the recorded hologram of the object in its initial state is reconstructed. The reconstructed object wave—E_o in Fig. 7.6—and the wave scattered on the altered object which is transmitted through the hologram, E_o^d in Fig. 7.6, interfere with one another. If the object is deformed or displaced due to stress a changing interference fringe pattern emerges. In this way geometric object alterations can be observed in real time with interferometric accuracy.

The hologram is illuminated by the deformed object wave as well as by the reference beam. For the amplitude of the transmitted wave the following is valid:

$$E_{\mathrm{t}} = t\left(E_{\mathrm{o}}^{\mathrm{d}} + E_{\mathrm{r}}\right), \tag{7.12}$$

where t is the amplitude transmission of the hologram, cf. (7.2); $E_{\mathrm{o}}^{\mathrm{d}}$ is the amplitude of the wave of the deformed object state, and E_{r} is the amplitude of the reference wave.

For simplification of further calculation it is accepted that the intensity of the object wave is small in comparison to that of the reference wave, i.e. $|E_{\mathrm{o}}|^2, |E_{\mathrm{o}}^{\mathrm{d}}|^2 \ll |E_{\mathrm{r}}|^2$. For the amplitude of the object wave $E_{\mathrm{o}}^{\mathrm{d}}$ transmitted from the hologram the following is valid, cf. first term in (7.3) with $E_{\mathrm{o}}^{\mathrm{d}}$ instead of E_{r}:

$$E_{\mathrm{t1}} = \left(1 - \varepsilon\tau \frac{1}{Z} |E_{\mathrm{r}}|^2\right) E_{\mathrm{o}}^{\mathrm{d}}. \tag{7.13}$$

For the reconstructed object wave E_{o} the following is valid, cf. second term in (7.3):

$$E_{\mathrm{t2}} = -\varepsilon\tau \frac{1}{Z} |E_{\mathrm{r}}|^2 E_{\mathrm{o}}. \tag{7.14}$$

Therefore, the interference pattern (with (7.13), (7.14) and (7.8)) results in:

$$\begin{aligned} I &= \frac{1}{Z}(E_{\mathrm{t1}} + E_{\mathrm{t2}})\left(E_{\mathrm{t1}}^* + E_{\mathrm{t2}}^*\right), \\ I &= \frac{1}{Z}|E_{\mathrm{o}}|^2\left[\left(1 - \varepsilon\tau\frac{1}{Z}|E_{\mathrm{r}}|^2\right)^2 + \varepsilon^2\tau^2\frac{1}{Z^2}|E_{\mathrm{r}}|^4 \ldots\right. \\ &\left. -\varepsilon\tau\frac{1}{Z}|E_{\mathrm{r}}|^2\left(1 - \varepsilon\tau\frac{1}{Z}|E_{\mathrm{r}}|^2\right)2\cos\phi\right]. \end{aligned} \tag{7.15}$$

Dark interference fringes occur if the phase ϕ is an integral multiple of 2π:

$$\phi = n\,2\pi, \quad n \text{ integer}.$$

The interrelationship between the phase ϕ and the geometric alteration of the object is explained in Sect. 7.3.1.

7.3 Interpretation and Evaluation

7.3.1 Sensitivity Vector

Holographic interferometry allows to record interference patterns of objects with scattering surfaces. The phase fronts of the waves interfering with one another have a similar structure, which are determined through the characteristics of the object surface. The following is valid for the phase difference at two corresponding

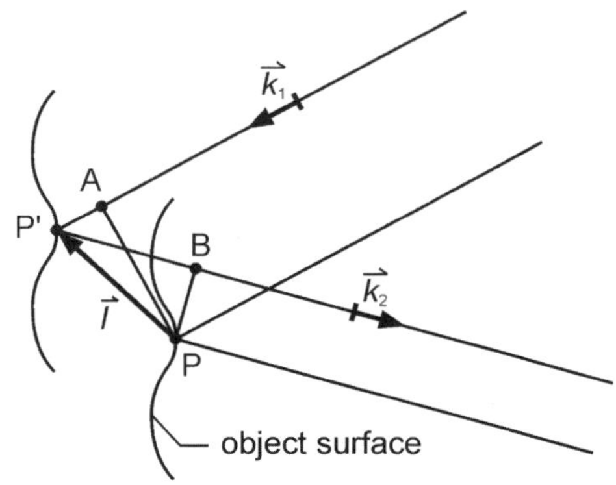

Fig. 7.7 Phase shift caused by a displacement of the object surface

points, P and P′, on the object surface in the original state and in the altered state, cf. Fig. 7.7:

$$\phi = \left(\vec{k}_1 - \vec{k}_2\right) \cdot \vec{l}, \tag{7.16}$$

where $\vec{k}_1, \vec{k}_2$ are the angular wavevectors of the incoming and scattered light and $\vec{l}$ is the displacement vector.

In (7.16) the scalar product is formed with the difference of the angular wavevectors $\vec{k}_1 - \vec{k}_2$ and the displacement vector $\vec{l}$. Only the projection of the displacement vector $\vec{l}$ in the direction of the vector $(\vec{k}_1 - \vec{k}_2)$ contributes to the phase difference. The difference of the wavevectors $\Delta\vec{k}$ is therefore denoted as the sensitivity vector because an object shift only in this direction leads to a phase difference.

For example, if we take the case that $\vec{k}_1$ is aligned anti-parallel to $\vec{k}_2$ and the object is shifted in the beam direction, i.e. $\vec{k}_1$ is parallel to $\vec{l}$, then (7.16) is simplified to: $\phi = (4\pi/\lambda)\, l$. For $l = n(\lambda/2)$, n integer, ϕ is a multiple of 2π. Between two neighboring interference fringes the phase varies at $\Delta\phi = 2\pi$, cf. (7.11), corresponding to an object shift in the direction of $\vec{k}_1$ respectively $-\vec{k}_2$ of $\lambda/2$.

7.3.2 Interference Fringe Patterns for Object Translation and Object Rotation

The interference pattern created by holographic interferometry depends on the deformation or shifting of the recorded object. In general a multitude of interference patterns of different shape and position are observed. The fringe pattern only indirectly yields the information about object alterations. To simplify the interpretation of the fringe patterns the following simple cases of object shifting should be looked at in connection with their interference fringe systems.

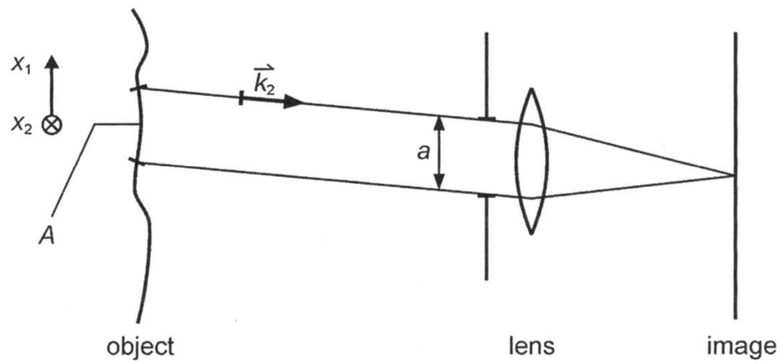

Fig. 7.8 Observation of a reconstructed object

The reconstructed wavefronts of a hologram are, in practice, generally recorded with a camera. Figure 7.8 shows the principle arrangement with a reconstructed object, lens and image plane.

The light in a diffraction limited point in the observation plane, cf. Sect. 2.3, comes from a finite surface element A of the scattering object surface. If the observation plane lies in the focal plane of the lens, the diameter of this surface element equals the aperture of the lens. This case is shown in Fig. 7.8. If this object, however, is imaged by the lens, then the observed surface element A corresponds to the Airy disk, cf. Sect. 2.3.2. This surface element A is denoted in the following as a resolution element. Next we discuss the recording of two mutually shifted resolution elements by holographic interferometry.

The phase differences between corresponding pairs of points of an undisplaced and a displaced resolution element are, in general, not constant but change depending on the position of the points. This means, for example, that the path difference does not amount to $\lambda/2$ for all pairs of points of a resolution element. In this case, no complete destructive interference appears and the fringe contrast is smaller than one, cf. Chap. 6. To achieve a high fringe contrast it is necessary that the phase ϕ in (7.16) varies only slightly within the resolution element. With x_i, $i = 1, 2$ the coordinates of a point within a resolution element A are marked, cf. Fig. 7.8. High fringe contrast is reached, if, with (7.16)

$$\begin{aligned} \mathrm{d}\phi &= \frac{\partial \phi}{\partial x_1}\delta x_1 + \frac{\partial \phi}{\partial x_2}\delta x_2, \\ \mathrm{d}\phi &= \sum_{i=1,2}\left(\frac{\partial}{\partial x_i}\left(\vec{k}_1 - \vec{k}_2\right)\cdot \vec{l}\delta x_i + \left(\vec{k}_1 - \vec{k}_2\right)\cdot \frac{\partial \vec{l}}{\partial x_i}\delta x_i\right) \end{aligned} \tag{7.17}$$

takes a minimum. Equation (7.17) describes the change of the phase within the resolution element caused by an incremental displacement $\delta x_1, \delta x_2$ of the points viewed. The maximum size of $\delta x_1, \delta x_2$ is limited by the size of the resolution element. To illustrate (7.17) two limiting cases are considered:

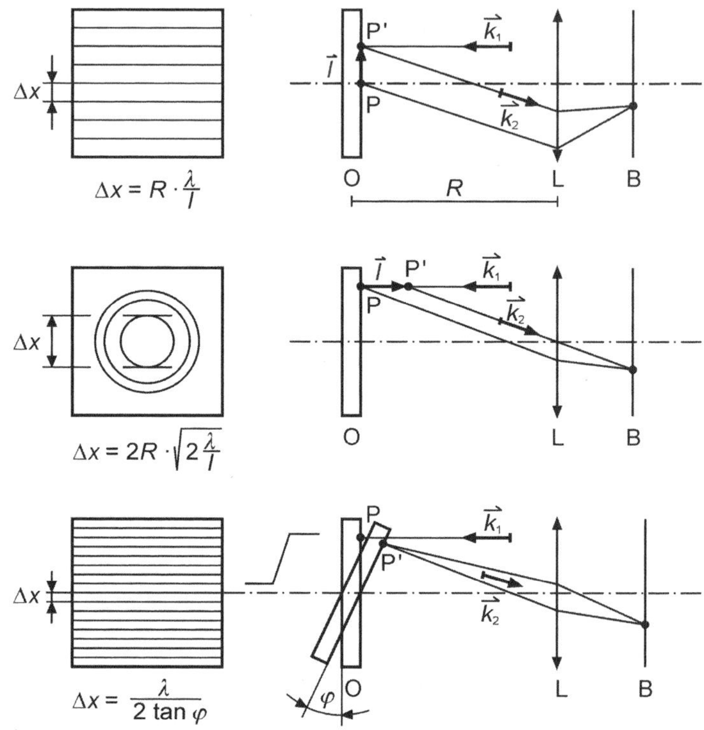

Fig. 7.9 Interference fringe patterns for different object changes, *top* in-plane shift, *middle* out-of-plane shift, *bottom* rotation. O object, L lens, B observation plane

(a) *Pure translation*, i.e. $\vec{l} = \text{const.}$ and $\partial\vec{l}/\partial x_i = \vec{0}$.

In this case, the second term of (7.17) equals zero and the first term should be minimized. This term equals zero, however, if $\vec{k}_1$ and $\vec{k}_2$ are constant. This is equivalent to an illumination of the object with a plane wave and observation of the interference fringes in the focal plane of a lens, i.e. $\vec{k}_2 = \text{const.}$, too. The lens is then focused at infinity. In this sense, the observed interference fringes are located at infinity.

(b) *Pure rotation*, i.e. $\vec{l} = \vec{0}$ and $\partial\vec{l}/\partial x_i \neq \vec{0}$.

The first term in (7.17) equals zero and the second term reduces, if δx_i is minimized. This is achieved with an image of the object surface, because then the resolution element assumes its smallest possible size. In this case the interference fringes are located on the object surface.

The cases considered are illustrated in Fig. 7.9. With a hologram produced by the double-exposure method, an image of the object O is reconstructed (hologram and reconstruction beam are not shown in Fig. 7.9). The reconstructed object is observed with a lens L in the plane B. Depicted are the characteristic interference fringe patterns and the image geometry for three different changes in the objects position between the two hologram recording times: (1) translation parallel to the object surface (top), (2) translation perpendicular to the object surface (middle) and (3) pure rotation (bottom). A point P of the object surface is located after the object displacement at position P′. This displacement is described with the translation through the displacement vector $\vec{l}$. The angular wavevector $\vec{k}_1$ characterizes the illumination wave of the object on the recording of the hologram, $\vec{k}_2$ the light scattered by the object: the object wave, cf. Sect. 7.3.1.

As shown above, the interference fringes for the pure object translation are located at an infinite distance. The observation plane B is therefore located in the focal plane of lens L, cf. Fig. 7.9, top and middle. In cases of an object displacement parallel to the object surface, an interference fringe pattern, as shown top left in Fig. 7.9, is observed. The distance Δx between two adjacent interference fringes in the object plane is given in Fig. 7.9. This distance is determined analogous to that of the distances between interference fringes on diffraction at a grating, cf. Sect. 2.3.3. The quantity R indicates the distance between object and lens.

For an object displacement perpendicular to the object surface a ring shaped interference fringe pattern occurs, analogous to the so-called interference fringes of equal inclination on a plane parallel plate [10]. The diameter of the innermost interference ring in the object plane is, likewise, shown in Fig. 7.9, middle. The derivation of this equation is referred to the literature [10].

In cases of a pure rotation of the object, as shown in the bottom section of Fig. 7.9, parallel interference fringes appear, located on the object. The object is imaged with the lens into the observation plane, cf. increased distance between lens and observation plane in Fig. 7.9, bottom right. A pattern of parallel interference fringes occurs where the spacing between the fringes depends on the angle of inclination φ and the wavelength λ. Assuming a distance between the object and the lens being large with respect to the lateral object dimensions—i.e. the wave vectors $\vec{k}_1$ and $\vec{k}_2$ are approximately antiparallel to each other (not shown in Fig. 7.9, bottom)—then the optical path difference for two adjacent fringes on the object is given by:

$$s_1 \tan \varphi - s_2 \tan \varphi = \frac{\lambda}{2}, \tag{7.17a}$$

where s_1, s_2 denotes the positions of two neighboring fringes relative to the axis of rotation being perpendicular to the plane of projection shown in Fig. 7.9, bottom right. Hence for the fringe spacing we obtain:

$$\Delta x = s_1 - s_2 = \frac{\lambda}{2 \tan \varphi}. \tag{7.17b}$$

To clarify the interpretation of interference patterns for holographic interferometry, the fringes located on the object are normally used. Moreover, the experimental set-up is chosen in such a way that global translations of the object between the recording times is to a large extent ruled out.

7.3.3 Phase-Shifting Method

Without further information, the interference fringe pattern of a holographic interferogram is not sufficient to work out the phase ϕ, cf. (7.10). This is, however, necessary to determine quantitatively a local object displacement, described through the displacement vector $\vec{l}$, see (7.16).

These displacements can be established with the aid of the phase-shifting method [11]. In Fig. 7.10 the construction for holographic interferometry with the phase shift method is schematically depicted.

In contrast to the usual holographic arrangement, here two reference beams with different propagation directions are applied. For a double-exposure recording, reference beam 1 is used for the first exposure and for the second exposure, reference beam 2. On reconstruction each reference beam produces two virtual images of the object, which according to the different propagation directions of the reference beams are laterally displaced against one another (for clarification this is not shown in Fig. 7.10). In total four reconstructed images are observed. Two of these four images completely overlap, so that they interfere with one another. This is the image of the object in recorded state 1 with reference beam 1 and the image of the object in recorded state 2 with reference beam 2.

The other pair of images partially overlap with this interferogram. If the angle between the propagation directions of the reference and reconstruction beams 1 and 2 is, however, chosen sufficiently large, these appear displaced further to the side and do not interfere with the interferogram.

The intensity distribution of the holographic interferogram relating to the object plane is described as follows:

$$J(x,y) = I_1 + I_2 + 2\sqrt{I_1 I_2}\cos\phi, \qquad (7.18)$$

with (x,y) coordinates in the object plane, $J(x,y)$ intensity distribution in the object plane and $I_{1,2}(x,y)$ intensity of the reconstructed object wave at the recording times 1, 2.

Equation (7.18) is a generalization of (7.10) for the case that $|E_o| \neq |E_o^d|$. The phase ϕ is via the equation (7.16) combined with the object shift $\vec{l}$. Equation (7.18) contains I_1, I_2, as unknowns as well as the phase ϕ. Hence with this equation not enough information is available in order to be able to determine the phase ϕ.

With the phase shift method, the phase of the wave used for reconstruction is shifted with a phase shifting element in a defined way. This is indicated schematically in Fig. 7.10 for reconstruction wave 1. A practical realization is introduced in Sect. 7.4. Through the additional phase shift, which is indicated by ϕ_i, a different interference fringe pattern develops, which is described as:

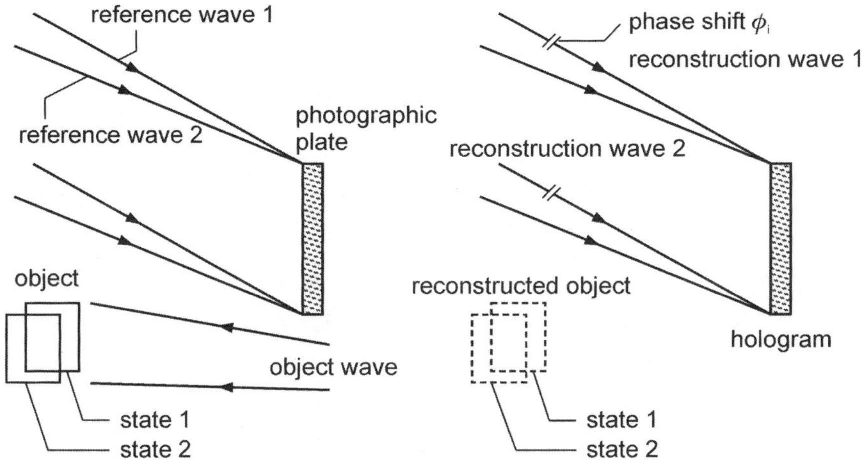

Fig. 7.10 Principle of phase-shifting holographic interferometry, *left* recording, *right* reconstruction

$$J_i(x, y) = I_1 + I_2 + 2\sqrt{I_1 I_2}\cos(\phi + \phi_i), \tag{7.19}$$

where ϕ_i is an additional phase shift in a reconstruction beam.

If three different phase shifts are introduced, i.e. ϕ_i, $i = 1$, 2 and 3, then there are three equations of the form (7.19). The respective interference patterns J_i are recorded, the ϕ_i are likewise known, so that three equations are available to determine the unknowns I_1, I_2 and ϕ. In order to record the interference patterns J_i, the object is reconstructed and observed e.g. with a video camera. For every phase shift ϕ_i, the respective video frame with the interference fringe pattern J_i is fed into an electronic memory and processed. On solving the equations of form (7.19) for the three recorded interference patterns, the following equation for the phase ϕ is obtained:

$$\tan\phi = \frac{(J_3 - J_2)\cos\phi_1 + (J_1 - J_3)\cos\phi_2 + (J_2 - J_1)\cos\phi_3}{(J_3 - J_2)\sin\phi_1 + (J_1 - J_3)\sin\phi_2 + (J_2 - J_1)\sin\phi_3}. \tag{7.20}$$

Due to the tangent function in (7.20) the phase ϕ can only be determined uniquely in the interval $-\pi/2 < \phi < +\pi/2$. The absolute phase differentiates itself from this value by a whole-number multiple of Π. Nevertheless, in order to determine the phase overcoming the irregularities at $\pm\pi/2$, the gradients of the phase function on both sides of an irregularity are compared for the computer aided evaluation. In this way, the phase $\phi = \phi(x, y)$ for the complete interferogram and with it the displacement vector $\vec{l}$ are calculated in the direction of the sensitivity vector, see Sect. 7.3.1.

7.4 Digital Holography

7.4.1 Fundamentals of Digital Holography

The rapid development of computer technology in recent decades has made it possible to perform recording and reconstruction of a hologram with a computer [11–18]. This topic is discussed in the following. We will not treat computer generated holography, which allows to produce artificial holograms by numerical methods; the interested reader is referred to the literature, e.g. [12–17].

Classical holography is, as discussed in Sect. 7.1, divided in two different steps:

Step 1: *Recording*

In classical holography, the interference pattern

$$I(x,y) \propto (E_{\mathrm{o}}(x,y) + E_{\mathrm{r}}(x,y))(E_{\mathrm{o}}^{*}(x,y) + E_{\mathrm{r}}^{*}(x,y)) \tag{7.21}$$

produced by the superposition of the object wave $E_{\mathrm{o}}(x,y)$ and the reference wave $E_{\mathrm{r}}(x,y)$ in the xy-plane of the hologram, is stored on a photographic plate. The developed plate is referred to as a hologram.

Step 2: *Reconstruction*

When the hologram is illuminated with the original reference wave $E_{\mathrm{r}}(x,y)$, the diffracted wave field $E_{\mathrm{d}}(x,y)$ immediately behind the hologram consists of three terms:

$$E_{\mathrm{d}}(x,y) = E_{1}(x,y) + E_{0}(x,y) + E_{-1}(x,y), \tag{7.22}$$

where the indices 1, 0, −1 denote the respective diffraction order. In the following we assume that the reconstruction is made under the same geometrical conditions as the recording, $E_{\mathrm{r}}(x,y)$ is constant and $|E_{\mathrm{o}}(x,y)| \ll |E_{\mathrm{r}}(x,y)|$. Then, cf. (7.3):

$$E_{1}(x,y) \propto E_{\mathrm{o}}(x,y), \tag{7.23}$$

$$E_{0}(x,y) \propto E_{\mathrm{r}}(x,y) \tag{7.24}$$

and

$$E_{-1}(x,y) \propto E_{\mathrm{r}}^{2}E_{\mathrm{o}}^{*}(x,y). \tag{7.25}$$

The wave $E_{1}(x,y)$ is proportional to the original object wave (corresponds to the second term in relation (7.3)). An observer perceives this wave as a virtual image, which is located exactly where the object was placed during the recording, see Fig. 7.11. This virtual image is indistinguishable from the original object. The reconstructed object wave corresponds to the $+1^{\mathrm{st}}$ order of the diffracted reference wave. The second wave $E_{0}(x,y)$ is the transmitted reference wave (zeroth

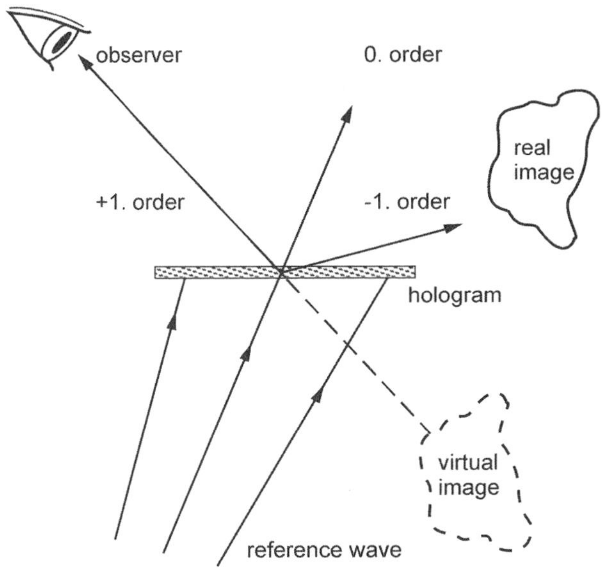

Fig. 7.11 Reconstruction of a hologram with −1., 0. and +1. diffraction orders

diffraction order, so called DC term; corresponds to the first term in relation (7.3)). The third wave $E_{-1}(x, y)$ is the so-called conjugate object wave (-1st diffraction order, twin image term; corresponds to last term in (7.3)). It is responsible for a distorted—because of the term E_r^2 and the conjugated field E_o^* in (7.25)—real image of the object.

Digital holography is very similar and thus also separated in two different steps:

Step 1: *Recording*

In digital holography the photographic film is replaced by a CCD or CMOS sensor with high spatial resolution, see Sect. 5.4.5. The holographic interference pattern is recorded electronically and stored—pixel by pixel—as so-called virtual hologram in digital form, e.g. on a hard drive of a personal computer or a memory stick. Alternatively, it is possible to produce a digital hologram by scanning and digitizing an analogue hologram, which was generated by the classical way.

Step 2: *Reconstruction*

In classical holography, the virtual and real images of an object are produced by diffraction of the reconstruction wave by the hologram. In digital holography the image of an object is calculated numerically by a computer program based on the digital stored data, i.e. the virtual hologram, and displayed on a computer screen.

Since only real images can be displayed on a screen, there are two options:

(a) The object wave, which corresponds to the virtual image, is computed in a first step. Then, with a virtual lens—corresponding to the eye of an observer—a real image is calculated numerically and displayed on a screen.

Fig. 7.12 Sketch of the object plane and the hologram plane

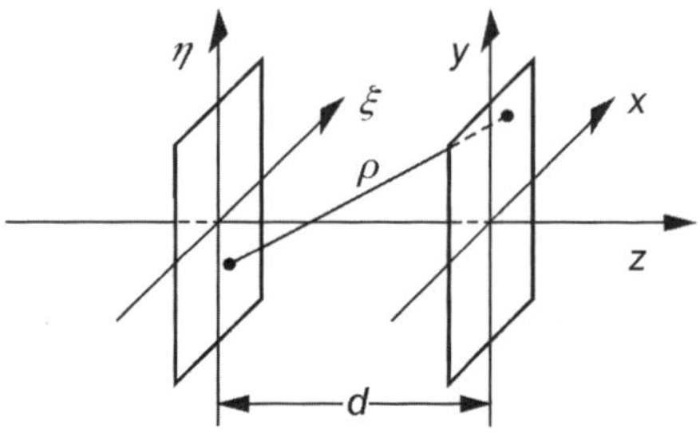

(b) Alternatively, the conjugate object wave, which corresponds to the real image of the object, can be calculated directly and displayed on a screen. Because the corresponding real image is distorted by the reconstruction wave—see (7.25) and cf. the explanation of the pseudoscopic distortion of the real image in Sect. 7.1—the diffraction of the conjugate complex reference wave $E_r^*(x, y)$ by the hologram is computed. In this case, an undistorted real image results. The relation between the stored intensity pattern $h(x,y)$ and the intensity distribution

$$I(\xi, \eta) \propto E(\xi, \eta) E^*(\xi, \eta) \tag{7.26}$$

of the image on the computer screen is given by the Fresnel-Kirchhoff integral [15]

$$E(\xi, \eta) \propto \iint_{x,y} h(x, y)\, E_r^*(x, y)\, \frac{e^{-i\frac{2\pi}{\lambda}\rho}}{\rho} \mathrm{d}x\mathrm{d}y \tag{7.27}$$

with

$$\rho = \sqrt{(x-\xi)^2 + (y-\eta)^2 + d^2}, \tag{7.28}$$

where d is the distance between the object (object plane) and the sensor (hologram plane), see Fig. 7.12; x and y or ξ and η are the coordinates of the hologram plane and the object plane, respectively, see Fig. 7.12. Equation (7.27) holds in the case that the observer looks perpendicularly to the hologram, i.e. in negative z-direction in Fig. 7.12.

It should be noted that—because of the diffraction of the conjugate complex reconstruction wave—the locations of the real and virtual image—compared to Fig. 7.11—have interchanged their positions. Therefore the real image coincidences with the object.

The numerical processing of the Fresnel-Kirchhoff integral (7.27) is very time-consuming. Therefore, approximations are made and the so-called convolution theorem is used, which is beyond the scope of this book.

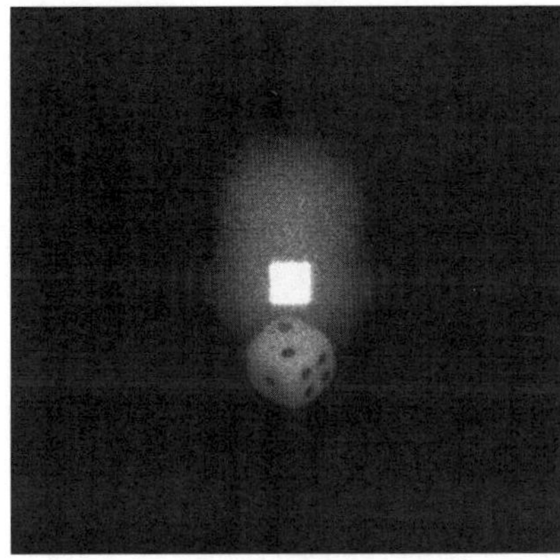

Fig. 7.13 Numerical reconstruction of a dice chosen as object [12]

As mentioned before, the diffracted light field, which is calculated with the Fresnel-Kirchhoff integral (7.27), corresponds to three different waves, i.e. object wave, transmitted reference wave and conjugate object wave. All these waves contribute, more or less, to the image displayed on the screen. Figure 7.13 shows a calculated image of a dice. The bright square in the center is the undiffracted reference wave, which is not related to the object (called also DC term). The twin image is out of focus and cannot be recognized. Both, the DC term and the twin image term, can be eliminated numerically.

The advantages of digital holography are diverse. A great advantage is the absence of time-consuming wet chemical processing connected with the photographic material development. Additional benefits are the possibilities to compensate numerically errors, to reduce numerically speckle noise and to remove numerically unwanted diffraction orders from the image plane. So sometimes a better resolution can be achieved compared to classical holography. With modern graphic cards parallel processing with several hundreds of processors can be implemented reducing calculation times by orders of magnitude. For a measuring field of 4×4 cm^2 the digital hologram enables a lateral resolution of 20 μm and an axial resolution of 1 mm [18]. With digital holographic microscopy 3D information of biological or technical objects can be extracted achieving spatial resolutions down to 300 nm by the use of oblique illumination and a wavelength of 193 nm [19].

7.4.2 Digital Holographic Interferometry

In classical holography the phase of the object wave or the conjugate object wave is not obvious. Therefore, a classical holographic interferogram is usually given by a single hologram, on which an object in two different states is stored. When the hologram is illuminated with the original reference wave, two interfering object waves and two interfering conjugate object waves occur, i.e., images with dark and bright fringes result, see Sects. 7.2 and 7.3.

In digital holography the complex field strength $E(\xi, \eta)$ of the real image of an object is numerically calculated, see (7.27). That is why the phase $\varphi(\xi, \eta)$ of the conjugate object wave can be calculated for any arbitrary object point (ξ, η) modulo π:

$$\varphi(\xi, \eta) = \arctan \frac{\mathrm{Im}E(\xi, \eta)}{\mathrm{Re}E(\xi, \eta)}. \tag{7.29}$$

If there are two different digital holograms, the phases $\varphi_1(\xi, \eta)$ and $\varphi_2(\xi, \eta)$ of each hologram can be determined and subtracted. So the phase shift

$$\Delta\varphi(\xi, \eta) = \varphi_2(\xi, \eta) - \varphi_1(\xi, \eta) \tag{7.30}$$

and thus the displacement vector—see (7.16)—of each object point (ξ, η) can be determined.

7.5 Set-up for Holographic Interferometry and Examples of Applications

In the following sections the set-up to perform double-exposure holographic interferometry will be described followed by examples of applications.

7.5.1 Measurement Set-up

Figure 7.14 shows the set-up of holographic interferometry using the double-exposure method. Radiation sources are lasers operating in continuous wave mode (CW) as well as pulsed mode provided that their coherence length is greater than the longest optical path difference between object and reference beam paths. CW lasers are used to record objects, that are at rest during exposure of the hologram. Rapidly changing objects are investigated with pulsed lasers.

Figure 7.14 shows a ruby laser generating pulses with a full width at half maximum (FWHM) of about 20 ns and radiant energies of the pulses of up to some joule. The pulse duration of the laser pulses determines the exposure time of the hologram recording. Due to the short pulse duration of the ruby laser the motion of even fast changing objects can be frozen for the holographic recording. Such object changes are occuring, e.g., while an object is oscillating. The ruby laser is operated in such a way, that two laser pulses are emitted with an adjustable interpulse separation of 1 μs–1 ms. The point in time for the exposure as well as the interpulse separation between both exposures are tuned for the studied state of motion or oscillation of the measuring object.

The laser beam is split by a wedge plate W into an object and a reference beam, see Fig. 7.14. The object beam is expanded by a diverging lens to illuminate larger

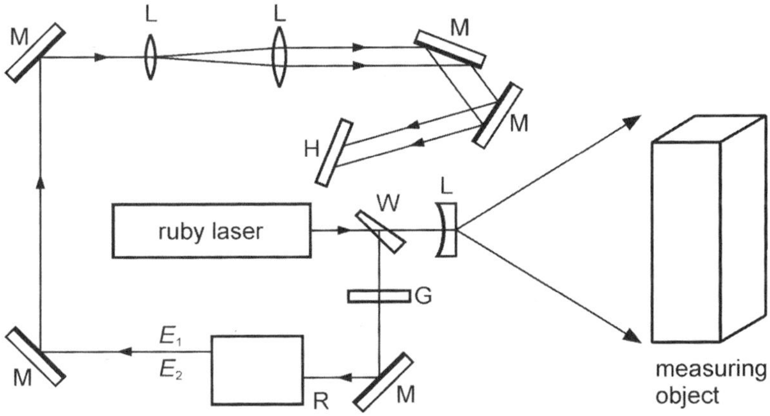

Fig. 7.14 Set-up for double-exposure holographic interferometry. W wedge plate, L lens, G grey filter, M mirror, R splitting of reference beam, cf. Fig. 7.15; H hologram

objects. The pulse energies of the ruby laser are sufficient to record objects with dimensions of about 1 m × 1 m.

A grey filter G attenuates the reference beam in order to adjust the irradiance in such a way that object and reference beam expose the hologram plate within the linear regime of the characteristic curve, cf. Sect. 5.4.1, Fig. 5.30. As a rule the ratio of irradiances of reference to object beam is chosen to be 5:1 up to 10:1. If the irradiances would be equal, then the interferences would lead to the largest variations of intensities and therefore the exposure would no longer be restricted to the linear range of the characteristic curve.

For a quantitative evaluation of holographic interferograms according to the phase-shifting method the reference beam is split into two reference beams whose propagation directions are inclined to each other, see Sect. 7.3.3, Fig. 7.10, left. The splitting of the reference beam R—cf. Fig. 7.14—is shown in Fig. 7.15. The direction of polarization is given by the electric field vector $\vec{E}$, cf. Sect. 2.2.5. The reference beam is guided through a Pockels cell which allows to change the direction of polarization within a short time. An electric voltage is applied to the electro-optic crystal causing a change of the polarization direction of the transmitted light, cf. Sect. 4.1.3. During the first laser pulse the Pockels cell is de-energized, i.e. the direction of polarization of $\vec{E}$ is not changed. This polarization state of the light beam exiting the Pockels cell is denoted with $\vec{E}_1$. A polarizing beam splitter reflects this beam to path 1.

Prior to the second laser pulse a voltage is applied to the Pockels cell. Thus the polarization is turned by 90° to the polarization direction $\vec{E}_2$. This laser pulse is transmitted by the polarizing beam splitter to path 2. With the help of a half wave plate this polarization direction is again rotated by 90° so that both reference beams have the same direction of polarization after the second beam splitter. This is a necessary condition to observe stationary interference patterns, cf. Chap. 6.

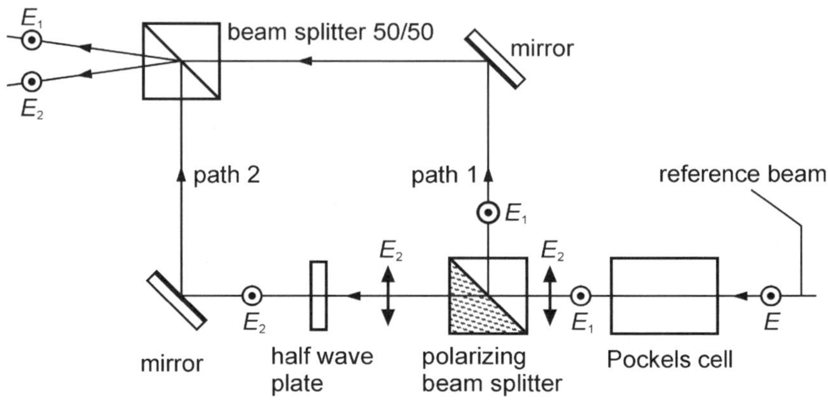

Fig. 7.15 Splitting of the reference beam into two reference beams inclined to one another for two laser pulses of the ruby laser; detailed illustration of the unit "splitting of reference beam", R, shown in Fig. 7.14

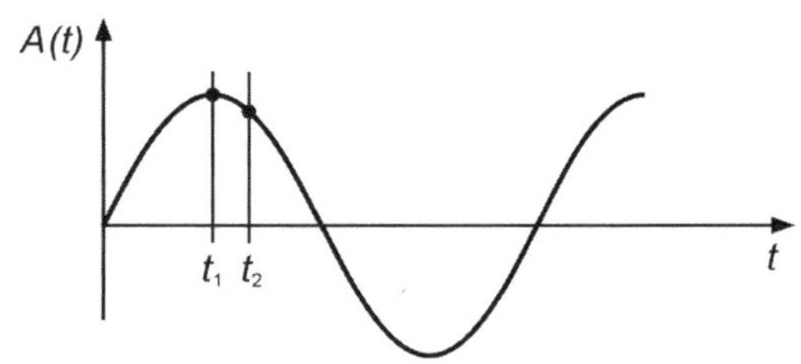

Fig. 7.16 Oscillation signal of a test object and recording times t_1, t_2 for holographic interferometry

The mirrors are aligned in such a way that the reference beams are inclined to one another, cf. Sect. 7.3.3. In Fig. 7.15 this splitting is shown exaggerated (top left), in Fig. 7.14 it is omitted to give a clearer picture of the beam path.

Both reference beams are expanded with a lens combination—as illustrated in Fig. 7.14—and lead via mirrors to the hologram plate H.

The exposure times and the temporal distance between both exposures are to be synchronized with the object movement. This is illustrated by means of Fig. 7.16 for the case of the investigation of an oscillating object. The oscillation of an object point is plotted as a function of time. For simplification, only the case of a harmonic oscillation is looked at here. This signal can be obtained, e.g. from the vibration excitation of a test object or through a sensor, such as e.g. a vibration pickup or a laser vibrometer. The times t_1 and t_2 mark the recording times relative to the zero crossing of the oscillation signal. The ruby laser is correspondingly triggered so that the first laser pulse is generated at time t_1 and the second laser pulse at time t_2. The position and distance of these times are chosen in such a way that holographic interferograms, which can be evaluated, could be recorded, i.e. a sufficient number of interference fringes appear over the object.

In Fig. 7.17 the reconstruction of the hologram with two reconstruction beams E_1, E_2 corresponding to both reference waves is shown (the mutual inclination between the beams is not reproduced in Fig. 7.17, it is generated through the tilting

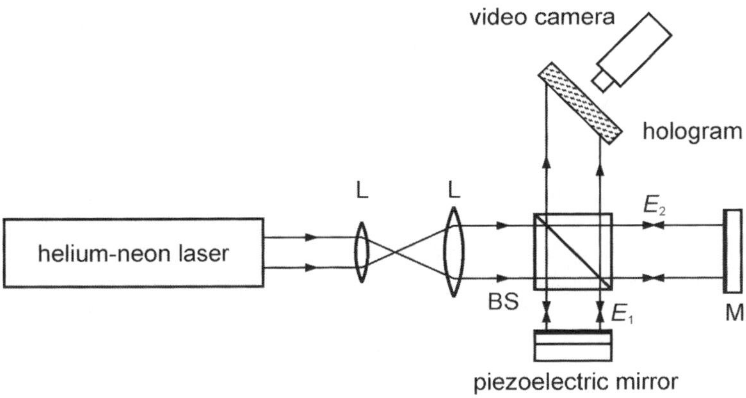

Fig. 7.17 Reconstruction of the hologram for the phase-shifting method. L lens, BS beam splitter, M mirror

of e.g. the mirror M). The phase of one reconstruction beam is altered by a piezoelectric adjustable mirror, in order to generate the single interferograms for the quantitative evaluation according to the phase-shifting method, cf. Fig. 7.10, right. These interferograms are fed into a computer system with a video camera.

7.5.2 Examples of Applications

From the very beginning holographic interferometry was applied as a tool in experimental physics to study e.g. plasma phenomena, to determine refractive index distributions and electron densities [20, 21].

In the following section application examples for holographic interferometry from the realm of vibration investigations are put forward where engineering tasks are in the focus. The reproduced holograms were generated with a set-up corresponding to that of Fig. 7.14.

In the first example the object under test is a loudspeaker housing, whose oscillation behavior was examined for definite frequency ranges. The objective is to obtain information about the spatial structure of the oscillations of the loudspeaker housing, in order to deduce the appropriate constructive measures to reduce unwanted sound emitters.

The loudspeaker housing has a height of about 1 m. For the experiments, it was excited over the built in loudspeaker system with a sinusoidal signal of definite frequency and amplitude. On the left in Fig. 7.18 the holographic interferogram of the loudspeaker housing is shown. On the front of the loudspeaker two distinct antinodes appear, and a torsion-like oscillation of the side wall is observed.

On the right in Fig. 7.18 the same loudspeaker housing is shown under the same excitation conditions after a constructive modification. The side wall was mechanically stiffened and additionally filled with an absorbing material. The

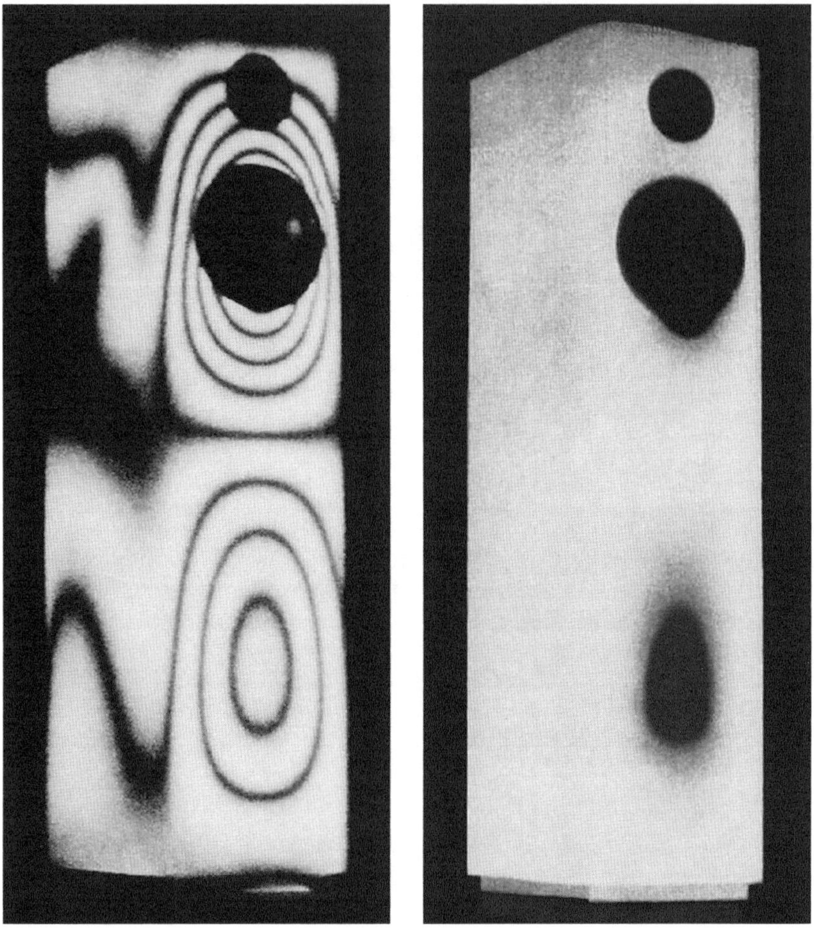

Fig. 7.18 Holographic interferograms of a loudspeaker housing with the same excitation conditions. At the front side of the cabinet the membranes of two dynamic speakers can be seen (two black circular areas in the upper third). *Left* original situation, *right* after constructive changes to the loudspeaker housing

unwanted oscillations of the casing could be substantially reduced so that the electroacoustic properties of the system were improved.

In Fig. 7.19 the holographic interferogram of an oscillating engine bonnet of a passenger car is reproduced [22]. The dimensions of the bonnet are approximately $0.9 \times 1.2\ m^2$. This hologram was evaluated according to the phase-shifting method. The determined absolute deformations of the bonnet between both recording points are shown in Fig. 7.20 as a function of the coordinates in the surface of the engine bonnet. Thus for every surface point of the object the deformation can be given. With the help of such investigations qualitative information on oscillation forms as well as quantitative data on spatial distributions of oscillation amplitudes of components or machines are obtained.

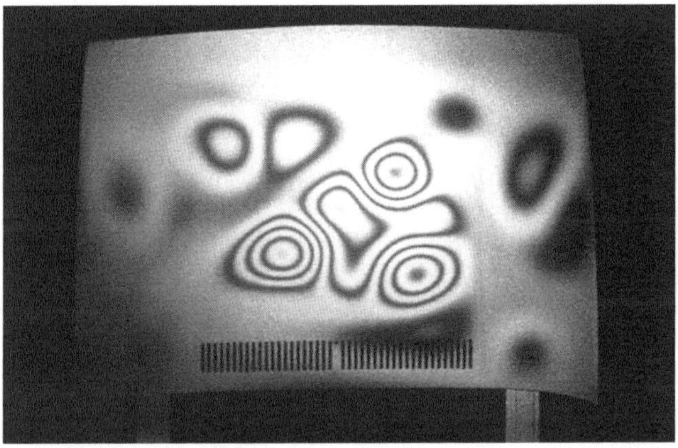

Fig. 7.19 Holographic interferogram of an oscillating engine bonnet

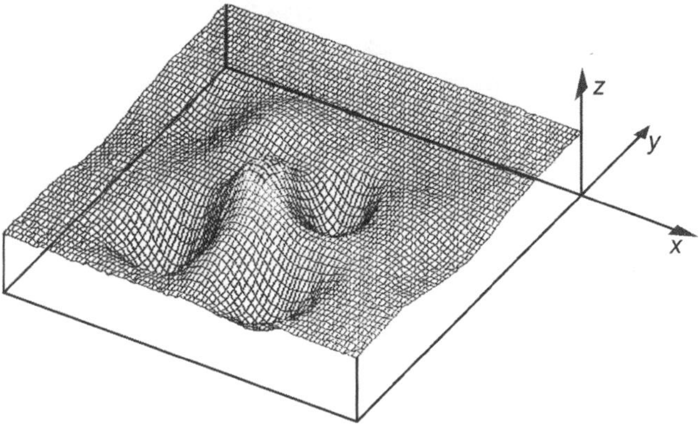

Fig. 7.20 Deformation of the engine bonnet determined from the holographic interferogram shown in Fig. 7.19

These test informations can also be used to calculate sound fields. The objective is to localize the sources of the sound, which cause e.g. unwanted noise stress for people in the vicinity of the oscillating component.

Extensions of holographic interferometry aim at measuring the three-dimensional displacement vector $\vec{l}$ for every point of an oscillating object [23, 24]. For this an optical set-up is used, which measures in three sensitivity directions simultaneously, cf. Sect. 7.3.1. The set-up shown in Fig. 7.14 records merely one sensitivity direction.

With a knowledge of the vector $\vec{l}$ for each point of the object surface the oscillation behaviour of an object can be described completely [24]. With the help of e.g. three different illumination directions and one observation direction three

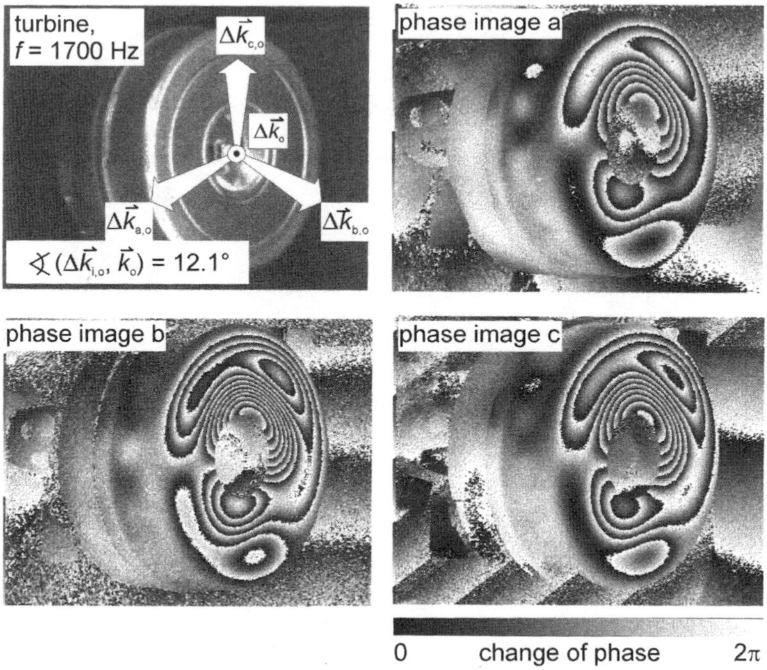

Fig. 7.21 Measurement of the mode of vibration of a turbine housing with three dimensional holographic interferometry; excitation frequency 1,700 Hz. *Top left* photograph of the turbine with the three sensitivity vectors $\Delta\vec{k}_{a,o}, \Delta\vec{k}_{b,o}, \Delta\vec{k}_{c,o}$ and the direction of observation $\vec{k}_o$ being perpendicular to the plane of projection. *Top right* and *bottom* phase images for the three sensitivity directions

different sensitivity vectors can be measured for each object point. Figure 7.21 shows a turbine housing of a vacuum cleaner motor with the three sensitivity vectors and the corresponding phase images for an excitation of the turbine with an eigenfrequency of 1,700 Hz [24]. Figure 7.22 displays the three-dimensional deformation fields of the mode of vibration deduced from the data of the phase images of Fig. 7.21.

The flexibility of a set-up for holographic interferometry can be increased if it is possible to guide the laser pulses via fiber optics [25]. For the reference beams monomode fibers must be used to preserve the spatial coherence across the hologram aperture for recording and reconstruction. The required energies of the reference beams are in the range of 10–100 μJ, which can be transmitted by fibers with small core diameters. For the object beams multimode fibers come into question, if the condition is fulfilled that a phase shift in the object beam path between the two hologram recordings is caused only by object displacements and not by changes of the mode structure within the optical fiber. Necessary object-beam energies amount to several joule, depending on object size. An approach to

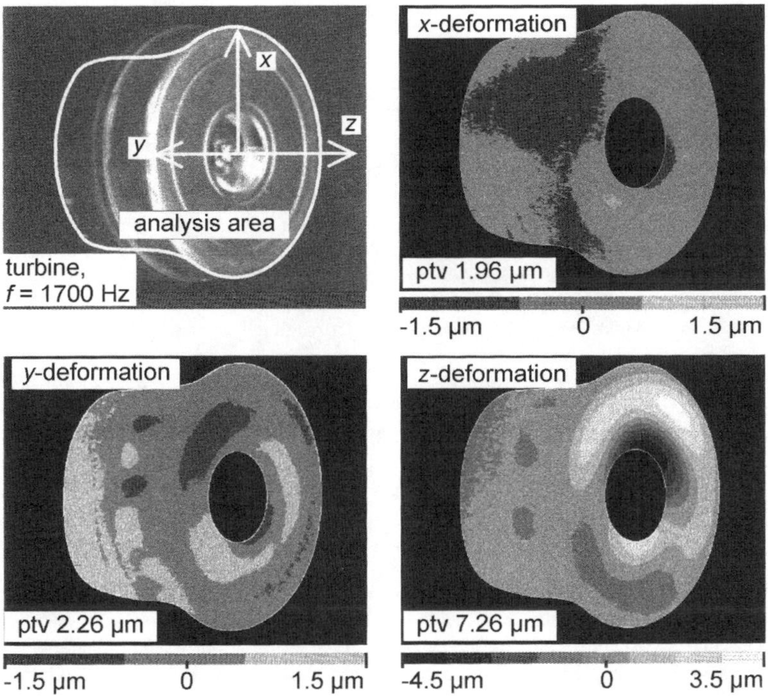

Fig. 7.22 Three dimensional deformation fields of the mode of vibration of a turbine housing based on the data of Fig. 7.21. *Top left* photograph of turbine with cartesian coordinate system. *Top right* and *bottom* x-, y- and z-deformation of the turbine housing at an excitation frequency of 1,700 Hz. ptv = peak-to-valley

deliver laser pulse energies of 300 mJ for a ruby laser through multimode fibers is based on pulse stretching in order to keep the peak irradiances below the damage thresholds of the fiber material [26].

References

1. D. Gabor, A new microscopic principle. Nature **161**, 777–778 (1948)
2. G. Saxby, *Practical Holography*, 3rd edn. (IOP Publishing Ltd, London, 2004). ISBN 0 7503 0912 1
3. A. Blumlein, Improvements in and relating to sound-transmission, sound-recording and sound-reproducing, Patent GB394325 (1933)
4. E. Hecht, *Optics*, 4th edn. (Addison Wesley, Reading, 2003). ISBN 0-321-18878-0
5. R. Collier, C. Burkhardt, L. Lin, *Optical Holography* (Academic Press, New York, 1971)
6. G. Groh, *Holographie: Physikalische Grundlagen und Anwendungen* (Verlag Berliner Union, Stuttgart, 1973). ISBN 3-408-53501-9
7. D. Halliday, R. Resnick, J. Walker, *Fundamentals of Physics*, 7th edn. (Wiley, New York, 2005). ISBN 0-471-21643-7

8. L. Heflinger, R. Wuerker, R. Brooks, Holographic interferometry. J. Appl. Phys. **37**, 642–649 (1966)
9. C. Vest, *Holographic Interferometry* (Wiley, New York, 1979)
10. D. Meschede (ed.), *Gerthsen Physik*, 24th edn. (Springer, Berlin, 2010) ISBN 978-3-642-12893-6
11. P. Hariharan, B. Oreb, T. Eiju, Digital phase-shifting interferometry: a simple error-compensation phase calculation algorithm. Appl. Opt. **26**, 2504–2506 (1987)
12. W.H. Lee, Computer-generated holograms: techniques and applications. Prog. Opt. **16**, 120–232 (1978)
13. O. Bryngdahl, F. Wyrowski, Digital holography—computer generated holograms. Prog. Opt. **28**, 1–86 (1990)
14. D. Schreier, *Synthetische Holografie* (VCH, Weinheim, 1984). ISBN 387664089X
15. S. Pasko, R.J. Wicki, Novel Fourier approach to digital holography. Opto-Electron. Rev. **10**(2), 89–95 (2002)
16. U. Schnars, W. Jueptner, *Digital Holography: Digital Hologram Recording, Numerical Reconstruction, and Related Techniques* (Springer, Berlin, 2010). ISBN 978-3642060182
17. N. Pandey, B.M. Hennelly, D.P. Kelly, Th.J. Naughton, Speed up of Fresnel transforms for digital holography using pre-computation. Proc. SPIE **7442**(744205), 1–11 (2009)
18. D. Carl, M. Fratz, A. Hofmann, H. Kock, Dreidimensionale Bilder auf einen Blick. Qualität und Zuverlässigkeit **55**, 36–38 (2010)
19. W. Osten, A. Faridan, P. Gao, K. Körner, D. Naik, G. Pedrini, A. Singh, M. Takeda, M. Wilke, Recent advances in digital holography. Appl. Opt. **53**, G44–G63 (2014)
20. F. Jahoda, R. Jeffries, G. Sawyer, Fractional-fringe holographic plasma interferometry. Appl. Opt. **6**, 1407–1410 (1967)
21. C. Tallents, M. Burgress, B. Luther-Davies, The determination of electron density profiles from refraction measurements obtained using holographic interferometry. Opt. Commun. **44**, 384–387 (1983)
22. C. Wanders, C. Haas, S. Lampe, R. Noll, Determination of absolute deformations of a vibrating automobile hood using holographic double-pulse interferometry, Laser in Engineering, ed. by W. Waidelich, 10th International Congress Series on Laser 91. (Springer-Verlag, Berlin Heidelberg, 1992), pp. 22–27. ISBN-13: 978-3-540-55247-5
23. R. Noll, Vibration analysis collects spatial information, Noise Vibr **24**(4), 12–14 (1993)
24. M. Sellhorst, H. Ostendarp, C. Haas, R. Noll, Dreidimensionale holografische Interferometrie und Modalanalyse, Laser in Engineering, ed. by W. Waidelich, 11th International Congress Series on Laser 93. (Springer-Verlag, Berlin Heidelberg, 1994), pp. 139–142. ISBN 978-3-540-57444-6
25. H. Nebeling, R. Noll, H. Ostendarp, M. Sellhorst, M. Weck, Modalanalyse mit Holografie—ein Verfahren zur hochauflösenden berührungslosen Analyse von Schwingungen (VDI Verlag, Düsseldorf, 1996) Nr. 532, 143 p.
26. S. Pflüger, M. Sellhorst, V. Sturm, R. Noll, Fiber-optic transmission of stretched pulses from a Q-switched ruby laser. Appl. Opt. **35**, 5165–5169 (1996)

Chapter 8
Speckle Metrology

Abstract The formation of speckles while illuminating an object with a laser beam is described. Different types of speckle interferometers are presented to measure out-of-plane, in-plane or derivatives of object displacements. The set-up of an electronic speckle interferometer is shown and examples of applications are given ranging from deformation analysis of stressed mechanical components, the detection of internal flaws in foam parts or tires to the analysis of stationary vibrations of components.

Speckle metrology comprises a number of different methods for measuring the geometric variation of objects with an optically rough surface. These methods use the so-called speckle effect which is discussed in Sect. 8.1. The areas of application of the speckle measuring methods, in particular of speckle interferometry, are comparable to those of holographic interferometry. Different types of object variations, for example dilatations, displacements or bulges can be determined quantitatively. Speckle interferometry allows to measure such object alterations with a video camera and electronic image processing, cf. Sect. 8.3. Electronic speckle interferometers distinguish themselves in practice through their simple and flexible operation.

8.1 Formation of Speckles

If a rough surface is illuminated with coherent laser light, this surface appears grainy to the observer. The light scattered from the surface varies in its spatial intensity distribution, bright and dark spots appear. This phenomenon is known as the speckle effect [1, 2]. The single bright or dark spots are called speckles. The speckles develop if the surface of the scattering object is rough as opposed to that of a mirror or a lens. A surface in optics is described as rough, if the roughness is at least of the order of the wavelength of the illuminating light. What is the reason for this statistically variable scattering intensity?

Each point of the surface which is illuminated by the incident laser light, scatters this light with a definite phase relative to the phase of the incoming wave.

A. Donges and R. Noll, *Laser Measurement Technology*, Springer Series in Optical Sciences 188, DOI: 10.1007/978-3-662-43634-9_8,

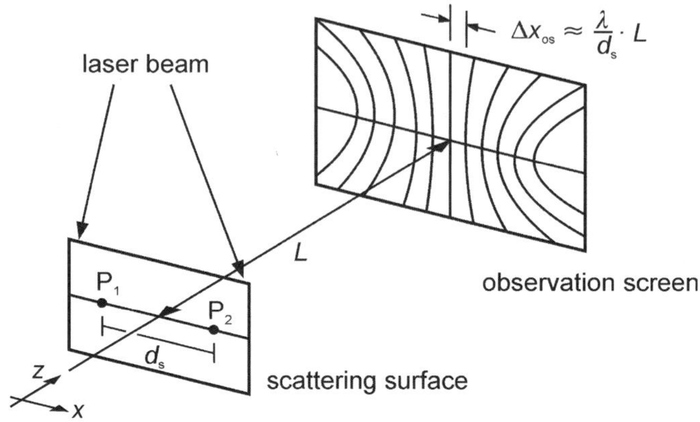

Fig. 8.1 Formation of speckles, interference pattern of two scattered waves emanating from the points P_1 and P_2 of a coherently illuminated surface

Due to the microscopic structure of the surface, which is characterized through the roughness, the optical path lengths between the various scatterers of that surface and an observation plane varies statistically and so do the phases of the waves arriving at a given point in the observation plane. In the following, two surface points are considered which are chosen in such a way that they scatter the incoming laser light and the waves emanating from these two points have equal phases (this is not a necessary assumption, it is just a simplification which leads to the central bright interference fringe on the observation screen in Fig. 8.1 lying on the z-axis). Figure 8.1 depicts this situation. A laser beam is directed onto a surface. An observation screen is mounted in the distance L to the scattering surface. Both of the waves emanating from P_1 and P_2 interfere and generate an interference pattern on the observation screen which is similar to that of the double slit, cf. Sect. 2.3.3. The interference fringes run approximately parallel to one another in the middle of the screen. For geometric quantities within the plane of the scattering surface the index "s" is used and for those in the observation screen "os". For the distance between two neighboring interference fringes in x-direction, the following is valid, cf. Sect. 2.3.3, for sufficiently large distances L:

$$\Delta x_{\mathrm{os}} \approx \frac{\lambda}{d_{\mathrm{s}}} L, \tag{8.1}$$

where λ is the wavelength, d_{s} the distance between P_1 and P_2 and L the distance between the scattering surface and the observation screen.

As well as the pair of points shown in Fig. 8.1, there are on a rough surface many more pairs of points with varying mutual distances and orientations. All these pairs generate their own interference pattern similar to that shown in Fig. 8.1. According to the size of the distance d_{s} between the points in compliance

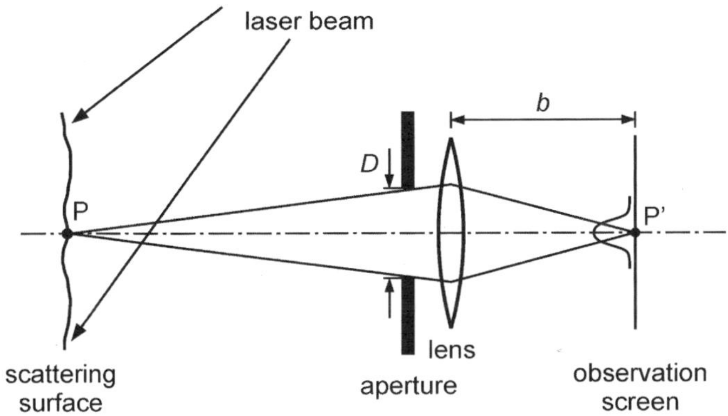

Fig. 8.2 Formation of speckles by imaging a coherently illuminated surface with a lens on an observation screen

with (8.1), each of these interference patterns have different fringe spacings or are each—according to the angular position of the point pair—twisted towards each other. As a result this statistical superposition of many interference patterns with different orientations and different fringe spacings leads to the formation of speckles on the observation screen.

The smallest typical dimension of a speckle can be estimated with the aid of equation (8.1). The smallest fringe spacing appears for two points which have the largest possible mutual distance d_s from one another. The largest possible distance is given through the size of the illuminated field. This so-called illumination aperture defines therefore the smallest dimension of the speckles in the observation plane.

In the following, the case of a lens being situated in the beam path, as shown in Fig. 8.2, is considered. With the lens the illuminated surface is imaged onto an observation screen. The aperture of the lens has a diameter D. The aperture plane can itself be regarded as a plane, on which a statistical wave field with spatially varying phase strikes. Similar to that already shown above, two points, which have the same phase, are again found in this plane. They generate an interference pattern in the observation plane. The superposition of many such interference patterns leads again to the formation of speckles. The smallest diameter arises analogous to (8.1):

$$\phi_{os} \approx \frac{\lambda}{D} b, \tag{8.2}$$

with D diameter of the aperture, b distance between lens or aperture plane and observation screen.

Apart from a constant numerical factor, cf. Sect. 2.3, the result (8.2) corresponds to the diameter of the so-called Airy disk. The Airy disk is the smallest possible area on which an object point can be imaged.

8.2 Speckle Pattern Photography

The different types of speckle pattern photography are introduced below [3–5]. They are based on a double exposure method, with which the geometric changes of the surface of an object, which appeared between both exposures, are registered. The imaging speckle photography is suitable for the determination of displacement gradients parallel to the surface of the test object, as they occur, for example, with the expansion or compression of a body. With the unfocused speckle photography the displacement gradients perpendicular to the object surface are obtained. These appear with the bulging or inclining of an object.

8.2.1 Imaging Speckle Pattern Photography

The principle layout for imaging speckle pattern photography is shown in Fig. 8.3. A laser beam illuminates the object. This object is imaged with a lens on a film plane. The speckle pattern of the object in state 1 is recorded in a first exposure on the film. The resolving power of the film must be so large that single speckles can be resolved. If necessary the diameter of the aperture of the imaging lens in Fig. 8.3 must be decreased, in order to increase the size of the speckles, cf. Sect. 8.1. After this illumination the position of the object is altered; in Fig. 8.3 a displacement in the plane of the object in negative x-direction is shown. The same speckle pattern exists in the film plane. It is, however, in contrast to state 1 now displaced, according to the magnification of the viewing system, in the positive x-direction. This speckle pattern in state 2 is recorded in a second exposure on the film. On the developed film two similar speckle patterns are recorded which are displaced against one another. Figure 8.3 shows for example the position of a speckle $\mathrm{P}'_{(1)}$ and $\mathrm{P}'_{(2)}$ for both exposures belonging to the same object point P in the two states 1 and 2 (the index for these two states is written as "(1)" or "(2)" for the points P′ in the film plane). The imaging of the object into the film plane can be described by the imaging equation, cf. (4.28):

$$\frac{1}{f} = \frac{1}{g} + \frac{1}{b}, \tag{8.3}$$

with the quantities object distance g, image distance b and the focal length of the lens f. For the magnification holds the relation:

$$\frac{B}{G} = \frac{b}{g}, \tag{8.4}$$

where B is the size of the image and G the size of the object.

The displacement between both speckle positions are determined with the layout shown in Fig. 8.4. A collimated laser beam is directed onto the developed

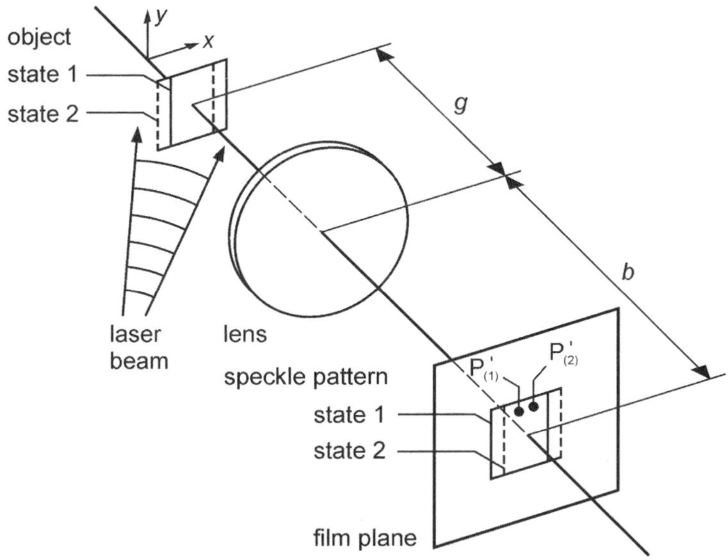

Fig. 8.3 Principle of speckle pattern photography, recording of the speckle pattern

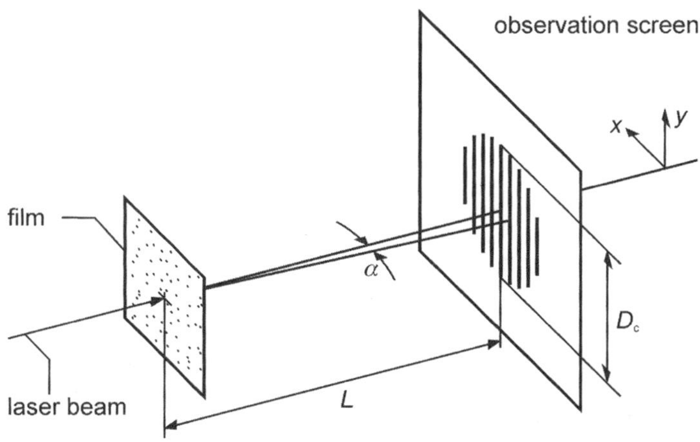

Fig. 8.4 Determination of the local object displacement from a speckle pattern photography. For clarity only a reduced number of speckles is shown on the film

film. The diameter of the laser beam in the film plane is chosen at least as large as the largest distance between a speckle pair $P'_{(1)}$ and $P'_{(2)}$. The illuminated speckles $P'_{(1)}$ and $P'_{(2)}$ represent two sources of equally phased coherent waves, which generate Young interference fringes. For the diagram Fig. 8.4, it was taken that the displacement between state 1 and state 2 is small compared with the distance L between the film and the observation plane. For this reason interference fringes essentially parallel to one another appear, as opposed to the situation shown in Fig. 8.1.

From the measured interference fringe spacing the displacement between both speckle positions $P'_{(1)}$ and $P'_{(2)}$ is calculated. It follows:

$$d_{\mathrm{fp}} = \frac{\lambda}{\Delta x_{\mathrm{os}}} L, \tag{8.5}$$

with Δx_{os} interference fringe spacing in the observation screen, L distance between film and observation plane, d_{fp} displacement between $P'_{(1)}$ and $P'_{(2)}$ in the film plane.

The displacement in the object plane d_{op} can then be determined with (8.5) and (8.4):

$$d_{\mathrm{op}} = d_{\mathrm{fp}} \frac{g}{b}. \tag{8.6}$$

Up to now it has been assumed that between both exposures the object experiences a pure translation in x-direction. The method described is also suitable for a spatially dependent object displacement in the xy-plane, like e.g. an expansion. For the evaluation in this case the film is scanned with the laser beam, in order to determine the local object displacement for each object point from the corresponding interference fringe system. The interference fringes are always aligned perpendicular to the direction of the local object displacement in the xy-plane.

The smallest still detectable displacement of the object depends on the size of the speckles in the film plane. For

$$d_{\mathrm{fp}} \geq d_{\mathrm{sp}}, \tag{8.7}$$

where d_{fp} is the speckle displacement in the film plane and d_{sp} the speckle size in the film plane, the displacement of the speckles can still be measured, but cannot in the reverse case. This limit can, too, be illustrated in another way. As shown in Fig. 8.4, the interference pattern in the observation plane is circularly limited. This circle, inside of which the interference fringes appear, is created by the diffraction of the laser light at individual speckles. The diffraction pattern is similar to that of a circular aperture, cf. Sect. 2.3. For the diameter of the circle in the observation plane D_{c}, the following is valid:

$$D_{\mathrm{c}} = 2.44 \frac{\lambda}{d_{\mathrm{sp}}} L, \tag{8.8}$$

see Fig. 8.4.

If the fringe spacing in the observation plane is larger than half the diameter $D_{\mathrm{c}}/2$, it can no longer be determined. From the equations (8.5) and (8.8), the following condition has to be fulfilled: $\Delta x_{\mathrm{os}} \leq D_{\mathrm{c}}/2$. Then, for the smallest resolvable object displacement holds the following relation:

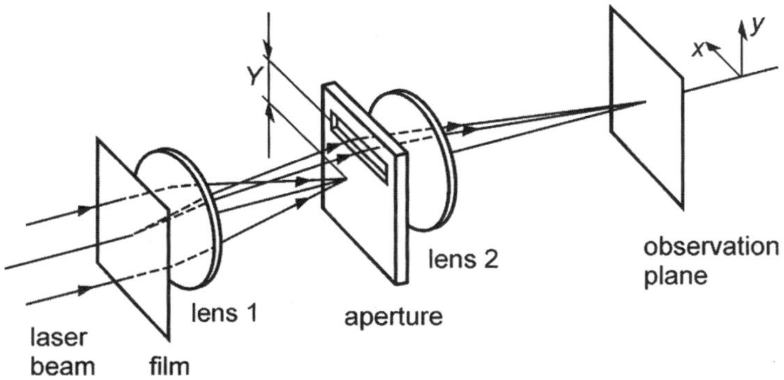

Fig. 8.5 Set-up to observe fringe patterns indicating locations of constant local object displacement

$$d_{\mathrm{fp}} \geq (1/1.22)\, d_{\mathrm{sp}}. \tag{8.9}$$

This relationship is identical to the inequation (8.7) with the exception of a constant numerical factor. The detection limit for speckle pattern photography is reached if the displacement of the speckle pattern is smaller than, or equal to, the size of the speckles. With equation (8.6), inequation (8.7) yields the condition for detectable object displacements for a given speckle size in the film plane d_{sp}:

$$d_{\mathrm{op}} \geq d_{\mathrm{sp}} \frac{g}{b}. \tag{8.10}$$

Up till now it has been shown how, for imaging speckle pattern photography, the exposed film is scanned point for point according to Fig. 8.4 in order to determine the respective local object displacement. With the arrangement shown in Fig. 8.5, points with the same object displacement can be displayed on the object which is imaged by lens 2 onto the observation plane. The coordinates of the object which is imaged into the film plane, cf. Fig. 8.3, are represented by $\left(x_{\mathrm{fp}}, y_{\mathrm{fp}}\right)$. On the film, a local displacement $d_y\left(x_{\mathrm{fp}}, y_{\mathrm{fp}}\right)$ of a speckle pattern was recorded in y-direction, e.g. with the arrangement according to Fig. 8.3. (A displacement in y-direction was chosen for the example in Fig. 8.5, in order that the beam paths could be shown more clearly). The displacement of the speckle pattern corresponds to a local object displacement in the y-direction, which should be determined.

The film is illuminated with an expanded laser beam. The laser beam is diffracted at the recorded speckle patterns, which are shifted against one another. An aperture is placed in the focal plane of lens 1 of Fig. 8.5. In the aperture plane an interference fringe pattern is formed due to the diffraction at the local speckles, which are shifted against each other on the film by $d_y\left(x_{\mathrm{fp}}, y_{\mathrm{fp}}\right)$. For a local displacement of the object in the y-direction, the interference fringes are oriented parallel to the x-axis. Only such interference fringes which run parallel to the

x-axis are allowed to pass through the slit shaped aperture. A second requirement is that the interference fringe distance must correspond to the distance of the slit in the y-direction from the axis of symmetry, (or an integral part thereof; see below). The zeroth order of the diffracted laser beam has a distance of $y = 0$ from the axis of symmetry and is therefore not allowed to pass through the aperture (cf. Sect. 2.3).

Lens 2 produces an image of the object recorded on the film in the observation plane (lens 1 is arranged directly behind the film and influences only slightly the imaging through lens 2). Due to the presence of the aperture the object points which contribute to the imaging are those which have a local displacement in the y-direction leading to an interference fringe pattern, the orientation and fringe spacing of which, is such that the diffracted beams of a diffraction order are allowed through by the aperture. In this way, those object points, at which a definite object displacement appeared in the y-direction, are imaged on the observation plane. The image of the object is covered by a fringe pattern, whereby the fringes correspond to points of constant local displacement in the y-direction.

If Y is the distance of the slit in the y-direction from the axis of symmetry, then for the interference fringes which contribute to the imaging, the following is valid:

$$(n\lambda/d_y)f_1 = Y, \tag{8.11}$$

with n diffraction order, $n = 0, 1, 2, \ldots$; d_y local displacement $d_y\left(x_{\mathrm{fp}}, y_{\mathrm{fp}}\right)$ of the speckle pattern in the film plane, f_1 focal length of lens 1 of Fig. 8.5.

Solving equation (8.11) for d_y gives:

$$d_y(x_{\mathrm{fp}}, y_{\mathrm{fp}}) = n\lambda f_1/Y. \tag{8.12}$$

The local displacement is a multiple of $\lambda f_1/Y$, if the corresponding interference fringes contribute to the imaging. We study two neighboring points $x_{1\mathrm{fp}}$ and $x_{2\mathrm{fp}}$ in the film plane, for which the local displacements

$$d_y(x_{1\mathrm{fp}}, y_{\mathrm{fp}}) = (\lambda f_1/Y), \tag{8.13}$$

$$d_y(x_{2\mathrm{fp}}, y_{\mathrm{fp}}) = 2(\lambda f_1/Y) \tag{8.14}$$

appear. The points $\left(x_{1\mathrm{fp}}, y_{\mathrm{fp}}\right)$, $\left(x_{2\mathrm{fp}}, y_{\mathrm{fp}}\right)$ where these displacements occur are imaged according to condition (8.11) into the observation plane. Furthermore, for simplification, it is assumed that lens 2 generates a 1:1 imaging of the film plane in the observation plane, i.e. the magnification is 1, cf. (8.4). The displacement gradient is determined from the distance between two neighboring fringes $x_2 - x_1 = -\left(x_{2\mathrm{fp}} - x_{1\mathrm{fp}}\right)$ in the observation plane, with (8.13) and (8.14):

$$\partial d_y/\partial x \approx (\lambda f_1/Y)/(x_2 - x_1). \tag{8.15}$$

The displacement gradients $\partial d_y/\partial y$, $\partial d_x/\partial x$, $\partial d_x/\partial y$ can be calculated completely analogously. The displacement gradient in the object plane is the same as the one given in (8.15) since numerator and denominator are transformed according to

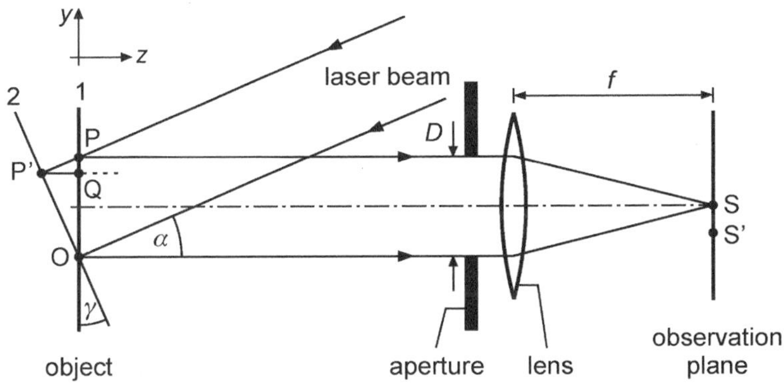

Fig. 8.6 Unfocused speckle pattern photography

(8.6). For the determination of the displacement gradients in the x-direction, the aperture in the arrangement according to Fig. 8.5 is simply rotated by 90°.

8.2.2 Unfocused Speckle Pattern Photography

In Fig. 8.6 the principle for unfocused speckle pattern photography is shown. An object is illuminated by a plane laser wave. A lens is situated in front of the object. The observation or film plane is arranged in the distance of the focal length. The laser light is scattered on the object in state 1. The scattering waves emanating from the points P and O lead to an interference pattern in the focal plane of the lens, which reaches a maximum at a given point S. Through a multitude of such scattering pairs of points, a speckle pattern, which is recorded on a film, cf. Sect. 8.1, is formed in the observation plane.

If the object in state 2 is inclined towards the original state, the phase relation between the points P′ and O changes. This phase change leads to a displacement of the observed interference maximum of S towards S′ in the observation plane. This phase change can be calculated as follows, see Fig. 8.6:

$$\overline{\mathrm{PP}'} = D \sin\gamma,$$

$$\overline{\mathrm{P}'\mathrm{Q}} = D \sin\gamma \cos\alpha, \tag{8.16}$$

where D is the diameter of the aperture and γ, α are the angles shown in Fig. 8.6.

If the phase of the scattering waves from P and O in state 1 is the same, then the phase from Q in state 2 is delayed with respect to the phase of the wave coming from O by the following amount:

$$\Delta\phi = (2\pi/\lambda)D(\sin\gamma + \sin\gamma\cos\alpha), \tag{8.17}$$

with $\Delta\phi$ phase difference between the scattering waves of Q and O, cf. Fig. 8.6. If the phase difference amounts to 2π, then the position of the interference pattern in the observation plane changes by just by one fringe spacing. For phase angles between 0 and 2π the interference pattern is displaced by the corresponding fraction of the fringe spacing. It is therefore valid:

$$\frac{\Delta\phi}{2\pi} = \frac{\overline{SS'}}{(\lambda/D)f}, \tag{8.18}$$

with $\overline{SS'}$ displacement of the interference pattern in the observation plane, f focal length of lens, $(\lambda/D)f$ interference fringe spacing in the observation plane, cf. (8.1). For sufficiently small angles γ using the approximation $\sin\gamma \approx \tan\gamma \approx \gamma$ the following is valid:

$$\gamma \approx \partial d_z/\partial y, \tag{8.19}$$

with d_z displacement of the object in z-direction, see coordinate system in Fig. 8.6, $\partial d_z/\partial y$ partial derivative of the displacement in z-direction with respect to y.

From (8.17), (8.18), (8.19) it follows that:

$$\overline{SS'} = (1 + \cos\alpha)f(\partial d_z/\partial y). \tag{8.20}$$

The displacement of the interference or speckle pattern can again be measured with the arrangement according to Fig. 8.4 or 8.5. With the help of equation (8.20) the gradient $\partial d_z/\partial y$ is calculated. Therefore unfocused speckle pattern photography offers the possibility to determine the displacement gradient of an object perpendicular to its surface. It is in this sense complementary to the imaging speckle pattern photography, with which the displacement gradient in the surface of an object is recorded, cf. Sect. 8.2.1.

8.3 Speckle Interferometry

Speckle interferometry distinguishes itself most markedly from speckle pattern photography in the following ways:

(a) With speckle interferometry a speckle pattern is observed, formed by the interference of the speckle pattern of a wave, scattered at the object, and a reference wave. The reference wave is for example a spherical wave or another speckle pattern.

(b) In speckle interferometry, two speckle patterns belonging to different object states—as described under (a)—are compared with one another. The observed fringes correspond to points at which the speckle patterns correlate. From the position and the structure of these fringes, the type and size of the spatial object change is determined.

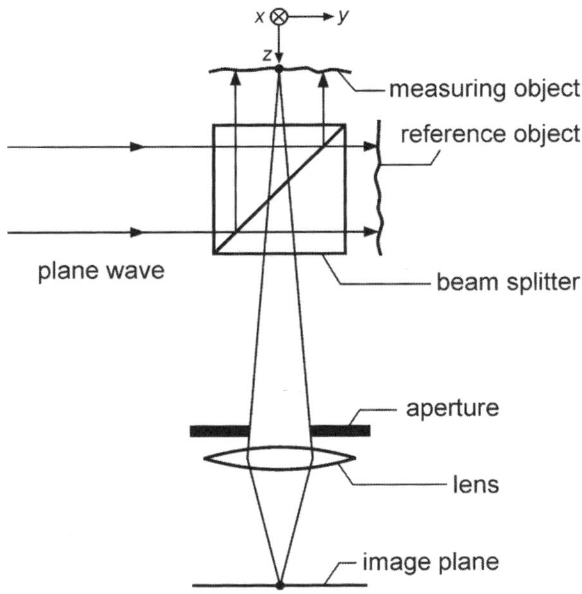

Fig. 8.7 Principle of speckle interferometry

8.3.1 Principle

The principle of speckle interferometry is represented by Fig. 8.7 [3, 4, 6]. Essentially, the design corresponds to that of a Michelson interferometer, cf. Sect. 6.1.2. Using a beam splitter, a plane wave is split into two. The waves illuminate a diffuse scattering surface of a test object and a reference object. Both objects are imaged, via the beam splitter and a lens, onto an image plane. The scattered waves from the test and reference objects interfere at this observation plane and a speckle pattern develops, as shown in Sect. 8.1, Fig. 8.2. The scattered waves are described in the image plane through their complex amplitudes, cf. Sect. 6.1:

$$\begin{aligned} E_\mathrm{m} &= |E_\mathrm{m}(x,y)|\exp(\mathrm{i}\vartheta_\mathrm{m}(x,y)), \\ E_\mathrm{r} &= |E_\mathrm{r}(x,y)|\exp(\mathrm{i}\vartheta_\mathrm{r}(x,y)), \end{aligned} \tag{8.21}$$

with $E_{\mathrm{m,r}}(x,y)$ complex amplitudes of the scattered waves from the test and reference objects in the observation plane, cf. Fig. 8.7, $\vartheta_{\mathrm{m,r}}(x,y)$ phases of the scattered waves in the observation plane.

Amplitudes and phases take on statistically varying values at different locations, as is characteristic of a speckle pattern. An interference pattern develops in the image plane, which itself becomes a speckle pattern. For the intensity distribution (the argument (x,y) is omitted for clarity) the following applies:

$$I_1 = \text{const.}(E_\mathrm{m} + E_\mathrm{r})\,(E_\mathrm{m} + E_\mathrm{r})^*,$$
$$I_1 = I_\mathrm{m} + I_\mathrm{r} + 2\sqrt{I_\mathrm{m} I_\mathrm{r}}\cos(\vartheta_\mathrm{m} - \vartheta_\mathrm{r}), \tag{8.22}$$

with $I_{\mathrm{m,r}}$ intensity of the scattered waves in the image plane.

A shift of the measuring object in Fig. 8.7 by a small amount in the z-direction gives rise to a phase difference between the waves E_m and E_r amounting to:

$$\phi(x, y) = (2\pi/\lambda)\,2d_z(x, y), \tag{8.23}$$

with λ wavelength of the laser light, d_z displacement of the test object in the z-direction, cf. Fig. 8.7. In general, this displacement would be locally different, i.e. $d_z = d_z(x, y)$ and also therefore $\phi = \phi(x, y)$. As a result of the phase difference ϕ, the interference pattern changes, as follows:

$$I_2 = I_\mathrm{m} + I_\mathrm{r} + 2\sqrt{I_\mathrm{m} I_\mathrm{r}}\cos(\vartheta_\mathrm{m} - \vartheta_\mathrm{r} + \phi). \tag{8.24}$$

The interference patterns I_1 and I_2 are equal if the following condition is fulfilled:

$$\phi = n\ 2\pi, \tag{8.25}$$

with n integer.

The degree of correlation between the two interference patterns can be observed as will now be described. The interference pattern I_1 is recorded on a photographic film. The developed film is brought back again to the position of the shot and if the test object has not changed, then the interference pattern I_1 will also have stayed the same. Only a very small amount of light is allowed through to the film, since the resulting dark areas will accurately overlap the bright speckles of the interference pattern I_1 where they exist. However, had the object changed locally, i.e. varying shifts occur in the z-direction at different object points $\mathrm{P}(x, y)$, then a fraction of light would have been allowed through. Dark fringes appear on the film in the image of the object. These occur in the areas where a phase difference of $\phi = n\ 2\pi$ exists. Bright fringes occur where $\phi = (2n + 1)\pi$ is valid. At these positions both interference patterns I_1 and I_2 do not overlap, i.e. in general dark regions do not appear on the film at the same places as bright speckles. Between two waves emanating from two neighboring dark fringes there is a path difference of λ, cf. (8.23).

Another variant of speckle interferometry following Fig. 8.7 is that both interference patterns, I_1 and I_2 are recorded on the same film. In areas in which there is no correlation, the film would be almost completely blackened. Regions which had not been exposed by speckle pattern I_1 would now be exposed by the bright speckles of the other speckle pattern. However, in areas in which the speckle patterns do show correlation, the film would be only partially exposed, as here, dark areas in both interference patterns lie at the same positions. The resulting

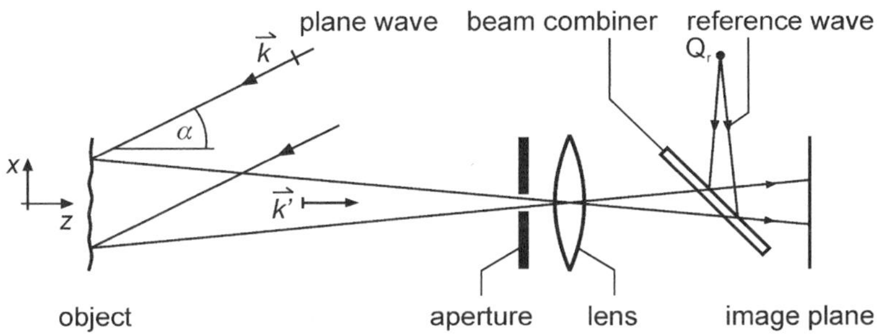

Fig. 8.8 Speckle interferometer to measure object displacements oriented mainly perpendicular to the object surface

fringes mark those object locations where a high correlation exists between the two speckle patterns. In this sense, fringes observed with speckle interferometry are called correlation fringes, in constrast to usual interference fringes.

8.3.2 Speckle Interferometer

The different variants of speckle interferometers distinguish themselves, among other things, by the way in which object displacements are measured. In the set-up shown in Sect. 8.3.1, Fig. 8.7 the object displacements were measured perpendicularly to the object surface—so-called "out-of-plane displacements". In general, the object displacement to be determined depends on the illumination and observation geometry. For the phase change of a scattered wave at a point the following equation is valid, in analogy to (7.16) and Fig. 7.7 in Sect. 7.3.1:

$$\phi = \left(\vec{k} - \vec{k}'\right) \cdot \vec{d}, \tag{8.26}$$

with $\vec{k}$ angular wave vector of the incident light, $\vec{k}'$ angular wave vector of the scattered light in the observation direction and $\vec{d}$ displacement vector.

Relationship (8.26) is a generalization of (8.23). In Fig. 8.7, $\vec{k}$ and $\vec{k}'$ are oriented parallel and antiparallel to the direction of the z-axis, so that, through a displacement of the test object, only the component of $\vec{d}$ in the z-direction causes a phase shift.

(a) Measurement of object displacements perpendicular to the object surface by speckle interferometry.

Figure 8.8 shows schematically how a speckle interferometer is used to measure displacements which are essentially perpendicular to the surface of an object. Using the speckle interferometer as illustrated in Fig. 8.8, a spherical wave is employed as a reference wave, in contrast to the speckle pattern used in the arrangement in Fig. 8.7.

The object is illuminated by a plane wave, which is inclined at an angle α to the z-axis. A lens projects the object onto the image plane. The reference wave is coupled with a beam combiner behind the lens. The aperture in front of the lens and the reference wave are both adjusted in such a way, so that the resulting speckle pattern can be resolved in the image plane, as is required by the selected recording medium—e.g. photographic film or a CCD camera. For the aperture this means that a sufficiently small diameter must be chosen, in order that the speckle dimensions are larger than the smallest still-resolvable structure of the recording medium in the image plane, cf. Sect. 8.1, equation (8.2).

The reference wave is projected initially from a point Q_r. The positions of Q_r and the beam combiner are adjusted so that the reference wave emanates from behind the beam combiner, as it would if it was being projected from the center of the aperture of the lens. In this case, the interference between the reference wave and the imaged speckle pattern leads to bright and dark speckles in the image plane, whose typical dimensions are larger than the smallest speckles caused by the aperture (for the development of the smallest speckles cf. Sect. 8.1). In this way, it can be guaranteed that the resolving ability of the recording medium is not exceeded.

For the angular wave vectors applied in equation (8.26) we have in the arrangement of Fig. 8.8:

$$\begin{aligned} \vec{k} &= (2\pi/\lambda)\,(-\sin\alpha\,\vec{e}_x - \cos\alpha\,\vec{e}_z), \\ \vec{k}' &\approx (2\pi/\lambda)\,\vec{e}_z, \end{aligned} \tag{8.27}$$

where $\vec{e}_x$, $\vec{e}_z$ are the unit vectors in the x-, z-directions.

The relationship given for $\vec{k}'$ applies only approximately if the width of the illuminated object is small in comparison to the distance of the object from the lens. For the displacement vector $\vec{d}$ applies:

$$\vec{d} = d_x\vec{e}_x + d_y\vec{e}_y + d_z\vec{e}_z. \tag{8.28}$$

With the equations (8.26) to (8.28) it follows that:

$$\phi = -(2\pi/\lambda)\,(\sin\alpha\,d_x + (1 + \cos\alpha)d_z). \tag{8.29}$$

For a sufficiently small angle α, the phase difference ϕ following (8.29) is essentially dependent on the object displacement in the z-direction. In the limiting case $\alpha = 0$ (8.29) yields again the expression $\phi = -(2\pi/\lambda)2d_z$ analogous to equation (8.23).

From the distance between neighboring correlation fringes Δx in the x-direction, the displacement gradient $\partial d_z/\partial x$ can be determined approximately. Therefore, we will consider two neighboring points (x_1, y) and (x_2, y) on the object surface. For the local displacements at these points, the following should apply:

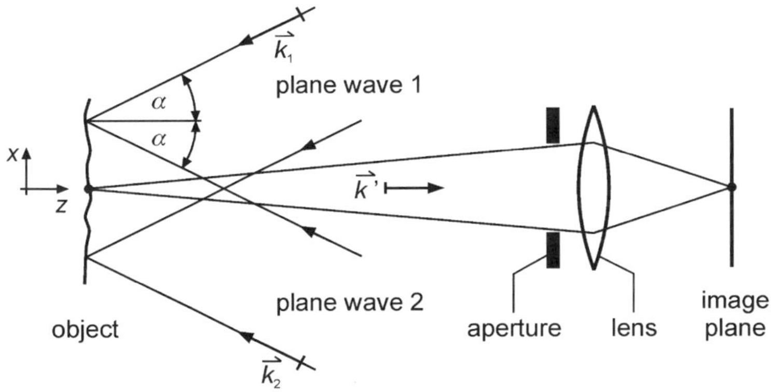

Fig. 8.9 Speckle interferometer to measure object displacements in the plane of the object surface

$$d_z(x_1, y) = \lambda/2, \quad d_z(x_2, y) = \lambda. \tag{8.30}$$

In this case, the phase difference, according to (8.23), between the neighboring correlation fringes, is just 2π. The displacement gradient $\partial d_z/\partial x$ can be expressed, using equations (8.30) in a first approximation, as a differential quotient:

$$|\partial d_z/\partial x| \approx \lambda/(2\Delta x), \tag{8.31}$$

with $\Delta x = |x_2 - x_1|$ fringe spacing in the object plane.

The speckle patterns in the image plane can be recorded on a film or with a CCD camera and evaluated, as previously described in Sect. 8.3.1.

(b) Measurement of object displacements in the object surface by speckle interferometry

Figure 8.9 represents schematically how a speckle interferometer can be used for the measurement of object displacements in the object surface—so-called "in-plane" displacements [7]. The object is illuminated by two plane waves which are inclined at an angle α to the z-axis, and whose angular wave vectors lie in the xz-plane. A lens projects the object onto the image plane.

For the angular wave vectors of the incidents waves holds, cf. Fig. 8.9:

$$\vec{k}_1 = (2\pi/\lambda)\,(-\sin\alpha\,\vec{e}_x - \cos\alpha\,\vec{e}_z), \quad \vec{k}_2 = (2\pi/\lambda)\,(+\sin\alpha\,\vec{e}_x - \cos\alpha\,\vec{e}_z). \tag{8.32}$$

For the angular wave vector in the observation direction, the following approximation is valid:

$$\vec{k}' \approx (2\pi/\lambda)\,\vec{e}_z,$$

cf. (8.27).

A shift of the object leads to a phase change of the scattered waves, relating to both waves $\vec{k}_1$ and $\vec{k}_2$, which, with equation (8.26), can be given by:

$$\begin{aligned}\phi_1 &= \left(\vec{k}_1 - \vec{k}'\right)\cdot\vec{d},\\ \phi_2 &= \left(\vec{k}_2 - \vec{k}'\right)\cdot\vec{d}.\end{aligned} \tag{8.33}$$

If the phase change ϕ at a point on the object surface caused by an object displacement in the field of the wave $\vec{k}_1$ is exactly as large as that in the field of the wave $\vec{k}_2$, or if it differs by a multiple of 2π only, then the speckle pattern in the image plane does not change. This implies that, at those object locations at which the following equation

$$\phi_1 = \phi_2 + n\,2\pi, \tag{8.34}$$

n integer, is satisfied, the speckle patterns agree locally and it is here that their correlation attains a maximum value. From equations (8.32), (8.33) and (8.34) follows

$$\phi_2 - \phi_1 = (2\pi/\lambda)2\sin\alpha\, d_x = -n2\pi. \tag{8.35}$$

Using (8.35) it follows that the object displacement in the x-direction d_x, for which the speckle patterns are identical, is given by:

$$d_x = -n\,\lambda/(2\sin\alpha). \tag{8.36}$$

The phase difference, which is calculated in (8.35), depends only on the x-component of the displacement vector $\vec{d}$. Hence the speckle interferometer from Fig. 8.9 is not sensitive to object displacements in the y- or z-direction. If the local displacement of the object in the x-direction takes the value of (8.36), then the corresponding local speckle patterns are correlated.

The result given in (8.36) can be explained also in another manner. The two incident waves shown in Fig. 8.9 are interfering, cf. Sect. 2.2.3. The waves described by the angular wave vectors $\vec{k}_1$ and $\vec{k}_2$ generate a fringe pattern on the object surface. The fringe spacing in x-direction is given by, cf. Sect. 7.1, equation (7.5):

$$a = \lambda/(2\sin\alpha), \tag{8.37}$$

with a fringe spacing, α angle of incidence, cf. Fig. 8.9.

If the object is locally displaced in the x-direction just by the amount of the fringe spacing or a multiple of that fringe spacing, then obviously all points of that region of the object surface are illuminated in the same way as before. That means, that the local phase of the incident waves has not changed or just changed by an integral multiple of 2π. The speckle patterns are correlated and correlation fringes appear.

From the distance between correlation fringes the displacement gradients within the object surface can be deduced. In the following we consider two neighboring points (x_1, y) and (x_2, y) on the object surface. For the local displacement at these points we take:

$$\begin{aligned} d_x(x_1, y) &= \lambda/(2\sin\alpha), \\ d_x(x_2, y) &= 2\lambda/(2\sin\alpha). \end{aligned} \tag{8.38}$$

In this case, according to (8.36), the speckle patterns are identical at the points (x_1, y) and (x_2, y). At these locations correlation fringes occur. Their mutual distance amounts to $\Delta x = |x_2 - x_1|$. The displacement gradient can be calculated approximately with equations (8.38) as a differential quotient:

$$|\partial d_x/\partial x| \approx \frac{|d_x(x_2, y) - d_x(x_1, y)|}{\Delta x} = \lambda/(2\sin\alpha\,\Delta x), \tag{8.39}$$

with $\Delta x = |x_2 - x_1|$ fringe spacing in the object plane.

With the help of (8.39) and the observed fringe spacing Δx the displacement gradient $\partial d_x/\partial x$ can be determined. In the same way the gradient $\partial d_x/\partial y$ is gained, with Δy being the fringe spacing in the y-direction. If in Fig. 8.9 the incident waves $\vec{k}_1$ and $\vec{k}_2$ are located in the yz-plane instead of the xz-plane then in an analogous way the gradients $|\partial d_y/\partial x|$ and $|\partial d_y/\partial y|$ can be determined.

(c) Measurement of the derivation of object displacements perpendicular to the object surface by speckle interferometry

Figure 8.10 shows a speckle interferometer which can be used to measure the derivation of object displacements perpendicular to the object surface [8, 9]. This variant is also described as shearing interferometer and the method is known as shearography. The characteristic feature of the shearing interferometer is the fact that it is sensitive to local changes of the object surface but it is insensitive to whole-body translations of the test object. For this reason, this type of interferometer has found widespread use.

The test object is illuminated by a laser beam. A lens projects the object surface onto the image plane. A beam splitter and mirrors 1 and 2 are located in the imaging ray path. At first, we will consider mirror 2 only. An object point P_1 is projected onto the point Q in the image plane by way of the beam splitter, mirror 2 and the lens. A second object point P_2, in the neighbourhood of point P_1, is projected likewise onto the point Q via the beam splitter and the slightly inclined mirror 1. The waves, resulting from the projection of the points P_1 and P_2 in the image plane, interfere at the point Q. For the calculation of the phase difference associated with a position

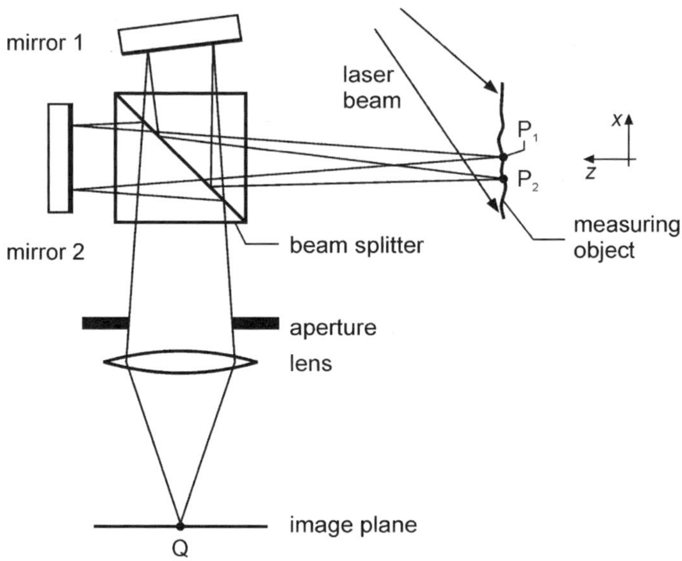

Fig. 8.10 Speckle interferometer to measure the derivative of object displacements oriented perpendicularly to the object surface

change of the object surface, it is assumed, referring to equation (8.26), that the point P_1 is displaced by $\vec{d}(x_{\mathrm{I}})$ and the point P_2 by $\vec{d}(x_{\mathrm{II}})$, where x_{I} and x_{II} are the x-coordinates of these points, see Fig. 8.10. Furthermore, P_1 lies close to P_2 which means that the wave vectors of the incident and scattered waves, $\vec{k}$ and $\vec{k}'$, for both points, are approximately the same. The phases of the waves scattered from points P_1 and P_2 are changed due to the object displacement as follows:

$$\begin{aligned} \phi_{\mathrm{I}} &= \left(\vec{k} - \vec{k}'\right) \cdot \vec{d}(x_{\mathrm{I}}), \\ \phi_{\mathrm{II}} &= \left(\vec{k} - \vec{k}'\right) \cdot \vec{d}(x_{\mathrm{II}}). \end{aligned} \tag{8.40}$$

For the difference of these phase changes, follows:

$$\phi_{\mathrm{I}} - \phi_{\mathrm{II}} = \left(\vec{k} - \vec{k}'\right) \cdot \left(\vec{d}(x_{\mathrm{I}}) - \vec{d}(x_{\mathrm{II}})\right). \tag{8.41}$$

We assume furthermore that the angular wave vectors $\vec{k}$ and $\vec{k}'$ are set opposite to each other and oriented parallel to the z-axis, cf. Fig. 8.10, in order to simplify (8.41):

$$\phi_{\mathrm{I}} - \phi_{\mathrm{II}} = (4\pi/\lambda)\,(d_z(x_{\mathrm{I}}) - d_z(x_{\mathrm{II}})), \tag{8.42}$$

where d_z is the z-component of the displacement vector $\vec{d}$.

Table 8.1 Properties of the different speckle interferometers; d_z refers to object translations perpendicular to the object surface and d_x to those lying within the object surface, cf. coordinate systems in Figs. 8.8, 8.9 and 8.10, s_x image shear in the x-direction

Speckle interferometer for measurement of object translations	Correlation fringes correspond to locations, where the following quantities are constant	Distance between neighboring correlation fringes in x-direction
Out-of-plane	d_z	$\frac{\lambda}{2}\frac{1}{\left\lvert\frac{\partial d_z}{\partial x}\right\rvert}$
In-plane	d_x	$\frac{\lambda}{2}\frac{1}{\sin\alpha\left\lvert\frac{\partial d_x}{\partial x}\right\rvert}$
Shearing	$\frac{\partial d_z}{\partial x}$	$\frac{\lambda}{2}\frac{1}{\left\lvert\frac{\partial^2 d_z}{\partial x^2}\right\rvert s_x}$

At the locations at which the difference of the phase changes is an integral multiple of 2π, the speckle patterns are identical. At these positions, correlation fringes occur.

With $\partial d_z/\partial x$, the derivative of the z-component of the displacement vector with respect to the x-coordinate is described. This derivative is given approximately by:

$$\frac{\partial d_z}{\partial x} \approx \frac{d_z(x_{\mathrm{I}}) - d_z(x_{\mathrm{II}})}{x_{\mathrm{I}} - x_{\mathrm{II}}}. \tag{8.43}$$

Inserting the relation (8.43) into (8.42) leads to:

$$\phi_{\mathrm{I}} - \phi_{\mathrm{II}} \approx \frac{4\pi}{\lambda}\frac{\partial d_z}{\partial x} s_x, \tag{8.44}$$

where $s_x = x_{\mathrm{I}} - x_{\mathrm{II}}$ is the image shear in the x-direction in the object plane.

If the phase difference $\phi_{\mathrm{I}} - \phi_{\mathrm{II}}$ is an integral multiple of 2π, then the speckle patterns will locally agree. The correlation fringes correspond to those locations at which $\partial d_z/\partial x$ is constant.

In analogy to the deduction in the previous sections (a) and (b), the curvature $\partial^2 d_z/\partial x^2$ can be calculated approximately from the fringe spacing:

$$\left\lvert\frac{\partial^2 d_z}{\partial x^2}\right\rvert \approx \frac{\lambda}{2}\cdot\frac{1}{s_x \Delta x}. \tag{8.45}$$

In Table 8.1 the properties of the described interferometer types are summarized.

8.3.3 Electronic Speckle Interferometry

With electronic speckle interferometry—abbreviated to ESPI—speckle patterns are recorded by CCD cameras instead of using photographic film, cf. Sects. 5.5.5 and 5.5.6) [10]. The principle of the various possible types of speckle interferometers which can be used for ESPI has already been explained in the previous section.

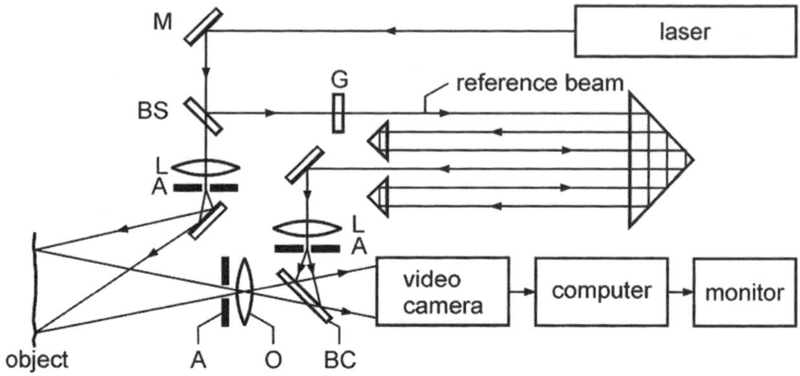

Fig. 8.11 Set-up of an electronic speckle interferometer. M mirror, BS beam splitter, L lens, G grey filter, A aperture, O objective, BC beam combiner. The video camera is operated without objective

The set-up of a typical electronic speckle interferometer, including the laser source and beam path, is represented by Fig. 8.11. It is based on the principle already shown in Fig. 8.8. A laser beam is split into a reference beam and an illumination beam. The illumination beam is directed at the object while the reference beam is passed through an optical delay line. This serves to equalize the different pathlengths of the reference and illumination beams, so that both are able to interfere, within the coherence length of the laser radiation, at the beam combiner located behind the objective.

The object is projected onto the light-sensitive detector with the aid of an objective. The reference wave is coupled into the beam path via a beam combiner and is superimposed with the wave scattered from the object. As was already described in Sect. 8.3.2, the reference wave and the aperture of the objective are chosen so that the speckle pattern can be resolved by the camera. The video signal is electronically processed and the image information is buffered. The resulting image data are subsequently displayed on a monitor.

In order to measure object displacements, the speckle patterns of the object in the initial state and in the final state are detected and stored. Following this, the video signals of both images are subtracted from one another and displayed again on the monitor. Another variant exists in the fact that the stored image is subtracted in real-time from the actual video signal of the camera. The regions of the image, in which the speckle patterns have not changed, yield a negligible difference signal, which appears as a dark zone on the monitor. If, however, the speckle pattern had changed locally, then a large difference signal would result. These regions appear on the monitor as bright areas.

In the following, the signal processing which accompanies electronic speckle interferometry should be considered closely. The intensity distribution of the displayed speckle patterns can be described by equations (8.22) and (8.24) from Sect. 8.3.1:

$$I_1 = I_m + I_r + 2\sqrt{I_m I_r}\cos(\vartheta), \tag{8.46}$$

$$I_2 = I_m + I_r + 2\sqrt{I_m I_r}\cos(\vartheta + \phi), \tag{8.47}$$

with I_m intensity distribution of the object wave in the image plane, I_r intensity distribution of the reference wave in the image plane, ϑ spatial statistically varying phase in the image plane, ϕ phase difference due to a local object displacement between the two recorded speckle patterns, cf. (8.23).

The voltage of the video signal is proportional to the intensity of the speckle pattern:

$$U \propto I, \tag{8.48}$$

where U is the voltage of the video signal.

For the subtraction of the video signals of both images, with equations (8.46) and (8.47), follows:

$$U = U_1 - U_2 \propto I_1 - I_2 \propto 4\sqrt{I_m I_r}\sin(\vartheta + \phi/2)\sin(\phi/2). \tag{8.49}$$

Equation (8.49) can attain both positive and negative values. Negative voltages would be represented on the monitor as dark regions. In order to prevent this, the signal is previously rectified. The brightness on the monitor is then proportional to the amount of U. This gives:

$$H = \beta\sqrt{I_m I_r \sin^2(\vartheta + \phi/2)\sin^2(\phi/2)}, \tag{8.50}$$

with H brightness on the monitor and β proportionality constant.

The distribution of brightness on the monitor is subject to statistical fluctuations, which occur because of the statistical spatial variation of the intensity I_m and of the phase ϑ. The phase difference ϕ changes only negligibly over the range of a few neighboring pixels of the CCD-chip. These statistical fluctuations can be eliminated using a spatial averaging process:

$$\langle H \rangle = \beta\left\langle\sqrt{I_m I_r}\right\rangle (2/\pi)\sqrt{\sin^2(\phi/2)}, \tag{8.51}$$

where the brackets "< >" denote the spatial averaging and $(2/\pi)$ is the average value of the function $\sqrt{\sin^2\vartheta}$ in the angle interval $[0, \pi]$.

The averaged brightness, dependent upon phase change, can attain different values. It follows that:

$$\begin{aligned} \langle H \rangle &= \beta\left\langle\sqrt{I_m I_r}\right\rangle (2/\pi), && \text{for } \phi = (2n+1)\,\pi,\ n = 0, 1, 2, \ldots, \\ \langle H \rangle &= 0, && \text{for } \phi = 2n\pi. \end{aligned} \tag{8.52}$$

Following equation (8.52), dark areas appear if the phase change between the recordings amounts to a multiple of 2π, cf. Sect. 8.3.1.

The phase cannot be unequivocally determined from the brightness using relationship (8.51). In order for it to be correctly determined, the phase shift method is again used, in analogy with holographic interferometry, cf. Sect. 7.3.3. In the latter, the additional phase change, which is necessary for the phase shift process, can be achieved through the reconstruction of the hologram, cf. Fig. 7.10, right. In contrast to this, the correlation fringe pattern in electronic speckle interferometry is generated after the subtraction of the second speckle pattern from the first. Therefore, an additional phase displacement cannot be introduced at this step, it is instead carried out when the speckle pattern I_1, is recorded [11]. Relationship (8.26) and the known phase lead to the calculation of the displacement field. An example using the method of subtraction of two speckle patterns and the obtained displacement field, in electronic speckle interferometry, is introduced in Sect. 8.4.

8.3.4 Time-Averaged Vibration Electronic Speckle Interferometry

The so-called time-averaging process is suitable for an investigation of vibrations with the help of speckle interferometry [12]. The set-up corresponds to Fig. 8.11. For simplification we assume that the observation is undertaken in a direction opposite to that of the illumination; this corresponds to the situation $\alpha = 0$ in Fig. 8.8. The object exhibits harmonic oscillation. A time-varying speckle pattern is the result at the input of the video camera. It is assumed, in the following, that the period of oscillation is small in comparison with the video camera's integration. This has the consequence that the speckle patterns—contrary to the case outlined in the previous section—are added on the camera detector. The oscillation at an object location is described through the following relationship:

$$s(t) = s_0 \sin \omega t, \tag{8.53}$$

with $s(t)$ local elongation, s_0 local amplitude, ω angular frequency of oscillation, t time.

If this elongation occurs in a direction parallel to the illumination and observation directions of the speckle pattern, then, for the associated phase displacement $\phi(t)$, the following is valid:

$$\phi(t) = \frac{4\pi}{\lambda} s(t). \tag{8.54}$$

The time-varying speckle pattern, with equation (8.47), is given by:

$$I(t) = I_m + I_r + 2\sqrt{I_m I_r}\cos\left(\vartheta + \frac{4\pi}{\lambda}s(t)\right). \tag{8.55}$$

These speckle patterns would be integrated within the integration time of the video camera, so for the video signal U we have:

$$U \propto \frac{1}{T}\int_0^T I(t)\mathrm{d}t, \tag{8.56}$$

where T is the integration time of the video camera.

Substitution of equation (8.55) in (8.56), assuming $T \gg 2\pi/\omega$, yields:

$$U = \gamma\left(I_m + I_r + 2\sqrt{I_m I_r}\cos\vartheta\, \mathrm{J}_0(4\pi s_0/\lambda)\right), \tag{8.57}$$

with γ proportionality constant, J_0 zeroth order Bessel function.

The features of the Bessel function will be discussed later. For reasons of the statistically varying phase ϑ in equation (8.57), a change of contrast is to be expected on the monitor, due to the object oscillation. To improve the monitor display, the following procedure is undertaken:

1. averaging of the signal in equation (8.57):

$$\langle U\rangle = \gamma(\langle I_m\rangle + \langle I_r\rangle),$$

2. subtracting the averaged signal from the actual signal value:

$$U - \langle U\rangle,$$

3. averaging the modulus of the signal difference, cf. (8.50):

$$\langle|U - \langle U\rangle|\rangle = \gamma\left(\sigma_m^2 + \sigma_r^2 + 2\langle I_m\rangle\langle I_r\rangle \mathrm{J}_0^2(4\pi s_0/\lambda)\right)^{1/2},$$

with $\sigma_m^2 = \langle I_m^2\rangle - \langle I_m\rangle^2$, $\sigma_r^2 = \langle I_r^2\rangle - \langle I_r\rangle^2$.

Hence we obtain for the brightness on the monitor:

$$\langle H\rangle = \gamma\left(\sigma_m^2 + \sigma_r^2 + 2\langle I_m\rangle\langle I_r\rangle \mathrm{J}_0^2(4\pi s_0/\lambda)\right)^{1/2}. \tag{8.58}$$

Figure 8.12 represents the function $\mathrm{J}_0^2(x)$, where $x = 4\pi s_0/\lambda$. If the Bessel function has its values such that a zero is assumed, dark fringes appear. The first four zeros are approximately:

$$\frac{4\pi}{\lambda}s_0 = 2.4;\ 5.5;\ 8.7;\ 11.4. \tag{8.59}$$

A dark correlation fringe corresponds to those locations which have constant local vibrational amplitude. For $s_0 = 0$, the Bessel function J_0 has its maximum value, i.e. following equation (8.58) the brightest regions exist at the nodal lines. Due to the features of the Bessel function, the contrast between the correlation fringes falls off with increasing amplitude.

8.4 Examples of Applications

The following examples relate to electronic speckle interferometry. The principle set-up used is shown in Fig. 8.11. Figure 8.13a shows a sketch of a holding and adjustment device of a laser mirror for a CO_2 high-power laser. A photograph of this component is reproduced in Fig. 8.13b. Due to the high thermal loading of the laser mirror, this holder also comprises a cooling cycle. The deformation of the holder caused by the pressure of the cooling water was investigated using electronic speckle interferometry. A speckle image is recorded, by the method described in Sect. 8.3.3, without the pressure loading of the cooling water. A second speckle image is taken after the pressure loading of the cooling water is applied. The subtraction of the two images leads to the fringe pattern represented in Fig. 8.13c. Figure 8.13d shows the result of the calculated displacement field in a pseudo-3D representation. Over the component coordinates (x, y) the displacement field is plotted in the z-direction in a magnified way. The maximum displacement over the diameter of the holding and adjustment device amounts to ca. 3 μm. There appears locally, in the region of the circular opening of the holding device for the mirror, an increased deformation.

Another application example is the inspection of foam parts, such as e.g. dashboards of modern automobiles or car door handles consisting of an aluminum or plastic supporting base and a moulded skin behind which a polyurethane foam is expanded. Inclusions of air in the foam are typical production flaws. They generate voids situated underneath the skin. The flaws become visible as blisters if the automobile is subjected to intense solar irradiation. The air trapped in the void is heated and expands. The blister appearing on the surface detracts from the visual appearance of the component. Such types of defects can be detected in a production line by speckle interferometry since even small deformations in the micrometer level can be made visible. Figure 8.14 illustrates the different types of stress applied for the inspection of materials and components. A heat load, mechanical force or change of ambient pressure can induce an expansion of the flaw in the elastic material leading to a slight deformation of the surface of the component.

Figure 8.15 shows schematically the sequence of material inspection with speckle interferometry. The specimen is positioned in the laser's field of irradiation, see Fig. 8.15a. Then the component to be inspected is put under stress so that a minute elastic deformation occurs, cf. Fig. 8.15b. The speckle interferometer detects the irregularities of the surface alteration caused by flaws in the material,

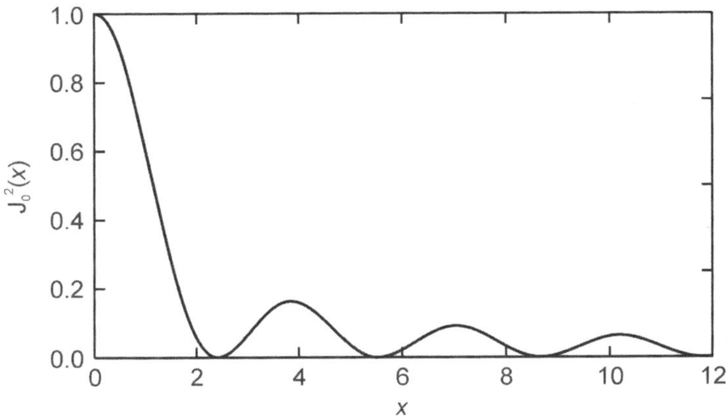

Fig. 8.12 The function $J_0^2(x)$

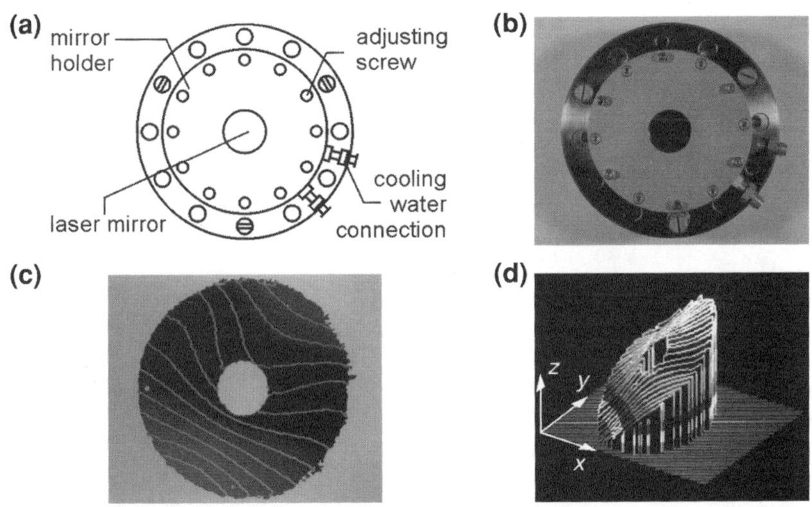

Fig. 8.13 **a** Illustration of the mirror holder of a high-power CO_2 laser mirror, **b** photograph of the mirror holder, **c** speckle interferogram under pressure load of the cooling water, **d** deformation field

cf. Fig. 8.15c. Finally, the measured signals are processed by a computer to yield an inspection decision and are documented in an inspection report, cf. Fig. 8.15d. Typical inspection times are in the order of a view seconds. The laser radiant flux necessary for the illumination of the component depends on the size and surface quality of the specimen. Light or white component surfaces scatter light better than dark or black surfaces. Therefore the necessary irradiance varies typically between 0.1 and 2 mW/cm^2.

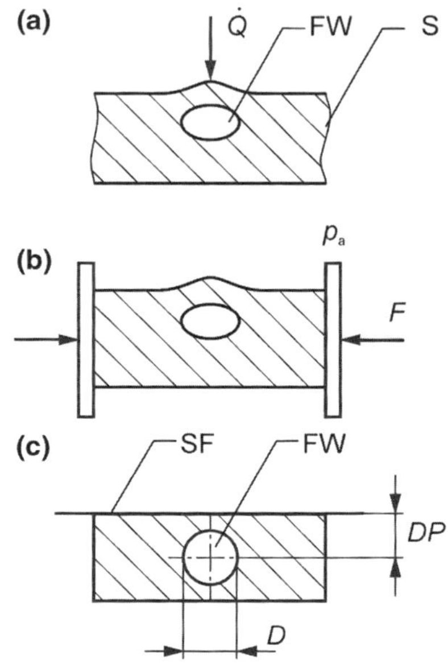

Fig. 8.14 Stress types applied to inspect internal flaws—here illustrated as void—in elastic materials with speckle interferometry: **a** thermal stress by means of a heat flow $\dot{Q}$, **b** mechanical stress by applying a force F or changing the ambient pressure p_a. Cross section **c** illustrates an equivalent void with spherical shape and the quantities *DP* and *D* to define the nondimensional parameter *P*, see text. FW flaw, S specimen, SF surface of measuring object, *DP* depth, *D* diameter

Figure 8.16 shows the determination of defects in a car door handle [13]. A view of a part of this door handle is shown in Fig. 8.16a. Essentially, the handle consists of a metal part, polyurethane foam and a film coating as described above. The handle is positioned in a pressure chamber and observed using the speckle interferometer according to Fig. 8.11. First, a speckle image of the unloaded handle is recorded. Then, it is exposed to a pressure change in the chamber, where upon the surface of the handle is deformed slightly in the regions of the voids. The resulting speckle pattern is compared with that obtained first. Figure 8.16b shows the result. The regions on the surface of the handle in which many correlation fringes are visible indicate the position of a defect.

With image processing methods, it is possible to automatically indicate voids by evaluation of the fringe information [13]. Figure 8.17 illustrates the geometrical quantities used for the evaluation of the image signals. A calibration process defines a limit *LT* to mark the found defect in such a way that the area of the marking corresponds to the area *A* of the flaw projected onto the surface of the specimen.

Figure 8.16c shows the result obtained using this method of examination. The detected flaw is indicated as black spot superimposed with the camera image of the car door handle.

The calibration is carried out with test pieces in which pre-defined voids are present. The material of these test pieces is chosen to have a similar deformation behaviour as the specimens to be inspected in the production line. Evaluating a sequence of speckle interferograms while reducing the pressure load step-by-step

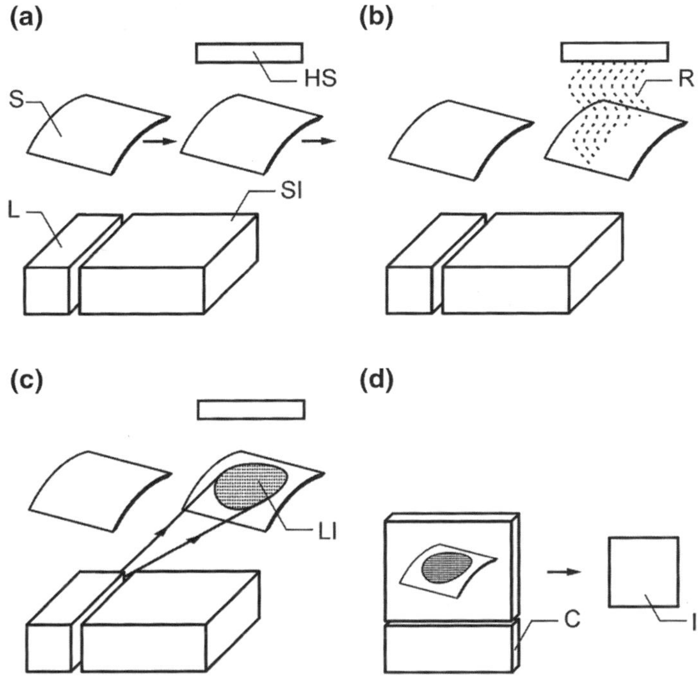

Fig. 8.15 Inspection sequence for flaw detection in foam parts with speckle interferometry. **a** positioning of the measuring object, **b** putting the sample under stress, **c** illumination of the workpiece with an expanded laser beam and measurement with a speckle interferometer, **d** signal evaluation, inspection decision and documentation. S specimen, HS heat source, L laser source, SI speckle interferometer, R infrared radiation, LI laser irradiation, C computer, I inspection report

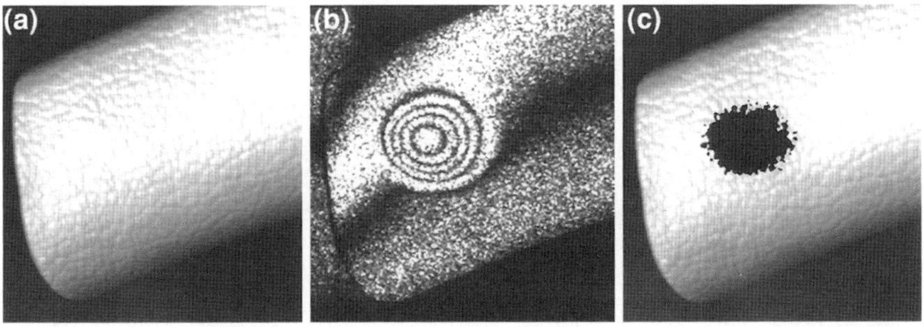

Fig. 8.16 **a** Partial view of a car door handle, **b** speckle interferogram of the handle with a pressure loading, **c** examination result following an automatic defect recognition and assessment

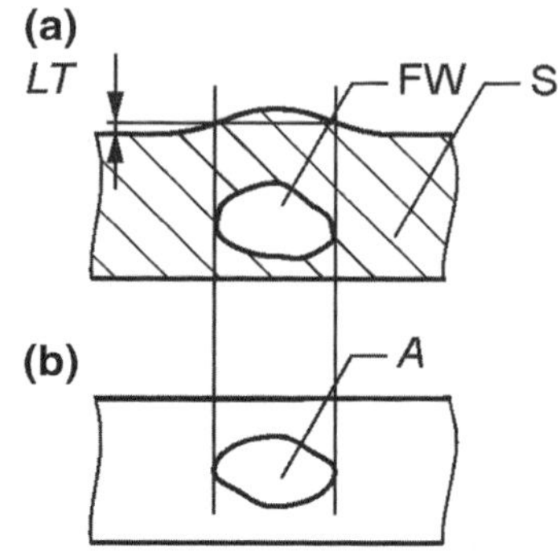

Fig. 8.17 Determination of the projected area of a flaw. **a** cross section of a workpiece with a flaw shown as void, **b** *top view* of the workpiece with the area of the void projected onto the surface. *LT* limit, FW flaw, S specimen, *A* projected area of the flaw at the surface of the workpiece

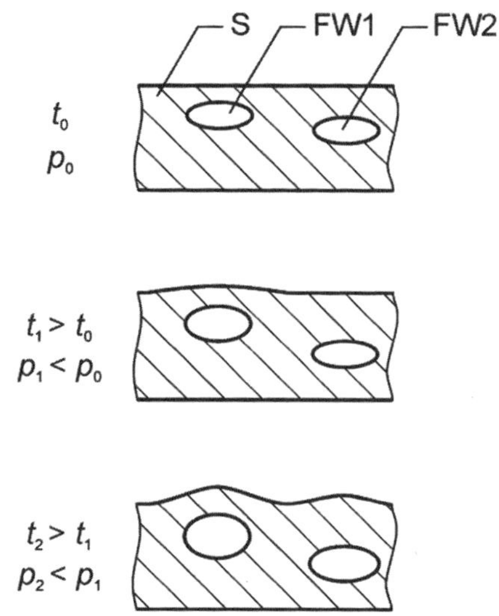

Fig. 8.18 Cross-sectional view of a workpiece with voids in different depths illustrating that a detectable surface deformation corresponding to a deeper lying flaw shows up at a further reduced ambient pressure. S specimen; FW1, FW2 flaws; t_0, t_1, t_2 time; p_0, p_1, p_2 ambient pressure

allows to indicate the depth of the voids below the component surface. Figure 8.18 shows as example a component with two voids at different depths. The component is put under stress by reduction of the pressure, i.e. from p_0 to p_1. Around flaw FW1 a deformation of the surface is already detectable. Flaw FW2 can only be seen under a larger underpressure load p_2. With an appropriate calibration of the inspection system the distance of a flaw from the surface can be estimated in this way.

The detection limit for flaws depends on the material properties, the position and form of the flaws as well as on the viewing angle of the speckle interferometer. In a simplified scheme, the form of the void is approximated by a sphere, see Fig. 8.14c. The diameter of the sphere is defined in such a way that the volume V of the void in question corresponds to the volume of the sphere with diameter D. The distance of the void from the center to the surface is denoted as DP, cf. Fig. 8.14c. A nondimensional parameter P is defined by $P = D/DP$. The results of

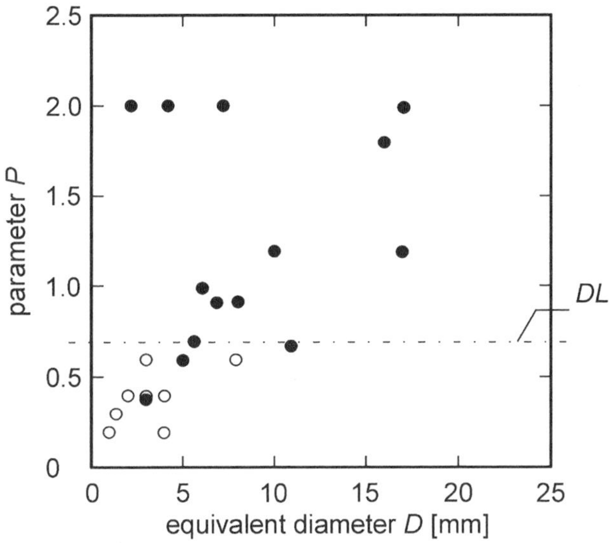

Fig. 8.19 Nondimensional parameter P as a function of the equivalent diameter D, cf. Fig. 8.14c. Filled circles show voids detected by speckle interferometry. DL detection limit

the test measurements are shown in Fig. 8.19. The parameter P is plotted as a function of the diameter D. The flaws detected by speckle interferometry are shown as filled circles.

The open circles are voids which could not be shown clearly. Small flaws with an equivalent diameter of 2 mm situated directly beneath the skin are detected corresponding to P-values of 2. If the voids are lying lower, corresponding to $P < 0.6$, they are no longer detected. The industrial requirement for this application is the detection of flaws of a diameter of $D > 8$ mm up to a depth of $DP = 9$ mm corresponding to $P = 0.9$. Figure 8.19 shows that all voids with $P > 0.7$ can be detected. The quantity $DL = 0.7$ is called detection limit.

The inspection of tires is another application field of shearing speckle interferometry, see Fig. 8.20 [14]. The tire is placed on a rotating disk. The speckle interferometer is placed inside the tire to be inspected. Step by step the whole internal surface of the tire is rastered while the tire undergoes a pressure load and is turned. The tire is placed inside a vacuum chamber to exert pressure differences in the order of a few 10 mbar to several 100 mbar [15]. Figure 8.21 shows a photograph taken during inspection of a tire with a shearing speckle interferometer, cf. Fig. 8.10.

The following materials can be inspected, e.g.: rubber compound, sandwich material, glass-reinforced plastic laminates, foam composites. Detected defects are for example debondings, material delaminations, cracks, impacts, indents, material inclusions.

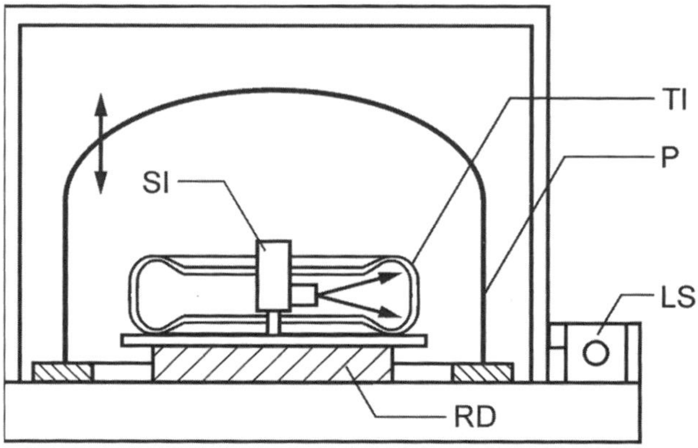

Fig. 8.20 Set-up of an inspection machine to detect flaws in tires. SI speckle interferometer, TI tire, P pressure chamber, LS laser source, RD rotating disk

Fig. 8.21 Inspection of the internal surface of tires with a shearing interferometer

Fig. 8.22 Fringe pattern of a harmonically oscillating membrane using the time-averaged ESPI method

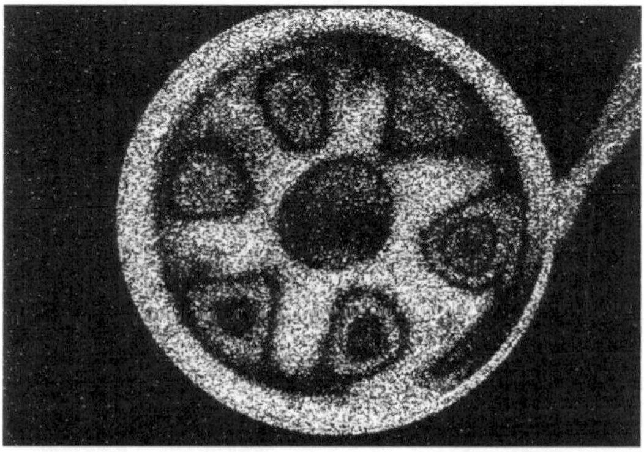

In the next example, a stationary vibration of a membrane was investigated using the time-averaged vibration electronic speckle interferometry method, cf. Sect. 8.3.4. Figure 8.22 shows the result. The bright regions correspond to the nodal lines, the dark regions to those locations with equal local vibrational amplitude. A six-fold symmetry of vibration is observed. With the change of excitation frequency and amplitude of the membrane, the different stationary forms of vibration can be observed online on the monitor. The frequencies of resonances and the associated vibration patterns are determined in this manner using the time-averaged ESPI method.

The magnitude of the local vibrational amplitude at the location of the dark fringes is calculated with equation (8.59), given in Sect. 8.3.4.

References

1. W. Goodman, Statistical properties of laser speckle patterns, in *Laser Speckle and Related Phenomena*, ed. by J. Dainty (Springer, Berlin, 1975)
2. K. Hinsch, Lasergranulation. Physik in unserer Zeit **23**, 59–66 (1992)
3. R. Erf (ed.), *Speckle Metrology*. (Academic Press, New York, 1978)
4. R. Jones, C. Wykes, *Holographic and Speckle Interferometry* (Cambridge University Press, Cambridge, 1989)
5. F. Chiang, in *Speckle Metrology*. Metals Handbook, 9th edn., vol. 17, Non-Destructive Evaluation and Quality Control (Metals park, Ohio, 1989)
6. H. Tiziani, Kohärent-optische Verfahren der Oberflächenmesstechnik. Tech. Mess. **58**, 228–234 (1991)
7. J. Leendertz, Interferometric displacement measurement on scattering surfaces utilizing speckle effect. J. Phys. E: Scient. Instr. **3**, 214–218 (1970)
8. J. Leendertz, J. Butters, An image-shearing speckle-pattern interferometer for measuring bending moments. J. Phys. E: Scient. Instr. **6**, 1107–1110 (1973)
9. Y. Hung, A speckle-shearing interferometer: a tool for measuring derivatives of surface dispalcements. Opt. Comm. **11**, 132–135 (1974)
10. J. Butters, J. Leendertz, Speckle pattern and holographic techniques in engineering metrology. Opt. Laser Technol. **3**, 26–30 (1971)
11. K. Creath, Phase-shifting speckle interferometry. Appl. Opt. **24**, 3053–3058 (1985)
12. P. Hariharan, *Optical Interferometry*, 2nd edn. (Academic Press, London, 2003)
13. R. Noll, M. Krauhausen, Zerstörungsfreie Prüfung von Verbundmaterialien mit dem Laser. Kunststoffe **83**, 116–120 (1993)
14. R. Noll, Speckle-Interferometrie, Praxis der Naturwissenschaften 7/43. Jg.(1994)37–44
15. http://www.steinbichler.co.uk/downloads-us/category/11-intact.html, September 25, 2013

Chapter 9
Optical Coherence Tomography: OCT

Abstract Optical coherence tomography (OCT) allows to measure geometric features inside translucent objects such as organic materials or organic tissues. The principle is based on interferometry. Time-domain and Fourier-domain methods are applied to make visible differences in the refractive index profile of the measuring object. We present the set-up of OCT sensors and show examples of applications such as 3D-imaging of an eye, B-scans of blood vessels or thickness measurements of multi-layer polymer films.

Optical coherence tomography (OCT) is an interferometric method to measure geometric features inside an object such as an organic tissue or an organic material like a polymer [1, 2]. The light used in OCT is able to propagate inside the material. According to the wavelength-dependent absorption and scattering properties the penetration depth is in the range of a few 100 μm up to several millimeters.

In Sect. 9.1 the principles of the two main variants of OCT methods will be presented. The set-up of OCT sensors is described in Sect. 9.2. OCT is applied as a medical imaging technique and here especially in ophthalmology. Industrial applications are found e.g. in non-destructive testing, thickness measurement of silicon wafers and polymer films, cf. Sect. 9.3.

9.1 Principle

The basic set-up for OCT is shown in Fig. 9.1. A light source, which emits light with a low coherence length, is coupled into a Michelson interferometer, cf. Sect. 6.1.2. The beam is split into reference and measuring beam. The mirror in the reference path can be shifted. The beam in the measuring path irradiates the measuring object and penetrates partially into the object. Inside the object the light is attenuated and scattered depending on the chosen wavelength and the properties of the organic object material. At interfaces between regions of different refractive indices the light is reflected. A part of this radiation propagates in the backward

A. Donges and R. Noll, *Laser Measurement Technology*, Springer Series in Optical Sciences 188, DOI: 10.1007/978-3-662-43634-9_9,

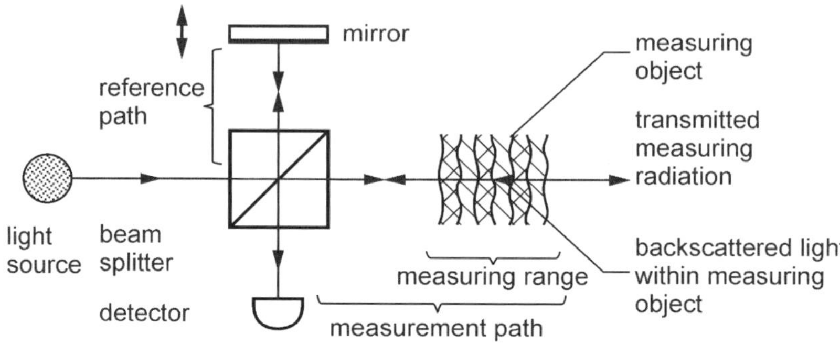

Fig. 9.1 Basic set-up for optical coherence tomography

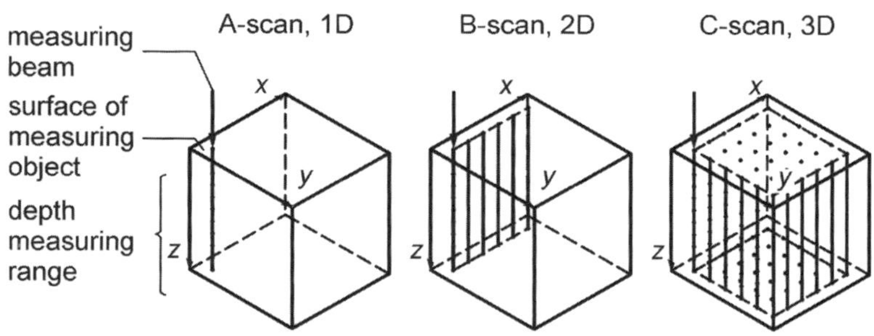

Fig. 9.2 Scan modes for optical coherence tomography

direction and another part—depending on the thickness of the object—is leaving the object as transmitted radiation on the backside.

Radiation from the reference and measuring path are combined at the beam splitter and the detector registers the interference signal, if the optical path difference between both signals is smaller than the coherence length of the applied light, cf. Sect. 2.4.2, Fig. 2.19. Since the coherence length was chosen to be small, for most scattering centers of the object this path difference will be greater than the coherence length. As a consequence only an incoherent superposition occurs. The reference mirror is now translated to match the reference path length to a defined measuring path length thus yielding a coherent superposition of both beams for a defined depth inside the measuring object. Now a high modulation degree of the superimposed beams in the sense of a high interference fringe visibility is detected. With a step-by-step translation of the reference mirror and its known position depth scans are generated.

Such depth scans are called A-scans a terminology taken over from ultrasonic methods [3]. They correspond to a one dimensional measurement. By a lateral relative motion between the measuring beam and the object two and three-dimensional images of translucent objects can be obtained. Figure 9.2 illustrates these A-, B- and C-scans.

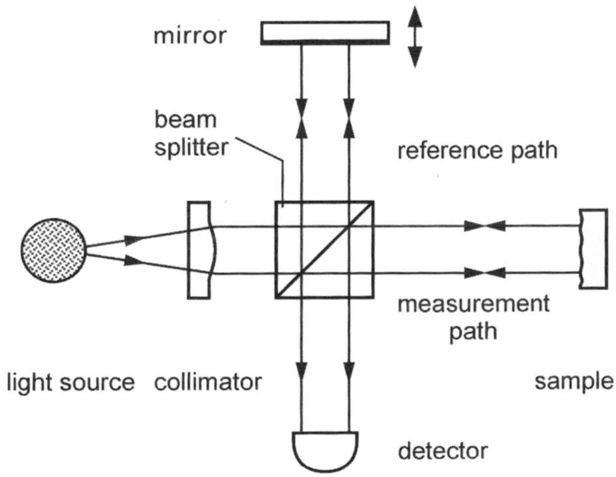

Fig. 9.3 Principle of time-domain OCT. The reference mirror is shifted while the detector registers the time-dependent interference signal, cf. Fig. 9.4

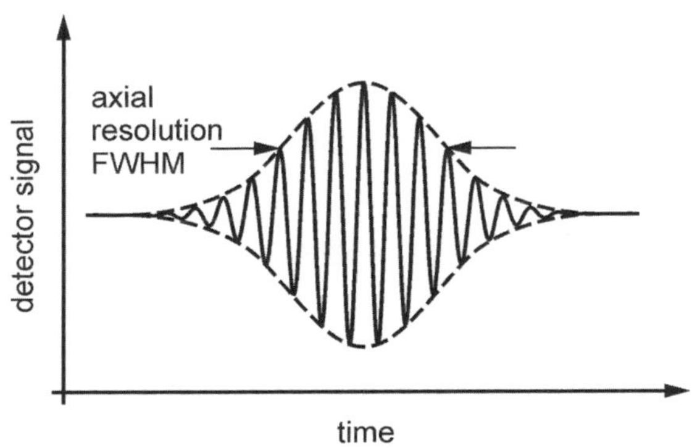

Fig. 9.4 Detector signal as a function of time for time-domain OCT and a single backscattering surface of the measuring object

9.1.1 Time-Domain OCT

Figure 9.3 shows the principle of the so-called *time-domain OCT* (TD-OCT) which is similar to the one described in the previous section. The light used has a spectrum with e.g. a full width at half maximum (FWHM) of $\Delta\lambda_{\mathrm{FWHM}} = 100$ nm. The reference path length is changed by shifting the mirror position, see double arrow in Fig. 9.3 and the detector registers the time-dependent interference signal.

This signal is shown in Fig. 9.4. The envelope of the interference pattern has a FWHM temporal width Δt—assuming a constant velocity v_{R} of the reference mirror—which follows the relation:

$$v_{\mathrm{R}} \cdot \Delta t \cdot 2 = L_{\mathrm{c}}, \tag{9.1}$$

where v_{R} is the velocity of the reference mirror and L_{c} is the coherence length. The left hand side describes the optical path difference occurring, when the reference mirror moves with v_{R} for a time Δt. The factor 2 takes into account that the light travels the reference path in forward and backward direction. The coherence length determines the axial resolution for the measurement of the distance to the sample.

The coherence time depends on the angular frequency width of the light source, cf. (2.29) in Sect. 2.4.2, $\Delta\omega T_c \approx \pi$ from which follows $2\pi\Delta f T_c \approx \pi$ and finally:

$$T_c \approx \frac{1}{2\Delta f}, \tag{9.2a}$$

where Δf is the frequency width of the light source. For the coherence length holds:

$$L_c = cT_c \approx \frac{c}{2\Delta f} = \frac{\pi}{\Delta k}, \tag{9.2b}$$

where

$$k = \frac{2\pi}{\lambda} \tag{9.3}$$

is the angular wave number and λ the wavelength. From $c = \lambda f$, cf. (2.4) in Sect. 2.1, follows $|\Delta f| = \frac{c}{\lambda^2}|\Delta\lambda|$ for sufficiently small values of Δf and $\Delta\lambda$, so the coherence length and axial resolution is given by:

$$L_c = \frac{1}{2}\frac{\lambda^2}{\Delta\lambda}. \tag{9.4}$$

For the following calculation we assume as origin for the considered waves the center of the beam splitter, cf. Fig. 9.3. Here the spatial phases of reference and measurement wave are set to be zero. This is an arbitrary assumption without loss of generality. Any other computation point could also be taken, e.g. the location of the detector. The only difference would be a constant spatial phase. Then for the electrical field of the reference wave reflected at the reference mirror with angular wave number k holds the following relation in complex notation:

$$E_r = E_0 e^{i(\omega t - 2kz_r)}, \tag{9.5}$$

with z_r describing the position of the reference mirror with respect to the center of the beam splitter. Assuming a reflection coefficient r_m for the measuring object, we have for the electrical field coming from the measuring arm:

$$E_m = r_m E_0 e^{i(\omega t - 2kz_m)}, \tag{9.6}$$

with z_m denoting the position of the object where the reflection takes place. The intensity of the resultant wave at the detector is:

$$I = I_m + I_r + 2\sqrt{I_m I_r}\cos[2k(z_r - z_m)]. \tag{9.7}$$

Relations (9.5)–(9.7) are generalized for a spectrum of angular wave numbers by replacing the amplitude E_0 by the spectral amplitude $E_{0,k}$ referring to angular wave numbers as follows:

$$E_0 \rightarrow E_{0,k}. \tag{9.8}$$

Equations (9.5) and (9.6) become then:

$$E_{\mathrm{r},k} = E_{0,k}\mathrm{e}^{\mathrm{i}(\omega t - 2kz_{\mathrm{r}})} \tag{9.9}$$

and

$$E_{\mathrm{m},k} = r_{\mathrm{m}}E_{0,k}\mathrm{e}^{\mathrm{i}(\omega t - 2kz_{\mathrm{m}})}, \tag{9.10}$$

where $E_{\mathrm{r},k}$ and $E_{\mathrm{m},k}$ are the complex spectral field strengths of both light waves—reference beam and measuring beam—at the detector, respectively.

Now we have to consider that not only a single reflection at the object surface but several reflections at different depths $z_{\mathrm{m},j}$ within the object occur. In the simple case of a set of discrete reflection layers, the reflection coefficient per unit length can be described as follows:

$$r_{\mathrm{m},z_{\mathrm{m}}}(z_{\mathrm{m}}) = \sum_{j=1}^{N} r_{\mathrm{m},j}\delta(z_{\mathrm{m}} - z_{\mathrm{m},j}), \tag{9.11}$$

where $r_{\mathrm{m},j}$ is the reflection coefficient at the position $z_{\mathrm{m},j}$ and $\delta(z_{\mathrm{m}} - z_{\mathrm{m},j})$ is the Dirac function. For the Dirac function holds:

$$f(a) = \int_{-\infty}^{+\infty} f(x)\delta(x-a)\,\mathrm{d}x, \tag{9.12}$$

with $f(x)$ arbitrary function (in a rigorous description $f(x)$ is a continuous compactly supported function), x integration variable, a constant. If x has the unit of a length and $f(x)$ has no dimension, then it follows from (9.12) that the unit of the Delta function is the inverse of the length unit. Hence $r_{\mathrm{m},z_{\mathrm{m}}}(z_{\mathrm{m}})$ in (9.11) is a reflection coefficient per unit length. Hence the index "z_{m}" was added to r_{m}.

Assuming $z_{\mathrm{m},j} > 0$ and using (9.11), (9.12) we obtain for the spectral electric field strength of the beam traveling backward to the beam splitter (9.10) the following expressions:

$$E_{\mathrm{m},k} = E_{0,k}\int_{0}^{\infty} \mathrm{i}(\omega t - 2kz_{\mathrm{m}})\left(\sum_{j=1}^{N} r_{\mathrm{m},j}\delta(z - z_{\mathrm{m},j})\right)\mathrm{d}z_{\mathrm{m}}, \tag{9.13}$$

$$E_{\mathrm{m},k} = E_{0,k}\sum_{j=1}^{N} r_{\mathrm{m},j}\mathrm{e}^{\mathrm{i}(\omega t - 2kz_{\mathrm{m},j})}. \tag{9.14}$$

For relation (9.14) we have implicitly assumed that the partial reflections or any other attenuations of the measuring radiation are small and can be neglected. This is not a necessary condition to apply OCT, but it simplifies the representation chosen here to a large extent.

With (9.9) and (9.14) the spectral electric field strength of the resulting wave at the detector is given by

$$E_k = E_{0,\,k}\mathrm{e}^{\mathrm{i}(\omega t - 2kz_\mathrm{r})} + \sum_{j=1}^{N} r_{\mathrm{m},\,j} E_{0,\,k}\mathrm{e}^{\mathrm{i}\left(\omega t - 2kz_{\mathrm{m},\,j}\right)} \tag{9.15a}$$

or

$$E_k = \left(\mathrm{e}^{-\mathrm{i}2kz_\mathrm{r}} + \sum_{j=1}^{N} r_{\mathrm{m},\,j}\mathrm{e}^{-\mathrm{i}2kz_{\mathrm{m},\,j}}\right) E_{0,\,k}\mathrm{e}^{\mathrm{i}\omega\mathrm{t}}. \tag{9.15b}$$

The corresponding time-averaged spectral intensity—referring to angular wave numbers—is

$$I_k \propto E_k E_k^* = E_{0,\,k}E_{0,\,k}^* \left(\mathrm{e}^{-\mathrm{i}2kz_\mathrm{r}} + \sum_{j=1}^{N} r_{\mathrm{m},\,j}\mathrm{e}^{-\mathrm{i}2kz_{\mathrm{m},\,j}}\right)\left(\mathrm{e}^{\mathrm{i}2kz_\mathrm{r}} + \sum_{n=1}^{N} r_{\mathrm{m},\,n}^*\mathrm{e}^{\mathrm{i}2kz_{\mathrm{m},\,n}}\right) \tag{9.16a}$$

or

$$I_k = I_{0,\,k}\left(\mathrm{e}^{-\mathrm{i}2kz_\mathrm{r}} + \sum_{j=1}^{N} r_{\mathrm{m},\,j}\mathrm{e}^{-\mathrm{i}2kz_{\mathrm{m},\,j}}\right)\left(\mathrm{e}^{\mathrm{i}2kz_\mathrm{r}} + \sum_{n=1}^{N} r_{\mathrm{m},\,n}^*\mathrm{e}^{\mathrm{i}2kz_{\mathrm{m},\,n}}\right) \tag{9.16b}$$

or

$$I_k = I_{0,k}\left(\begin{array}{c} 1 + \sum_{j=1}^{N} r_{\mathrm{m},j}\mathrm{e}^{-\mathrm{i}2k(z_{\mathrm{m},j}-z_\mathrm{r})} + \sum_{n=1}^{N} r_{\mathrm{m},n}^*\mathrm{e}^{\mathrm{i}2k(z_{\mathrm{m},n}-z_\mathrm{r})} + \sum_{j=1}^{N} r_{\mathrm{m},j}r_{\mathrm{m},j}^* \cdots \\ \cdots + \sum_{v=1}^{N}\sum_{w=1,v\neq w}^{N} r_{\mathrm{m},v}r_{\mathrm{m},w}^*\mathrm{e}^{-\mathrm{i}2k(z_{\mathrm{m},v}-z_{\mathrm{m},w})} + \sum_{v=1}^{N}\sum_{w=1,v\neq w}^{N} r_{\mathrm{m},v}^*r_{\mathrm{m},w}\mathrm{e}^{\mathrm{i}2k(z_{\mathrm{m},v}-z_{\mathrm{m},w})} \end{array}\right). \tag{9.16c}$$

Assuming

$$r_{\mathrm{m},j} = r_{\mathrm{m},j}^* = \sqrt{R_{\mathrm{m},j}}, \tag{9.17}$$

with $r^*_{\mathrm{m},j}$ complex conjugate reflection coefficient, $R_{\mathrm{m},j}$ reflectivity in the measurement path at location j we obtain:

$$I_k = I_{0,k}\left(\begin{array}{c} 1+\sum_{j=1}^{N} R_{\mathrm{m},j} + 2\sum_{j=1}^{N}\sqrt{R_{\mathrm{m},j}}\cos(2k(z_{\mathrm{m},j}-z_{\mathrm{r}}))\ldots \\ \ldots + 2\sum_{v=1}^{N}\sum_{w=1,v\neq w}^{N}\sqrt{R_{\mathrm{m},v}}\sqrt{R_{\mathrm{m},w}}\cos(2k(z_{\mathrm{m},v}-z_{\mathrm{m},w})) \end{array}\right). \tag{9.18}$$

The term $I_{0,k}\left(1+\sum_{j=1}^{N} R_{\mathrm{m},j}\right)$ in (9.18) describes the background part of the detected signal, which does not depend on the path difference of the two interferometer arms. The term $2I_{0,k}\sum_{j=1}^{N}\sqrt{R_{\mathrm{m},j}}\cos(2k(z_{\mathrm{m},j}-z_{\mathrm{r}}))$ is the *interference term* or *cross correlation term* depending on the path difference. This term contains the depth information $z_{\mathrm{m},j}$ to be gained. The last term $2I_{0,k}\sum_{v=1}^{N}\sum_{w=1,v\neq w}^{N}\sqrt{R_{\mathrm{m},v}}\sqrt{R_{\mathrm{m},w}}\cos(2k(z_{\mathrm{m},v}-z_{\mathrm{m},w}))$ is called *auto-correlation term* which depends on path differences between different depths within the measuring object.

For time-domain OCT the signal S caused by the spectral intensity (9.18) is detected as a spectrally integrated signal to which all angular wave numbers of the light source contribute:

$$S(z_{\mathrm{r}}) = \eta\int_0^{\infty} I_k \,\mathrm{d}k = \eta I, \tag{9.19}$$

where η is the detector constant—assuming a linear detector response—and I the intensity of the superimposed reference and measurement beams. The measured signal is a time-dependent signal described by the relation $z_{\mathrm{r}} = v_{\mathrm{r}}t$, cf. Fig. 9.4.

For Fourier-domain OCT the position of the reference mirror is kept constant, i.e. $z_{\mathrm{r}} = \mathrm{const.}$, while the signal is detected spectrally resolved, thus we have:

$$S_k = \eta I_k(k, z_{\mathrm{r}} = \mathrm{const.}). \tag{9.20}$$

In the following Sect. 9.1.2 this method will be presented in more detail.

In order to evaluate relation (9.19), we consider the spectrum of the radiation source to have a maximum at the angular wave number k_0 and describe this spectrum in a simplified way by a Gaussian function:

$$I_{0,k} = \frac{I_0}{\sqrt{\pi}\Delta k}e^{-\left[\frac{(k-k_0)}{\Delta k}\right]^2}, \tag{9.21}$$

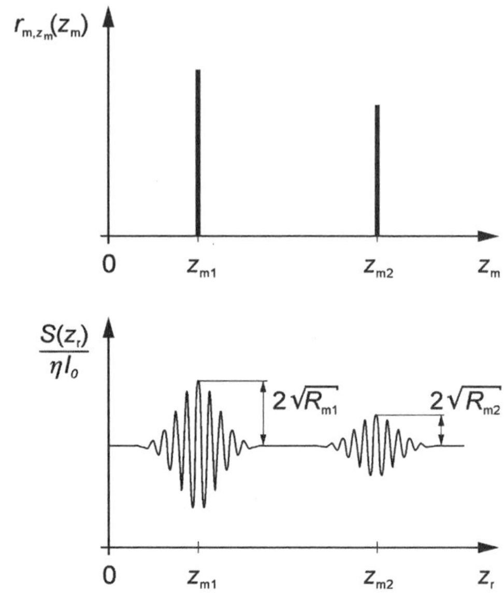

Fig. 9.5 *Top* reflection coefficient per unit length, cf. (9.11), as a function of the distance z_{m}. *Bottom* detector signal for time-domain OCT as a function of the position of the reference mirror z_{r}, cf. (9.23), (9.24)

where $\Delta k \ll k_0$ is a measure of the width of the emission spectrum of the light source and

$$I_0 = \int_0^\infty I_{0,k} \mathrm{d}k, \tag{9.22}$$

where I_0 is the intensity of the beam. Putting (9.21) in (9.18), (9.19) then leads after some calculation steps to:

$$I(z_{\mathrm{r}}) = I_0 \left(1 + \sum_{j=1}^{N} R_{\mathrm{m},j} \right) + 2I_0 \sum_{j=1}^{N} \sqrt{R_{\mathrm{m},j}} \cos\left(2k_0 \left(z_{\mathrm{r}} - z_{\mathrm{m},j}\right)\right) \mathrm{e}^{-\left(z_{\mathrm{r}} - z_{\mathrm{m},j}\right)^2 \Delta k^2}. \tag{9.23}$$

For the derivation of (9.23) the autocorrelation term in (9.18), i.e. the last term, was not taken into account. Assuming small reflection coefficients—as we did so far—this term scales with a product of two small reflection coefficients and is therefor small compared to the preceding terms. The derivation of (9.23) is described in A.3.

In relation (9.23) again the first term describes the background. The second term is the sum of products of an oscillating function—the cosine function—and an envelope of Gaussian shape. Figure 9.5 illustrates the meaning of (9.23) taking as example two depths of the measuring object where backscattering occurs with different reflection coefficients. At position z_{m1} the reflection coefficient is higher than at position z_{m2}. Close to these positions of the reference mirror—see bottom graph—the detector signal shows an oscillation due to the interference within the coherence length. On a base line given by the first term of (9.23) an oscillating signal with a Gaussian envelope appears whose amplitude is proportional to the reflection coefficient, i.e. the square root of the respective reflectivity R_{m1} or R_{m2}. The measuring result

$$S(z_r) = \eta I(z_r) \tag{9.24}$$

can be interpreted as a convolution of the reflection profile with the coherence function of the light of the source. The carrier frequency of the observed oscillation is determined by k_0, the central angular wave number of the light source spectrum, cf. (9.21), which appears in the argument of the cosine function in (9.23).

Due to this carrier frequency signal processing can be accomplished by lock-in techniques. Finally, the distances z_{m1} and z_{m2} are determined as well as the strength of the backscattered signal at these locations. In this way a depth scan, see Fig. 9.2, is generated.

9.1.2 Fourier-Domain OCT

Starting point of *Fourier-domain OCT* (FD-OCT) is (9.20). The signal is detected spectrally resolved as a function of the wavelength by use of a spectrometer based on a dispersing prism or a grating, cf. Sect. 2.3.4. Figure 9.6 shows the simplified set-up with a fixed reference mirror and a spectrometer with detector array (details of the spectrometer are omitted here for better clarity). In a first step we transform the measured spectrum $S_\lambda(\lambda)$ to the equivalent representation $S_k(k)$ by use of relation (9.3). The spectrum $S_k(k)$ is a function of the angular wave number k and is proportional to the spectral intensity I_k, see (9.20).

The principal characteristics of the detected spectrum is shown in Fig. 9.7. The spectrum has a bell-shaped envelope, whose bandwidth is equal to the bandwidth of the light emitted by the source. This envelope is modulated. For a given single scattering surface of the measurement object a simple modulation occurs as shown schematically in Fig. 9.7. The modulation contains the information about the distance to the scattering structure of the object, see (9.18).

To extract this information the signal $S_k(k)$—which is proportional to the spectral intensity with respect to angular wave number—is processed by an inverse Fourier transformation to go from the k-space to the z-space:

$$\tilde{S}(z) = \frac{1}{2\pi} \int_{-\infty}^{+\infty} S_k e^{ikz} dk. \tag{9.25}$$

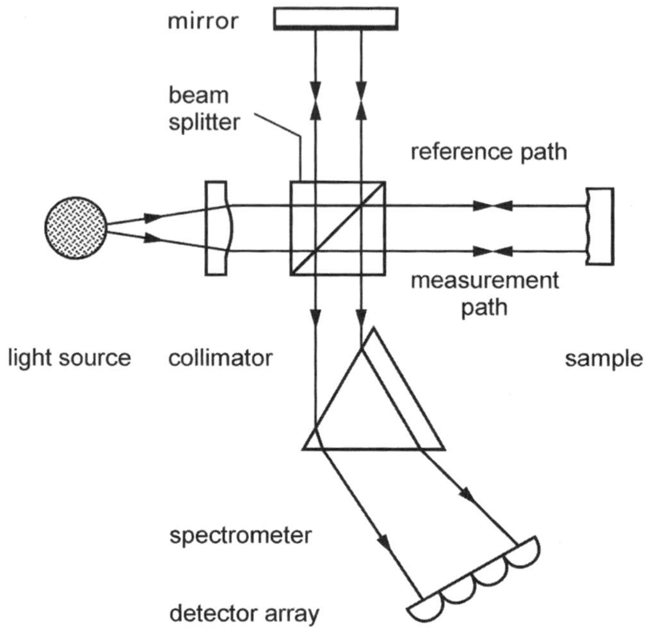

Fig. 9.6 Principle of Fourier-domain OCT. The reference mirror is fixed. The superimposed signals from reference and measurement beam are detected by a spectrometer

Fig. 9.7 Detector signal versus angular wave number gained with a Fourier-domain OCT. The spectrum has a bell-shaped envelope with a width corresponding the bandwidth of the light source

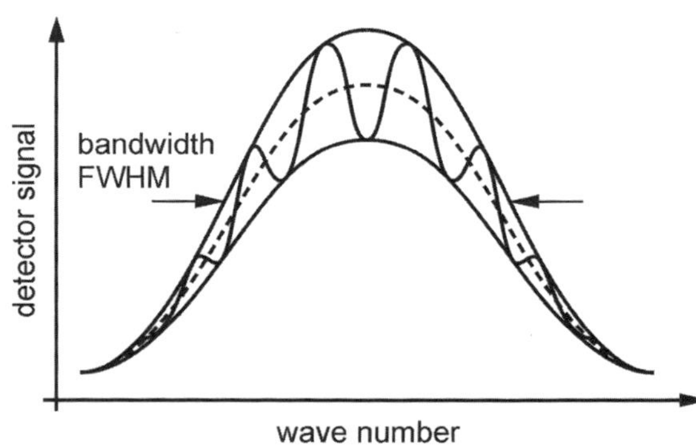

Executing this integration requires the inverse Fourier transform of the spectrum of the light source $I_{0,k}$, cf. (9.21), multiplied by the detector constant η, which is given by:

$$\tilde{f}(z) = \frac{\eta}{2\pi} \int_{-\infty}^{+\infty} \frac{I_0}{\sqrt{\pi}\Delta k} \mathrm{e}^{-\left[\frac{(k-k_0)}{\Delta k}\right]^2} \mathrm{e}^{\mathrm{i}kz} \mathrm{d}k = \frac{\eta I_0 \mathrm{e}^{\mathrm{i}k_0 z}}{2\pi} \mathrm{e}^{-\frac{z^2 \Delta k^2}{4}}. \qquad (9.26)$$

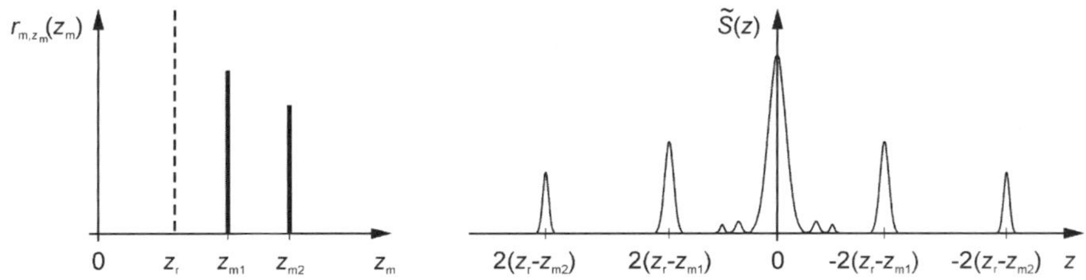

Fig. 9.8 *Left* reflection coefficient per unit length, cf. (9.11), as a function of the distance z_m. *Right* inverse Fourier transform according to (9.27) of the detected spectrum $S_k(k)$ of a Fourier-domain OCT

Formula (9.26) is derived in Appendix A.4. This function has its maximum at $z = 0$. The full width at half maximum (FWHM) is given by:

$$\Delta z_{\mathrm{FWHM}} = \frac{4\sqrt{\ln 2}}{\Delta k} \approx \frac{\pi}{\Delta k}. \tag{9.26a}$$

With (9.2b) it follows that

$$\Delta z_{\mathrm{FWHM}} \approx L_c. \tag{9.26b}$$

Inserting (9.18) to (9.20) and (9.25) and some calculation steps, we finally obtain:

$$\tilde{S}(z) = \tilde{f}(z)\left(1 + \sum_j R_{m,j}\right) + \left(\sum_j \sqrt{R_{m,j}}\left(\tilde{f}(z + \Delta z_j) + \tilde{f}(z - \Delta z_j)\right)\right) + \cdots, \tag{9.27}$$

with the abbreviation $\Delta z_j = 2(z_r - z_{m,j})$. The autocorrelation term is again not considered here (this term is indicated by "…" in (9.27)). Relation (9.27) is derived in Appendix A.5. Figure 9.8 illustrates this result. In the example shown—see the graph in Fig. 9.8, left—two positions exist where a backscattering occurs: z_{m1} and z_{m2}. Furthermore we assume that $z_r < z_{m1} < z_{m2}$ holds, then the Δz_j-values are negative. The inverse Fourier transform of the measured spectrum shows five peaks, cf. Fig. 9.8, right.

The central peak corresponds to the first term in (9.27), which is proportional to $\tilde{f}(z)$ and attains its maximum at $z = 0$, cf. (9.26). The two peaks on the negative z-axis belong to the terms $\tilde{f}(z - 2(z_r - z_{m,1}))$ and $\tilde{f}(z - 2(z_r - z_{m,2}))$ of (9.27) for $j = 1, 2$. On the positive z-axis the two peaks correspond to $\tilde{f}(z + 2(z_r - z_{m,1}))$ and $\tilde{f}(z + 2(z_r - z_{m,2}))$. From e.g. the two peak positions on the positive axis the distances z_{m1} and z_{m2} to be measured are deduced for a given position of the reference arm z_r.

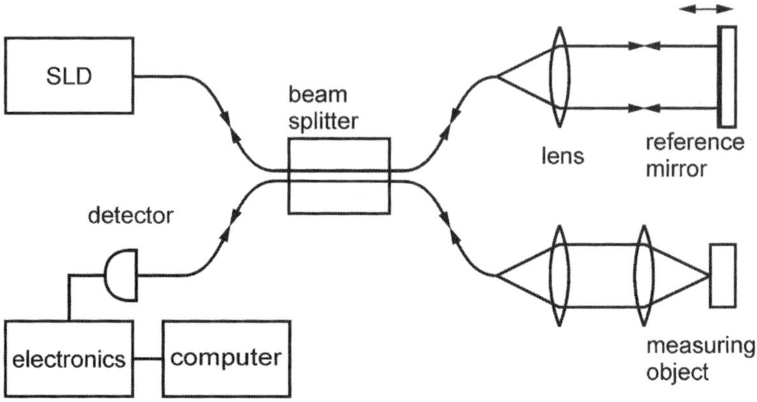

Fig. 9.9 Set-up of a time-domain OCT sensor, SLD superluminescence diode

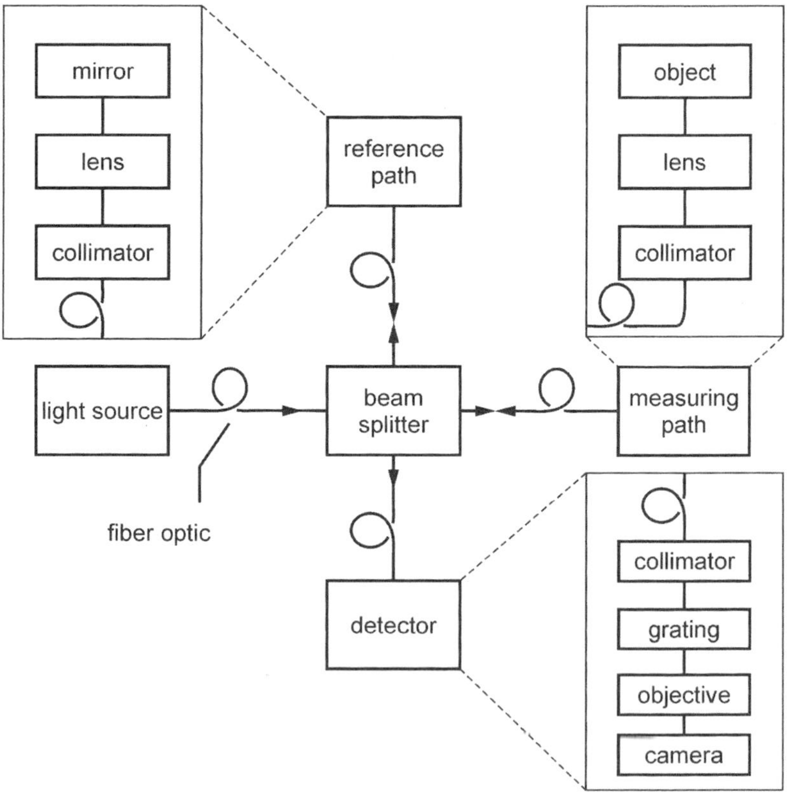

Fig. 9.10 Block diagram of a Fourier-domain OCT sensor

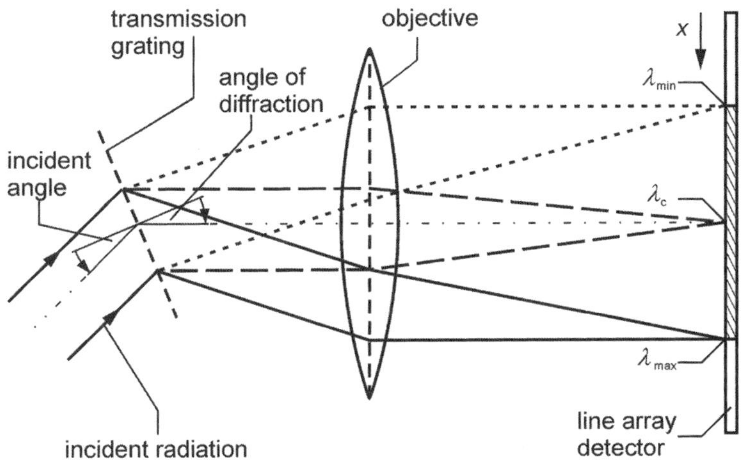

Fig. 9.11 Principle of a spectrometer for Fourier-domain OCT

9.2 OCT Sensors

The set-up of a time-domain OCT sensor is shown in Fig. 9.9. Instead of free beam paths fiber optics are used to guide the light from the light source, to split the light into reference and measurement path, to combine the backpropagating signals and finally to transmit the superimposed light beams to the detector. Typically as light source a superluminescent diode is applied. Various methods are used to change the optical path length of the reference arm, such as piezoelectric transducers to shift the reference mirror or fiber stretchers [1].

Figure 9.10 shows a block diagram of a fiber-based Fourier-domain OCT sensor. The detector block comprises a spectrometer with fiber optics, collimators, grating, objective and camera.

A possible spectrometer set-up illustrates Fig. 9.11 making use of the light diffraction by a grating, cf. Sect. 2.3.4. The collimated incident light irradiates a transmission grating and is diffracted depending on the wavelength. In the focal plane of the objective a position sensitive detector is placed, such as those described in Sect. 5.4.5.

The raw data captured for a given distance of the measuring object is displayed in Fig. 9.12a as a function of the wavelength, which is the primary measured quantity with a spectrometer. In a first data processing step the background is subtracted to yield the interference signal with high modulation degree, cf. Fig. 9.12b. A close look shows that the distance between two neighboring minima is not constant but shows a chirp, i.e. for increasing wavelength this distance increases as well.

To understand this characteristic we assume a given path difference between reference and measurement arm of Δs. A minimum occurs if this path difference

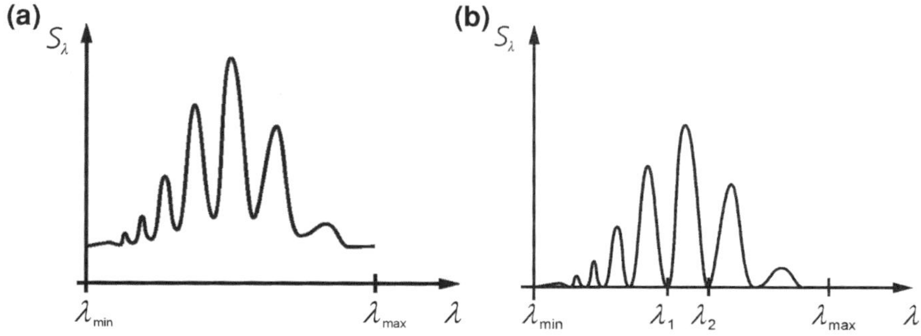

Fig. 9.12 **a** Raw data of the detected spectrum as a function of the wavelength for a Fourier-domain OCT sensor, **b** Spectrum after background subtraction. Two neighboring minima are considered at the wavelengths λ_1 and λ_2, see text

corresponds to a phase difference of 180° for a given wavelength. Hence, for the minimum at λ_1 in Fig. 9.12b the following condition holds:

$$N_1\lambda_1 + \frac{\lambda_1}{2} = \Delta s, \tag{9.28}$$

where N_1 is an unknown integral number. For the minimum at λ_2 we have:

$$N_2\lambda_2 + \frac{\lambda_2}{2} = \Delta s. \tag{9.29}$$

Since $\lambda_2 > \lambda_1$ is assumed, cf. Fig. 9.12b, then from (9.28), (9.29) follows: $N_2 < N_1$. For neighboring minima the smallest distance of integral numbers is just 1, i.e. $N_2 = N_1 - 1$. Inserting this relation in (9.29), then subtracting (9.28) from (9.29) yields:

$$\Delta s = \frac{\lambda_1\lambda_2}{\lambda_2 - \lambda_1}. \tag{9.30}$$

With the abbreviation $\Delta\lambda = \lambda_2 - \lambda_1$ and two wavelengths being close to each other, i.e. $\lambda \approx \lambda_1 \approx \lambda_2$, follows from (9.30):

$$\Delta\lambda \approx \frac{\lambda^2}{\Delta s}, \tag{9.31}$$

which is consistent with the observation, that the wavelength distance between two neighboring minima increases with the wavelength for a given path difference Δs. If the spectrum is plotted as a function of the angular wave number, then the angular wave number spacing between two minima Δk is independent of k, which can easily be shown with the help of $k = 2\pi/\lambda$ and the differential $|\Delta k| \approx \frac{k^2}{2\pi}\Delta\lambda$.

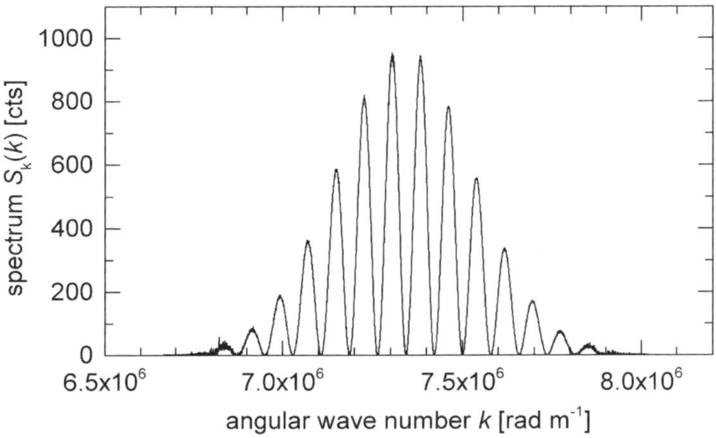

Fig. 9.13 Spectrum of the interfering signals for a Fourier-domain OCT as a function of the angular wave number

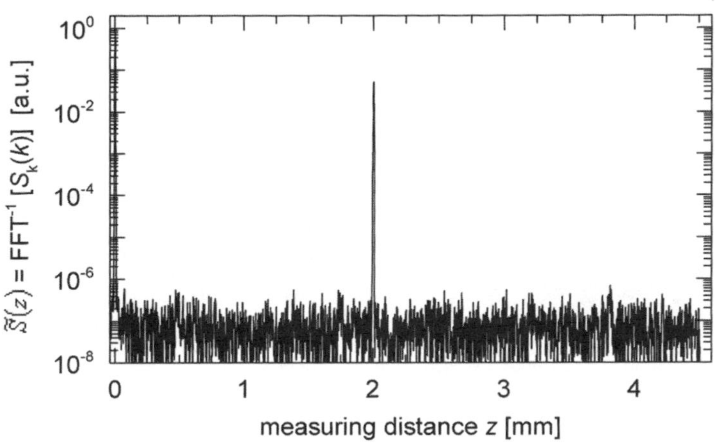

Fig. 9.14 Inverse Fourier-transformed (FFT^{-1}) spectrum of Fig. 9.13 as a function of the distance z

Figure 9.13 shows a detected spectrum as a function of the angular wave number k for a single back-scattering surface in the measuring path.

Figure 9.14 gives the result of the spectrum shown in Fig. 9.13 after inverse Fourier transformation where the ordinate has a logarithmic scale. At a distance of $z = 2$ mm a distinct peak occurs being many order of magnitude above the noise level.

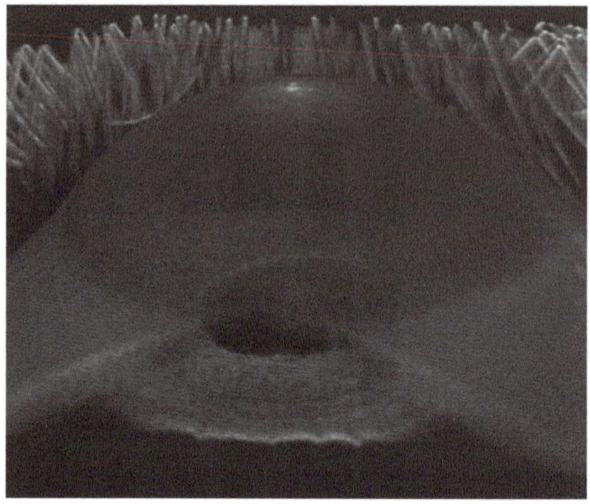

Fig. 9.15 3D-image of the whole anterior segment of an eye taken with a fast OCT system acquired in 0.8 s [4]

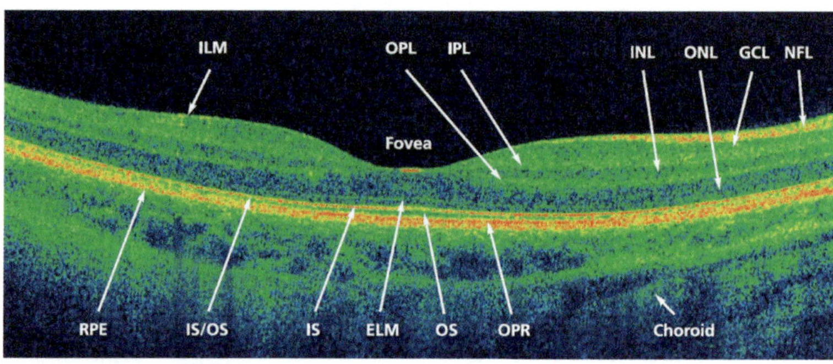

Fig. 9.16 B-scan of the retina of a healthy eye taken with a FD-OCT system. ILM inner limiting membrane, OPL outer plexiform layer, IPL inner plexiform layer, INL inner nuclear layer, ONL outer nuclear layer, GCL ganglion cell layer, NFL nerve fiber layer, RPE retinal pigment epithelium, IS/OS interface between IS and OS, IS photoreceptor inner segment, ELM external limiting membrane, OS photoreceptor outer segment, OPR outer photoreceptor/RPE complex [8]

9.3 Examples of Applications

Most dominant application fields of the various OCT methods are found in medicine as e.g. ophthalmology or diagnostics of coronary atherosclerosis [4–7]. Figure 9.15 shows the anterior of an eye with cornea and iris in a C-scan, cf. Fig. 9.2 [4]. The number of voxels contributing to this 3D-image amounts to $1000 \times 985 \times 560$, they were measured with a fast swept source OCT system within 0.8 s. The diameter of a human cornea amounts to about 11 mm.

FD-OCT is used as a tool for noninvasive in vivo analysis of retinal tissue for diagnosis and management of retinal disease and glaucoma. The main advantage

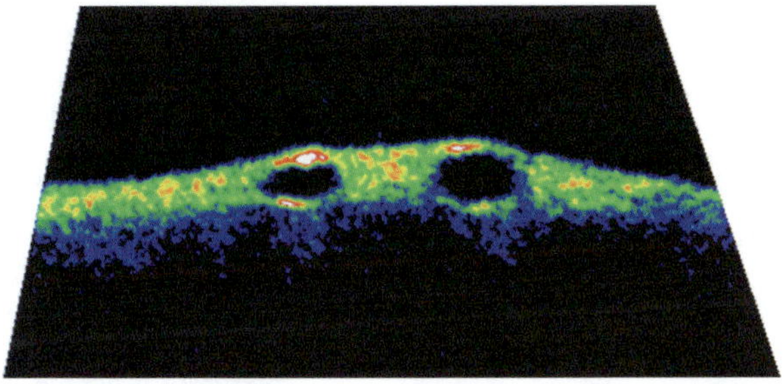

Fig. 9.17 B-scan of blood vessels of the heart of a pig taken with a FD-OCT system

Fig. 9.18 Multi-layer film tube produced in a blown-film plant with inline monitoring of the thicknesses with OCT [11]

compared to TD-OCT is the high number of scans acquired per second ranging up to 27,000 A-scans/s. This feature allows for three-dimensional imaging and reduces the possibility of artifacts due to eye movement during the data acquisition. Measuring objects are e.g. the cross-section anterior segment and 3D-views of the retina. Figure 9.16 shows a B-scan of a healthy eye indicating the different layers and structures that can be resolved by OCT and used to diagnose different diseases [7, 8]. Handheld ultrahigh speed swept source OCT instruments were developed using microelectromechanical systems (MEMS) for beam scanning [9]. With such a system fundus images are generated in a field of view of 6×6 mm^2. The axial scan rate amounts to 350 kHz and the achieved resolution in tissue is 10 μm.

Figure 9.17 shows a B-scan of blood vessels of the heart of a pig taken with a FD-OCT system [10]. The blood vessels were purged with NaCl solution. The vessels have a diameter of about 0.8 mm. With the help of such images coagulation processes of blood vessels can be monitored inline.

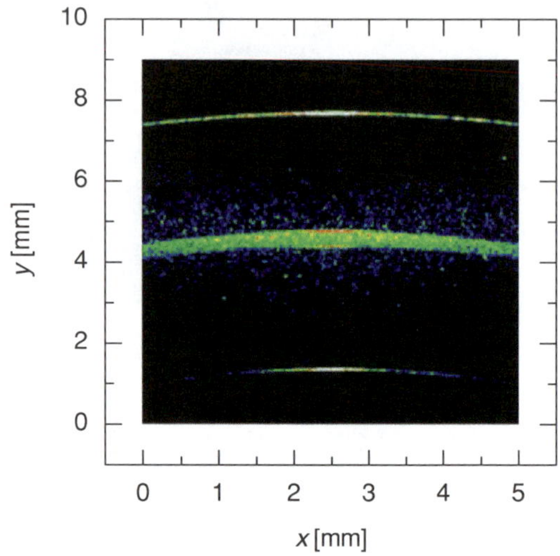

Fig. 9.19 B-scan measured with FD-OCT of a multi-layer PET preform

Industrial applications of OCT are found for inline quality control tasks of polymer products such as the measurement of the thicknesses of multi-layer polymer films. These multi-layer films comprise various functional layers to optimize their properties with respect to mechanical stability, UV resistance and diffusion barriers. Figure 9.18 shows a multi-layer film tube produced in a blown-film plant with an installed inline monitoring system based on OCT [11]. The measuring range of the applied FD-OCT sensor amounts to 5.0 mm, the linearity error is less than 5.0 μm.

The result of a B-scan OCT measurement of a multi-layer PET preform product is shown in Fig. 9.19 yielding the thicknesses of the three different layers [12]. In this case the thickness of the internal diffusion barrier layer amounts to 270 μm. Resolvable differences of refractive indices causing sufficient backscattering go down to 0.002.

In the field of laser machining OCT is studied to measure the hole depth in real-time during laser drilling [13].

References

1. W. Drexler, J. Fujimoto (eds.), *Optical Coherence Tomography* (Springer, New York, 2008)
2. B. Bouma, G. Tearney (eds.), *Handbook of Optical Coherence Tomography* (M. Dekker Inc., New York, 2002)
3. R. Cobbold, *Foundations of Biomedical Ultrasound* (Oxford University Press Inc., New York, 2007)
4. A. Fercher, W. Drexler, C. Hitzenberger, T. Lasser, Optical coherence tomography—principles and applications. Rep. Prog. Phys. **66**, 239–303 (2003)

5. W. Wieser, T. Klein, D. Adler, F. Trépanier, C. Eigenwillig, S. Karpf, J. Schmitt, R. Huber, Extended coherence length megahertz FDML and its application for anterior segment imaging. Biomed. Opt. Express **3**, 2647–2657 (2012)
6. H. Bezerra, M. Costa, G. Guagliumi, A. Rollins, D. Simon, Intracoronary optical coherence tomography: a comprehensive review. Cardiovasc. Intervent. **2**, 1035–1046 (2009)
7. B. Lumbroso, M. Rispoli. *Guide to interpreting spectral domain optical coherence tomography* (I.N.C. Innovation-News-Communication, J. Allyn Inc, Dublin, California, 2009)
8. Zeiss wall poster *Exploring the Eye with Cirrus HD-OCT*, http://www.meditec.zeiss.com/C125679E00525939/ContainerTitel/CirrusOCT/$File/wall-poster.pdf. Accessed 1 Nov 2013
9. C. Lu, M. Kraus, B. Potsaid, J. Liu, W. Choi, V. Jayaraman, A. Cable, J. Hornegger, J. Duker, J. Fujimoto, Handheld ultrahigh speed swept source optical coherence tomography instrument using a MEMS scanning mirror. Biomed Opt Express **5**, 293–311 (2013)
10. Fraunhofer Institute for Laser Technology, Aachen (2011)
11. S. Hölters, C. Farkas, R. Fleige, A. Lenenbach, R. Noll, Low coherence interferometry for the inline measurement of translucent multilayer structures, 10th IMEKO Symposium Laser Metrology for Precision Measurement and Inspection in Industry (LMPMI) 2011. VDI-Berichte Nr. **2156**, 161–168 (2011)
12. S. Hölters, J. Overbeck, L. Ederleh, W. Michaeli, A. Lenenbach, R. Noll, Precise measurement technology for complex multilayer films. Coating **9**, 6–12 (2010)
13. P. Webster, L. Wright, K. Mortimer, B. Young, B. Leung, J. Yu, J. Fraser, Automatic real-time guidance of laser machining with inline coherent imaging, J. Laser Appl. **23**, 1–6 (2011)

Chapter 10
Laser Triangulation

Abstract Laser triangulation is a method to measure the absolute distance to an object. The principle of triangulation is presented including the extensions to 2-D and 3-D measurements. We discuss the characteristic curve of laser triangulation, the implications of the laser beam propagation and the properties of the object surface. Examples of applications are given ranging from thickness measurements of rolled sheets, flatness measurements of heavy plates, coordinate-measuring machines to the straightness and profile measurement of rails.

Laser triangulation is a method of measurement, in which the absolute distance of a point on the surface of an object is determined from a reference plane. Triangulation is, in this sense, a point orientated method, similar to a mechanical probe. The distance determined corresponds to a one-dimensional measurand. A generalization of the laser triangulation method is the so-called *laser light section technique*. In this, a light line is projected onto the object to be measured. This line is observed from another direction and, by measuring the line as it progresses, the two-dimensional contours of the object section in its path can be determined. There is an easy transition between the two methodical approaches. For example, it is feasible to carry out the measurement of two-dimensional contours through the combination of laser triangulation and beam scanning, the latter being achieved by means of e.g. a rotating mirror or a translation stage.

For non-contacting distance and contour measuring, different types of laser triangulation sensors are available commercially. As a rule, they are designed as compact devices, which have themselves opened up an abundance of applications in the fields of quality assurance and process control. In the following, the principle of laser triangulation will be represented first.

A. Donges and R. Noll, *Laser Measurement Technology*, Springer Series in Optical Sciences 188, DOI: 10.1007/978-3-662-43634-9_10,

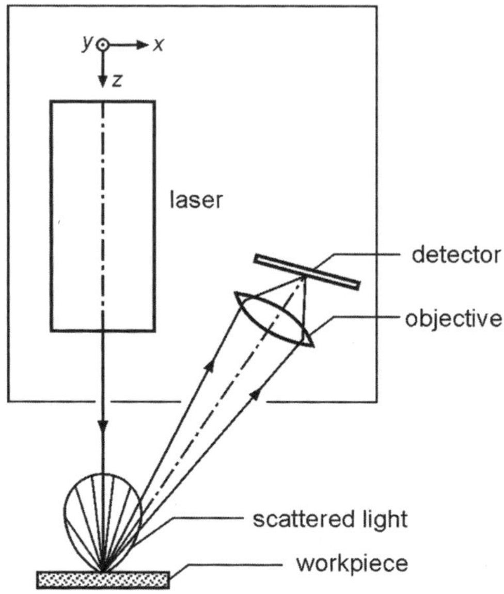

Fig. 10.1 Principle of laser triangulation

10.1 Principle

Figure 10.1 shows the principle of laser triangulation [1]. A laser beam is directed at a workpiece. Light is scattered at the spot at which the beam irradiates the measuring object. This scattering of light is then used to observe the intersection point. With an ideally specular object surface, no diffuse light scattering would appear but only a directed reflection, cf. Sect. 3.2.3, Figs. 3.8, 3.9. Such objects are not suited to measurement by laser triangulation. In the general case, the scattering behaviour depends on the material properties of the workpiece, its object surface structure or texture and the wavelength of the laser. The scattering is described by an angular intensity distribution, as is shown schematically in Fig. 10.1.

The light spot at the intersection point of the beam is imaged, by the use of an objective on a detector, through a given angle to the incident direction. This angle, with respect to the scattering characteristics, must be chosen so that sufficient intensity can be registered from the object. The position-sensitive detector delivers a signal dependent on the location of the imaged light spot. With respect to the detector, a lateral effect diode, CCD or CMOS line array would be used in preference, cf. Sect. 5.4.

The location of the light spot imaged onto the detector depends on the position of the workpiece in the z-direction, see coordinate system in Fig. 10.1. From the measured location of the image spot and the known imaging geometry and incident direction, the distance of the test object in the z-direction can be determined from the reference plane, e.g. the xy-plane in Fig. 10.1. Figure 10.2 shows the extension of the one dimensional distance measurement to contour (two

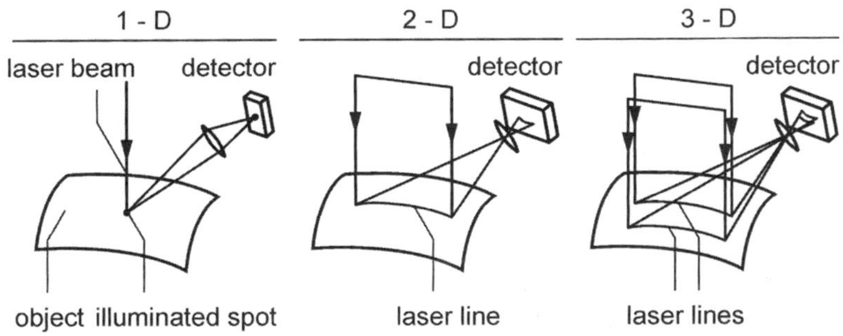

Fig. 10.2 Variants of laser triangulation methods. *Left* distance measurement with a collimated laser beam, cf. Fig. 10.1, *center* laser light section for the measurement of contours and profiles, *right* projection of several laser lines to measure the shape of an object

dimensional, 2-D) and shape (three dimensional, 3-D) measurements of objects [2]. For a 2-D measurement of contours, the laser beam is formed to a laser line. This line is imaged under a known angle on a 2-D position-sensitive detector, for example, a CCD or CMOS array, cf. Sect. 5.4.5. This type of laser triangulation is called *laser light section method.*

For triangulation sensors, semiconductor lasers are usually deployed. They are operated in a continuous or pulsed manner. The radiant flux of the semiconductor laser is controlled electrically. In this way it is possible to compensate for strong fluctuations of the scattered radiation at the test object, which appear as a result of differing surface properties. Typical wavelength used are in the red, and near-infrared spectral range as, e.g., 660, 670, 685, and 780 nm. The radiant flux varies between 1 and 100 mW depending on the mean measurement distance, the measurement range, the wavelength, and the specific application.

Besides the radiant flux important characteristics of the laser source are the beam quality—cf. Sects. 2.6.3 and 2.7—and the pointing stability. As a rule, a spectral laser line filter is required in front of the detector. This filter transmits the scattered laser light and reduces disturbing light from the surroundings.

10.1.1 Scheimpflug Condition

For 1-D laser triangulation the objective and detector must be arranged relative to the incident direction of the laser beam so that, for various distances of the test object, the light spot is focused as sharply as possible in the plane of detection. The detector is therefore inclined relative to the optical axis of the objective. In Fig. 10.3 the required geometrical arrangement is illustrated in more detail.

To simplify the representation, the coordinate system is chosen so that the v-axis runs parallel to the optical axis of the objective and the u-axis in the objective plane is oriented towards the z-axis. The origin of the uv-coordinate system lies in the middle of the objective. The laser beam directed at the workpiece is inclined to

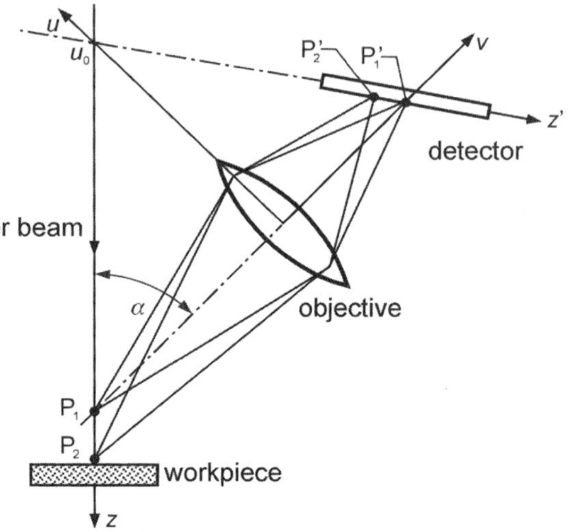

Fig. 10.3 Imaging geometry of 1-D laser triangulation

the v-axis, as is represented in Fig. 10.3. If the laser beam meets the test object at a point P_1, then the corresponding light spot would be observed at a point P_1' in the plane of detection. This applies equally to the points P_2 and P_2'. Corresponding to the fact that the point P_2 lies further away from the objective than the point P_1, the distance of the point P_2' to the objective is shorter than that for P_1'. The detector must obviously be inclined to the v-axis—i.e. at an angle different from 90°—in order to achieve sharp imaging. In the following we will define how large the value chosen for this inclination should be.

The laser beam in the incident direction is described by the following linear equation:

$$u = m_{\mathrm{L}} v + u_0, \tag{10.1}$$

with u, v coordinates, see Fig. 10.3, m_L slope of the line, u_0 position of the laser beam at $v = 0$. Given (v, u) as the coordinates of a point P on the workpiece, the coordinates (v', u') of the corresponding image point P′ are yielded from the following relationships:

$$v' = \frac{v}{v/f + 1}, \tag{10.2}$$

$$u' = \frac{v'}{v} u, \tag{10.3}$$

with f focal length of the objective. Relationship (10.2) follows from the thin lens equation $1/s_{\mathrm{i}} + 1/s_{\mathrm{o}} = 1/f$, with s_{i} image distance, s_{o} object distance; (10.3) comes from the expression $B/G = s_{\mathrm{i}}/s_{\mathrm{o}}$ where B is the image size and G is the object size, cf. Sect. 8.2.1, equations (8.3), (8.4). With equations (10.1), (10.2), and (10.3), the undashed coordinates can be eliminated, in order to derive the equation of the curve on which the image point P′ is found. The result is again a linear equation:

$$u' = (m_L - u_0/f)v' + u_0. \tag{10.4}$$

For a sharp imaging of the points P_1 and P_2, in Fig. 10.3, the detector is aligned along this straight line. This is equally valid for all other points along the incident direction of the laser beam. For $v' = 0$, the line cuts the u-axis at $u' = u_0$, i.e. at the same position as the incident laser beam, cf. (10.1). The inclination of the detector is then chosen so that the incident direction itself, the u-axis in the objective plane and the straight line, on which the detector is aligned (that which was described by the (10.4)), intersect at one point $(0, u_0)$. This rule is known as the *Scheimpflug condition.*

10.1.2 Characteristic Curve of a Triangulation Sensor

In the following, the characteristic curve of a triangulation sensor will be calculated. It describes the relationship between the position of the light spot on the test object and the location of the imaged light spot at the detector.

The location of the light spot on the straight line described by equation (10.1) determines unequivocally the position of the imaged light spot. For the following representation, the coordinate axes, z and z', given in Fig. 10.3, will be used, these running parallel to the lines described by (10.1) and (10.4), respectively. The coordinate origin of these axes lies, in each case, on the v-axis in Fig. 10.3.

A light spot at the position z would be projected onto the detector plane at the corresponding point z' (the derivation of this equation is not given):

$$z'(z) = m_L f \sqrt{1 + (m_L - \frac{u_0}{f})^2} \cdot \frac{(\frac{z}{\sqrt{1+m_L^2}})}{(\frac{z}{\sqrt{1+m_L^2}} + \frac{u_0}{m_L} - f)(m_L - \frac{u_0}{f})}. \tag{10.5}$$

For $z = 0$ equation (10.5) yields $z' = 0$ as it should be as a consequence of the defined origins of the z and z' coordinates. Figure 10.4 represents the characteristic curve of a triangulation sensor, described by equation (10.5), for randomly chosen parameters u_0, m_L and f. The triangulation sensors' measuring range in the z-direction of –120 mm to +120 mm would, in this case, be imaged onto a detector length in the z'-direction of approximately 20 mm. This length corresponds, for example, to the length of a CCD line array with 2048 elements, cf. Sect. 5.5.6.

For reasons of the special arrangement with the Scheimpflug condition, there exists, in general, no linear relationship between the coordinates z and z', as is evident from equation (10.5) and Fig. 10.4. For the limiting case, when u_0 tends to infinity (considering this limiting value, the object distance s_o has to be kept constant, i.e. $m_L = u_0/s_o$ and therefore m_L runs as well to infinity), z' is linearly dependent upon z.

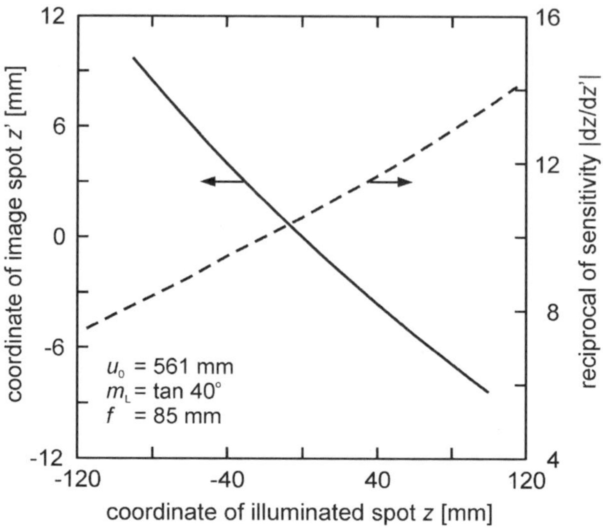

Fig. 10.4 Coordinate of image spot z' and reciprocal of the sensitivity $|dz/dz'|$ as a function of the coordinate z of the illuminated spot for a given laser triangulation set-up

In this case, in Fig. 10.3, the lines (10.1) and (10.4) are aligned parallel to the u-axis corresponding to the usual imaging geometry. In practice, this case is not relevant in laser triangulation, since then the scattered light would have to be observed in a direction perpendicular to the normal of the object surface, cf. Figs. 10.1 and 10.3.

As a consequence of the generally non-linear characteristics shown by the curve of equation (10.5), a displacement of the beam intersection point by an amount dz, leads to a displacement of the corresponding image point by dz', this being dependent on the position of the intersection point. In order to describe this dependency, we consider the derivation of the function $z' = z'(z)$ from (10.5) with respect to z: dz'/dz. The modulus of this derivation is referred to as sensitivity. The reciprocal sensitivity is applied to the right hand side ordinate in Fig. 10.4. This sensitivity can also be thought of as a differential imaging scale: a distance increment dz at a position z is imaged onto a distance increment dz'. The reciprocal sensitivity, in the example shown, varies within the measuring range from 7.5 to 14. Hence, the sensitivity of a triangulation sensor is not constant within its measuring range. It is lower, the further is the test object away from the sensor.

10.2 Influencing Quantities in Laser Triangulation

With respect to the performance characteristics of laser triangulation, e.g. the uncertainty and precision of measurement as well as the resolving power, the following influencing quantities have to be considered:

- laser beam propagation,
- properties of the object surface,
- imaging aberrations,
- detector and signal evaluation and
- atmospheric conditions.

Of course these items are as well of importance for other laser measuring methods as those described in Chaps. 6–14. They are considered in the following in an exemplary way for the case of laser triangulation.

10.2.1 Laser Beam Propagation

Laser beam propagation and beam profile are determined generally by the type of laser source used and the beam shaping optics. For simplification, in the following representation, we will assume that the laser beam can be described by a Gaussian beam, cf. Sect. 2.6.1. Figure 10.5 indicates the characteristic quantities of a Gaussian beam.

In order to achieve high resolving power in triangulation, the diameter of the laser beam at the workpiece should be kept as small as possible. Therefore, the laser beam is generally focused with a lens. We assume that the beam waist diameter of the laser beam in front of the lens amounts to $2w$, cf. Fig. 10.5. We also assume that the distance of this beam waist from the focusing lens is as large as the focal length of the lens. For the diameter of the laser beam at the focus, the following is valid, cf. Sect. 4.2.2:

$$2w_0 = \frac{2\lambda}{\pi w} f_F, \tag{10.6}$$

with λ wavelength of the laser source, $2w$ beam waist diameter of the laser beam in front of the focusing lens, $2w_0$ diameter of the laser beam at the focus and f_F focal length of the focusing lens.

The Rayleigh length z_R indicates the distance along the beam, from the focusing position, at which the beam diameter is larger than the focus diameter by a value $\sqrt{2}$, see Fig. 10.5. For the Rayleigh length holds, c.f. Sect. 2.6.1:

$$z_R = \frac{\pi}{\lambda} w_0^2, \tag{10.7}$$

where z_R denotes the Rayleigh length.

In Fig. 10.6, the Gaussian beam diameter $2w_0$ is plotted as a function of $2z_R$, for varying wavelengths λ, according to equation (10.7). At a given Rayleigh length, the beam diameter is reduced, the shorter is the wavelength. The curves shown are valid for an ideal Gaussian beam. For real laser beams, the beam diameter at a given Rayleigh length is always larger than the values shown in Fig. 10.6.

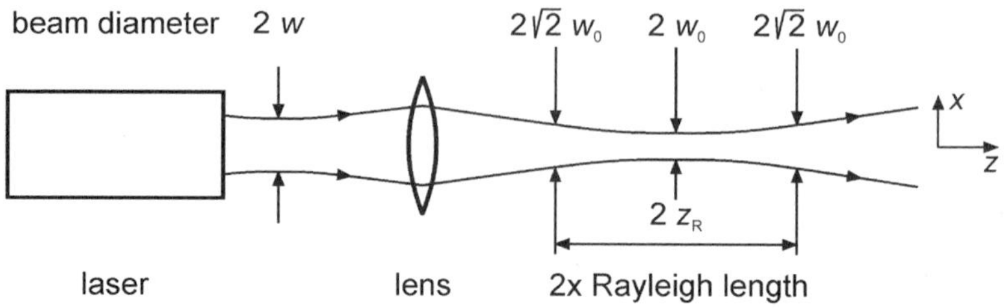

Fig. 10.5 Laser beam propagation of a Gaussian laser beam (diagram is not to scale)

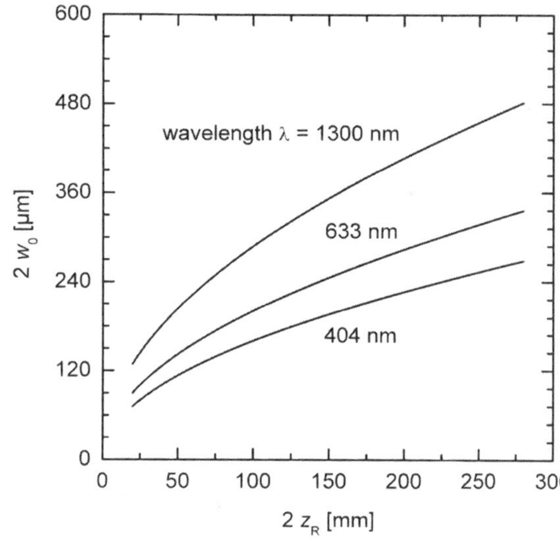

Fig. 10.6 Waist diameter of a Gaussian beam as a function of the Rayleigh length, cf. relation (10.7)

It is frequently the case with a triangulation sensor, that the beam diameter within the measuring range should be as small as possible. The beam would then be formed in such a way, that:

$$2z_{\mathrm{R}} = M_z, \tag{10.8}$$

with M_z measuring range of the triangulation sensor in the z-direction, cf. Figs. 10.1, 10.3.

The beam waist can also be fixed, within a given measuring range, by relationship (10.7). As a numerical example, $M_z = 240$ mm will be chosen, cf. Fig. 10.4, with $\lambda_{\mathrm{HeNe}} = 633$ nm. Using (10.8) and (10.7), this yields $2w_0 = 311\,\mu\mathrm{m}$. At the limits of the measuring range the respective beam diameter amounts to $2\sqrt{2}w_0 = 440\,\mu\mathrm{m}$.

In the following, it will be estimated, using the given expressions, at least how many measuring positions can be separately distinguished within the measuring range. For the sake of simplicity and a rough estimate, we will limit ourselves to the case represented in Fig. 10.7. The beam waist of the laser beam finds itself

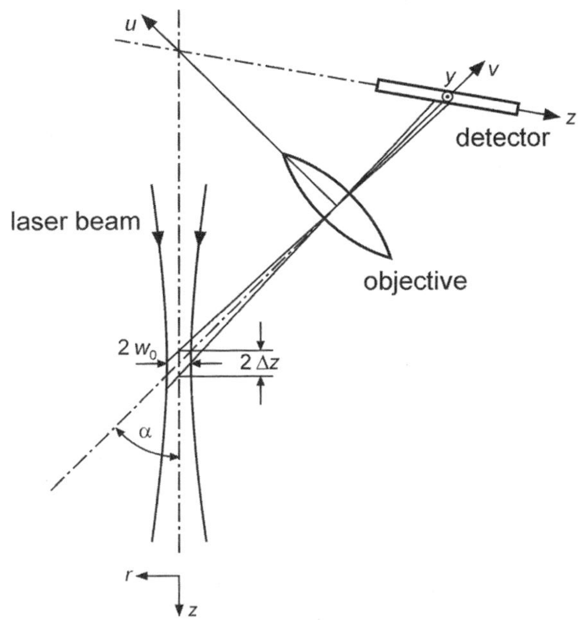

Fig. 10.7 Determination of the resolution in the z-direction

situated at the location $z = 0$ on the v-axis. The surface of the test object is oriented perpendicularly to the z-axis at the location $z = 0$ (not shown in Fig. 10.7). By observation of the light spot through an angle, $\alpha \neq 90°$, the finite diameter $2w_0$ leads to an uncertainty in the determination of the position of the light spot on the z-axis of $\pm\Delta z$, where:

$$\Delta z = w_0/\tan\alpha, \tag{10.9}$$

with α angle of the incident laser beam to the v-axis.

At other test object locations in the measuring range, another Δz value would result, due to the different beam diameters and directions of observation.

With the help of equation (10.9), the smallest achievable triangulation resolving power can be estimated roughly. To simplify the calculation, we neglect the fact that Δz can attain different values within the measuring range. Furthermore, we also assume that the light spots from two measuring positions can still be distinguished if the distance between their centers amounts to Δz. With these simplifying assumptions the number of distinguishable measuring positions within the measuring range $M_z = 2z_R$ in the z-direction can be estimated using the following equation:

$$N = 2z_R/\Delta z = (\pi w_0/\lambda)\tan\alpha. \tag{10.10}$$

For the numerical example given above and a value $\alpha = 40°$, cf. Fig. 10.4, the smallest achievable resolving power is then $N = 1295$. By the use of triangulation sensors, which determine the center of gravity of the projected light spot, resolutions of up to $N = 4000$ can be realized, cf. Sect. 10.2.4.

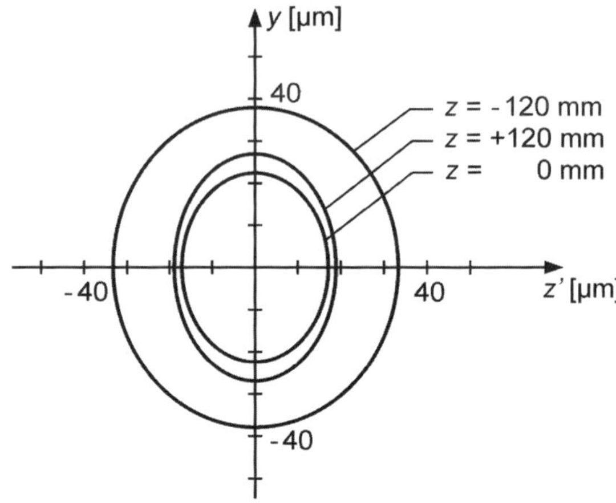

Fig. 10.8 Form of the imaged light spots in the plane of detection for different test object locations in the measurement range; for the position of the y and z'-coordinates, see Fig. 10.7

Figure 10.8 illustrates, for the numerical example which had already been selected for Fig. 10.4, the form of the imaged light spot in the plane of detection. The position of the y-coordinate is shown in Fig. 10.7. For an improved overview, the projected light spots are all shifted to the coordinate origin $z' = 0$, for three different test object z-positions, respectively. Due to the observation geometry, the projected light spot has an elliptical form. Between the test object positions $z = 0$ and $z = +120$ mm, the size of the projected light spots varies only negligibly. The reason for this is illustrated with the help of Fig. 10.7. The light spot has a larger diameter at an object location $z = 120$ mm, than at $z = 0$. On the other hand, the size of the light spot at $z = 120$ mm is reduced more during its imaging onto the plane of detection, than at $z = 0$, cf. Fig. 10.4. On the grounds of these opposing relationships, the effects partially compensating each other, the size of the imaged light spots does not vary substantially.

The situation for the position of the light spot at $z = -120$ mm is somewhat different. Here, a larger light spot diameter is linked with a smaller imaging reduction. The consequence of this clearly leads to an increase in the size of the imaged light spots, as is represented in Fig. 10.8. In Sect. 10.2.4 we will consider the relevance of the size of the projected light spots on their detection in more detail.

10.2.2 Properties of the Object Surface

The properties of the object surface to be measured, together with the beam profile of the laser source, influence the accuracy in the laser triangulation measurement. This is illustrated in Fig. 10.9a–c. For the following consideration, we would assume, as previously stated, that the laser beam has a Gaussian intensity profile, see Fig. 10.9a. The laser beam is projected perpendicularly onto a plane object

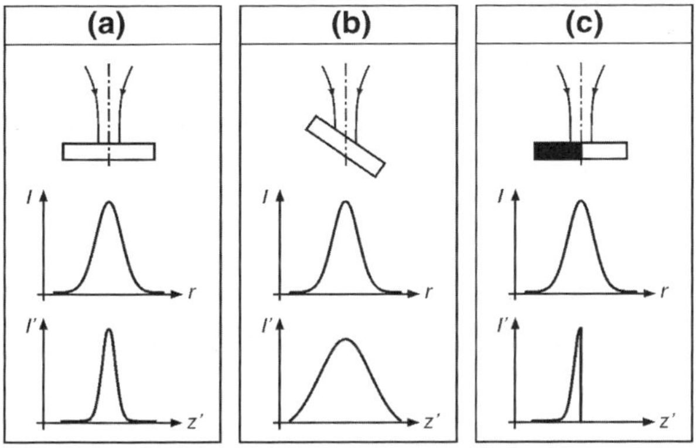

Fig. 10.9 Influence of the object surface on the intensity distribution of the imaged light spot. **a** laser beam irradiates a homogeneously scattering object oriented perpendicularly to the axis of propagation of that beam (dot and dash line), intensity distribution as a function of the radius on the object surface $I = I(r)$, corresponding intensity distribution $I' = I'(z')$ in the detector plane; **b** object is inclined towards the direction of the detector; **c** object oriented perpendicularly to the incoming beam and an inhomogeneous scattering surface

surface, and an uniform scattering behaviour is assumed. The beam profile $I(r)$ describes the intensity distribution at the point of intersection with the object surface (for the coordinate r, see Fig. 10.7). Figure 10.9a, represents schematically the intensity distribution $I' = I'(z')$ of the imaged light spot in the plane of detection. Due to the scale of imaging and the direction of observation, the half-width value of this intensity distribution is smaller than that of $I(r)$. If the object surface is inclined with respect to the incident direction—as is represented in Fig. 10.9b—then a wider intensity distribution results in the detector plane.

In both cases, the intensity distribution of the imaged light spots is symmetrical. If, for example, the position of the light spot were evaluated through a center of gravity determination, then the inclination of the object would have no influence on the result.

Figure 10.9c illustrates the case in which the object surface is oriented perpendicularly to the incident beam, the surface having non-uniform scattering properties. Indeed, one half of the object completely absorbs the incident radiation, whereas the other half scatters the beam. The intensity distribution $I' = I'(z')$ is now asymmetrical. With a center of gravity determination, this distribution would lead to a result being obtained which, in comparison to that in case (a), would be different. The resulting deviation depends on the diameter $2w$ of the incident intensity distribution. With such strongly varying scattering properties of the test object, an error is to be expected being of the same order of magnitude as the beam half-width value.

If the laser beam meets an object edge, then a similar effect occurs. This case corresponds to the combination of situations (b) and (c), sketched in Fig. 10.9. The intensity distribution $I' = I'(z')$ becomes asymmetrical and consequently a measurement error arises.

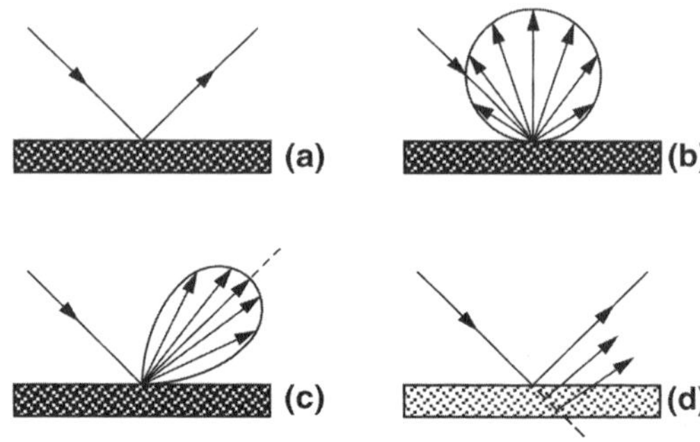

Fig. 10.10 Scattering properties of different test objects. **a** specular surface, **b** uniformly scattering surface, **c** directed scattering lobe, **d** laser beam enters a translucent object and is scattered underneath the object surface

In the following, the scattering characteristics of the object surface will be examined more closely. They are influenced by the material properties and the structure of the surface. Figure 10.10 shows schematically the different possible cases. Figure 10.10a represents an object with a specular surface. The incident laser beam undergoes a reflection, whose direction depends on the relative orientation of the laser beam and the object surface. In general, the reflected light would not fall on the aperture of the objective, so no signal is measured.

In Fig. 10.10b the ideal situation is described, in which there is an uniform scattering at the object surface in all spatial directions. This case is described by Lambert's emission law where the radiant intensity is proportional to the cosine of the angle between the direction of observation and the normal vector of the surface element, cf. Sect. 3.2.3, formula (3.9). Such objects can be measured with laser triangulation at any spatial orientation. We had always assumed this to be the case in the previous sections. The relevant case, in practice, is that of a scattering distribution, whose maximum is oriented along a considered direction of reflection, as is illustrated by the dotted line in Fig. 10.10c. The angular half-width value of this scattering distribution depends on the surface structure of the object. As a rule, a higher-valued surface roughness leads to a scattering distribution of increased angular width. If a preferred direction arises in the objects' surface structure—as, for example, with rolled sheet metal, turned or ground workpieces—then the resulting scattering distribution does not show rotational symmetry. The scattering distribution's angular half-width values differ, in this case, in two directions which are perpendicular to each other.

The known angular half-width values of the object's scattering distribution to be measured, determines the maximum permissible angle through which the object surface may be inclined to the incident direction of the laser beam and with which enough scattering light remains for detection purposes.

In Fig. 10.10d the laser beam partially propagates into the object. This occurs, for example, with organic materials such as plastics and as well with glass, foam or liquids. The laser beam is partially scattered in the region of the propagation path. In this way, the observed light spot is modified which influences the accuracy in measuring its position in the detector plane and has to be considered in the evaluation algorithms.

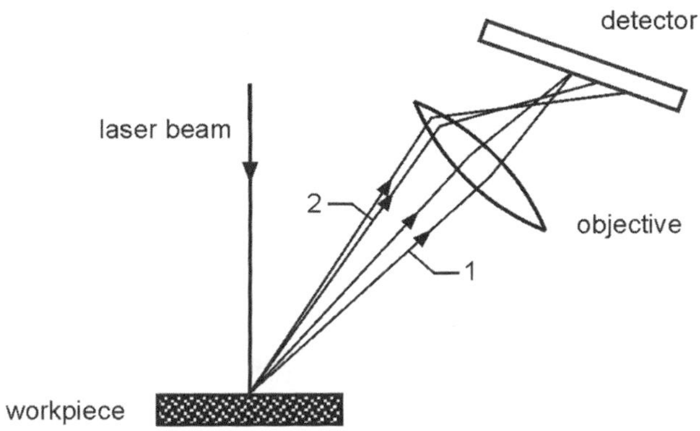

Fig. 10.11 The influence of the objectives' spherical aberration with non-uniform illumination of the aperture on the position of the imaged bundles of rays

10.2.3 Imaging Aberrations

The geometrical arrangement of laser beam, test object, objective and detector was already presented for laser triangulation in Sect. 10.1. In the following, the aspects affecting the measuring accuracy shall be concentrated on.

The objective used in laser triangulation for the observation of the light spot is, in general, a site which can generate errors. The so-called aberrations appear, for example, as coma, astigmatism, distortion and spherical aberrations [3]. The spherical aberrations describe the effect of the beam being more strongly refracted at the edge of the objective, than in the vicinity of the optical axis. The effects of this on laser triangulation are explained in Fig. 10.11 for the case of a non-uniform illumination of the objective aperture. Here, two bundles of rays, (1) and (2), within the angular light distribution scattered at the object, will be considered. The first bundle of rays, (1), is found close to the optical axis of the objective. The rays are imaged onto a point in the plane of detection. The second bundle of rays, (2), emanating from exactly the same light spot on the workpiece, see Fig. 10.11, but encounters instead, the outer region of the objective. Due to the spherical aberration, the rays are not imaged onto the same location in the plane of detection as those in the first bundle, (1), and, as is shown schematically in Fig. 10.11, a displacement appears. The spherical aberration leads to the light spot's image being distorted on one side. This phenomenon is described as a coma.

With an object scattering which corresponds to that shown in the diagram in Fig. 10.10c the following situation can occur. Due to the scattering distribution and the orientation of the object surface, only a part of the objective's aperture is illuminated. For example, in Fig. 10.11, only the radiation of the second bundle, (2), is available for imaging. Similarly, another angular orientation at the same object z-position would image only the radiation relating to the first bundle, (1), onto the plane of detection. The consequence is that there appears in the detector plane two imaged light spots in different positions, even though there was no spatial variation of the z-position. The measurement result depends on the angular orientation of the object surface. With unknown orientation, the measurement uncertainty is determined by the magnitude of the spherical aberration. If the size of the objectives' aperture were reduced, then so would be the influence of the spherical aberration. However, the light intensity available at the detector plane also falls off accordingly.

10.2.4 Detectors and Signal Evaluation

In laser triangulation, the following types of detectors are used: lateral effect diodes, CCD-line arrays, CMOS-line arrays. For the functional details of these detectors, refer to Sects. 5.4.5 and 5.4.6. The work of the detectors is that of delivering a signal from which the position of the imaged light spot can be gained. As a rule, the center of gravity of the intensity distribution in the plane of detection would be determined in this way. Table 10.1 compares the characteristics of these detectors with respect to laser triangulation.

The lateral effect diode integrates over the intensity distribution and thus directly delivers a signal, dependent on the center of gravity of the intensity distribution. With CCD or CMOS arrays, the first spatial momentum of the intensity distribution must first be calculated from the pixel values. However, for reasons of asymmetrical intensity distribution—as shown in Fig. 10.9—possible errors in the determination of the center of gravity cannot be established with lateral effect diodes.

The effect of multiple scattering on the measurement result is shown in Fig. 10.12. If the workpiece to be measured has an unevenly formed surface, then multiple scattering could arise. Beam (1) in Fig. 10.12 is scattered once at the object surface before meeting the objective. Beam (2) is scattered twice at the object surface. Both beams are imaged at different locations in the detector plane. Multiple scattering leads to a change in the intensity distribution, so that the center of gravity of this distribution no longer reproduces the exact position of the primary light spot on the workpiece. With a lateral effect diode, multiple scattering is not registered. The measurement result is therefore false. With a CCD-line array, the individual pixel values can be compared with given threshold values to discriminate the generally weaker multiple scattering intensity. Only those beams scattered once at the workpiece are used in the determination of the first spatial momentum of the intensity distribution.

In general, lateral effect diode signals are evaluated with an analog electronic system. The output currents are first converted into voltages by means of an amplifier. These voltages are bandwidth limited in order to suppress the influence of the noise of the electronic components. With operational amplifiers, difference and summation signals are formed, from which the center of gravity of the intensity distribution for the lateral effect diode results, cf. Sect. 5.4.6. With lateral

Table 10.1 Comparison of lateral effect diode, CCD-line arrays, CMOS-line arrays for 1-D laser triangulation

	Lateral effect diode	CCD	CMOS
Center of gravity	Direct	To be calculated	To be calculated
Intensity distribution	Not measurable	Evaluation possible	Evaluation possible
Multiple scattering	Not measurable	Discrimination possible	Discrimination possible
Measuring frequency	20 kHz–10 MHz	1 kHz–70 kHz	300 Hz–8 kHz
Number of pixels	N/A	2048 × 1	2048 × 32
Pixel size [μm]	N/A	10 μm × 10 μm	5.5 μm × 5.5 μm
Active detector area	5 mm × 2 mm	20 mm × 10 μm	11.3 mm × 176 μm

The numerical values are valid for typical detectors

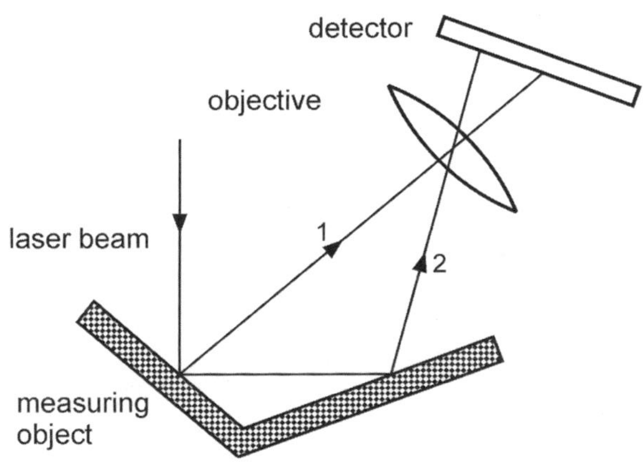

Fig. 10.12 Multiple scattering on uneven workpiece surfaces in laser triangulation

effect diodes, for reasons of the analogous evaluation, measurement frequencies of up to 10 MHz can be realized. A triangulation sensor with a lateral effect diode is therefore suitable for use in those measurement procedures which require a high measurement rate.

A CCD-line array consists of a row of light-sensitive elements connected with a shift register, cf. Sect. 5.4.5. Charges, which are generated on these elements during an integration phase, are led over a transfer gate, parallel to the shift register, at a given point in time. The charges are read out from the shift register in series. The video signal of the CCD-line array can be amplified and digitized. The intensity distribution resulting from the pixel values is evaluated by a processor in order to determine the first spatial momentum of the intensity distribution.

Another possibility exists for the evaluation of the CCD-line array video signal, in which the signal leads to a comparator. The output of the comparator controls a counter. This counter is first started with the charge transfer from the CCD-line array to the shift register. It is stopped when the video signal overshoots a given comparator threshold. The counter-state is therefore proportional to the position of a certain pixel value on the CCD-line array. This evaluation method is generally faster than that previously given above. However, one disadvantage is that the

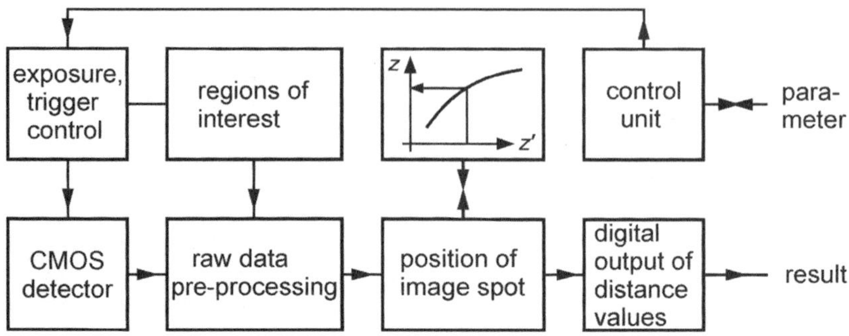

Fig. 10.13 Block diagram of the signal processing of laser triangulation sensors

form of the intensity distribution cannot be evaluated. With CCD-line arrays, measuring frequencies in the range 1–70 kHz can be achieved.

Figure 10.13 shows the concept of signal evaluation of a triangulation sensor equipped with a CMOS detector [4]. The sensor comprises a control unit which can be parametrized from a master computer. This allows for an exact triggering of the measuring process with a low jitter with respect to other time signals of external process sequences. Regions of interest (ROI) in the detector signal of the CMOS array can be defined to eliminate e.g. interfering reflexes as those shown exemplarily in Fig. 10.12. After pre-processing the position of the image spot is determined and finally a digital output is generated to transmit the result.

In the following, the effects of the speckles on the detection of the signal will be considered. Within the imaged light spot, as is represented in Fig. 10.8, for example, there exists no regular intensity distribution. A statistical intensity modulation arises as a result of the speckle effect, cf. Sect. 8.1. The size ϕ_{sp} of the speckles in the image plane depends on the objective-image plane distance and the aperture of the objective as well as on the wavelength of the laser beam, cf. equation (8.2):

$$\phi_{sp} \approx (\lambda/D)s_i, \tag{10.11}$$

where D is the diameter of the objective's aperture and s_i is the distance between the objective and the image plane.

The distance s_i is determined through the use of equation (10.4), whereby for $u' = 0$, $v' = s_i$. In the numerical example, accompanying Fig. 10.4, we have $s_i = 97.4$ mm. For an objective focal length to aperture ratio, such that $f/D = 2.8 : 1$, it follows that for $f = 85$ mm: $D = 30.4$ mm. From (10.11), it then results in an average speckle size, in the middle of the image plane, of $\phi_{sp} = 2\,\mu$m. In this case, with a CCD-line array and a pixel size of approx. 10 μm × 10 μm, some 25 speckles appear on one element. Signal noise occurs because the individual number of speckles per pixel is subject to statistical fluctuations. If a Poisson distribution is assumed, then the relative fluctuation in the number of speckles per pixel, N_{sp}, amounts to $\Delta N_{sp}/N_{sp} = 1/\sqrt{N_{sp}}$ [5]. For the chosen numerical

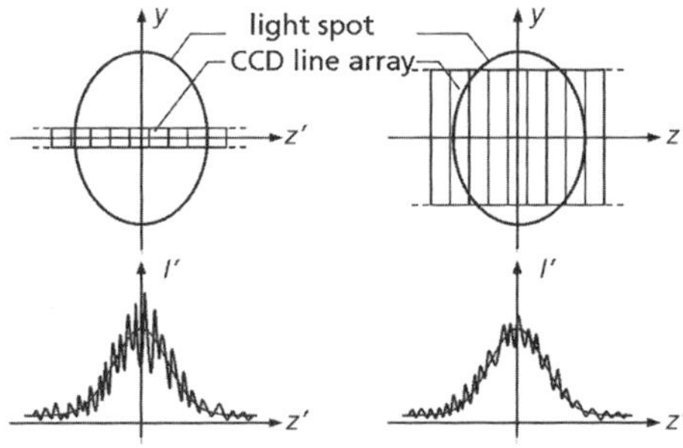

Fig. 10.14 Schematic illustration of the intensity distribution influenced by speckle noise with different CCD pixel geometries

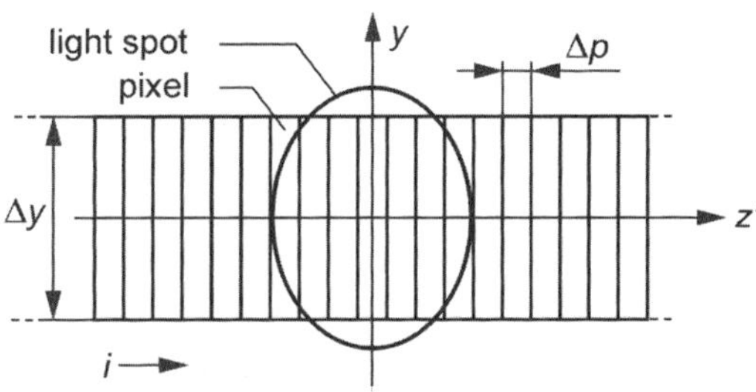

Fig. 10.15 Detection of the imaged light spot on a linear detector array with pixels of width Δp and height Δy

example, it follows that $\Delta N_{sp}/N_{sp} = 0.2$. As a consequence of the fluctuation in the individual pixel values, the intensity distribution in the plane of detection is noisy and, for this reason, there appears an additional error in the determination of the center of gravity.

Figure 10.14 shows schematically the intensity distribution in the detection plane for two differently-sized active detection areas of a CCD-line array, with an identically sized light-spot. In Fig. 10.14, left, the detector possesses an extension in the y-direction of 10 μm and pixels with a quadratic geometry. As has already been stated, in this case there are only relatively few speckles per pixel available. Those intensity distributions measured with a linear CCD-sensor, $I' = I'(z')$, indicate strong fluctuations due to the speckle noise. In Fig. 10.14, right, a CCD with non-quadratic pixels is represented. The extension in the y-direction amounts to a multiple of the extension of a pixel in the z'-direction, for example. The resulting intensity distribution shows reduced fluctuations because it is now integrated over a larger number of speckles in the y-direction. The location of the light-spot in the image plane, with respect to the z'-coordinate, can be determined more exactly in this way.

In the following we consider in more detail the effect of the speckles on the uncertainty of the determination of the center of gravity of the imaged light spot. Figure 10.15 shows again the imaged light spot and the detector array where the single photo-sensitive elements i, also called pixels, have a height of Δy and a width of Δp. The radiant energy Q_i received by an element i is given by the product of the irradiance, the area of the pixel and the exposure time.

The position of the imaged light spot in terms of the center of gravity of the counter variable i is calculated as follows:

$$i_{\mathrm{cg}} = \frac{\sum_{i=1}^{m} iQ_i}{\sum_{i=1}^{m} Q_i}, \tag{10.12}$$

with m total number of pixels of the detector array. For the radiant energy received by a pixel we write:

$$Q_i = Q_{i0} + \Delta Q_i, \tag{10.13}$$

where Q_{i0} is the radiant energy received by a pixel i if no speckles would occur and ΔQ_i is the change of this radiant energy caused by the speckles. Due to energy conservation we have $\sum_{i=1}^{m} \Delta Q_i = 0$. For the relative standard deviation of this energy fluctuation we define:

$$\delta_i = \frac{\Delta Q_i}{Q_{i0}}. \tag{10.14}$$

For the sum of all energy portions we set:

$$Q = \sum_{i=1}^{m} Q_i. \tag{10.15}$$

With (10.13), (10.14) and (10.15) inserted in (10.12) the center of gravity of the light spot can be written as follows:

$$\begin{aligned} i_{\mathrm{cg}} &= \frac{1}{Q}\sum_{i=1}^{m} i(Q_{i0} + \Delta Q_i) = \frac{1}{Q}\sum_{i=1}^{m} iQ_{i0}(1+\delta_i) \\ &= \frac{1}{Q}\sum_{i=1}^{m} iQ_{i0} + \frac{1}{Q}\sum_{i=1}^{m} iQ_{i0}\delta_i = i_{\mathrm{cg0}} + \frac{1}{Q}\sum_{i=1}^{m} iQ_{i0}\delta_i, \end{aligned} \tag{10.16}$$

where i_{cg0} denotes the center of gravity of the light spot if no fluctuations due to speckles are present. Let us now take N measurements of this light spot. The average of these measurements is calculated in the following way:

$$\langle i_{\mathrm{cg}} \rangle = \frac{1}{N}\sum_{j=1}^{N} \left(i_{\mathrm{cg0}} + \frac{1}{Q}\sum_{i=1}^{m} iQ_{i0}\delta_i \right)_j. \tag{10.17}$$

For many measurements with $N \gg 1$ the sum over the second term in the brackets of (10.17) will be zero. Hence we have for (10.17):

$$\langle i_{\mathrm{cg}} \rangle = \frac{1}{N} \sum_{j=1}^{N} i_{\mathrm{cg0}} \big|_j = i_{\mathrm{cg0}}. \tag{10.18}$$

The variance of the quantity i_{cg} is calculated with:

$$s_{\mathrm{cg}}^2 = \left\langle \left(i_{\mathrm{cg}} - i_{\mathrm{cg0}} \right)^2 \right\rangle = \frac{1}{N-1} \sum_{j=1}^{N} \left(i_{\mathrm{cg}} - i_{\mathrm{cg0}} \right)^2 \Big|_j = \frac{1}{N-1} \sum_{j=1}^{N} \left(\frac{1}{Q} \sum_{i=1}^{m} i Q_{i0} \delta_i \right)^2_j,$$

$$s_{\mathrm{cg}}^2 = \frac{1}{N-1} \frac{1}{Q^2} \sum_{j=1}^{N} \left(\sum_{i=1}^{m} i^2 Q_{i0}^2 \delta_i^2 + \sum_{i=1}^{m} \sum_{k=1, k \neq i}^{m} (i Q_{i0} \delta_i)(k Q_{k0} \delta_k) \right) \Bigg|_j. \tag{10.19}$$

Since the δ_i and δ_k are uncorrelated the double sum vanishes. Hence we obtain for (10.19):

$$s_{\mathrm{cg}}^2 = \frac{1}{N-1} \frac{1}{Q^2} \sum_{j=1}^{N} \left(\sum_{i=1}^{m} i^2 Q_{i0}^2 \delta_i^2 \right) \Bigg|_j. \tag{10.20}$$

For simplification we assume that all δ_i have the same value, i.e.:

$$\delta_i = \delta. \tag{10.21}$$

Using (10.21) we rewrite (10.20) as follows:

$$s_{\mathrm{cg}}^2 = \frac{1}{N-1} \delta^2 N \frac{\sum_{i=1}^{m} i^2 Q_{i0}^2}{Q^2} \approx \delta^2 \frac{\sum_{i=1}^{m} i^2 Q_{i0}^2}{Q^2}. \tag{10.22}$$

In order to evaluate this relation we assume that the energy portions Q_{i0} can be described by a Gaussian function assuming—without loss of generality—that this function has a maximum at $i = 0$:

$$Q_{i0} = \frac{\Delta p Q}{a \sqrt{\pi}} \mathrm{e}^{-\frac{i^2 \Delta p^2}{a^2}}, \tag{10.23}$$

where Δp is the width of a pixel in z'-direction, cf. Fig. 10.15; $a = l_{\mathrm{FWHM}} / \left(2\sqrt{\ln 2} \right)$ and l_{FWHM} is the full width at half maximum of the light spot described by (10.23). For the further calculation steps we approximate the sum in (10.22) by an integral assuming $\Delta p \ll a \ll L$, where L is the length of the detector chip: $L = m\Delta p$. Integrating (10.23) over i from minus to plus infinity just gives Q. With the taken assumptions we rewrite (10.22):

$$s_{\mathrm{cg}}^2 = \delta^2 \frac{\Delta p^2}{a^2 \pi} \int_{-\infty}^{\infty} i^2 \mathrm{e}^{-2 \frac{i^2 \Delta p^2}{a^2}} \mathrm{d}i = \frac{\delta^2 a}{\sqrt{32\pi} \Delta p}. \tag{10.24}$$

The measuring uncertainty in the determination of the position of the light spot due to the speckle formation can now be calculated:

$$\Delta z'_{\mathrm{cg}} = \sqrt{s_{\mathrm{cg}}^2} \cdot \Delta p = \frac{\delta \sqrt{a \Delta p}}{(32\pi)^{1/4}}. \tag{10.25}$$

Relation (10.25) shows that in the case of an absent speckle effect, i.e. $\delta = 0$, the measuring uncertainty vanishes, which is intuitively evident. To estimate the relative standard deviation of the energy portions due to the speckle effect, we calculate the number of speckles in a pixel:

$$N_{\mathrm{sp}} = \frac{\Delta p \Delta y}{\frac{\pi}{4} \phi_{\mathrm{sp}}^2}, \tag{10.26}$$

with ϕ_{sp} size of the speckles cf. (10.11). As discussed above, assuming a Poisson statistics we set:

$$\delta = \frac{1}{\sqrt{N_{\mathrm{sp}}}}. \tag{10.27}$$

Inserting (10.27), (10.26), (10.11) in (10.25) then yields:

$$\Delta z'_{\mathrm{cg}} = \frac{\sqrt{\pi}}{(32\pi)^{1/4}} \frac{\lambda s_{\mathrm{i}}}{D} \sqrt{\frac{a}{\Delta y}}. \tag{10.28}$$

Formula (10.28) shows that the uncertainty is improved if the size of the pixels in the y-direction, cf. Fig. 10.15, is increased. Obviously this is only the case as long as the number of speckles captured by the pixel is increasing according to (10.26). If Δy becomes greater than l_{FWHM} the number of pixels no longer increases. To estimate the minimal achievable uncertainty we therefore take $\Delta y \approx l_{\mathrm{FWHM}} = 2\sqrt{\ln 2}\, a$; inserted in (10.28) yields:

$$\Delta z'_{\mathrm{cg}} = \frac{\sqrt{\pi}}{2(32\pi)^{1/4}} \frac{1}{\sqrt{2}(\ln 2)^{1/4}} \frac{\lambda s_{\mathrm{i}}}{D} = \frac{1}{2^{3/2} 32^{1/4}} \left(\frac{\pi}{\ln 2}\right)^{1/4} \frac{\lambda s_{\mathrm{i}}}{D} = \kappa \frac{\lambda s_{\mathrm{i}}}{D}, \tag{10.29}$$

where the constant is $\kappa \approx 0.22$.

In the next step we transform the uncertainty $\Delta z'_{\mathrm{cg}}$ from the detector plane to the object space. Remembering the geometry shown in Fig. 10.3 and using the lens equation we get:

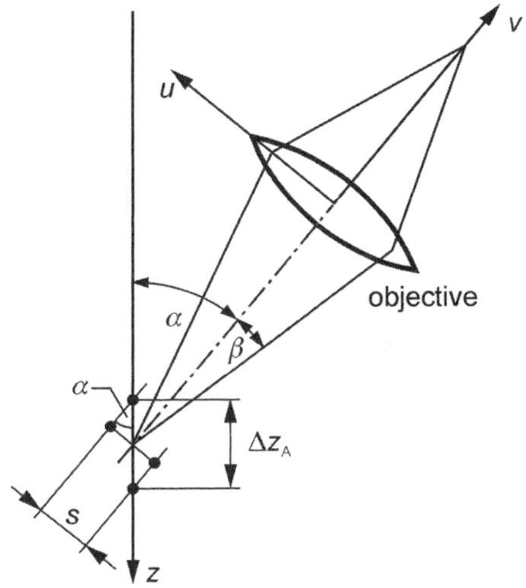

Fig. 10.16 Observation of two object points at a mutual distance s with the objective in the triangulation geometry and projection of this distance to the z-axis

$$\frac{\Delta z'_{cg}}{\Delta z \sin \alpha} = \frac{s_i}{s_o}, \tag{10.30}$$

where s_i is the distance between the objective to the detector plane and s_o is the distance between the objective and the measuring object, cf. Sect. 10.1.1. The half opening angle β of the objective when looking from the measurement location towards the objective and assuming $D/(2s_o) \ll 1$, is given by, see Fig. 10.16:

$$\tan \beta = \frac{D}{2s_o} \approx \sin \beta. \tag{10.31}$$

Inserting (10.30), (10.31) into (10.29) then yields as theoretical lower limit of the measuring uncertainty for laser triangulation sensors [6]:

$$\Delta z = \kappa \frac{\lambda}{2 \sin \alpha \sin \beta}. \tag{10.32}$$

According to (10.32) the measuring uncertainty can be reduced for smaller wavelengths, greater triangulation angles α and greater apertures of the objective. For the numerical example chosen for Fig. 10.4, $\lambda = 633$ nm, $\alpha = 40°$, $\sin \beta = 0.023$, we get as estimate of the minimum achievable uncertainty: $\Delta z = 4.8$ μm. Practically a further increase of the triangulation angle is limited by the reduced intensity of the scattered light, cf. Figs. 10.1, 10.10a–c. A further increase of the numerical aperture of the objective is restricted by the available construction space of the sensor and the consequential increase of the size of the observation window in the sensor housing, which has to be kept clean in an industrial environment.

It is interesting to see, that (10.32) is identical, with exception of a constant factor, with the limit to resolve two adjacent points of an object in microscopy as described by Abbe. According to Abbe two points with a mutual distance s can be distinguished, if the following relation holds, see Fig. 10.16 [3]:

$$s = \frac{\lambda}{2 \sin \beta}. \tag{10.33}$$

If we project the distance s—which is running parallel to the u-axis—onto the z-axis, we receive $\Delta z_A = s / \sin \alpha$. The resolvable distance according to Abbe (10.33) becomes then:

$$\Delta z_A = \frac{\lambda}{2 \sin \alpha \sin \beta}. \tag{10.34}$$

Relation (10.34) shows the same dependencies on the wavelength, the triangulation angle α and the numerical aperture as does (10.32). So Abbe's interpretation of using at least the $\pm 1.$ diffraction orders of the smallest object detail to resolve this detail described with (10.33), (10.34) is equivalent to the derivation shown here considering the influence of the speckle effect—which is a diffraction effect—to estimate the uncertainty in the determination of the position of the imaged light spot.

10.2.5 Atmospheric Influences

The propagation of the irradiated and scattered light with laser triangulation is influenced by variations in the refractive index of the surrounding atmosphere. These appear particularly in the environment of hot workpieces, which should be measured. Figure 10.17 shows how a laser beam would be deflected due to the presence of a refractive index gradient in a region with the extension l in the z-direction. The angle of deflection α_x, valid under the assumption that $|\alpha_x| \ll 1$, is given by [6]:

$$\alpha_x = \int_0^l \frac{1}{n} \frac{dn}{dx} dz, \tag{10.35}$$

where α_x is the angle of deflection in the x-direction, n the index of refraction; x, z the coordinates, see Fig. 10.17; dn/dx the refractive index gradient in the x-direction and l the length of the range in the z-direction in which the gradient dn/dx is present.

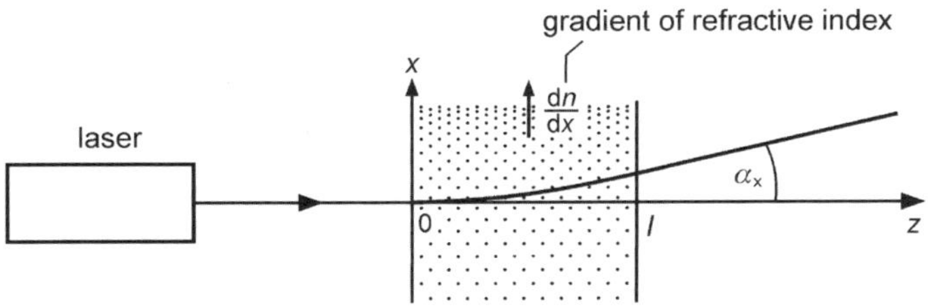

Fig. 10.17 Beam deflection in the atmosphere due to a refractive index gradient in the x-direction

Relationship (10.35) can be derived from Snell's law. The integration follows over the beam path in the z-direction. As is shown in Fig. 10.17, the beam is deflected in the direction of the more dense medium.

The refractive index gradient is produced essentially as a result of pressure and temperature gradients in the atmosphere. The following is approximately valid (cf. equation (6.15) in Sect. 6.2.3):

$$\mathrm{d}n/\mathrm{d}x \approx \Delta n/\Delta x = \left((0.27\,\mathrm{m/hPa})\frac{\Delta p}{\Delta x} - (0.92\,\mathrm{m/K})\frac{\Delta T}{\Delta x}\right)10^{-6}/\mathrm{m}, \qquad (10.36)$$

with $\Delta p/\Delta x$ difference quotient of the pressure in the x-direction and $\Delta T/\Delta x$ difference quotient of the temperature in the x-direction.

Equation (10.36) implies that with constant temperature and an increase in pressure, the refractive index increases similarly. Using the general gas equation, an increase in pressure at constant temperature indicates that the gas density has increased. A stronger refraction ensues at a higher density. On the other hand, an increase in temperature at constant pressure results in a decrease in the refractive index. In this case, the density of the gas is lower, as is the refraction.

If $1/n(\mathrm{d}n/\mathrm{d}x)$ is independent of z, then equation (10.36) can be simplified: $\alpha_x = 1/n(\mathrm{d}n/\mathrm{d}x)l$. The angle of deflection increases proportionally with the distance along the beam path. For the beam position in the x-direction along a path z, the following is approximately valid:

$$x \approx \frac{1}{2n} \cdot \frac{\mathrm{d}n}{\mathrm{d}x} \cdot z^2. \qquad (10.37)$$

In Fig. 10.18, equation (10.37) is represented for various pressure and temperature gradients.

In the presence of static pressure and temperature gradients, the beam does not linearly propagate but, instead, assuming a constant refractive index, the beam describes a parabolic trajectory, following equation (10.37). In this way, a systematic measuring error arises with laser triangulation. These influential factors have to be considered, particularly when hot workpieces are to be measured.

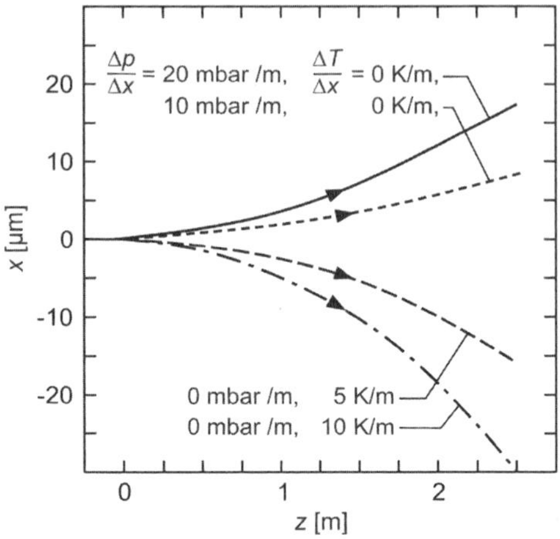

Fig. 10.18 Beam deflection in x-direction according to (10.37) as a function of the propagation path in z-direction for different pressure and temperature gradients

10.3 Triangulation Sensors for Contour Measurement

The principle of triangulation sensors to measure distances was shown in Fig. 10.1. The sensor comprises mainly the components: laser source, objective and detector. If a contour or profile of an object is to be measured, corresponding to the 2-D measurement illustrated in Fig. 10.2, middle—then the laser beam is formed to a line projected onto the measuring object. An alternative is to use a 1-D triangulation sensor with a collimated laser beam which is moved across the object by a linear axis or the object is moved relative to the sensor.

Figure 10.19 shows the principle set-up of a 2-D triangulation sensor based on laser light section. By use of a cylindrical lens the laser beam is formed to a line which is projected onto the measuring object. Figure 10.20 represents a photograph of a laser light section sensor for profile and contour measurements [1]. Typical technical data are: measuring range 130 mm × 105 mm, mean measuring distance 272 mm, detector array 1280 × 1024 pixels, measuring frequency 200 Hz, spatial precision 130 µm, temporal precision 65 µm, linearity 100 µm (the last three characteristics are defined in [2]).

Another possibility for contour measurements is the combination of a triangulation sensor with moving mirrors [8, 9]. Figure 10.21 shows schematically the set-up of such a scanning sensor. The illuminating laser beam is guided via a rotating mirror onto the measuring object. The scattered light is collected by another rotating mirror which is aligned coplanar on the same axis as the first one and directed to the objective. Again laser beam, objective and detector are arranged according to Scheimpflug's condition, cf. Sect. 10.1. The detector signal depends only on the perpendicular distance between the rotating axis and the respective illuminated spot on the measuring object. The angular position of the axis is measured by an angular transmitter. In this way the scanned contour of an object is measured in polar coordinates.

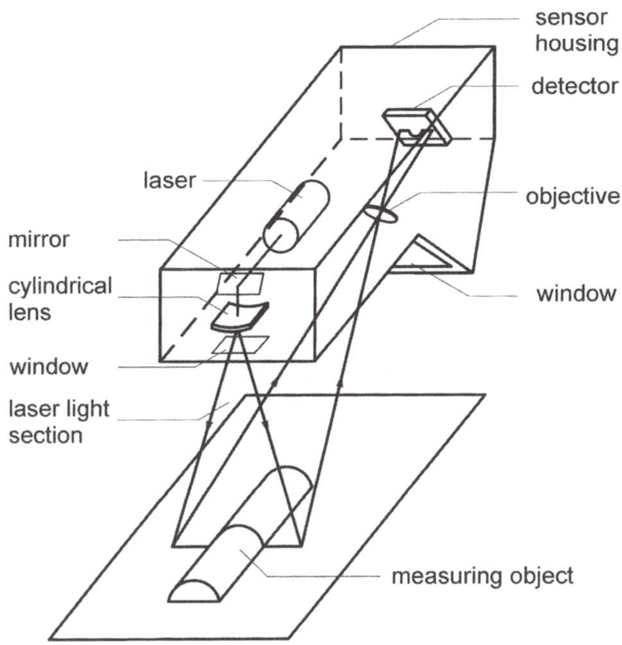

Fig. 10.19 Set-up of a 2-D triangulation sensor. The projection of a laser line onto a measuring object and observation of this line under an angle to measure a profile is also referred to as laser light section method

Fig. 10.20 Laser light section sensor for profile and contour measurements, *left* object with projected laser line, *right* triangulation sensor

Another type of triangulation sensor to measure contours is shown in Fig. 10.22 [10]. The laser beam is guided via two lenses and a rotating polygon mirror onto the workpiece. The lenses are designed and arranged in such a way, that the rotation of the polygon mirror leads to a parallel translation of the illuminating beam in the object space. This kind of beam propagation is called telecentric.

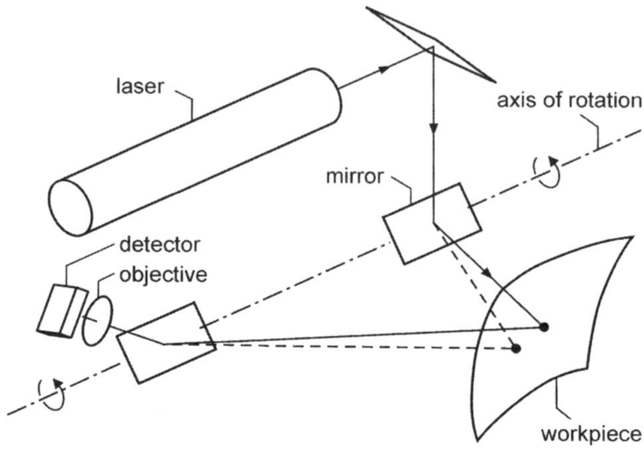

Fig. 10.21 Triangulation sensor for contour measurement with rotating mirrors on a common axis

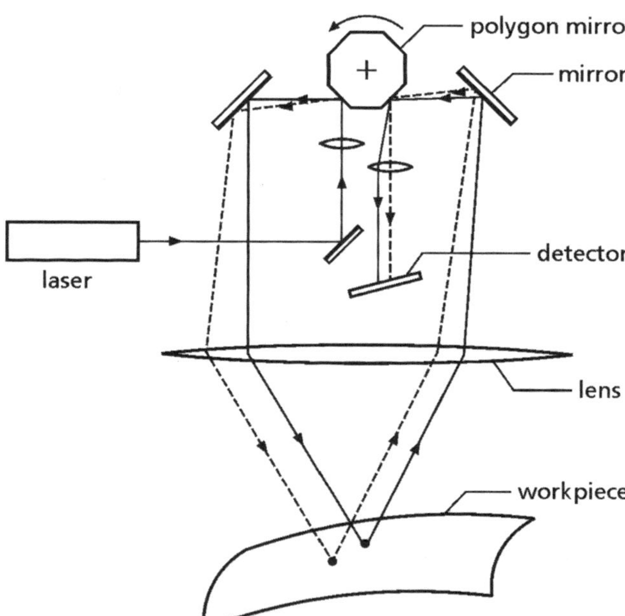

Fig. 10.22 Triangulation sensor for contour measurement with telecentric optical path and a polygon mirror

All illuminating light beams run parallel to each other in the object space. The scattered light of the light spot is also imaged onto the detector with a telecentric beam propagation via the polygon mirror. This enables all illuminating light beams and the corresponding optical axes of the imaging apparatus to intersect at the same angle in the object space. In this way, a high degree of position-independent accuracy in contour measurement is attained.

The detector is inclined in order that a sharp imaging be retained for different light spot positions, corresponding to the Scheimpflug condition. The detector signal is a function of the perpendicular distance between the plane, formed by the lens, and the light spots on the test object. In the perpendicular direction, the angular position of the polygonal mirror determines the location of the light spot. In this sense, the triangulation sensor, such as that in Fig. 10.22, measures the workpiece contour in cartesian coordinates.

10.4 Examples of Applications

Triangulation sensors are deployed for a variety of applications in process control and quality assurance. They measure geometric quantities of moving objects. Figure 10.23a–d shows exemplarily some application cases of triangulation sensors for distance measurements.

In Fig. 10.23a two sensors are mounted on a translation stage and guided along a profile to be measured. With the known distance between the guiding rails and the two distances measured with the triangulation sensors the profile of the measuring object is determined. In another application the sensors are mounted fixed and measure the straightness of rails passing the sensors continuously [1]. Figure 10.23b shows an example to measure internal dimensions of objects [11].

The imbalance of a rotating workpiece can be measured with the set-up shown in Fig. 10.23c, where one sensor measures the radial runout and the other the axial runout. Similar set-ups are used to inspect the geometry of various types of shafts and shaft-like components, as e.g. cam shafts, drive shafts or crank shafts [1]. However, due to the continuously increasing requirements with respect to the accuracy for the inspection of high-precision shafts such as ground shafts, a principle limit of the triangulation principle is reached, cf. Sect. 10.2.4. For measurement uncertainties in the lower micrometer and especially sub-micrometer range absolute measuring interferometric methods have significant advantages [12].

Figure 10.23d shows schematically an array of triangulation sensors for the inline control of the thickness of rolled sheet metal [13]. On two sides of a passing sheet, at opposite positions, several sensors are mounted. The sensors are triggered synchronously. From the distance signals gained the thickness of the sheet for each measuring position is determined.

The flatness of rolled heavy plates is measured with a variant of the laser triangulation principle as illustrated schematically in Fig. 10.24 [14]. Three triangulation sensors, M1 to M3, are mounted 5 m above a roller table to measure the flatness of moving hot rolled heavy plates. Each sensor projects 10 laser beams onto the surface of the passing measuring object. Within each sensor the row of illuminated spots is imaged onto two CCD detector arrays using the Scheimpflug condition in such a way, that the set of 2×5 spots across the width of the rolled

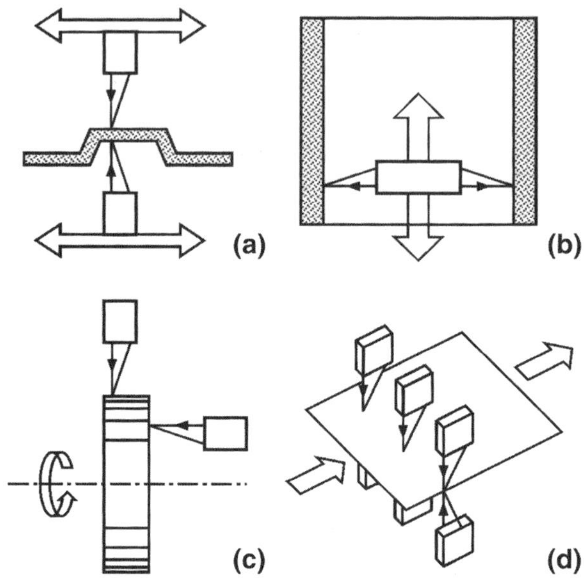

Fig. 10.23 Examples of applications of triangulation sensors. **a** measurement of the profile of an object, **b** measurement of inner dimensions of objects, **c** measurement of the radial and axial runout of a rotating workpiece, **d** measurement of the thickness of sheets at various lateral positions. Arrows indicate direction of movement

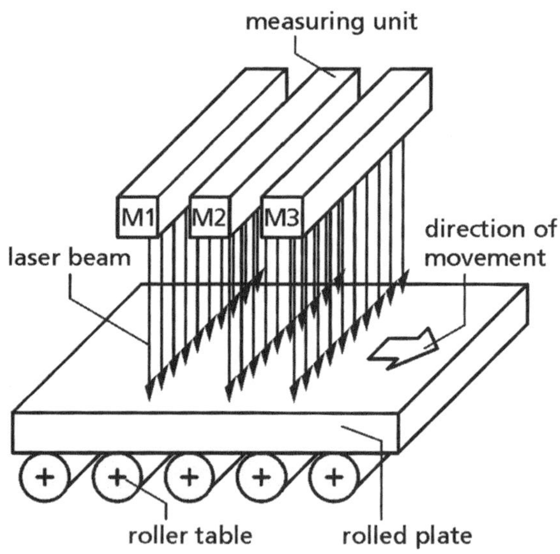

Fig. 10.24 Measurement of the flatness of heavy plates with laser triangulation sensors

plate is imaged sharply onto the respective detector plane. The array of measuring spots covers an area of 1500 mm × 4500 mm [15]. The measuring range in z-direction, cf. Figure 10.3, amounts to 400 mm. The measuring frequency is 100 Hz. The three triangulation sensors are synchronized so that all thirty distances are measured simultaneously. By this approach it is possible to determine the flatness of the rolled plate and to eliminate artefacts induced by irregular

Fig. 10.25 Flatness measurement system based on laser triangulation in a rolling mill

movements of the plate itself while transported on the roller table. Hot sheet metal with temperatures up to 900 °C can be measured. Due to the heat radiation of the hot plate the surrounding atmosphere grows warm and convection currents are induced with accompanying gradients of the refractive index, cf. Sect. 10.2.5. At plate temperatures of 900 °C the measuring uncertainty caused by this effect is of the order of ±2 mm. The measuring result is evaluated in real-time and the recognized flatness defects are displayed in the control room of the rolling mill as a basis for feedback and optimization actions to a preceding water cooling device.

Figure 10.25 shows a photograph of the installed system in the rolling mill. The sensors are mounted at the top of a steel scaffolding above the roller table. On the roller table a hot heavy plate is passing.

Figure 10.26 presents some examples of flatness topologies of heavy plates measured with the system shown in Figs. 10.24 and 10.25. The flatness deviations are shown enlarged to recognize more easily the type of flatness defect occurring. The colors represent the amount of deviation from an ideal plane. The flatness defects are automatically classified in terms of defect types and their global and local quantification is determined. The raw data from the multi-point triangulation is processed as follows: pre-processing to eliminate sheet movement, assessment of global structure, extraction of local features, classification of pattern features by use of a neuro-fuzzy algorithm.

Figure 10.27 shows an application of a triangulation sensor in a coordinate-measuring machine. The machine has e.g. two linear axes and one axis of rotation at which the sensor is mounted. The specimen—in the shown example a section of an H-girder—is placed on the measuring table. A computer controls the axes and moves the sensor around the contour of the specimen. From the positions of the axes and the distances measured with the triangulation sensor the contour of the specimen is determined. Due to the contactless measurement the measuring speed

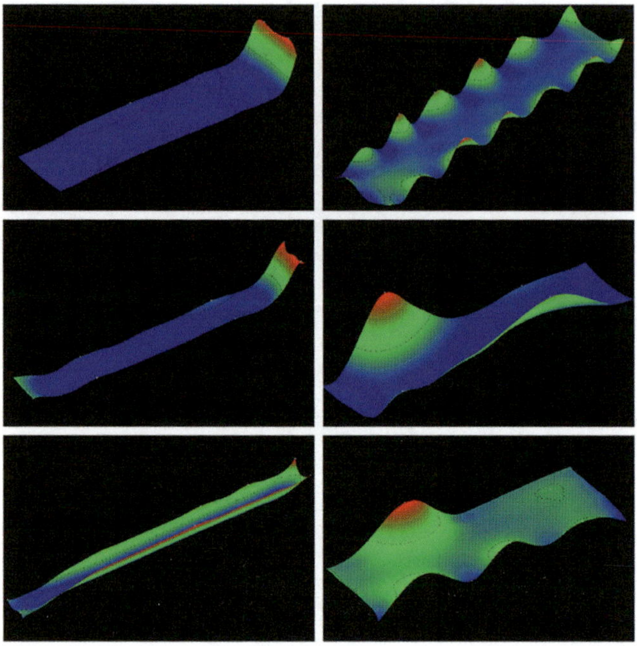

Fig. 10.26 Flatness topologies of heavy rolled plates measured with laser triangulation

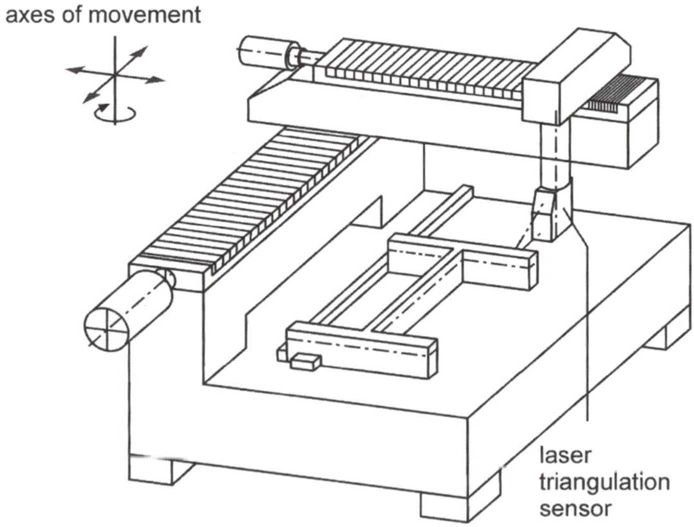

Fig. 10.27 Coordinate-measuring machine with a laser triangulation sensor

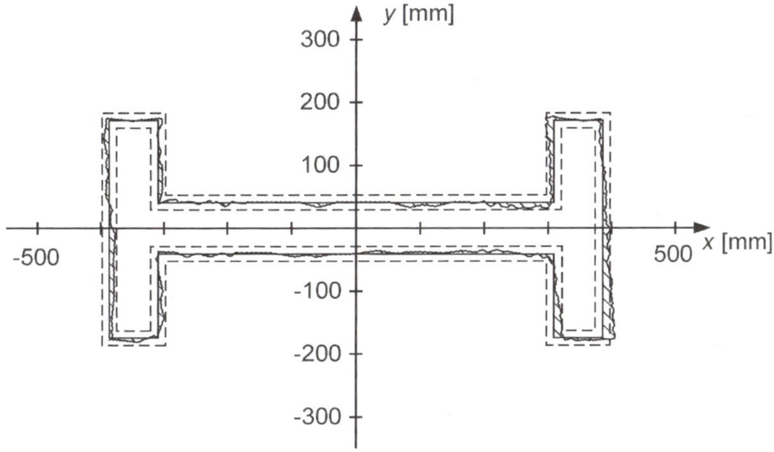

Fig. 10.28 Measurement report of an H-girder

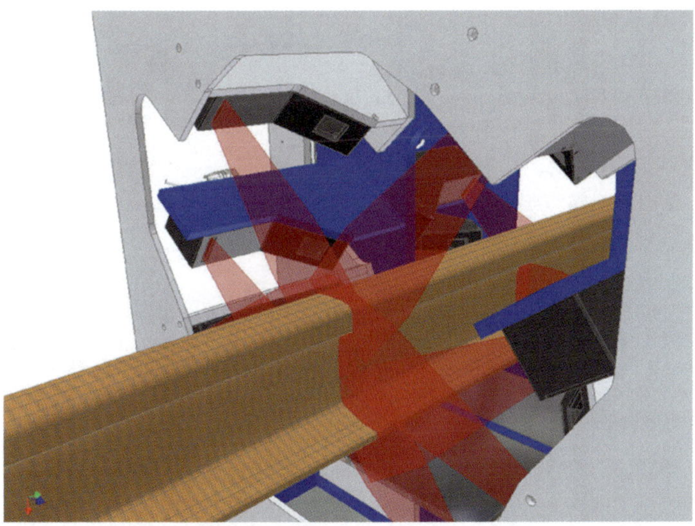

Fig. 10.29 Three dimensional CAD model of a laser measuring unit to measure the cross-sectional profile of rails. The measuring object shown is the high-speed rail UIC 60

is much faster than with a conventional coordinate-measuring machine working with tactile sensors.

Figure 10.28 shows the measurement report documenting the deviations from the desired contour.

The noncontact measurement with laser triangulation sensors allows for the inspection of product shapes and geometries without the need to take the specimens out of the production line. Figure 10.29 shows a set-up to measure the cross-section of rails with an array of ten laser light section sensors mounted around the rail passing through the measuring rig [1]. The light section sensors generate a laser line over the entire circumference of the rail profile. To avoid overlaps between the laser lines, different wavelengths are used and they are triggered with a time shift. With this set-up the contour of the rails is measured automatically along the whole length of 120 m of each produced rail piece in a production line.

With a laser light section sensor the microtopology of weld seams can be measured to gain a fingerprint of the surface structure, which is the result of statistical processes occurring during welding. In this way weld seams can be used to identify components or casks in a unique way [16].

References

1. A. Lamott, R. Noll, Laser triangulation, in *Tailored Light 2, Laser Application Technology*, Chapt. 19.1, ed. by R. Poprawe (Springer, Berlin Heidelberg, 2011), pp. 473–486. ISBN 978-3-642-01236-5
2. DIN 32 877: 2000-08, Optoelectronic measurement of form, profile and distance, Beuth Verlag GmbH, Berlin, Germany, August 2000
3. L. Bergmann, C. Schäfer, *Lehrbuch der Experimentalphysik.* (*Band III*: *Optik*), ed. by H. Gobrecht (Walter de Gruyter, Berlin, 1978)
4. R. Noll, Online Lasermesssysteme für die Stahl- und Automobilindustrie, VDI-Berichte Nr. 2011, ISBN 978-3-18-092011-5, 2008, 3–12
5. I. Bronstein, K. Semendjajew, *Taschenbuch der Mathematik* (Verlag Harri Deutsch, Thun, 1980)
6. R. Noll, W. Brandenburg, H. Aehling, S. Seiwert, Laser-based inspection of product shapes, Proc. 3. Int. IMEKO Symposium, Heidelberg, VDI-Bericht Nr. 1118, ISBN 3-18-091118-2, 1994, 1–6
7. H. Wolter, *Handbuch der Physik. Band 24*, ed. by S. Flügge, (Berlin, Springer-Verlag, 1956)
8. H. Schüssler, Industrielle Lasermesstechnik und ihr Einsatz in Produktion und Entwicklung, Automobil-Industrie 28, 475–485 (1983)
9. G. Seitz, H. Tiziani, R. Litschel, 3-D-Koordinatenmessung durch optische Triangulation, in *Laser-Technologie und Anwendungen*, Jahrbuch, 1st. edn., (Essen, Vulkan-Verlag, 1987)
10. G. Bickel, G. Häusler, M. Maul, Three-dimensional inspection of large objects, in *Optical Metrology*, ed. by O. Soares (Martinus Nijhoff Publishers, Dordrecht, 1987)
11. R. Noll, Online-Prüfung von Bändern und Rohren, Bänder Bleche Rohre, 11/12, S. 72–75 (2008)
12. R. Noll, S. Hölters, J. Kämmerling, A. Lenenbach, Laserabstandssensoren messen die Dicke von Walzbändern, Qualität und Zuverlässigkeit 58(10), 44–47 (2013)
13. R. Noll, M. Krauhausen, Online laser measurement technology for rolled products. Ironmaking & Steelmaking **35**, 221–227 (2008)
14. M. Sellhorst, R. Noll, Triangulationsverfahren, patent DE 195 32 767, granted 28 Dec 2000
15. H. Kirsch, B. Bödefeld, R. Wester, R. Noll, J. Treuil, P. Autesserre, J. Guisti, R. Fournier, Automatic recognition and quantification of flatness defects at heavy plates after a water cooling treatment for online optimisation of the cooling process, final report, EUR 20503, ECSC project no. 7210-PR/024, 2002
16. K. Rudolf, R. Noll, B. Richter, Verification of Pollux cask seals by laser techniques, in *Proceedings 19th Annual Symposium on Safeguards and Nuclear Material Management,* (Montpellier, 1997), pp. 635–639

Chapter 11
Laser Doppler Methods

Abstract Laser Doppler methods are used to measure the state of motion of objects and particles. Laser vibrometers and laser anemometers make use of this effect. Their interferometric principle is explained which allows to determine velocities and to discriminate as well the direction of movement. Signal processing schemes are described such as frequency demodulation, digital signal processing, tracker and photon correlators. Examples of applications as the measurement of the three dimensional vibration state of car bodies and flow fields in wind tunnels are presented.

Laser Doppler methods are, as a rule, point-oriented measuring methods, with which the state of motion of objects or individual particles can be investigated. The basis for this method is the Doppler effect which describes the shift in frequency of an electro-magnetic wave, scattered at a moving object. The velocity of the object can be determined from this frequency shift. In the following, the Doppler effect will be explained and, based on this, the methods of laser vibrometry and laser Doppler anemometry are presented. These methods are used to investigate component vibrations as well as in flow measuring technology.

11.1 Doppler Effect

The Doppler effect describes the change in frequency of a wave directed at an object, as a result of the motion of the object. The change in frequency of the scattered wave is measured, in order to determine the velocity of an object point.

Figure 11.1 represents the scattering of an electromagnetic wave at a moving object. The laser beam, directed at the object, is described by a plane wave, which propagates in the direction of the unit vector, $\vec{e}_{\mathrm{i}}$. The coordinate in the direction of this unit vector is denoted by z. A point on the surface of the scattering object, at a certain point in time, is given the position P(0). This point travels with a velocity $\vec{v}$.

A. Donges and R. Noll, *Laser Measurement Technology*, Springer Series in Optical Sciences 188, DOI: 10.1007/978-3-662-43634-9_11,

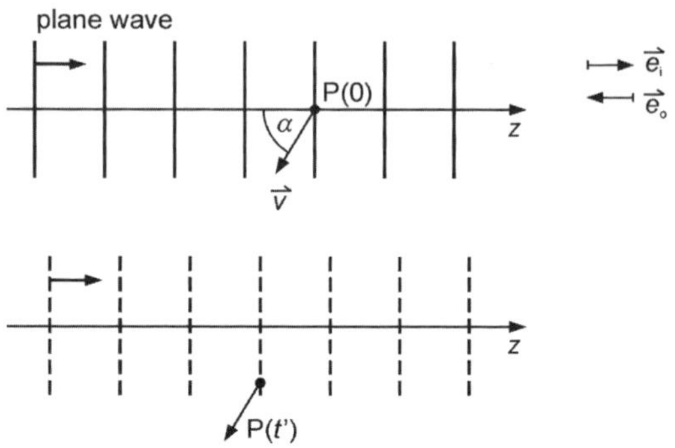

Fig. 11.1 Scattering of an electromagnetic wave at a moving object point P

In Fig. 11.1, above, a wavefront just reaches the position P(0). In Fig. 11.1, below, the point has travelled to the position P(t'). The incident wave has propagated further in the meantime, as is represented by the dashed lines in Fig. 11.1, below. An observer at the point P, in the system at rest of point P, measures a wave frequency of $f' = 1/t'$. Thus, for this frequency, cf. Sect. 3.6, we have:

$$f' = f + (v/\lambda)\cos\alpha, \tag{11.1}$$

with f frequency of the incident wave, f' frequency of the wave measured by the observer at P, v absolute value of velocity $\vec{v}$, α see Fig. 11.1, above.

Normally, equation (11.1) would be written in terms of the angular frequencies and the angular wave number, whereby:

$$\omega' = \omega + kv\cos\alpha, \tag{11.2}$$

with $\omega' = 2\pi f'$, $\omega = 2\pi f$, $k = 2\pi/\lambda$.

The light of angular frequency ω' is scattered at the point P on the surface of the object. In the following, we will assume for the laboratory system that the scattered light is observed in the direction of the unit vector $\vec{e}_o$, a direction which exactly opposes the incident direction of the radiation, cf. Fig. 11.1, below. An observer in the laboratory system observes an angular frequency other than ω', because the point P moves towards him.

Analogous to equation (11.2), it shows that for the frequency ω'', measured by the observer in the laboratory system, the following is valid:

$$\omega'' = \omega' + k'v\cos\alpha, \tag{11.3}$$

where ω'' is the angular frequency of the scattered wave measured in the laboratory system, k' the angular wave number of the wave measured by the observer at P.

The quantities ω' and k' can be eliminated using (11.2), (11.3) and $\omega' = k'c$. Under the assumption that the velocity v of the object point is much smaller than

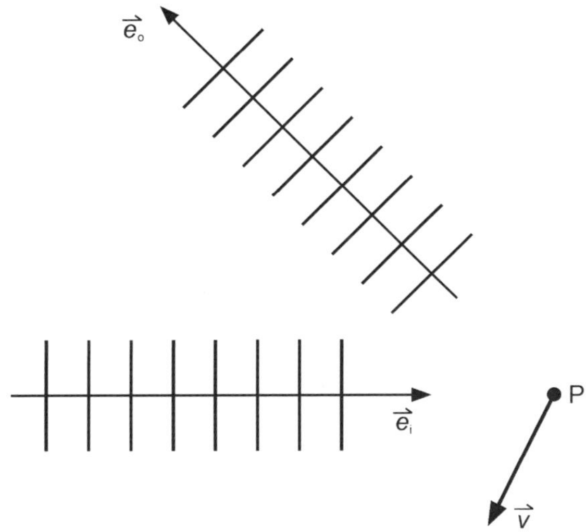

Fig. 11.2 Scattering of a wave in the incident direction $\vec{e}_i$ at a moving object point P and observation of the scattered wave in direction $\vec{e}_o$ in the laboratory system

the speed of light c, it follows, for the angular frequency of the scattered wave observed in the laboratory system, that:

$$\omega'' = \omega + 2kv\cos\alpha. \tag{11.4}$$

The difference, $\Delta\omega = \omega'' - \omega$ or $\Delta f = f'' - f$, is defined as the Doppler shift. For example, for $\lambda = 633\,\text{nm}$, $\alpha = 0°$, v = 10 m/s, the Doppler shift amounts to $\Delta f = 31.6\,\text{MHz}$.

In the derivation of equation (11.4), it was assumed that the scattered light is observed in a direction opposing that of the incident irradiation. Figure 11.2 shows the general case of observing the scattered wave from a direction which differs from that of the incident radiation. Analogous to the derivation of (11.4), it can be shown that, in this case, the following is valid:

$$\omega'' = \omega + k\left(\vec{v}\cdot\vec{e}_o - \vec{v}\cdot\vec{e}_i\right), \tag{11.5}$$

with $\vec{v}$ velocity vector, $\vec{e}_o$ unit vector in the observation direction, $\vec{e}_i$ unit vector in the incident direction. Equation (11.4) is given once more if $\vec{e}_o = -\vec{e}_i$ and $\vec{v}\cdot\vec{e}_o = v\cos\alpha$ are substituted into equation (11.5).

11.2 Laser Vibrometer

The term laser vibrometer is derived from the main use of this device, namely the investigation of vibrations of component parts. Laser vibrometers are a particular type of interferometer, designed for the measurement of vibrations.

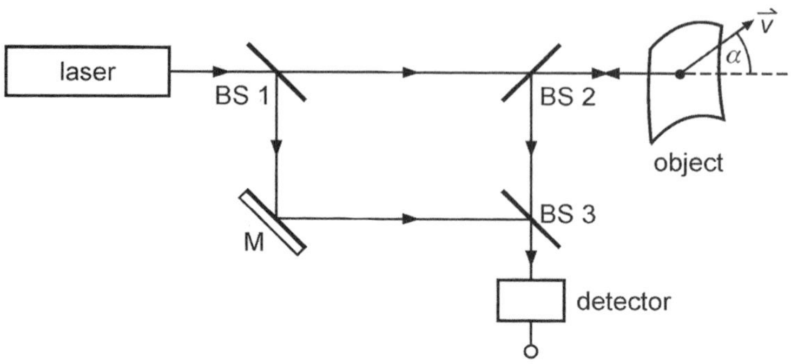

Fig. 11.3 Principle of a laser vibrometer. BS beam splitter, M mirror

11.2.1 Principle

Figure 11.3 shows the principal set-up of a laser vibrometer. The laser beam is split up into a measuring beam and a reference beam by the beam splitter BS 1. The measuring beam is directed at the vibrating object, whereupon it is reflected. The Doppler shifted wave, backscattered in the direction of the incident radiation, is guided, via the beam splitter BS 2, to the detector where it is superimposed with the reference beam. The electrical field strength of these two waves, at the location of the detector, is given by, cf. Sect. 6.1.1:

$$\begin{aligned}\vec{E}_{\mathrm{r}} &= \vec{E}_{\mathrm{r}0}\,\mathrm{e}^{\mathrm{i}(\omega t+\phi_{\mathrm{r}})},\\ \vec{E}_{\mathrm{m}} &= \vec{E}_{\mathrm{m}0}\,\mathrm{e}^{\mathrm{i}(\omega'' t+\phi_{\mathrm{m}})},\end{aligned}\tag{11.6}$$

with $\vec{E}_{\mathrm{r}}$, $\vec{E}_{\mathrm{m}}$ electrical field strength of the reference and measuring beams at the detector; $\vec{E}_{\mathrm{r}0}$, $\vec{E}_{\mathrm{m}0}$ real amplitudes of the reference and measuring beams; ω, ω'' angular frequencies, see (11.4); ϕ_{r}, ϕ_{m} phases of the reference and measuring beams at $t = 0$.

The intensity reaching the detector amounts to:

$$\begin{aligned}&I \propto (E_{\mathrm{r}} + E_{\mathrm{m}})\,(E_{\mathrm{r}} + E_{\mathrm{m}})^*,\\ &I = I_{\mathrm{r}} + I_{\mathrm{m}} + 2\sqrt{I_{\mathrm{r}} I_{\mathrm{m}}}\cos[(\omega'' - \omega)\,t + \phi_{\mathrm{m}} - \phi_{\mathrm{r}}],\end{aligned}\tag{11.7}$$

where I_{r}, I_{m} are the intensities of reference and measuring beam at the detector.

The intensity I is a periodic function, dependent on the argument of the cosine function of (11.7). The intensity lies in the following range: $I_{\mathrm{r}} + I_{\mathrm{m}} - 2\sqrt{I_{\mathrm{r}} I_{\mathrm{m}}} \leq I \leq I_{\mathrm{r}} + I_{\mathrm{m}} + 2\sqrt{I_{\mathrm{r}} I_{\mathrm{m}}}$. The contrast of the interference with respect to time, cf. Sect. 6.1.2, is largest when $I_{\mathrm{r}} = I_{\mathrm{m}}$ is valid, as is illustrated in Fig. 11.4.

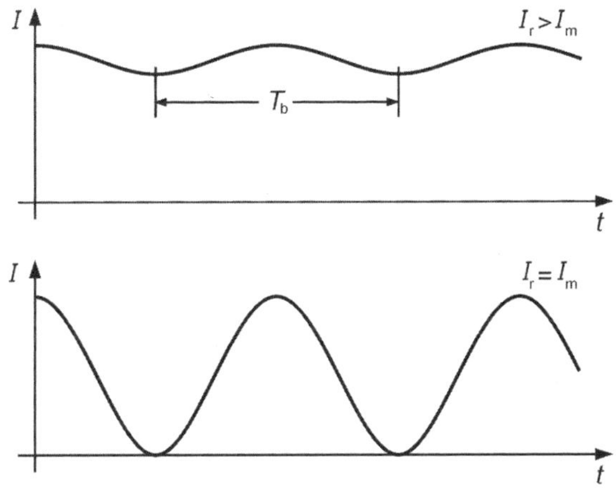

Fig. 11.4 Intensity as a function of time according to relation (11.7) for the cases $I_r > I_m$ *(above)* and $I_r = I_m$ *(below)*

The distance, with respect to time, between two minima or two maxima on the curve in Fig. 11.4 is given by the beat period, $T_b = 2\pi/|\omega'' - \omega|$. Substituting $\omega'' - \omega$ with equation (11.4) in equation (11.7) leads to:

$$I = I_r + I_m + 2\sqrt{I_r I_m} \cos\left[2kv\,(\cos\alpha)\,t + \phi_m - \phi_r\right]. \tag{11.8}$$

Therefore, it follows for the beat period that:

$$T_b = 2\pi/|2kv\cos\alpha|. \tag{11.9}$$

If the beat period T_b is measured, the component of the object velocity in the direction of the measuring beam, cf. dashed line in Fig. 11.3, can be determined using equation (11.9).

The beat period of the signal does not depend on the positive or negative nature of the velocity, but merely on its modulus. It is therefore impossible to distinguish whether the object is brought closer or whether it is further distanced. In order to remedy this situation, the angular frequency of the reference beam is lowered by a known angular frequency ω_0. This shift in frequency can be achieved, for example, using an acousto-optic modulator, cf. Sect. 4.1.1, placed in the path of the reference beam. The following is then valid for the reference beam omitting the vector character:

$$E_r = E_{r0} e^{i(\omega t - \omega_0 t + \phi_r)}, \tag{11.10}$$

with ω_0 fixed angular frequency shift of the reference beam. Analogous to (11.8), this reference wave yields:

$$I = I_r + I_m + 2\sqrt{I_r I_m} \cos\left[2kv\,(\cos\alpha)\,t + \omega_0 t + \phi_m - \phi_r\right] \tag{11.11}$$

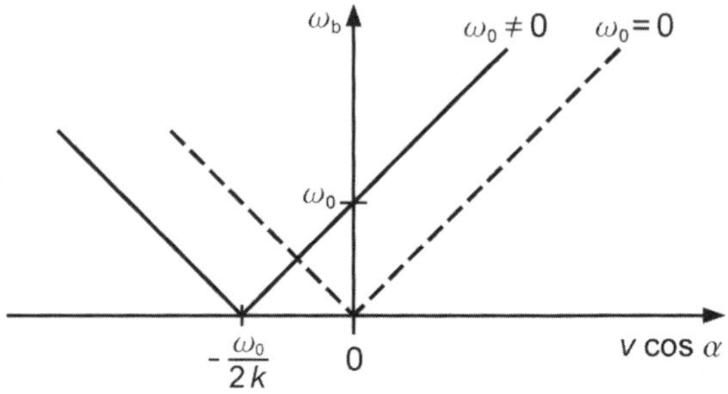

Fig. 11.5 Angular beat frequency ω_b as a function of the velocity component $v \cos \alpha$. ω_0 angular frequency shift of the reference beam

and the beat period is then given by:

$$T_b = 2\pi / |2kv \cos \alpha + \omega_0|. \tag{11.12}$$

In Fig. 11.5 the angular beat frequency $\omega_b = 2\pi/T_b$ is shown for both cases, i.e. equation (11.9) and equation (11.12), as a function of the velocity component $v \cos \alpha$. The dashed curve plotted for $\omega_0 = 0$ illustrates, that it is not possible to deduce the direction of the velocity component for a given beat frequency ω_b. At $\omega_0 \neq 0$ this is clearly possible, if the following condition holds:

$$v \cos \alpha > -\omega_0/(2k). \tag{11.13}$$

The angular frequency shifts used in practice typically amount to $\omega_0 = 2.5 \times 10^8\,\text{rad s}^{-1}$. Using equation (11.13) and a laser wavelength of $\lambda = 633\,\text{nm}$, the velocity component in the direction of the measuring beam is such that: $v \cos \alpha > -12.7\,\text{m/s}$. With respect to the further processing of the signal which oscillates with an angular beat frequency ω_b, the velocity range for an unequivocal detection of the velocity direction is, as a rule, established as follows:

$$0 \leq |v \cos \alpha| \leq \omega_0/(2k). \tag{11.14}$$

At object component velocities in the interval given in (11.14), angular beat frequencies $\omega_b = 2kv \cos \alpha + \omega_0$ of the intensity signal (11.11) appear in the range $0 \leq \omega_b \leq 2\omega_0$. The frequency-modulated detector signal is demodulated, in order to generate a voltage signal which is proportional to the velocity component $v \cos \alpha$.

11.2.2 Signal Processing

To illustrate the signal processing method, we will consider, in the following, the simplified case in which the harmonic oscillation of an object is investigated using a vibrometer. The frequency spectrum of the detector signals depends on the range of variation of the object point velocity, at which the measuring beam is directed. For the harmonic oscillation of an object point P with an angular frequency ω_p:

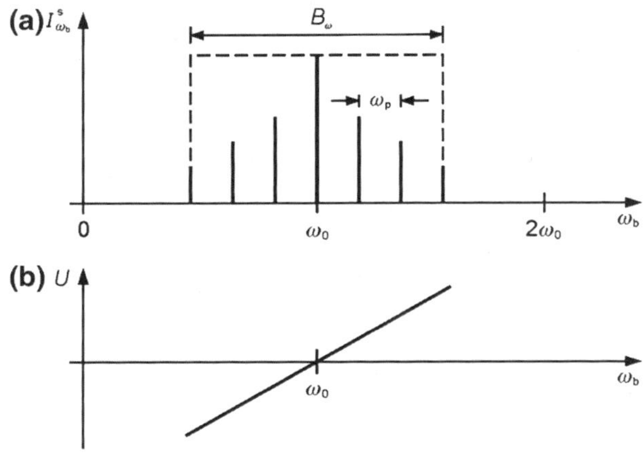

Fig. 11.6 Angular frequency spectrum of the signal of a laser vibrometer (**a**) and idealized response curve for demodulation (**b**)

$$\vec{v} = \hat{\vec{v}} \cos(\omega_P t + \phi_P) \tag{11.15}$$

is valid, whereby $\hat{\vec{v}}$ is the vectorial amplitude of the velocity, ω_p is the angular frequency of oscillation of the object point P and ϕ_p the phase of the oscillation of the object point for $t = 0$.

Equation (11.15) leads to a time-dependent angular frequency, cf. (11.11), which is described by (omitting again the vector character):

$$\omega(t) = 2k\hat{v}\cos(\omega_p t + \phi_p)\cos\alpha + \omega_0. \tag{11.15a}$$

The peak angular frequency deviation described by equation (11.15a) is $\Delta\omega = 2k\hat{v}\cos\alpha$. The time-dependent phase to be inserted in equation (11.11) is calculated with:

$$\varphi(t) = \int_0^t \omega(t)\,\mathrm{d}t = \frac{\Delta\omega}{\omega_p}\sin(\omega_p t + \phi_p) + \omega_0 t. \tag{11.15b}$$

Inserting the phase $\varphi(t)$ into (11.11) then yields:

$$I = I_r + I_m + 2\sqrt{I_r I_m}\cos\left[(\Delta\omega/\omega_p)\sin(\omega_p t + \phi_p) + \omega_0 t + \phi_m - \phi_r\right]. \tag{11.16}$$

According to (11.16), the intensity is frequency-modulated with the angular frequency of the component oscillation ω_p at the point P. The quantity

$$\eta = |\Delta\omega|/\omega_p = 2k\hat{v}|\cos\alpha|/\omega_p \tag{11.17}$$

is denoted as the modulation index [1]. The angular frequency spectrum of the intensity, according to (11.16), consists of discrete lines, equally distanced in frequency by ω_p, as is shown in Fig. 11.6a. The intensities of the individual lines of this line spectrum depend on the size of the modulation index.

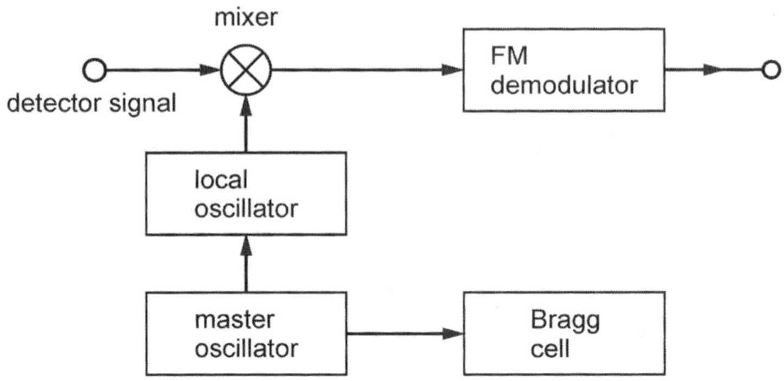

Fig. 11.7 Block diagram of the analog signal processing of a laser vibrometer

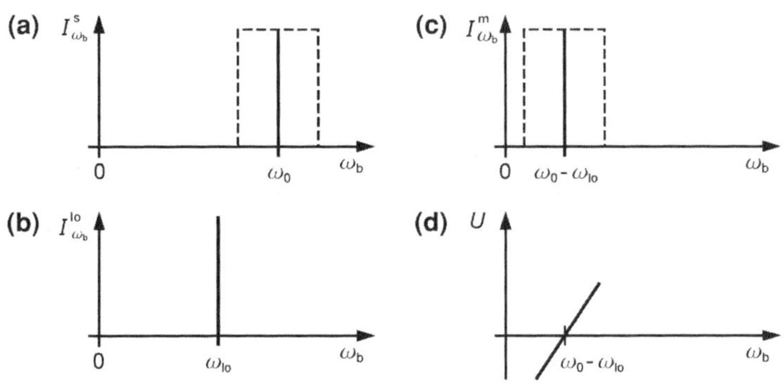

Fig. 11.8 Illustration of signal spectrum (**a**), spectrum of local oscillator (**b**), spectrum after mixing (**c**), response curve for demodulation (**d**)

The angular spectrum bandwidth B_ω for the harmonic frequency modulation is given by [1]:

$$B_\omega = 2(1 + \eta)\,\omega_P. \tag{11.18}$$

A voltage signal, proportional to the signal frequency, is generated via a demodulator. Figure 11.6b represents schematically a characteristic line of a demodulator $U = U(\omega_b)$. This characteristic line should be as linear as possible within the frequency spectrum, in order that there be a linear dependency between the voltage signal U obtained and the momentary object velocity.

Figure 11.7 shows a block diagram representing laser vibrometry with analog signal processing [2]. The detector signal is first superimposed, in a mixing stage, with the signal of a local oscillator. Figure 11.8 shows the relevant frequency spectra for the signal (a) and the local oscillator (b). Due to the mixing stage, the spectrum is displaced in the direction of lower frequencies, see Fig. 11.8c. In this frequency region, steeper characteristic lines are able to be realized for the demodulator, so that vibrations with lower amplitudes of velocity can also be measured.

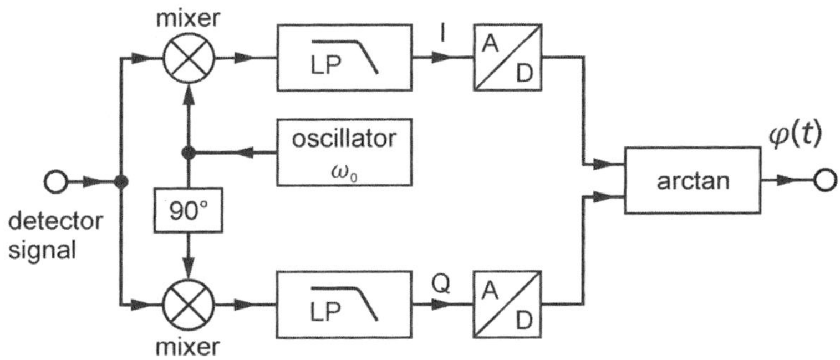

Fig. 11.9 Block diagram of the signal processing of a laser vibrometer using digital methods. LP low pass filter; I, Q quadrature signals, A/D analog-to-digital converter

The frequency of the local oscillator in Fig. 11.7 is derived from a master oscillator. This, in turn, forms the reference frequency for the drive of the Bragg cell and, in so doing, avoids any drift between the frequencies ω_{lo} and ω_0. A drift would otherwise cause an undefined position in the transformed frequency spectrum $I^{\mathrm{m}}_{\omega_{\mathrm{b}}}$ to appear, cf. Fig. 11.8c.

After the mixing stage, the signals are filtered and directed to the frequency demodulator. This is carried out by a FM-discriminator, for example [3]. With these methods, velocity measurement ranges of between 1 μm/s and 10 m/s can be achieved at object frequencies f_{p} of 1–10^6 Hz.

Apart from using frequency demodulation for signal evaluation, there is also the possibility, at low object frequencies, to measure the motion of an object via the path change between the object and the vibrometer. This path change, in the set-up according to Fig. 11.3, leads to a migration of interference fringes over the detector. Two consecutive interference fringes correspond to a path change of $\lambda/2$ in the direction of the beam. The interference fringes are counted electronically and the amplitude or velocity of the object's vibration is then determined. High absolute precision is feasible because these methods relate directly to stable laser wavelengths.

An alternative signal processing is based on digital signal processors (DSP) which allow to achieve nearly a real-time decoding. Instead of counting passing fringes which is limited to phase information in discrete increments of 2π, a continuous phase measurement is strived for. The methodical approach is illustrated in Fig. 11.9 [4]. Two so-called quadrature signals are generated having a relative phase relation of 90°, cf. Sect. 6.2.1. The measuring signal according to (11.11) is a frequency modulated signal due to the phase change caused by the object velocity. Omitting the DC term in (11.11) and assuming for simplicity $\cos\alpha = 1$ and $\phi_{\mathrm{m}} = \phi_{\mathrm{r}} = 0$, the voltage signal of the detector in Fig. 11.3 generated by the interference term can be written as follows:

$$u(t) = 2\xi\sqrt{I_\mathrm{r}I_\mathrm{m}}\cos\left(\varphi(t) + \omega_0 t\right), \tag{11.19}$$

$$\varphi(t) = 2kvt = \frac{4\pi s}{\lambda}, \tag{11.20}$$

where ξ is a detector constant, assuming a linear response, and $s = vt$. This frequency modulated signal with the carrier angular frequency ω_0 (11.19) is mixed by a cosine and sine signal as illustrated in Fig. 11.9 to the base band. These signals are described by:

$$u_{\cos}(t) = \hat{u}\cos\left(\omega_0 t\right), \tag{11.21}$$

$$u_{\sin}(t) = \hat{u}\sin\left(\omega_0 t\right), \tag{11.22}$$

with $\hat{u}$ voltage amplitude. The output signal of a mixer—as shown in Fig. 11.9—is proportional to the product of the signals fed to that mixer, i.e.:

$$u_\mathrm{I}(t) = \kappa u(t)\cdot u_{\cos}(t), \tag{11.23}$$

where κ is a constant characterizing the mixer. With (11.19), (11.21) inserted in (11.23) and using trigonometric product-to-sum identities, we obtain:

$$u_\mathrm{I}(t) = 2\kappa\xi\hat{u}\sqrt{I_\mathrm{r}I_\mathrm{m}}\left[\cos\varphi(t) + \cos\left(\varphi(t) + 2\omega_0 t\right)\right], \tag{11.24}$$

and in a similar way:

$$u_\mathrm{Q}(t) = 2\kappa\xi\hat{u}\sqrt{I_\mathrm{r}I_\mathrm{m}}\left[-\sin\varphi(t) + \sin\left(\varphi(t) + 2\omega_0 t\right)\right]. \tag{11.25}$$

After a low pass filter, see Fig. 11.9, the signal oscillating with $2\omega_0$ can be eliminated:

$$\tilde{u}_\mathrm{I}(t) = \hat{u}_\mathrm{I}\cos\varphi(t), \tag{11.24a}$$

$$\tilde{u}_\mathrm{Q}(t) = -\hat{u}_\mathrm{Q}\sin\varphi(t), \tag{11.25a}$$

where $\tilde{u}_\mathrm{I}(t)$, $\tilde{u}_\mathrm{Q}(t)$ denotes the low pass filtered signal of (11.24), (11.25) and $\hat{u}_\mathrm{I}$, $\hat{u}_\mathrm{Q}$ are the respective amplitudes. After analog-to-digital conversion a series of voltages is obtained $\tilde{u}_\mathrm{I}(t_k)$, $\tilde{u}_\mathrm{Q}(t_k)$ at discrete time steps t_k with $k = 1, 2, \ldots$. With this data the phase can be calculated:

$$\varphi(t_k) = \arctan\left\{\frac{-\tilde{u}_\mathrm{Q}(t_k)}{\tilde{u}_\mathrm{I}(t_k)}\right\} + m\pi, \tag{11.26}$$

where m is an integer. We finally obtain the displacement, cf. (11.20):

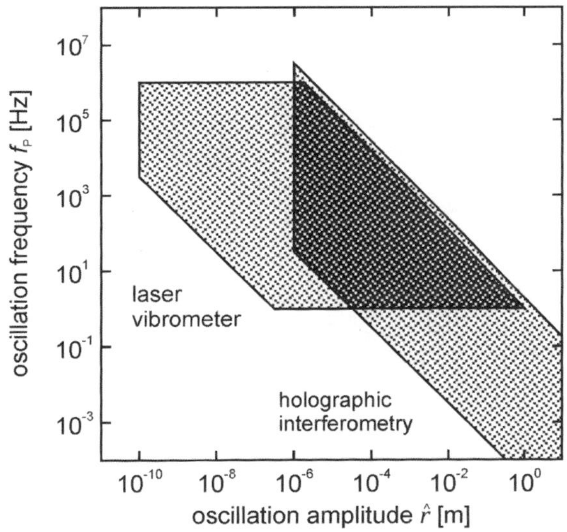

Fig. 11.10 Typical measuring range of laser vibrometry and holographic double-pulse interferometry for harmonic oscillations

$$s(t_k) = \frac{\lambda}{4\pi}\varphi(t_k). \tag{11.27}$$

Instead of generating the quadrature signals in an analog way by the two mixers shown in Fig. 11.9, the detector signal can be first mixed down to an intermediate frequency carrier (IF) which is then directly digitized. The I and Q signals are generated numerically by mixing the signal with a numerical oscillator tuned to the IF carrier. Then—as described before—the arctangent operation is carried out to get the phase. The time derivative of the phase yields the velocity of the object, cf. (11.20), (11.27).

11.2.3 Measuring Range of Laser Vibrometers

The typical measuring range of a laser vibrometer is reproduced in Fig. 11.10 [5]. The representation relates to an object's harmonic oscillation. The component of vibration at a measurement point P in the incident direction of the laser beam of the vibrometer, is described by the following relationships, cf. (11.15):

$$\begin{aligned} r &= \hat{r}\sin\left(\omega_\mathrm{p} t + \phi_\mathrm{p}\right), \\ v &= \hat{r}\omega_\mathrm{p}\cos\left(\omega_\mathrm{p} t + \phi_\mathrm{p}\right), \end{aligned} \tag{11.28}$$

with r displacement of the object point P and $\hat{r}$ amplitude of the oscillation of object point P. The vector character of the displacement is omitted here, since only the movement in the direction of the incident laser beam is considered.

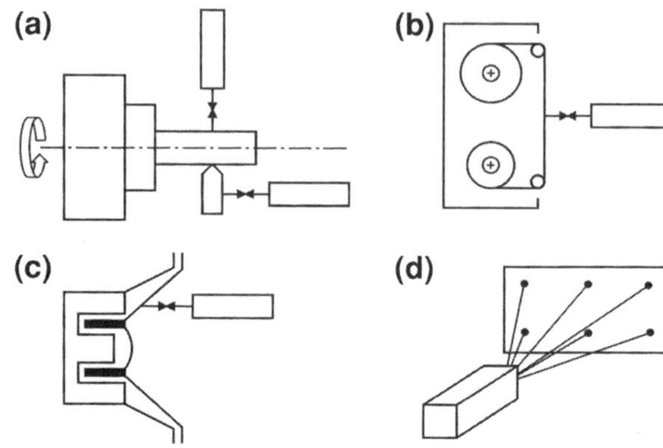

Fig. 11.11 Application examples of laser vibrometers. **a** tools and rotating spindles, **b** moving magnetic tapes, **c** membranes of loudspeakers, **d** vibration analysis of extended objects by scanning of the measuring laser beam

The diagram in Fig. 11.10 represents the measuring range of laser vibrometers and has frequency of oscillation and amplitude of oscillation on its coordinate axes. The measuring range extends itself in both coordinate directions over many orders of magnitude, so that, in practice, almost all relevant vibrations can be measured. This is especially applicable to acoustic vibrations in the 10 Hz–20 kHz range. The holographic double-pulse interferometry measurement range, using a ruby laser, cf. Chap. 9, for the investigation of vibrations, is represented in Fig. 11.10 for comparison. The lower limit at vibration amplitudes of 1 μm results from the condition that, for a practicable evaluation of the interferometric hologram, at least one interference fringe must exist. For amplitudes greater than ca. 1 μm, the measurement ranges of the laser vibrometer and holographic interferometry overlap to a large extent.

To compare laser vibrometry and holographic interferometry in the investigation of component vibrations, it must be considered that laser vibrometry is a point-oriented measuring process. Holographic interferometry, however, allows for a 2-dimensional survey of the vibration states. In the case of stationary vibrations laser vibrometers combined with galvo-scanners are able to measure oscillations of a set of surface points, cf. Sect. 11.2.4 [6–8].

11.2.4 Examples of Applications

Figure 11.11 shows examples of the applications of laser vibrometers [9–12]. In Fig. 11.11a, laser vibrometers have been installed in order to localize unwanted vibrations in machine tools. In particular, these measurements can be carried out on moving spindles and tools [13]. Through vibration analysis, damage to bearings can be recognized in good time [14].

Laser vibrometers are used in industry to investigate computer hard drives with an arrangement comparable to that in Fig. 11.11a. For example, the relative movement between the rotating disk and the combined write-read heads can be measured [5].

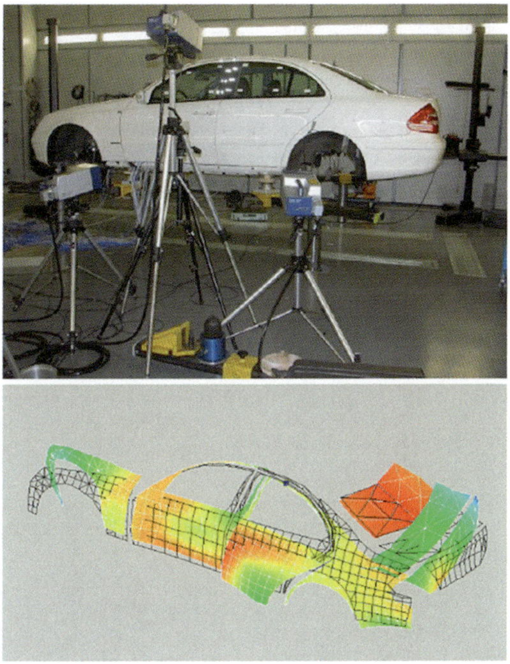

Fig. 11.12 *Above* set-up with three laser vibrometers to analyze the three dimensional vibration state of a car body, *bottom* measuring results. Foto and measuring results by courtesy of Polytec GmbH, Germany

Figure 11.11b shows an example for the measurement of vibrations on moving magnetic tapes as used for mass data storage archives. Unwanted tape motions influence the reproduction of the stored information. A contact measurement is excluded here. Using laser vibrometry, it has been found that the motion of the tapes lies typically in the sub-micrometer range, cf. Fig. 11.10.

Figure 11.11c shows a laser vibrometer which can be used for the measurement of vibrations of the membrane of a loudspeaker. A similar application case is the study of vibrational modes of violines [15]. In such structures of low mass, the fixings of conventional mechanical vibration sensors would distort the actual vibration movement.

Figure 11.11d shows a laser vibrometer combined with a rotating mirror or galvoscanner for the purpose of beam deviation [6]. In this way, it is possible to scan different points on a stationary object in sequence. At every measurement point, the amplitude and phase of the oscillation are measured, relative to a reference signal. The reference signal can, for example, be generated by the excitation signal of the component parts' oscillation. A computer evaluates the data and links it with the test object geometry in a grid model, in order to obtain a clear representation of the vibration state.

Projecting three beams of three laser vibrometers from different directions in space on the same surface point of an oscillating object allows to detect the Doppler shifted signals in these three directions. Thus the three dimensional oscillation elongation is measured, i.e. the vectorial displacement according to (11.15).

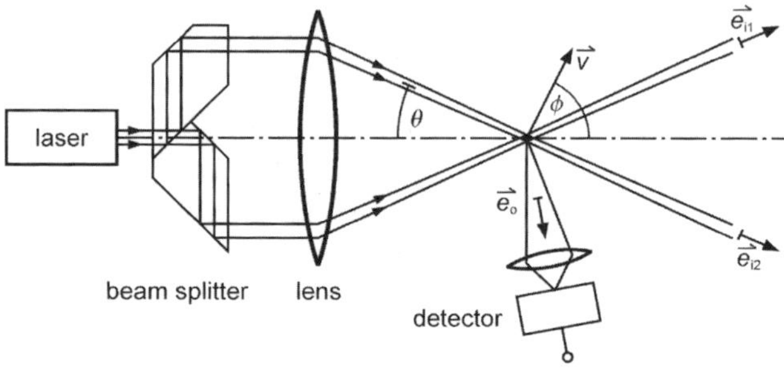

Fig. 11.13 Set-up of a differential Doppler anemometer

With such an arrangement the time expenditure for a modal analysis can be considerably decreased. Typical areas that can be scanned have sizes up to 1 m^2 with a working distance of 0.4 m. Vibration frequencies range up to 80 kHz with surface velocities from 0 to 10 m/s.

Figure 11.12, above, shows a set-up with three laser vibrometers to analyze the vibrational state of a car body. In this way vibration simulations can be verified with experimental data. Figure 11.12, bottom, shows the result with an exaggerated presentation of the three dimensional elongations of different parts of the car body with respect to their position of rest.

11.3 Laser Doppler Anemometer

Laser Doppler anemometers are primarily installed for use in flow measuring technology [16, 17]. The abbreviation LDA is frequently used for these systems. The LDA exploits the Doppler effect of scattering particles which then flow with the fluid to be investigated. An example of such particles is small droplets in a gas stream. Laser Doppler anemometry is a point-oriented method of measurement, with which the velocity of a stream is measured.

11.3.1 Principle

The design of a LDA relates in principle to that of a laser vibrometer, as is represented in Fig. 11.3. However, this design has a particular disadvantage for the measurement of the scattering signal from individual moving particles. The backscattered intensity is very small because, in general, only few particles are found in the test volume defined by the laser beam. As is shown in Fig. 11.4, Sect. 11.2.1, the contrast of the interference on the detector between the measuring beam and the reference beam is, in this case, small.

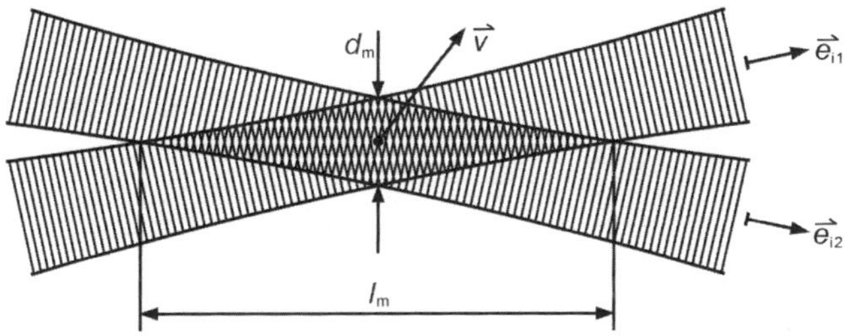

Fig. 11.14 Two-beam interference and measuring volume for differential Doppler anemometry with diameter d_m and length l_m

This disadvantage is avoided with a so-called differential or two-beam LDA. The principal set-up is represented in Fig. 11.13. A laser beam is split into two parallel beams by way of a beam splitter. These beams are focused with a lens. The overlapping region of these two beams on the optical axis of the LDA defines the measuring volume. The incident directions of both beams in the measurement volume are described—in a way analogous to Fig. 11.2—by the unit vectors $\vec{e}_{i1}$ and $\vec{e}_{i2}$. A particle moves through the measuring volume with a velocity $\vec{v}$. The light scattered at this particle is recorded via a detector in the direction $\vec{e}_o$.

The frequency of the scattered light can be calculated using equation (11.5). Under consideration of the two incident directions, it follows that, for the Doppler-shifted angular frequencies of the scattered light:

$$\begin{aligned} \omega_1'' &= \omega + k(\vec{v}\cdot\vec{e}_o - \vec{v}\cdot\vec{e}_{i1}), \\ \omega_2'' &= \omega + k(\vec{v}\cdot\vec{e}_o - \vec{v}\cdot\vec{e}_{i2}), \end{aligned} \tag{11.29}$$

with ω, k angular frequency and angular wave number of the incident laser light; $\vec{e}_{i1}$, $\vec{e}_{i2}$ unit vectors in the incident direction, cf. Fig. 11.13; $\vec{e}_o$ unit vector in the direction of observation, $\vec{v}$ velocity of a particle in the measuring volume.

The electrical field strength of the scattered waves interfere at the detector. For the intensity registered by the detector, the following is valid:

$$I = I_1 + I_2 + 2\sqrt{I_r I_m}\cos\left[\left(\omega_1'' - \omega_2''\right)t + \phi_1 - \phi_2\right], \tag{11.30}$$

where I_1, I_2 are the intensities of the two beams at the detector; ω_1'', ω_2'' the Doppler-shifted angular frequencies, see (11.29) and ϕ_1, ϕ_2 the zero phases of the two beams. Substituting equation (11.29) in (11.30) and using the angles θ and ϕ shown in Fig. 11.13, the angular beat frequency ω_b is given by:

$$\omega_b = \left|\omega_1'' - \omega_2''\right| = 2kv|\sin\phi\sin\theta|, \tag{11.31}$$

with v absolute value of the velocity vector $\vec{v}$; θ, ϕ see Fig. 11.13.

Thus, the registered frequency contains only information about the modulus of the velocity component perpendicular to the optical axis of the LDA system, see dash dot line in Fig. 11.13. However, the positive or negative nature of this velocity component cannot be established, due to the fact that (11.31) yields the same beat frequency for both ϕ and $\phi + \pi$. Hence, for the identification of this component direction, the frequency of one of the incident beams is shifted, as was already described for the laser vibrometer in Sect. 11.2.1.

In relation (11.31) the direction of observation is not included. This means that the signal frequency does not depend on the direction of observation, see $\vec{e}_o$ in Fig. 11.13. The receiving solid angle from the detector can therefore be chosen to be as large as possible in order that a sufficient amount of scattered light is registered. The scattered light is imaged onto the detector by a lens, for example. Over and above that, the direction of observation relative to the measuring volume can be chosen freely in as much as that it is determined from the direction-dependent scattering properties of the particle, cf. Sect. 3.4.

Equation (11.31) can also be clearly derived in the following manner. The two beams, inclined at an angle θ to the optical axis, lead to a two-beam interference in the region of overlap. Figure 11.14 shows an enlarged section of Fig. 11.13 from the overlapping region of the two beams. For the distance between neighboring interference fringes, the following is valid, cf. Sect. 7.1.1, Fig. 7.4:

$$d_i = \frac{\lambda}{2\sin\theta}, \tag{11.32}$$

with d_i perpendicular distance of the interference fringes, λ wavelength of the laser light, θ angle, see Fig. 11.13.

A particle propagates through this periodic intensity distribution. The intensity of the light scattered at the particle would likewise fluctuate periodically. For the frequency of this light scattering variation, with (11.32), the following is valid:

$$f_s = v|\sin\phi|/d_i, \tag{11.33}$$

where $v\sin\phi$ is the velocity component perpendicular to the interference fringes. With $\omega_s = 2\pi f_s$ and $k = 2\pi/\lambda$, it follows from (11.32) and (11.33) that:

$$\omega_s = 2kv|\sin\phi\sin\theta|. \tag{11.34}$$

This result agrees with equation (11.31). In Fig. 11.15 a LDA signal is represented schematically as a function of time. The scattering signal oscillates with the angular frequency ω_s. The beam profile of the laser beams determines the signal envelope. The depth of modulation depends on the size of the scattering particles, their concentration in the measuring volume and on the background scattering present in the flowing medium.

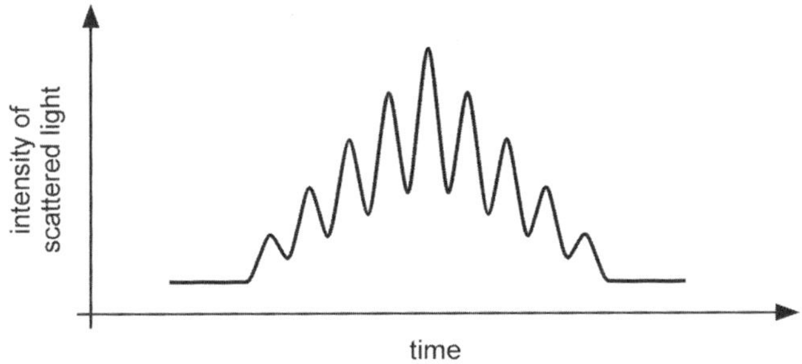

Fig. 11.15 Schematic LDA signal of a scattering particle passing the measuring volume

The dimensions of the measuring volume of a differential LDA are marked on the diagram in Fig. 11.14. If both component beams are described by a Gaussian beam, then, the following is approximately valid, cf. Sect. 2.6:

$$d_{\mathrm{m}} = \frac{4\lambda f}{\pi(2w)\cos\theta}, \tag{11.35}$$

with d_{m} diameter of the measuring volume for differential LDA, f focal length of the focusing lens of the two beams, see Fig. 11.13; $2w$ beam diameter of each beam in front of the focusing lens.

The length of the measuring volume l_{m} amounts to:

$$l_{\mathrm{m}} = \frac{d_{\mathrm{m}}}{\tan\theta}. \tag{11.36}$$

With the arrangement following that of Figs. 11.13, 11.14, only one velocity component of a particle can be measured perpendicular to the interference planes. In order to determine two velocity components, perpendicular to one another, a further two-beam interference is generated, arranged perpendicularly to the first. For this, another laser wavelength would be chosen in order that, using filters, the scattering signals belonging to different velocity components can be distinguished from one another.

The laser light scattering depends particularly on the size of the particles contained in the fluid to be investigated. The following cases are to be distinguished:

$$\begin{aligned} &(\mathrm{a})\, d \ll \lambda,\ I_{\mathrm{s}} \propto d^6, \\ &(\mathrm{b})\, d > \lambda,\ I_{\mathrm{s}} \propto d^2, \end{aligned} \tag{11.37}$$

where d is the diameter of the particle, λ the wavelength of the laser source and I_{s} the intensity of the scattered light.

Case (a) is described by the so-called Rayleigh scattering, cf. Sect. 3.4.1. The scattering intensity in different spatial directions corresponds to a dipole

Table 11.1 Cutoff-frequency of different combinations of flowing media and particles

Fluid	Particles	d [mm]	f_{max} [Hz]
Water	Sand	0.5	0.16
Air	Water droplet	7×10^{-3}	8×10^2
Air	Oil droplet	1×10^{-3}	1.2×10^4
Water	Hydrogen bubble	0.2	20

characteristic and is proportional to d^6. The reason for this is that the individual waves scattered by the atoms, from which the particles consist, are superimposed in-phase, approximately. The number of atoms in a particle is proportional to d^3. If E is the electrical field strength of the scattered wave of an individual atom, then, for the total scattered intensity, with in-phase superposition of the scattered waves of all atoms in the particle, the following is valid: $I_s \propto (d^3E)^2 \propto d^6$.

In case (b), Mie scattering is present, cf. Sect. 3.4.2. The scattering intensity scales with the cross-sectional area of the particle. The intensity scattered from the particle is strongly directionally dependent. For example, for a water droplet with $d = 1$ µm, the ratio of the scattering intensity in the forward direction—i.e. parallel to the direction of the incident laser radiation—to that in the reverse direction amounts to 10^3: 1. In order to obtain a scattering intensity which is as high-valued as possible, it is therefore appropriate to arrange the detector in Fig. 11.13 in the direction of the angle bisector of $\vec{e}_{i1}$, $\vec{e}_{i2}$.

According to (11.37), it is advantageous to use large scattering particles so that the scattering intensity increases. On the other hand, the scattering particles of a stream flow together better, the smaller they are. In practice, a compromise is reached.

One measure for the flow of particles in a stream is the maximum frequency of fluctuation at which the particles still follow the stream. In Table 11.1, these cut-off frequencies f_{max} are represented for various particles and flow media.

11.3.2 Signal Processing

To process the signal of a Doppler anemometer, the following devices are usually employed: tracker, counter and photon-correlator.

It is understood that a so-called tracker implies a frequency demodulator. The principle is represented in Fig. 11.16. The detector signal of the anemometer contains a constant part and a sinusoidal part, cf. (11.30). The sinusoidal part contains information about the velocity of the scattering particle at the angular beat frequency ω_b, cf. (11.31). Therefore, the constant part is separated from the signal first, using a highpass filter. Joined subsequently in series to the filter is a phase detector, in which the angular frequency of the signal is compared to the angular frequency ω_{VCO}. This reference frequency is generated by a VCO, a "voltage controlled oscillator".

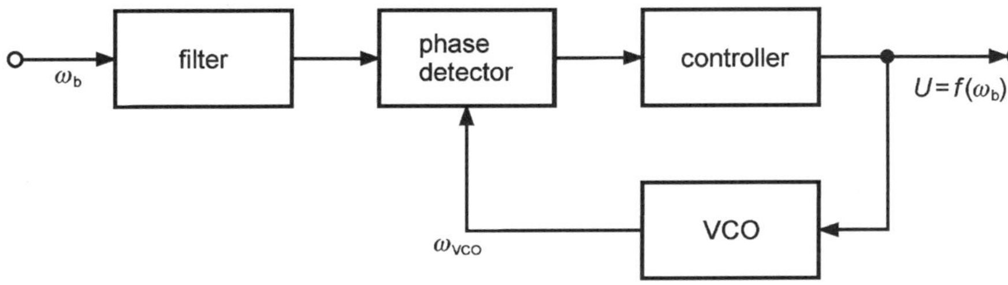

Fig. 11.16 Principle of LDA signal processing with a tracker

Fig. 11.17 Principle of LDA signal processing with a counter

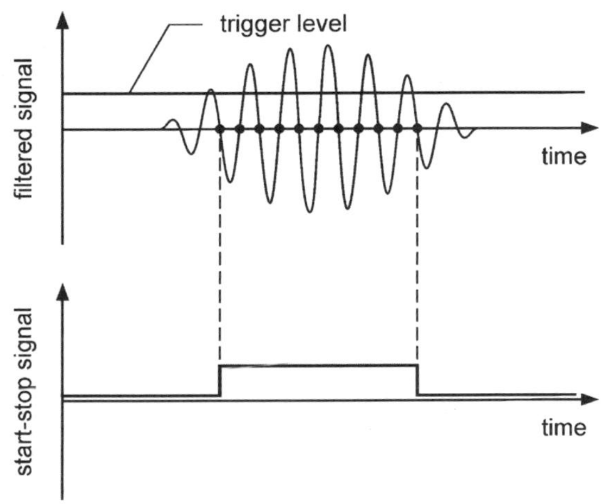

If the signal angular frequency and ω_{VCO} do not agree, the phase detector registers an increasing phase difference between the two signals. This is sent to a controller, which delivers a voltage controlling the frequency of the VCO. The angular frequency ω_{VCO} is followed until it agrees with the signal frequency of the anemometer. The output signal of the controller is therefore a function of the angular frequency ω_b. The linearity of the characteristic line of the demodulation, described by this function, depends in particular on the properties of the voltage-frequency-characteristic line of the VCO.

A prerequisite for frequency demodulation with a tracker is that a continuous input signal should be supplied. This signal is, for example, the result of the superposition of a series of single input signals, as the one shown in Fig. 11.15. A tracker can also be used if high concentrations of scattering particles are present in the medium to be investigated. It is suitable for processing noisy signals up to a frequency of ca. 15 MHz.

The use of a counter for signal processing is explained by Fig. 11.17. Figure 11.17, above, shows a filtered scattering signal containing only the sinusoidal part, cf. Fig. 11.15. If the signal increases above a selected trigger level, the resulting number of zero intersections is counted. The signal frequency is

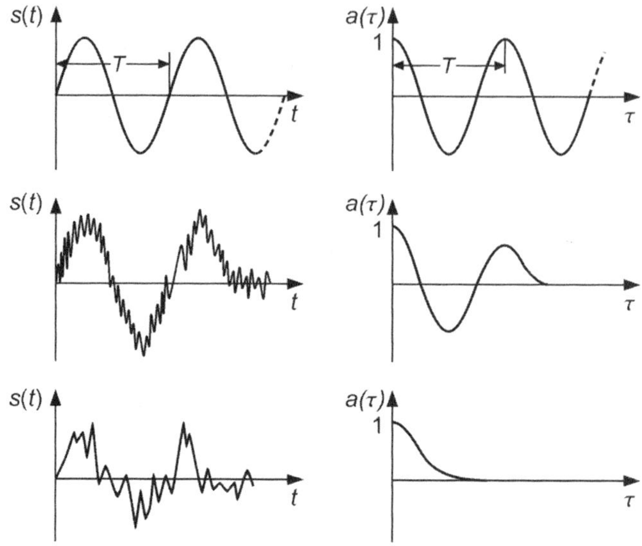

Fig. 11.18 Various signal forms $s(t)$ and corresponding auto-correlation functions $a(\tau)$

determined from this value per unit time. For the time measurement, the pulses of a high frequency reference oscillator are counted. The time measurement is started at the first zero intersection and stopped by the last, see Fig. 11.17, below.

With the counter method, in contrast to the tracker, individual signals can also be evaluated at lower scattering particle concentrations. With the prerequisite of having a sufficiently high signal quality, signal frequencies up to 150 MHz can be processed.

For very noisy signals, photon correlators are employed. In order to evaluate the signal, a so-called auto-correlation function is formed. This delivers a measure of the presence of regular structures in the signal—e.g. an oscillation with a definite period. The auto-correlation function $A(\tau)$ is defined as follows:

$$A(\tau) = \lim_{T\to\infty} \frac{1}{T} \int_0^T s(t)s(t+\tau)\mathrm{d}t, \tag{11.38}$$

with T limit of integration, $s(t)$ signal as a function of time, τ displacement in time.

For the normalized auto-correlation function, the following is valid:

$$a(\tau) = A(\tau)/A(0). \tag{11.39}$$

This yields a value $a(\tau) = 1$ for $\tau = 0$. Figure 11.18 illustrates the relationship between various signals and the relevant normalized auto-correlation functions. With a sinusoidal signal $s(t)$, $a(\tau)$ has a periodic progression with an equal period of oscillation. This is also valid for a sinusoidal signal which has been

Table 11.2 Measuring range of laser Doppler anemometers with different signal processing systems

	Tracker	Counter	Photon correlator
Frequency range	< 15 MHz	1 kHz–150 MHz	1 kHz–50 MHz
Velocity	< 75 m/s	5 mm/s–750 m/s	5 mm/s–250 m/s
Main field of application	Fluid flows	Gas flows, flames, ultrasonic flows	Gas flows
Concentration of scattering particles	High	Low	Low
Focal length		120–2,500 mm	
Fringe spacing d_i		1–10 μm	
Measuring volume			
Diameter d_m		20–300 μm	
Length l_m		100 μm–10 mm	

Velocity values given refer to a fringe distance of $d_i = 5\,\mu m$, cf. (11.32). The dimensions of the measuring volume are illustrated in Fig. 11.14

superimposed with a noise signal. For a sinusoidal signal of finite length with a superimposed noise signal, as is represented in Fig. 11.18, center, $a(\tau)$ also progresses periodically, but with decreasing amplitude.

In Fig. 11.18, below, an irregular signal is represented. The auto-correlation function falls off strongly in this instance for small time intervals τ and shows no periodic behaviour whatsoever. In such signals, there is no periodic structure present and no information to be given regarding the beat frequency.

Photon correlation is used in LDA for frequencies up to 50 MHz. Since it is necessary to form an average value, according to (11.38), it is not possible to carry out a real-time analysis in this procedure, in contrast to the methods previously mentioned. The average value must technically be formed within a finite time period. The limit of the integration T in (11.38) is so chosen, that $T \gg 1/\omega_b$ is valid, whereby ω_b is the angular beat frequency, cf. (11.31). Photon correlation is used primarily in the investigation of air streams with low concentrations of scattering particles.

11.3.3 Measuring Range of Laser Doppler Anemometers

Typical measuring ranges of laser Doppler anemometers with different signal processing systems have been summarized in Table 11.2 [16]. Furthermore, the principle areas of application of each, and data for the geometry of the measurement, has also been given.

The focal distances given in the table for the focusing lens of the LDA, cf. Fig. 11.13, establish, among other things, the minimum working distance required between the LDA system and the measuring location.

HeNe lasers with powers up to 10 mW are predominantly employed as sources of radiation, as are argon lasers in the power range 3–10 W. For compact LDA

Table 11.3 Scattering particles for LDA measurements

Disperse particles	Silicon carbide	Titanium dioxide	Polystyrene latex	Oil/water droplets
Shape	Irregular	Irregular	Spherical	Spherical
Density [g/cm^3]	3.2	4.1	1.05	≤1
Mean diameter [μm]	1.5	0.2	1.0	1–10
Standard deviation [%]	1.4	–	1	–
Fluid	Water	Gas	Gas	Gas

sensors, semiconductor lasers would be used, such as gallium–aluminum–arsenide lasers (GaAlAs) for example, with wavelengths of 780–850 nm and a power up to 100 mW.

Flow accessibility is considerably improved using fiber optics for the beam propagation of the laser and scattered light. So, for example, compact measuring heads can be placed directly in a flow stream without giving rise to large disturbances.

In Table 11.3, the data of exemplary scattering particle types used in LDA measurements is summarized. In general, the scattering particles are not equally sized, but there exists a size distribution. The ability of the particles to follow a flowing stream depends on their particle size. A monodispersal size distribution is therefore strived for, in order to achieve defined measurement conditions.

11.3.4 Examples of Applications

LDA systems are predominantly employed for the investigation of a flowing medium [17–22]. Various examples of the application of LDA systems are represented in Fig. 11.19.

Figure 11.19a shows the application of a LDA to the measurement of volumetric flows [23]. Conventional methods such as the use of inductive volumetric flow measuring devices and Coriolis mass flow measuring devices reach their measuring limits for very small volumetric flows. In this region, LDA flow measurement proves to be particularly advantageous.

The measurement beams of the LDA are directed into a transparent fluid through a glass pipe. The laser light is scattered at the particles found in the flowing medium and registered by a detector in the forward direction, cf. Sect. 11.3.1. Under the prerequisite that a fully developed velocity profile should exist, the volumetric flow is calculated from the pipe cross-section and the measured velocity. This system would be used in the lacquer industry, for example. The mixing ratio of lacquer and hardening agent should remain constant during a lacquering process. For this, the volumetric flows can be measured to a higher accuracy under the operating conditions.

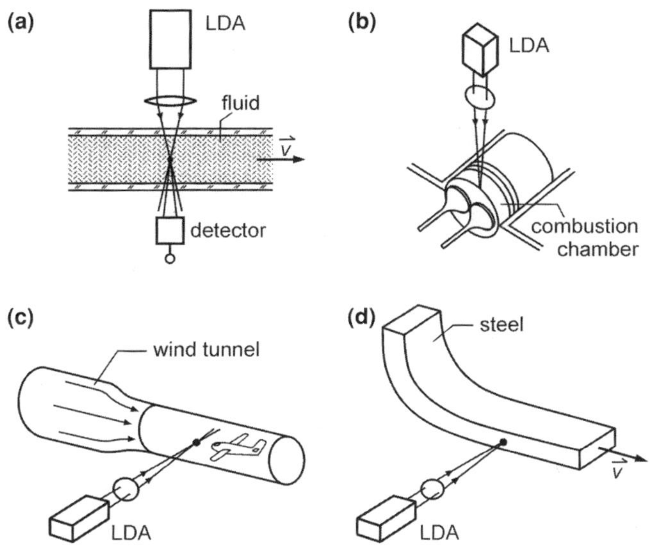

Fig. 11.19 Applications of laser Doppler anemometers. **a** volumetric flows, **b** combustion chamber, **c** wind tunnel, **d** casted steel bar

Figure 11.19b shows the application of a LDA for flow measurement in the combustion chamber of a combustion engine [24]. The aim of such investigations is to optimize the ignition process and the mixing behaviour. The laser beams are guided into the combustion chamber via a window. The scattered light is observed in the reverse direction. Titanium dioxide particles (TiO_2) are used for the LDA velocity measurement prior to the ignition of the mixture. For measurements during the combustion process, highest data rates are achieved using scattering particles of zirconium tetra fluoride (ZrF_4).

In Fig. 11.19c, the application of a LDA to a wind tunnel is shown [16]. With low velocity channels, the aerodynamics of vehicles and buildings can be investigated. In high velocity channels, flight objects are tested. Suitable conditions exist if the wind tunnel has an open measuring section, otherwise the laser radiation must be directed into the measurement volume via a window. For the measurement of spatial flow fields the LDA is displaced relative to the test object. One of the main points of the investigations is stall, characterized by high degrees of turbulence and unstationary behaviour. Figure 11.20 illustrates the measurement of the flow field in the vicinity of a passenger car using a 1: 5 car model in a wind tunnel.

Figure 11.19d shows an application of a LDA in which the test object is not a fluid, but a solid body [25, 26]. Micro-structures on the surface of the test object act as scattering centers here. In the example presented, the velocity of a continuously casted steel bar is measured. In this way, the casting process can be monitored and controlled. The steel bar exists in the red hot state, so that the non-contact measurement process presents particular advantage. Technically, the LDA

Fig. 11.20 Measurement of the flow field with LDA at a car model in a wind channel. Foto by courtesy of Dantec

device can be made to be resistant to the harsh operating conditions of a steelworks.

Other test objects in the steel production and processing industry whose velocity can be measured using LDA systems are, e.g. ingots, profiles, pipes, bands, sheets, wires and foils.

References

1. O. Zinke, H. Brunswig, *Lehrbuch der Hochfrequenztechnik* (Springer, Berlin, 1965)
2. P. Buchhave, Laser Doppler velocimeter with variable optical frequency shift. Opt. Laser Technol. **7**, 11–16 (1975)
3. H. Meinke, F. Gundlach, *Taschenbuch der Hochfrequenztechnik*. 5. Aufl., (Springer, Berlin, 1992)
4. M. Johansmann, G. Siegmund, M. Pineda, Targeting the limits of laser Doppler vibrometry, in *Proceedings of the International Disk Drive Equipment and Materials Association*, 2005, pp. 1–12
5. A. Lewin, F. Mohr, H. Selbach, Heterodyn-Interferometer zur Vibrationsanalyse. Tech. Mess. **57**, 335–345 (1990)
6. B. Stoffregen, Flächenabtastendes Laser-Doppler-Schwingungsanalysesystem. Tech. Mess. **51**, 394–397 (1984)
7. G. Righini, A. Tajani, A. Cutolo (eds.), *An Introduction to Optoelectronic Sensors*. Series in Optics and Photonics, vol. 7 (World Scientific Publishing, Hackensack, 2009)
8. B. Halkon, S. Rothberg, Vibration measurements using continuous scanning laser Doppler vibrometry: theoretical velocity sensitivity analysis with applications. Meas. Sci. Technol. **14**, 382 (2003). doi:10.1088/0957-0233/14/3/318
9. B. Junge, Experiences with scanning laser vibrometry in automotive industries, in *Proceedings of SPIE 2358, First International Conference on Vibration Measurements by Laser Techniques: Advances and Applications*, vol. 377 (1994), doi:10.1117/12.185348
10. P. Castellini, G. Revel, E. Tomasini, Laser Doppler vibrometry: a review of advances and applications. Shock Vib. Dig. **30**, 443–456 (1998)

11. A. Stanbridge, D. Ewins, Modal testing using a scanning laser Doppler vibrometer. Mech. Syst. Signal Process. **13**, 256–270 (1999)
12. P. Castellini, M. Martarelli, E. Tomasini, Laser Doppler vibrometry: development of advanced solutions answering to technology's needs. Mech. Syst. Signal Process. **20**, 1265–1285 (2006)
13. K. Tatar, M. Rantatalo, P. Gren, Laser vibrometry measurements of an optically smooth rotating spindle. Mech. Syst. Signal Process. **21**, 1739–1745 (2007)
14. S. Aye, Statistical approach for tapered bearing fault detection using different methods, in *Proceedings of World Congress on Engineering 2011*, vol. III, ISBN 978-988-19251-5-2, 2112-2115
15. G. Bissinger, D. Oliver, 3-D laser vibrometry on legendary old Italian violins. Sound Vib. **41**(7), 10–14 (2007)
16. B. Ruck, *Lasermethoden in der Strömungsmesstechnik* (AT-Fachverlag, Stuttgart, 1990)
17. F. Durst, A. Melling, J. Whitelaw, *Principles and Practice of Laser-Doppler Anemometry* (Academic Press, London, 1976)
18. B. Ruck, Laser Doppler anemometry—a nonintrusive optical measuring technique for fluid velocity. Part. Charact. **4**, 26–37 (1987)
19. C. Tropea, Laser Doppler anemometry—recent developments and future challenges. Meas. Sci. Technol. **6**, 605–619 (1995)
20. N. Lawson, The application of laser measurement techniques to aerospace flows, in *Proceedings of the Institute of Mechanical Engineering*. Part G, J. Aerosp. Eng. **218**, 33–57 (2004)
21. M. Campbell, J. Cosgrove, C. Greated, S. Jack, D. Rockliff, Review of LDA and PIV applied to the measurement of sound and acoustic streaming. Opt. Laser Technol. **32**, 629–639 (2000)
22. T. Charret, S. James, R. Tatam, Optical fibre laser velocimetry: a review, Meas. Sci. Technol. **23**(032001), (2012), 32 pp.
23. Strömungsmessung mit dem Laser, Info-Börse Laser 1990, Ed. VDI Technologiezentrum Physikalische Technologien, Düsseldorf
24. P. Witze, *Flow Lines* (Publishing TSI Incorporated, USA, 1989)
25. W. Schwenzfeier, F. Kawa, Laseroptisches Geschwindigkeitsmessen beim Stranggiesen von Stahl. Laser u. Optoelektronik **18**, 29–32 (1986)
26. X. Lan, J. Khodadadi, F. Shen, Evaluation of six k–ε turbulence model predictions of flow in a continuous casting billet-mold water model using laser Doppler velocimetry measurements. Metall. Mater. Trans. B **28B**, 321–332 (1997)

Chapter 12
Confocal Measurement Systems

Abstract We give an overview of confocal microscopy, discuss the physical principle and the achievable resolution. Different kinds of scanners are presented and some typical applications. In the second section we show, that a confocal measurement system can also be used to measure the surface profile of non-transparent objects. Finally, we discuss optical disc scanning systems.

The principle of *confocal microscopy* was first described in a U.S. patent in 1957, which was awarded to Minsky [1]. At that time—the laser was not yet invented—sufficient strong light sources were not available. It took 10 years more to make and to publish confocal microscopy recordings of sufficient image quality. In the 1970s, the development of confocal microscopes was strongly pushed, so that since the mid-1980s, advanced systems are available [2–15].

Compared to conventional microscopes, a confocal laser scanning microscope (CLSM) has higher contrast, better resolution in space and an enhanced depth of focus. Due to these advantages, it has been established in widespread fields, as e.g. in biology and medical research. But also in industrial applications, e.g. roughness measurements, measurements of three-dimensional micro-hardness indentations or coating thickness measurements, CLSM has gained great importance. Numerically most used is the principle of CLSM in DVD players or DVD-ROM drives [16–27].

12.1 Confocal Microscopy

12.1.1 Geometric Optical Analysis

Let us consider the configuration sketched in Fig. 12.1. A laser beam, e.g. from an argon ion laser or a HeNe laser, is focused at a pinhole. This illuminated pinhole acts, approximately, like a single-point source P.[1] The point source P is imaged via

[1] *Note* If the diameter of the focus of the laser beam is less than the diameter of the pinhole, the pinhole lying at P can be omitted.

A. Donges and R. Noll, *Laser Measurement Technology*, Springer Series in Optical Sciences 188, DOI: 10.1007/978-3-662-43634-9_12,

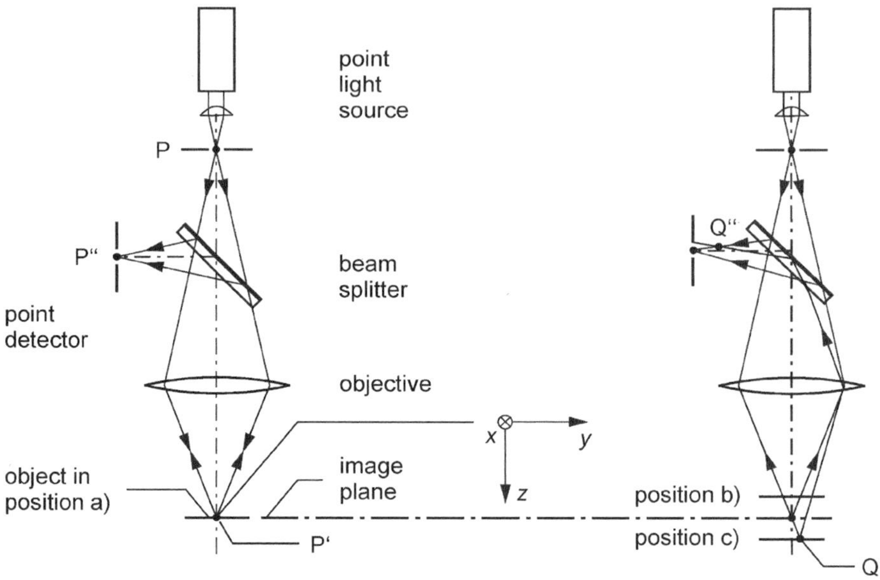

Fig. 12.1 Principle of confocal microscopy. Three object positions are considered—(*a*)–(*c*)—to illustrate that only a scatterer at point P′ is confocal to the point light source P. *xyz* coordinate system with origin at P′

a beam splitter with an imaging system such as an objective of a microscope with high numerical aperture (*NA*), in the image point P′ inside a partly transparent measuring object. Scattered and reflected laser light (wavelength λ_l) as well as fluorescent light[2] (wavelength $\lambda_f > \lambda_l$) returns to the objective and propagates, due to the beam splitter, towards the single-point detector. The single-point detector is approximately realized by a small pinhole, which reduces the active area of the used detector, usually a photomultiplier tube or an avalanche photodiode. The electrical signal received is processed in a computer.

In case of *fluorescence confocal laser scanning microscopy*, a dichroic beam splitter or an additional spectral filter is used, cf. Sect. 14.7. So only fluorescence light with wavelength λ_f reaches the detector.

In the further discussion, we have to distinguish two cases:

- The scattered light starts in the point P′ cf. Fig. 12.1, left, *position a)*. Then, a portion of this light is imaged at the single-point detector P″ and can be detected. We emphasize that in this case both points, P and P″, lie in the image plane of the point P′, i.e., the optical path length between P and P′ equals the path length between P′ and P″. This is referred to as confocal.

[2] There are materials that show inherently fluorescence. Others need to be colored with appropriate dyes.

- Next we assume that the scattered light origins at an out-of-focus point $Q' \neq P'$ of the measuring object, cf. Fig. 12.1, right, *position c)*. In this case, the point Q' is also mapped by the imaging system, but in an image point $Q'' \neq P''$. Q'' does not coincide with the single-point detector at P'', i.e., the points Q'' and P are not confocal. That is why the pinhole of the point detector suppresses most of the light from Q' and the following detector receives nearly no light. An analogous situation occurs for an out-of-focus point in the object plane at position b) in Fig. 12.1, right.

We summarize this geometric optical analysis: The signal of the single-point detector at P'' corresponds to the light, which is scattered in the image point P' of the single-point light source P. The contributions of other scattering points $Q' \neq P'$ to the signal can be neglected.

With a suitable scanning process, the measuring point P' can be moved through the measuring object. So, pixel-by-pixel and line-by-line, slice images $I(x,y,z)$ of the measuring object at fixed depths (z = const.) can be gained, cf. coordinate system shown in Fig. 12.1. It is a great advantage—compared to classical microscopy—that object points, which lie outside the image plane of the imaging system, do not contribute to these slice images. A three-dimensional dataset can be obtained by recording a greater number of slice images at different depths z. The scanning process is realized in different ways, e.g. by displacement of the measuring object, by moving the imaging system in the direction of the optical axis or by using special deflection systems.

12.1.2 Resolution

The resolution is defined as the minimum distance $\Delta r_{\min}$ between two object points that barely can be perceived as separate points. We distinguish axial and lateral resolution, whether the two points lie on a line parallel, i.e. parallel to the z-axis in Fig. 12.1, or in a plane perpendicular—xy-plane in Fig. 12.1—to the optical axis, respectively.

In the previous section we assumed that the points P and P' are mapped by the objective into image points P' and P'', respectively. This statement needs to be corrected. Due to diffraction at the aperture of the objective, the points P and P' are imaged into small, comet-like and rotationally symmetric volumes, which we call in the following *Airy-volume*. These volumes are described by the so-called 3-dimensional point-spread function (3D-PSF). The full widths at half maximum (FWHM) of such an Airy-volume—perpendicular and parallel to the optical axis—are given by [13]:

$$D_{\mathrm{FWHM,\,lateral}} = 0.51\,\frac{\lambda}{NA} \quad \text{(in the image plane)} \tag{12.1}$$

and

$$D_{\mathrm{FWHM,\,axial}} = 0.88\frac{\lambda}{n - \sqrt{n^2 - NA^2}} \quad \text{(on the optical axis)}, \tag{12.2}$$

with NA numerical aperture, n refractive index of the medium between objective and object, e.g. immersion oil. For $NA^2 \ll n^2$, equation (12.2) simplifies to:

$$D_{\mathrm{FWHM,\,axial}} = 1.8\frac{n\lambda}{NA^2} \quad \text{(on the optical axis)}. \tag{12.3}$$

In these equations λ is the wavelength of the diffracted light and NA the numerical aperture of the objective, cf. Sect. 4.3.1, equation (4.58). NA is defined by

$$NA = n \sin \sigma, \tag{12.4}$$

where σ is the half-angle of the maximum cone of light that can enter the lens.

(a) Great diameter of the pinhole

If there is no pinhole at P″ (or the pinhole has a great diameter), all the scattered light from the Airy-volume in the object is gathered by the detector. The resolutions are therefore given by the FWHMs of the Airy-volume caused by the illuminating laser light in the object. Thus [13]:

$$\Delta r_{\mathrm{min,lateral}} \approx 0.51\frac{\lambda_{\mathrm{l}}}{NA} \tag{12.5}$$

and

$$\Delta r_{\mathrm{min,axial}} \approx 0.88\frac{\lambda_{\mathrm{l}}}{n - \sqrt{n^2 - NA^2}}, \tag{12.6}$$

see Fig. 12.2, or for $NA^2 \ll n^2$:

$$\Delta r_{\mathrm{min,axial}} \approx 1.8\frac{n\lambda_{\mathrm{l}}}{NA^2}. \tag{12.7}$$

These equations can be applied, as long as the pinhole diameter D_{ph} at P″ satisfies the relation:

$$D_{\mathrm{ph}} > 1.2\frac{\lambda_{\mathrm{l}}}{NA}. \tag{12.8}$$

(b) Small diameter of the pinhole

With small diameters of the pinhole at P″ the resolutions can be improved (to the disadvantage of the intensity). Some considerations, which we do not discuss in the following, provide [13]:

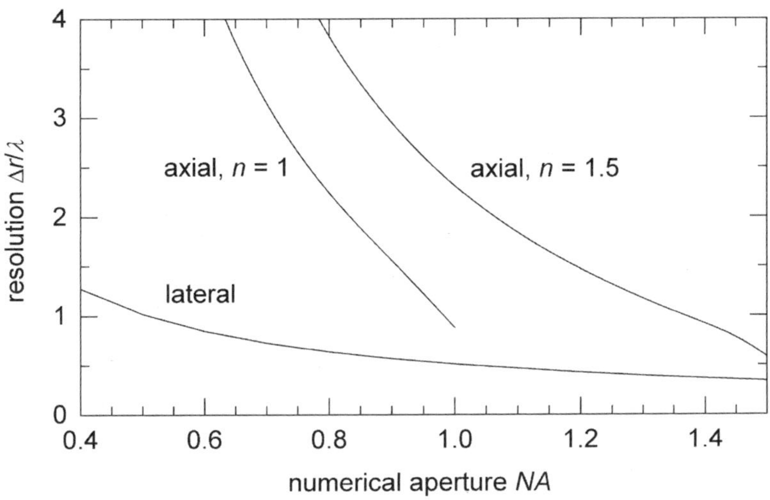

Fig. 12.2 Lateral and axial resolution as a function of the numerical aperture for the case (a) of a great pinhole diameter and the relations (12.5), (12.6), see text

$$\Delta r_{\text{min,lateral}} \approx 0.37 \frac{\bar{\lambda}}{NA} \tag{12.9}$$

and

$$\Delta r_{\text{min,axial}} \approx 0.64 \frac{\bar{\lambda}}{n - \sqrt{n^2 - NA^2}} \tag{12.10}$$

or

$$\Delta r_{\text{min,axial}} \approx 1.3 \frac{n\bar{\lambda}}{NA^2} \quad \left(\text{for } NA^2 \ll n^2\right). \tag{12.11}$$

These equations hold for:

$$D_{\text{ph}} < 0.31 \frac{\bar{\lambda}}{NA}. \tag{12.12}$$

In this case the fluorescence wavelength λ_{f} contributes to the result, so a medial wavelength $\bar{\lambda}$, which is defined as:

$$\bar{\lambda} = \sqrt{2} \frac{\lambda_{\text{f}} \cdot \lambda_{\text{l}}}{\sqrt{\lambda_{\text{f}}^2 + \lambda_{\text{l}}^2}}, \tag{12.13}$$

has to be inserted in (12.9) to (12.12).

Table 12.1 Comparison of the resolutions of a conventional and a confocal microscope considering fluorescent objects. In addition, the thickness of a slice image is given [13]

	Conventional microscopy	Confocal microscopy $D_{ph} > 1.2\frac{\lambda_l}{NA}$	Confocal microscopy $D_{ph} < 0.31\frac{\bar{\lambda}}{NA}$
Lateral resolution	$0.51\frac{\lambda_f}{NA}$	$0.51\frac{\lambda_l}{NA}$	$0.37\frac{\bar{\lambda}}{NA}$
Axial resolution	$\frac{n\lambda_f}{NA^2}$	$0.88\frac{\lambda_l}{n-\sqrt{n^2-NA^2}}$	$0.64\frac{\bar{\lambda}}{n-\sqrt{n^2-NA^2}}$
		$1.8\frac{n\lambda_l}{NA^2}\quad(NA^2 \ll n^2)$	$1.3\frac{n\bar{\lambda}}{NA^2}\quad(NA^2 \ll n^2)$
Thickness of slice image	Not defined	$\sqrt{\left(0.88\frac{\lambda_f}{n-\sqrt{n^2-NA^2}}\right)^2+\left(\frac{\sqrt{2}nD_{ph}}{NA}\right)^2}$	$0.64\frac{\bar{\lambda}}{n-\sqrt{n^2-NA^2}}$
		$\sqrt{\left(1.8\frac{n\lambda_f}{NA^2}\right)^2+\left(\frac{\sqrt{2}nD_{ph}}{NA}\right)^2}\quad(NA^2 \ll n^2)$	$1.3\frac{n\bar{\lambda}}{NA^2}\quad(NA^2 \ll n^2)$

In case of a non-fluorescent object, the detected light has the wavelength λ_l of the laser radiation. Therefore, the medial wavelength $\bar{\lambda}$ has to be replaced in the equations (12.9)–(12.12) by λ_l.

The choice of an appropriate pinhole diameter is critical. On the one hand, a smaller diameter improves the discrimination of depth, while on the other hand, the detectable intensity is reduced. Usually a pinhole diameter

$$D_{ph} \approx 1.2\frac{\bar{\lambda}}{NA} \tag{12.14}$$

is chosen.

In Table 12.1, the results are summarized and compared to conventional microscopy. In addition, the thickness of a slice image is given [13].

12.1.3 Scanners

To obtain a three-dimensional dataset of the measuring object, the image point P′ must be guided in a suitable manner through the object, i.e., all three dimensions—x, y, z—are to be scanned sequentially. This can be achieved in various ways. It is distinguished between

- object scanners,
- objective scanners,
- beam scanners

and combinations thereof. The beam scanners are divided in

- single-point scanners,
- multi-point scanners and
- slit scanners.

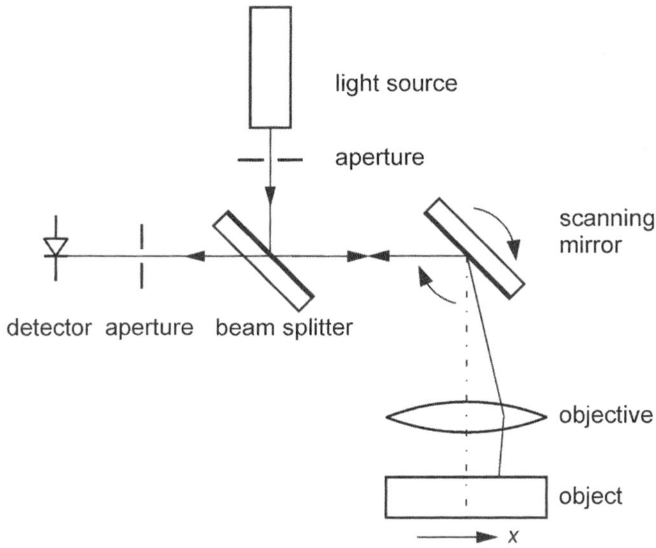

Fig. 12.3 Principle of a beam scanner

At a pure object scanner, the object is moved in all three spatial directions. This approach has the advantage of a simple design with only a static, centrally symmetric laser beam. The scanning speed is limited by the inertia of the object and the interaction with the immersion oil. Similar is the situation with a pure objective scanner, in which the entire optical system is moved. As actuators in object scanners and objective scanners serve galvanometers, piezoelectric crystals or geared electric motors.

At beam scanners, the illuminated point P′ is moved through the object by means of movable mirrors or acousto-optic deflectors, cf. Sect. 4.1.1, Fig. 4.8. In addition to single-point scanners, also multi-point scanners and slit scanners are used. In this case, a plurality of object points can be illuminated simultaneously and be evaluated with a plurality of single-point detectors. Multi-point scanners and slit scanners allow—compared to single point scanners—higher frame rates, but at a lower signal-to-noise ratio. This is due to the fact that each detector captures to a small extent also light of neighboring areas. Figure 12.3 shows, as an example, the principle of a one point scanner, which scans the x-direction fast, while the other two directions are scanned by slow object movement.

Typical sectional images in confocal microscopy consist of 1024 × 1024 pixels with a data depth of 8 bits (256 levels of gray). To scan this, a fast xy-scanner needs about 1 s. Depending on the layer thickness, usually between 16 to several 100 slice images are recorded. The corresponding requirements on the storage capacity of the computer system are high, besides the needs for the controlling of the scanning device and the graphical display. The development of confocal microscopy was, therefore, closely linked to the development of powerful computers.

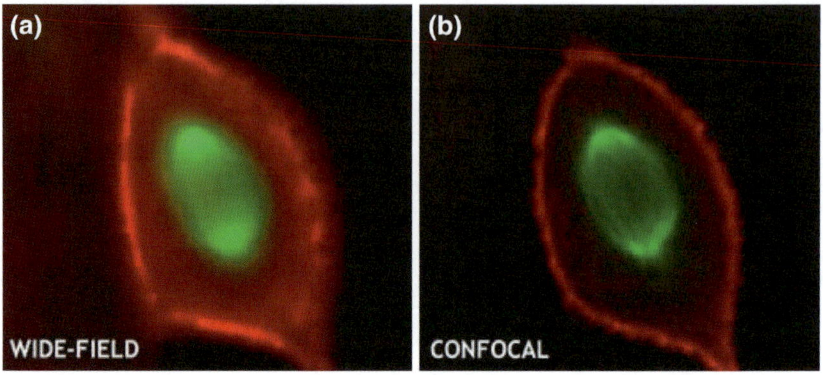

Fig. 12.4 Fluorescence microscopical image of cleavage in meta-/anaphase [13]. **a** widefield image, **b** confocal image. Pictures are provided by courtesy of Zeiss

Fig. 12.5 Cell imaging with confocal microscopy: nuclei (*red*), mitochondria (*green*), actin filaments (*blue*). *Source* Fraunhofer ILT

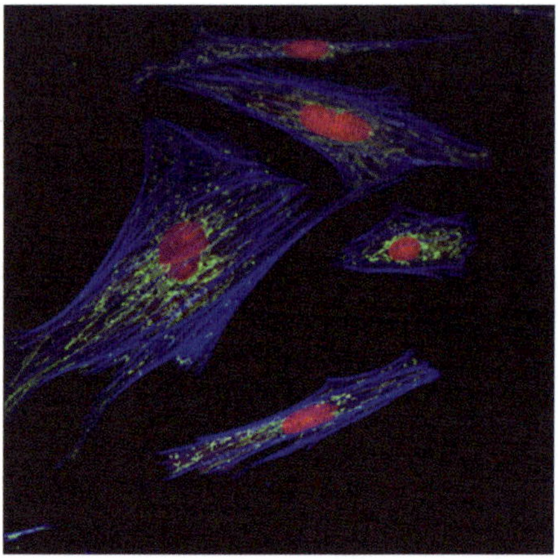

12.1.4 Typical Applications of Confocal Microscopy

(a) Biology and medicine

The advantage of confocal microscopy is demonstrated in Fig. 12.4. This figure presents the image of a cell labelled with two different fluorophores, cf. Sect. 14.3. The cell is in the meta-/anaphase of cleavage. The plasma membrane (in red) and the spindle apparatus (in green) of this cell are marked with fluorescent anaphylactics. Figure 12.4a shows an image generated with a conventional microscope. The membrane coloring of the cell border can be seen covering nearly the whole picture. The spindle fibres (middle of the cell) are blurred, but of notable intensity in the centromere region. Figure 12.4b shows the same picture taken with a confocal microscope. The outer plasma membrane is defined clearly—optical slice by suppressing the light outside the focal area—and the fibers of the spindle apparatus are identifiable. Another example of cell imaging with confocal microscopy is given by Fig. 12.5.

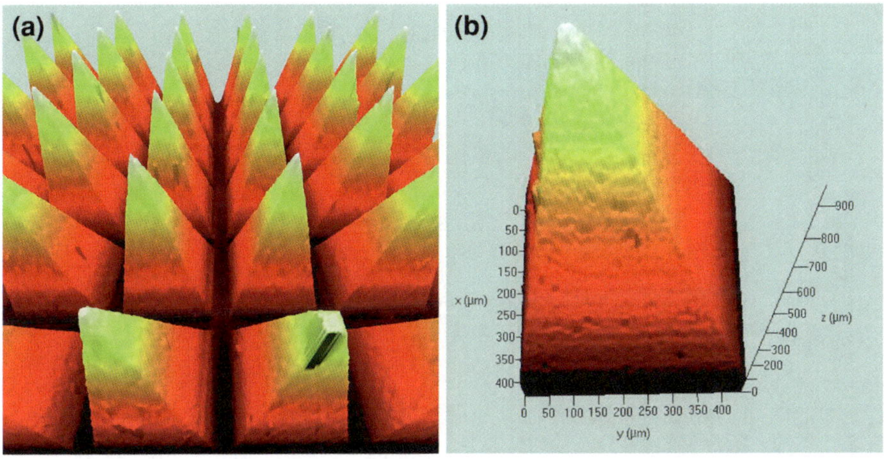

Fig. 12.6 Form electrode cut in sintered tungsten-copper. *Sample* Prof. Uhlmann, Technische Universität Berlin, Institut für Werkzeugmaschinen und Fabrikbetrieb, Germany. Pictures are provided by courtesy of Zeiss

(b) Materials science

Another example to underline the power of confocal microscopy is shown in Fig. 12.6. This image shows a form electrode cut in sintered tungsten-copper. Remarkable is the measured depth of 1 mm, which is achieved by the digital combination of single optical slices to an overall image.

12.2 Profilometer

Confocal microscopy offers the opportunity, to make sectional images at different depths of the object. An essential pre-requisite for this is the transparency of the measuring object. But even with non-transparent objects, the principle of confocal microscopy is used with success.

For the measurement of surfaces, the illumination spot is continuously moved along the surface of the opaque object, see Fig. 12.7. Thereby the movement of the imaging system along the optical axis—this is the z-axis in Fig. 12.7—is controlled, so that the detection signal for each position (x, y) becomes maximum. In such an auto-focus system, each illuminated surface point is always in the image plane of the single-point light source, i.e., the objective follows any change in the height of the object surface.

Thus the z-position of the objective represents the surface profile of the measuring object, see $z = z(x)$ curve shown in Fig. 12.7. The movement of the objective is detected with a secondary measuring system, in most cases based on an inductive measuring principle.

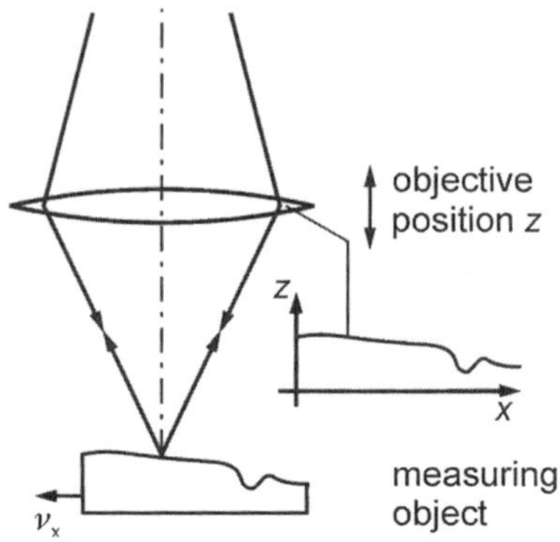

Fig. 12.7 Principle of confocal surface measurements. The z-position of the objective follows the surface profile

In addition to auto-focus systems, there are also measuring systems operating without a control loop. For a plurality of points (x, y) these systems measure the detector signal at different z-positions of the objective. Since the detector signal for each measuring point is maximized, when the image point Q′ of the single-point source lies on the surface, it is possible to measure a surface profile.

In contrast to the established tactile surface measurement systems, the non-contact laser scanning method allows shorter measuring times, also with soft and sensitive materials.

The next section is addressed to a particular type of profilometer, the so-called optical disc scanning systems.

12.3 Optical Disc Scanning Systems

12.3.1 Optical Disc

The Compact Disc (CD) was developed as a permanent and inexpensive mass storage medium by Philips and Sony and was introduced in the market in the year 1982. The information on the CD is stored physically in microstructures, so-called pits, which are indentations aligned in a spiral track running from the center of the CD to the outside. The area between two neighboring pits is referred to as land. Due to the wavelength of the semiconductor laser ($\lambda = 780$ nm) used to read the CD, the storage capacity of a conventional CD (diameter: 120 mm) is limited to about 0.7 GByte. This corresponds to a length of the spiral track of about 5.5 km.

Since about 1996, a further development of the CD, the Digital Versatile Disc (DVD), was introduced, having a storage capacity of about 4.7 GByte. This was achieved by use of a shorter laser wavelength (about 640 nm), which allowed to reduce the pit structure and hence to increase the storage capacity. In this case, the length of the spiral track was extended to 12 km. Additionally, some DVDs have two storage layers on top of each other—thereof one is semi-transparent—allowing to double the capacity (double-layer DVD).

Table 12.2 Geometrical and optical data of CD, DVD and BD

	CD	DVD	DVD (double layer)	BD
Length of a pit or land [μm]	0.83–3.1	0.40–1.9	0.44–2.1	0.13–0.6
Depth of a pit [μm]	0.12	0.11	0.11	0.15
Width of pit [μm]	0.60	0.32	0.32	0.13
Pitch of the spiral [μm]	1.6	0.74	0.74	0.32
Wavelength [nm]	780	650	650	405
NA	0.45	0.60	0.60	0.85
d_f [μm]	1.04	0.65	0.65	0.29

NA numerical aperture, d_f focus diameter

Fig. 12.8 Images of the pit structure of **a** CD, **b** DVD and **c** Blu-ray Disc taken by a scanning electron microscope [26]

In 2006, a new format called Blu-ray Disc (BD), designed by Sony, Philips, and Panasonic, was released as the successor to DVD. Because of its shorter wavelength of 405 nm, the capacity was further increased to 25 GByte for a single layer with a track length of 27 km. A dual layer Blu-ray Disc can store 50 GByte.

In the digital future so-called surface recording discs will be developed, where the data are stored directly at the surface and are read by a near-field optic [23]. This disc shall store 100 GByte.

The geometrical and optical data of a CD, DVD and BD are summarized in Table 12.2. Figure 12.8 shows images of these storage media taken with a scanning electron microscope (SEM).

12.3.2 Principle of Optical Disc Scanning

The information stored on an optical disc is read without mechanical contact and wear-free by using an optical scanning system. The microscopic size of the pits requires sophisticated optical and electronic systems. An optical scanning system must follow the track of the rotating optical disc and discriminate pits from lands. This problem was solved by laser measurement technology. In the following we discuss the principle of optical disc scanning. Numerical values stated refer to DVDs.

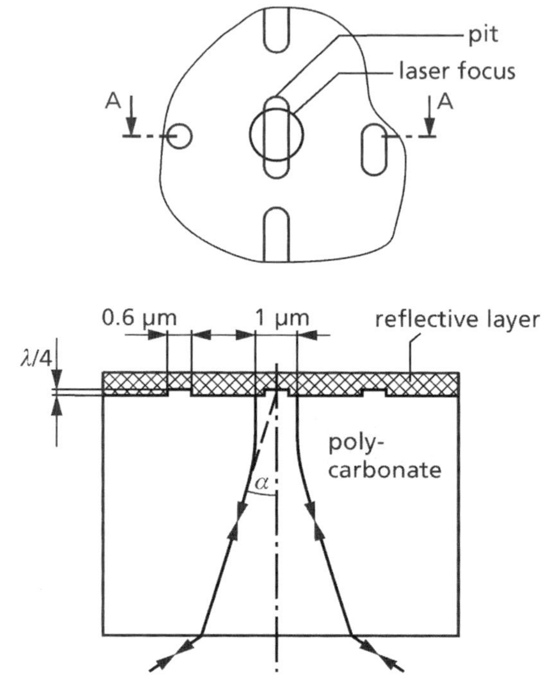

Fig. 12.9 *Top* view of an optical storage medium showing a detail with a pit and the laser focus. *Bottom* sectional view, cf. *A–A* in top graph, of an optical storage medium showing the propagation of the laser beam reading the pit and land structure

A laser beam with a wavelength of about 650 nm is focused on the reflecting surface of the optical disc, see Fig. 12.9. The focus diameter amounts to about 0.65 µm. The reflected light is detected by a photodiode. We have to distinguish two cases:

- The focused laser beam strikes on a land: The incident light is reflected and detected by the photodiode, i.e., a reflection at land corresponds to a clear detector signal.
- The focused laser beam irradiates a pit: Since the focus diameter is approximately twice as large as the pit width, the light is reflected at two levels of height. That is why the reflected light splits in two beams with a phase-shift of

$$\Delta\varphi = 2\pi \frac{2nh}{\lambda_0}, \tag{12.15}$$

where h is the depth of the pit, n the refractive index of the sealing plastics and λ_0 the wavelength of the laser radiation in vacuum. With typical values, the phase shift has an amount of about $\Delta\varphi \approx \pi$. So, destructive interference occurs and the detector signal nearly vanishes.

An optical scanning system must satisfy highest standards, according to the autofocus and the tracking abilities. In practice, two conceptions of scanning systems were introduced. Firstly, we first consider the so-called single-beam system.

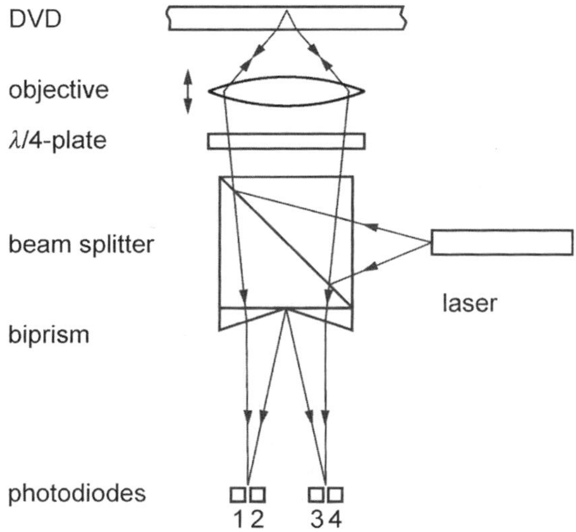

Fig. 12.10 Schematic representation of the single-beam system to read-out an optical storage medium

12.3.3 Optical Pick-up Systems

The principle of a single-beam pick-up system is sketched in Fig. 12.10. An intensity-stabilized and linearly polarized laser beam is deflected completely by a polarization sensitive beam splitter. With an objective, the laser beam is focused onto the information level of the DVD. Since the reflected light has passed twice the quarter-wave plate, its polarization direction is rotated by 90°, cf. Sect. 4.1.2. Therefore, the reflected light propagates without deflection through the beam splitter towards the direction of the photodiodes. In front of the photodiodes, there is a beam splitter, e.g. a bi-prism, which divides the beam into two partial beams.

If the system is perfectly aligned, i.e. the focus of the incident beam is exactly centered on the track of the DVD, then the two measuring foci lie exactly between the photodiodes D1 and D2 or D3 and D4, respectively, see Fig. 12.11a. All diodes will receive the same amount of light.

If, however, the distance between the DVD and the objective is too large or too small, the illuminated spots in the detector plane become blurred and their distances are, compared to the aligned state, smaller or greater, respectively, see Fig. 12.11b and c. The two pairs of diodes will no longer be illuminated uniformly. The difference signal (D1 + D4) − (D2 + D3) is, therefore, a measure for the defocus. It is used as a control signal for the coil, which moves the objective inductively.

With the four photodiodes, the track deviation can be detected. If the illumination focus is not centered on the track, a pair of diodes—e.g. D1 and D2—receives more light than the other, e.g. D3 and D4 [22]. That is why for the so-called radial-error the signal (D1 + D2) − (D3 + D4) serves as control signal. With this signal, the movement of the swing arm is controlled, on which the whole optical measuring system, as shown in Fig. 12.10, is mounted.

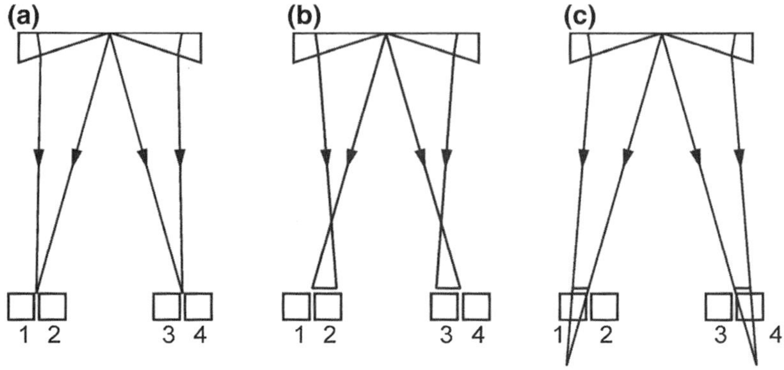

Fig. 12.11 Positions of the measuring spots in relation to the photodiodes D1–D4 **a** correctly focused **b** distance from DVD to objective too large **c** distance from DVD to objective too small

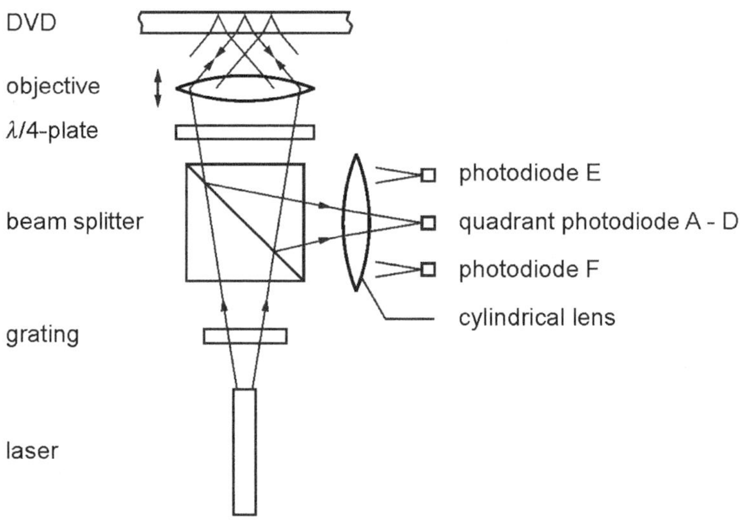

Fig. 12.12 Schematic set-up of a triple-beam system

The sum of all photodiode signals (D1 + D2 + D3 + D4) represents the actual signal to read and recover the stored data.

Finally, we point to the triple-beam system, which is widely used. Compared to the single-beam system, the laser beam is, in front of the beam splitter, divided through a diffraction grating in three sub-beams. Accordingly, in the optical disc plane and the detector plane, there are three focused beams, the central laser focal spot and two weaker foci on both sides, respectively, see Fig. 12.12. The central spot generated by the main beam supplies both, the digital information signal and the control signal for the focusing. The secondary spots—also called idler beams—are required for tracking.

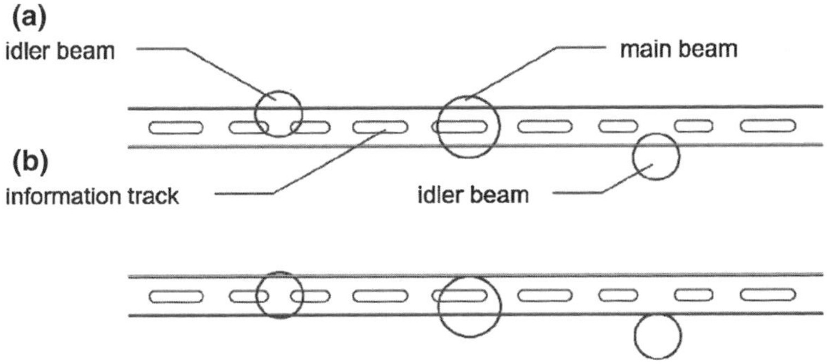

Fig. 12.13 Position of the three illumination spots **a** correctly adjusted **b** out of adjustment

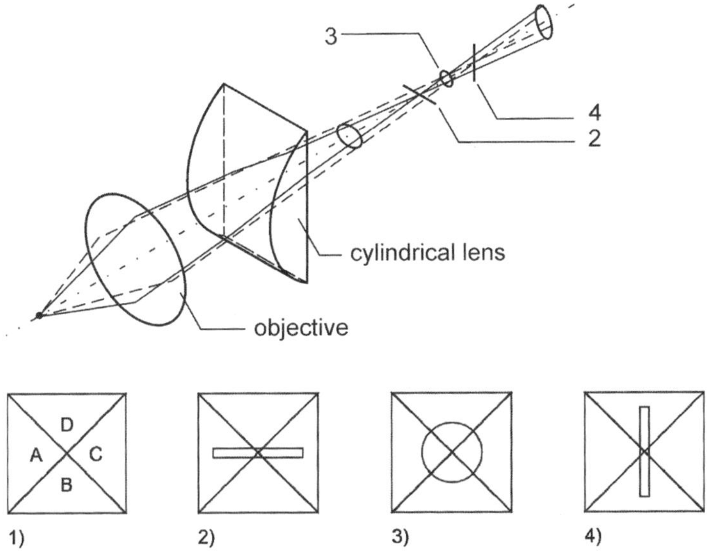

Fig. 12.14 Principle of the autofocus control of a triple-beam pick-up system. Top: shape of the beam cross-section at different positions *2–4* along the optical axis. Bottom: *1)* arrangement of the detector fields *A–D* of the four-quadrant diode, *2)–4)* schematic shape of the beam cross-section at the positions *2–4*, cf. top, in relation to the four detector fields

Let us assume that the main beam falls centered on the information track. So—with correct alignment of the system—both idler beams also touch the track, but are shifted slightly to the left or right, see Fig. 12.13a. The photodiodes E and F, which detect the light from the two sub-beams, cf. Fig. 12.12, yield the same signals. But if the main beam is not centered to the track as shown in Fig. 12.13b, the two detector signals are different. That is why the difference of the two photocurrents (E–F) serves as a signal for the tracking control.

The main beam is incident on the four-quadrant diode, i.e. the photodiodes A to D in Fig. 12.12, cf. Sect. 5.4.6. This beam provides the focus error signal to control the distance between the objective and the optical disc. The additive introduced cylindrical lens causes an astigmatism, which is of great importance. The astigmatic beam path is sketched in Fig. 12.14. It illustrates that the light distribution on the four-quadrant diode depends on the position of the objective. The objective is controlled by the signal (A + C) − (D + B). The sum of all signals (A + B + C + D) provides the information-carrying data signal.

References

1. M. Minsky, *Microscopy Apparatus*, U.S. Patent 3.013.467, applied for a patent 1957, granted 1961
2. T. Schwendt, R. Noll, Confocal microscopy, in *Tailored Light 2, Laser Application Technology*, Chapt. 19.4, ed. by R. Poprawe (Springer, Berlin Heidelberg, 2011), pp. 509–515. ISBN 978-3-642-01236-5
3. R.L. Price (ed.), W.G. Jerome (ed.), *Basic Confocal Microscopy*. (Springer, Heidelberg, 2011). ISBN 978-0387781747
4. J. Pawley, *Handbook of Biological Confocal Microscopy* (Springer, Heidelberg, 2006). ISBN 978-0387259215
5. W. Lukosz, Optical systems with resolving powers exceeding the classical limit. J. Opt. Soc. Am. **56**, 1463–4172 (1966)
6. R.H. Webb, Confocal optical microscopy. Rep. Prog. Phys. **59**, 427–471 (1996)
7. A. Donges, Grundlagen der konfokalen Laserscanning Mikroskopie. Laser J. **1**(7), 32–33 (2007)
8. W. Denk, Two photon laser scanning microscopy. Science **248**, 73–76 (1990)
9. Multiphoton Laser Scanning Mikroskopie, http://www.zeiss.de
10. C.H. Zander, J. Enderlein, R.A. Keller, *Single Molecule Detection in Solution* (Wiley-VCH, Berlin, 2002). ISBN 3-527-40310-8
11. T. Wilson, *Confocal Microscopy* (Academic Press, London, 1990). ISBN 978-0127572703
12. C. Sheppard, D. Shotton, C. Shepard, *Laser Scanning Microscopy (Microscopy Handbooks)* (Springer, Heidelberg, 1997). ISBN 978-0387915142
13. S. Wilhelm, B. Gröbler, M. Gluch, H. Heinz, *Confocal Laser Scanning Microscopy-Principles,* http://zeiss-campus.magnet.fsu.edu/referencelibrary/pdfs/ZeissConfocal Principles.pdf. Accessed 31 Oct 2013
14. http://www.itg.uiuc.edu/technology/atlas/microscopy/
15. M. Müller, *Introduction to Confocal Microscopy* (SPIE Press, Bellingham, 2006). ISBN 9780819460431
16. R. Leach, *Optical Measurement of Surface Topography* (Springer, Heidelberg, 2011). ISBN 978-3642120114
17. K. Stout, L. Blunt, *Three-Dimensional Surface Topography* (Penton, London, 2000). ISBN 9781857180268
18. K. Freischlad, Optical surface profiling: profilometer advances benefit surface analysis, film-thickness measurement. Laser FocusWorld **46**(1), 77–81 (2010)
19. K. Pohlmann, *Compact-Disc-Handbuch: Grundlagen des digitalen Audio, technischer Aufbau von CD-Playern, CD-I, Photo-CD* (IWT-Verlag München, 1994). ISBN 3-88322-500-2
20. K. Clements, *Understanding and Servicing CD Players* (Elsevier Science, Oxford, 1998). ISBN 0-7506-0934-6
21. K. Pohlmann, *The Compact Disc Handbook*, 2nd edn. (A-R Editions, Middleton, 1992)

22. J. Braat, P. Dirksen, A. Janssen, Diffractive read-out of optical discs, chap. 4, in *Optical Imaging and Microscopy*, ed. by P. Török, F.-J. Kao. Springer Series in Optical Sciences, vol. 87 (Springer, Berlin, 2007), pp. 71–111
23. F. Zijp, *Near-Field Optical Data Storage* (Technische Universiteit Delft, Proefschrift, 2007)
24. G. Winkler, *Tonaufzeichnung digital—moderne Audiotechnik mit Computerhilfe* (Elektor Verlag, Aachen, 1990). ISBN 9783921608876
25. A. Donges, Wie funktionieren CD bzw. CD-Player? Praxis der Naturwissenschaften-Physik **6**(48), 13–16 (1999)
26. Fraunhofer Institute for Laser Technology, Aachen, Sept 2008
27. E. Meinders, A. Mijiritskii, L. van Pieterson, M. Wuttig, *Optical Data Storage—Phase-Change Media and Recording* (Springer, New York, 2006). ISBN 978-1-4020-4216-4

Chapter 13
Laser Spectroscopy

Abstract Laser spectroscopy utilizes the specific properties of atoms and molecules to gain information about the chemical composition of the test object. The principle of laser material analysis is described as well as the important underlying physical processes. The evaluation of the emitted spectra yields the composition of the material. Examples of applications for mix-up detection, material-specific recycling and inline process control tasks are presented. Light detection and ranging—LIDAR—is a spectroscopic method for the remote analysis of the composition of gases in the atmosphere. The working principle and the methods for the signal evaluation are presented. Examples of applications are described such as measurements of atmospheric gas constituents, aerosol particles, atmosphere dynamics and organic pollutions in water. Coherent anti-Stokes Raman spectroscopy—CARS—is based on the non-linear interaction of laser light with matter. By this, information about the temperature and concentration of molecules in gas atmospheres is gained. Examples of applications are combustion processes such as Diesel and Otto engines, gas discharges, graphite furnaces or novel types of microscopy to make visible cellular structures of living cells.

Laser spectroscopy comprises those measurement methods, employing the specific properties of atoms or molecules and their interaction with laser radiation, to gain information concerning the state of the test object. As test objects a wide range of materials in their aggregate states come into question. The most important measurement characteristics are material composition, element or molecular concentration, temperature and pressure.

Laser radiation enables a series of innovative spectroscopic measurements to be undertaken, which prior to its discovery could not be realized. These are now predominantly used in the area of basic research. This chapter brings into focus those laser spectroscopic processes which have already entered into a technical application stage.

A. Donges and R. Noll, *Laser Measurement Technology*, Springer Series in Optical Sciences 188, DOI: 10.1007/978-3-662-43634-9_13,

13.1 Laser Material Analysis

Laser material analysis is a method in which the elements contained within a material can be determined in their corresponding concentrations. In the literature, this method is described as "laser-induced breakdown spectroscopy", abbreviated to LIBS [1–4]. Laser material analysis is a non-contact fast analysis method for which there are a series of applications in the areas of process control and quality assurance.

13.1.1 Principle

In Fig. 13.1, the principle of laser material analysis is represented. The laser beam is focused on the material to be investigated, cf. Fig. 13.1a. Due to natural absorption effects, the material is locally heated and, at sufficiently large laser irradiance, a fraction is vaporized, cf. Fig. 13.1b, c. The laser radiation is partially absorbed in the material vapor and this vapor can reach such high temperatures that a fraction of the atoms attains higher energy states by collisions with accelerated electrons. When the energy of the atoms is sufficiently large, ionization occurs. The system of neutral particles, ions and electrons is defined as a plasma, cf. Fig. 13.1d.

One way in which the excited atoms or ions present in the laser-induced plasma give up their energy of excitation is via the process of spontaneous emission, cf. Sect. 3.3. The frequency spectrum of this emission is characteristic for the composition of the material under investigation, see Fig. 13.1e, and the radiation is detected and evaluated. On progression, the plasma starts to decay and a crater forms as a result of the vaporized material. The diameter of this crater lies typically in the range of a few 10 μm, cf. Fig. 13.1h.

Figure 13.2, above, shows, as an example, an emission spectrum of a laser-induced plasma. In the spectrum, individual emission lines appear and these are characteristic of the various elements contained in the material under investigation. Each line represents a particular element and the height of the line is proportional to the concentration of this particular element appearing in the sample. This spectral line intensity is also influenced by a series of other factors, which will be discussed in Sect. 13.1.2. As a rule, the line intensity—which is the integral of the spectral line intensity in a wavelength interval enclosing the line—of an element j, is measured with respect to a reference line r [2]. A dominant element would generally be chosen for the reference line in the material to be investigated. In the following we use the term *line intensity* always in the sense of a spectral line intensity being integrated over the line profile.

In order to gain quantitative results, it is necessary to calibrate the method and this is carried out using reference samples whose composition is known. In Fig. 13.2, below, a schematic calibration curve is shown, indicating the ratio of the

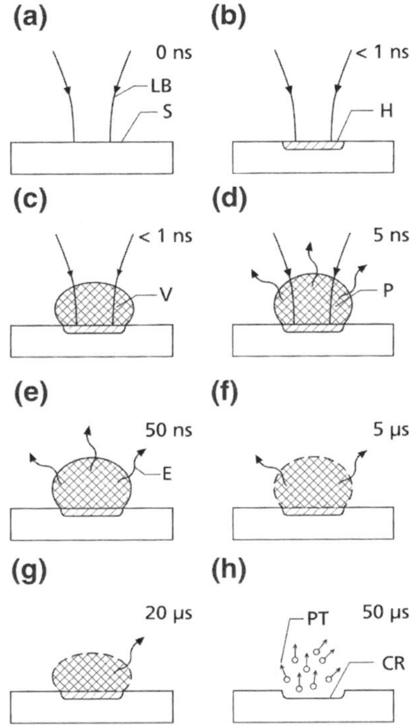

Fig. 13.1 Principle of laser material analysis based on laser-induced breakdown spectroscopy (LIBS) shown in phases (**a**)–(**h**). LB incident laser beam, S sample or measuring object, H region of energy deposition, V material vapor, P plasma, E element specific emission, CR crater, PT particulates. The times given depict the temporal evolution after start of irradiation of the laser pulse

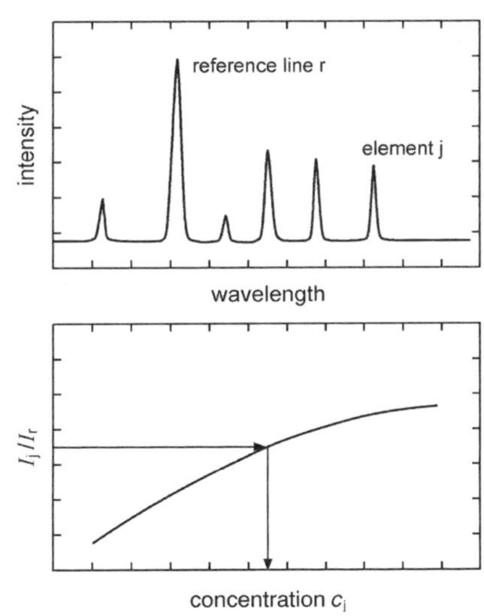

Fig. 13.2 Emission spectrum (*above*) and calibration function (*below*) in laser material analysis

intensity of the element I_j—also called analyte—to a reference line I_r, as a function of the known element concentration c_j from the set of calibration samples (for higher analyte concentrations the calibration curve is usually determined as a

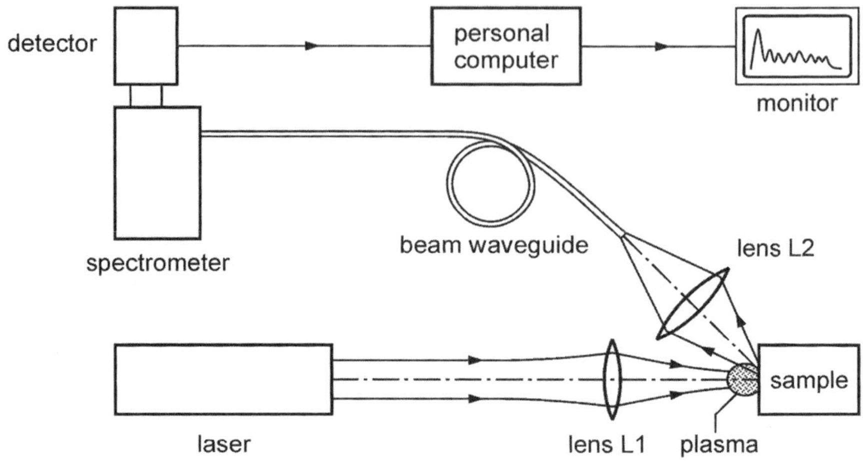

Fig. 13.3 Set-up for laser material analysis based on laser-induced breakdown spectroscopy

function of the concentration ratio c_j/c_r). In general, the calibration curve shows a non-linear progression. The concentration of an element in an unknown sample can then be determined using the measured line ratio and the calibration curve. The inverse function of the calibration curve, i.e. $c_j = f(I_j/I_r)$, is called analysis function.

Basically the spectrochemical method described above goes back to the pioneering work of Kirchhoff and Bunsen who were the first using a spectrometer to observe the emission of various elements excited in a Bunsen burner [5]. They were able to allocate distinct spectral lines to the elements potassium, sodium, lithium, strontium, calcium and barium. With this method they discovered "new" elements giving them the names caesium and rubidium.

Figure 13.3 shows a typical set-up used in laser material analysis [6]. A Q-switched laser can be installed, for example, as a source of laser radiation, cf. Sect. 2.7. The generated laser pulses have a pulse width of 5–100 ns and pulse energies ranging between 100 μJ and 2 J. With that, in the focus of the lens L1, irradiances of more than 10^6 W/cm^2 are reached, as is generally necessary for material vaporization, cf. Fig. 13.1.

The plasma radiation is collected by the lens L2 and a fiber optic guides the radiation to a spectrometer. Two methods can be used to record the spectrum. One possibility is to place individual detectors at those positions in the spectrometer at which the emission lines of the elements of interest, the analytes, are found. Photomultipliers are usually installed as detectors. An advantage of this method is that it is only the spectral information necessary for the analysis which is recorded and evaluated.

In the second method, a whole spectral region is registered using a diode array, cf. Sect. 5.4.5. The spectrum represented in Fig. 13.2, above, has been recorded in this

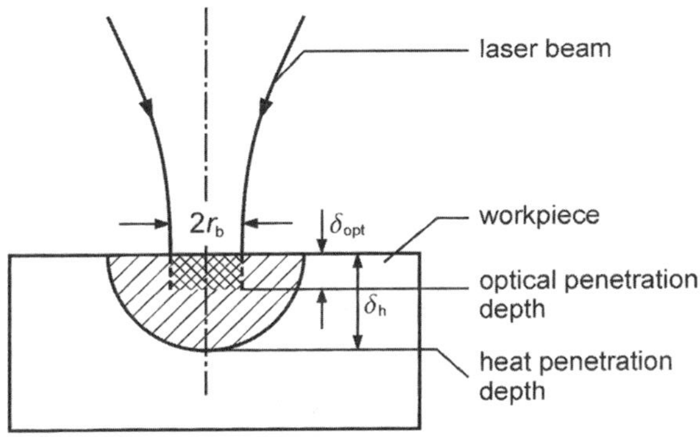

Fig. 13.4 Parameters of the interaction of laser radiation with a solid sample

way. The signals registered by the photomultipliers or the diode arrays are digitized and evaluated in a computer. The data evaluation is explained in Sect. 13.1.5.

13.1.2 Vaporization and Plasma Formation

In laser material analysis, the material is vaporized and excited to the point of radiative emission, cf. Fig. 13.1. For an effective vaporization the measuring object is heated locally by the focused laser radiation to the boiling or disintegration temperature. The fraction of incident radiation energy absorbed by the workpiece contributes to this. The degree of heating is influenced by the heat conduction in the material which can allow some of the absorbed energy to be removed from the coupling zone. Figure 13.4 shows schematically a few of the quantities which are characteristic for the laser-material interaction: diameter of the laser beam at the surface of the measuring object $2r_b$, optical penetration depth δ_{opt}, heat penetration depth δ_h.

The optical penetration depth indicates the path taken through the measuring object, along which the intensity of the laser radiation falls to approximately 37 % of the value at the surface of the measuring object. Its magnitude depends on the absorption properties of the material. For metals, with laser wavelengths in the visible and near infrared regions of the spectrum, δ_{opt} amounts typically to 10^{-6}–10^{-5} cm.

The heat penetration depth indicates the path taken by the heat energy following the onset of the incident laser irradiation. For the heat penetration depth as a function of the time t the following is valid assuming a point source at the surface of a semi-infinite object [7]:

$$\delta_h = 2\sqrt{\kappa t}, \tag{13.1}$$

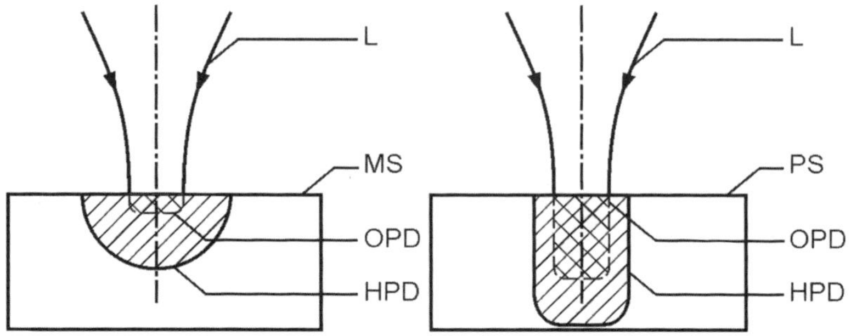

Fig. 13.5 *Left* The optical penetration depth is significantly smaller than the heat penetration depth; *right* the optical penetration depth is in the order of or greater than the heat penetration depth. L laser beam, MS metal sample, OPD optical penetration depth, HPD heat penetration depth, PS polymer sample

where κ is the thermal diffusivity, $[\kappa] = \text{cm}^2/\text{s}$.

The heating up of the workpiece can be described by the heat conduction equation. The characteristics of the interaction between laser beam and solid sample is illustrated in Fig. 13.5.

Analytical solutions are given only for certain cases. In the following, the cases which will be considered are:

(a) *surface absorption*, $\delta_{opt} \ll \delta_h$

The optical penetration depth is much smaller than the heat penetration depth. This is the case, for example, with steel for $t > 10^{-9}$ s ($\kappa_{steel} = 0.04$ cm²/s). In the heat conduction equation, the source term is represented as a surface source. This source term, with the laser irradiance following a Gaussian profile, is given by, cf. Sect. 4.2:

$$Q(r,t) = AI_0 e^{-\frac{2r^2}{r_b^2}}, \tag{13.2}$$

with A absorptance, I_0 irradiance of the laser radiation in the beam center at $r = 0$, r_b beam radius, cf. Fig. 13.4.

In this case, the solution of the heat conduction equation for the temperature progression at the workpiece surface and in the center of the laser beam is [7]:

$$T = \frac{AI_0 r_b}{\sqrt{2\pi}\kappa\rho\, c} \arctan \frac{\sqrt{8\kappa t}}{r_b}, \tag{13.3}$$

where T is the temperature at the workpiece surface in the center of the laser beam, κ is the thermal diffusivity, ρ is the density, c is the specific heat capacity and t is the time.

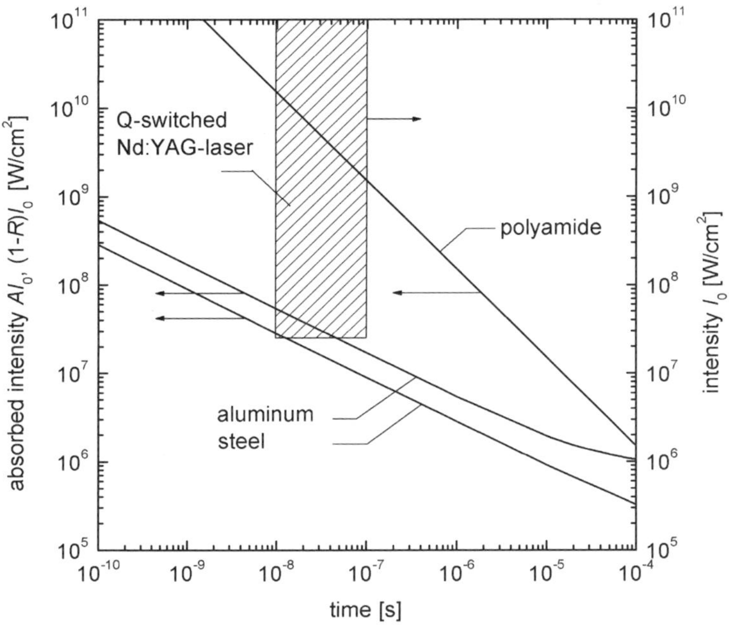

Fig. 13.6 Absorbed intensity as a function of time necessary to reach evaporation or decomposition temperature for various materials. The *hatched box* marks the area of irradiances accessible with commercial Nd:YAG lasers. The parameters used for calculation of the curves are given in the text

Using equation (13.3), for a certain absorbed laser irradiance, the time at which the temperature at the workpiece surface would reach the boiling temperature and intensified vaporization would begin, can be determined. In Fig. 13.6, the relationship for the absorbed irradiance is shown graphically as a function of time for steel and aluminum, according to equation (13.3). The heat conductivity and the thermal diffusivity are also functions of the temperature. To simplify the calculation of these curves average values were taken for the conductivity and diffusivity in the temperature range from 300 K to the evaporation temperature. The used parameters are: (a) steel, average heat conductivity $\langle \kappa \rho\, c \rangle = 0.27\ \mathrm{W/(cm\ K)}$, average thermal diffusivity $\langle \kappa \rangle = 0.06\ \mathrm{cm^2/s}$, $T = T_\mathrm{v} = 2{,}900$ K; (b) aluminum, $\langle \kappa \rho\, c \rangle = 1.79\ \mathrm{W/(cm\ K)}$, $\langle \kappa \rangle = 0.64\ \mathrm{cm^2/s}$, $T_\mathrm{v} = 2{,}700$ K; the beam radius amounts in both cases $r_\mathrm{b} = 100\ \mu\mathrm{m}$). The absorbed irradiance necessary to reach evaporation temperature decreases with longer irradiation times.

The laser irradiance at the workpiece surface has to be selected according to the absorptance, in order that the vaporization threshold is reached. Typical absorptances for steel and aluminum amount to 0.03 and 0.1. With Q-switched lasers and pulse lengths of between 10 and 100 ns, irradiances over 10^7 W/cm^2 are necessary to vaporize steel or aluminum. The irradiance regime accessible with commercial Q-switched lasers is shown in Fig. 13.6, see hatched box.

(b) *volume absorption* $\delta_{opt} \gg \delta_h$

The optical penetration depth is considerably large in comparison to the heat penetration depth. This situation is particularly relevant to plastics, for example. The optical depth of penetration for polymers, such as polyamide, amounts to ca. 3.5 mm for Nd:YAG laser wavelengths [8]. Using a thermal diffusivity of $\kappa_{polyamid} = 1.3 \times 10^{-3} \text{cm}^2/\text{s}$ [9], it follows, with equation (13.1), that for a time $t < 20$ s the optical depth of penetration is larger than the heat penetration depth.

If heat conduction is considered negligible, the increase in temperature can then be calculated directly from the locally absorbed laser intensity. The intensity profile through the absorbing medium is given by:

$$I = (1 - R)I_0 e^{-\frac{z}{\delta_{opt}}}, \tag{13.4}$$

with R reflectivity, I_0 incident irradiance at the workpiece surface, z coordinate in the direction of the laser beam; $z = 0$ refers to the workpiece surface.

The fraction of incident irradiance entering the workpiece is given by the factor $(1 - R)$ in equation (13.4). Contrary to case a) described above, the surface absorption A cannot be taken to represent the total absorption, because in the case of volume absorption, the absorption which takes place is not only limited to the surface. For the absorbed energy per unit volume after a time t, equation (13.4) gives:

$$-\frac{dI}{dz}t = (1 - R)\frac{I_0}{\delta_{opt}} e^{-\frac{z}{\delta_{opt}}} t. \tag{13.5}$$

This energy per unit volume leads to the volume element being heated up by a temperature difference ΔT and:

$$\Delta q = \rho c \Delta T, \tag{13.6}$$

where Δq is the thermal energy per unit volume, ρ is the density, c is the specific heat capacity and ΔT is the increase in temperature.

With $-(dI/dz)\,t = \Delta q$ and equations (13.5) and (13.6) follows at the workpiece surface ($z = 0$):

$$(1 - R)I_0 = \rho c \Delta T \delta_{opt}/t. \tag{13.7}$$

If the temperature difference ΔT was set as being the difference between the temperature at which disintegration and vaporization of the plastic occurs and the initial temperature of the plastic, then the required irradiances can be determined using equation (13.7), as a function of the time of irradiation. Figure 13.6 shows the curve for polyamide calculated using equation (13.7). Generally, the required beam irradiances are higher than those for metals.

13.1.3 Vaporized Material

In the following, the amount of material vaporized by a laser pulse will be estimated. A simplified energy balance will be considered, for which it will be assumed that the absorbed laser energy contributes exclusively to the vaporization of the material (specific melt enthalpy is neglected):

$$AW_{\text{Laser}} = \rho V(E_{\text{v}} + c\Delta T) \tag{13.8}$$

where A is the absorptance, W_{Laser} is the irradiated laser pulse energy, ρ is the density, V is the ablated volume, E_v is the specific evaporation enthalpy, c is the specific heat capacity and ΔT is the temperature difference between boiling temperature and room temperature.

Vapor recondensation is not considered in equation (13.8) and neither are any heat losses due to conduction or melt ejection. Furthermore, for the sake of simplicity, it was disregarded that the quantities A, ρ and c are generally not constant during the heating up and vaporization phases and the values assumed for the calculations were taken in their initial state at room temperature. A numerical example, using equation (13.8), yields an estimation of the removed mass of a steel workpiece by a laser pulse with $AW_{\text{Laser}} = 10$ mJ, such that: $\rho V = 1.6 \times 10^{-6}$ g ($E_{\text{v}} = 6 \times 10^3$ J/g, $c = 0.51$ J/(gK), $\Delta T = 2{,}600$ K). Experimentally determined mass ablations lie in the range of micrograms per pulse depending on the pulse energy, pulse width, burst mode, focusing conditions, ambient atmosphere and target material.

The resulting vapor density can be estimated using the Clausius-Clapeyron equation [10]. For information, typical numerical values will be merely specified. For absorbed beam irradiances of 10^6–10^7 W/cm^2, the vapor densities range between 10^{18} and 10^{20} cm^{-3}. Due to the high temperatures involved, there always exists a few free electrons which take up energy from the radiation field before losing it in collisions with the vaporized atoms. These collisions between electrons and atoms are a prerequisite for the electrons taking up energy from the oscillating electrical field of the laser beam. Collision-free electrons, on the other hand, would take up and release energy periodically, experiencing no net energy input. The electrons quickly reach high mean energies being sufficient to partially ionize the atoms of the material vapor. In this manner, a plasma is generated in which atoms and ions are excited and emit radiation, cf. Fig. 13.1d.

The plasma state can be described using balance considerations for particle density, momentum and energy. Calculation models show that, within a few nanoseconds of the onset of the incident radiation, electron energies in the range of 0.25–1 eV are attained [11]. Electron densities increase, due to processes of collision and ionization, to values between 10^{16} and 10^{18} cm^{-3} [6].

13.1.4 Time-Resolved Spectroscopy

The plasma generated in laser material analysis exists only for a short time, as does the radiation emitted by the excited atoms and ions in the plasma. The time-dependent line emission of particles for a transition between the energy levels j and i can be described as follows [12]:

$$\varepsilon_{ji}^{\nu,\mathrm{d}\Omega}(t) = \frac{1}{4\pi}\Gamma(\nu)\cdot A_{ji}\cdot h\nu_{ji}\cdot N_{\mathrm{a}}^{z}(t)\cdot\frac{g_j}{u_{\mathrm{a}}^{z}(T(t))}\cdot\mathrm{e}^{-\frac{E_j}{kT(t)}}, \tag{13.9}$$

with $\varepsilon_{ji}^{\nu,\mathrm{d}\Omega}(t)$ line emission coefficient of the transition $j \to i$, $\Gamma(\nu)$ line profile as a function of the frequency, A_{ji} Einstein coefficient for spontaneous emission for the transition from the upper level j to the lower level i, h Planck constant, $h\nu_{ji}$ photon energy of the transition from $j \to i$, cf. Sect. 3.3; N_{a}^{z} density of species a with charge z, g_j statistical weight of upper level j, $u_{\mathrm{a}}^{z}(T)$ partition function of species a with charge z, E_j energy of the upper level j, and T temperature. The unit of $\varepsilon_{ji}^{\nu,\mathrm{d}\Omega}(t)$ is W/(m^3 sr Hz). The emission coefficient specifies how much power per unit volume, per unit solid angle and per unit frequency interval would be emitted for the transition considered. The quantities A_{ji} and $h\nu_{ji}$ are particle specific and can be quoted from tables or data bases [13–16]. The number density $N_{\mathrm{a}}^{z}(t)$ of the particles is influenced by the momentary state of the plasma. The particles are excited, among other things, through collisions to the energy level j. The number of these collisions depends on the particle density, the electron density and the temperature. Since these parameters change during the lifetime of the plasma, $N_{\mathrm{a}}^{z}(t)$ is also a function of time. The total power emitted by a transition $j \to i$ from a plasma can be estimated by (13.9) if the plasma is optically thin, i.e. the radiation is not again absorbed inside the plasma volume. In case of a homogeneous plasma, where the density N_{a}^{z} and the temperature T are constant inside the plasma volume, the emission coefficient is multiplied with the plasma volume to yield the power per solid angle and frequency interval. In a more rigorous treatment the inhomogeneous distribution of densities and temperatures inside the plasma has to be considered. Furthermore the so-called radiation transport equation has to be solved taking into account emission and absorption processes along the path of propagation of a radiation bundle through the plasma [17].

To estimate the lifetime of the plasma, it is assumed that the plasma geometry can initially be represented by a sphere. In this sphere, there is a hot gas, the constituent particles of which move with an average thermal speed v_{th}. As a result of this non directional movement, the sphere begins to "diffuse". A measure of the time taken for this to occur is:

$$\tau = \frac{d_{\mathrm{Plasma}}}{v_{\mathrm{th}}}, \tag{13.10}$$

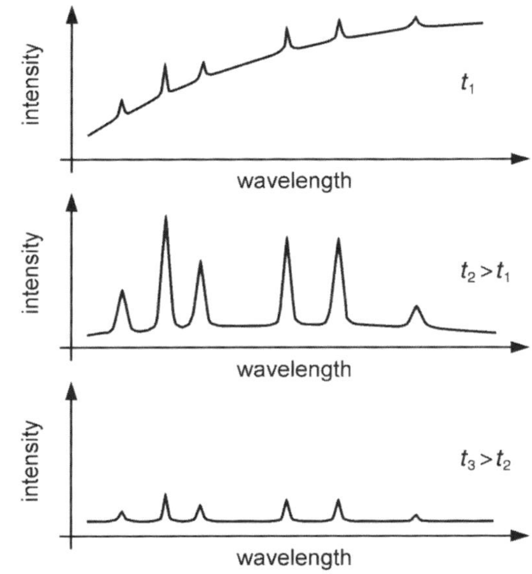

Fig. 13.7 Emission spectra of laser-induced plasma at different delay times after laser irradiation

with τ plasma lifetime, d_{Plasma} plasma diameter and v_{th} average thermal speed of the atoms and ions in the plasma.

For a plasma consisting of atoms and ions of iron at a temperature of $T = 9{,}000$ K and a diameter $d_{\text{Plasma}} = 2$ mm, the lifetime of the plasma is calculated using (13.10), to be: $\tau = 1.1$ µs. During this short period, the particles emit radiation which can be evaluated in laser material analysis.

Figure 13.7 shows schematically the emission spectra of the laser-induced plasma at three different time points following the onset of the incident laser radiation [18]. At the time t_1, the plasma emits predominantly a continuous spectrum caused by free-free and free-bound transitions of electrons. This phenomenon is called Bremsstrahlung. Only small peaks of the line intensities of atoms and ions are visible, and the ratio of the peak intensity of an emission line to the neighboring intensity of the spectral continuum emission is low. At the instant of time t_2, the plasma has expanded and has cooled down. The intensity of the line emission, as well as the ratio peak intensity to continuous background, increases significantly. By the time t_3 is reached, the plasma temperature decreases further and the emission intensities have decreased.

With time-resolved spectroscopy, radiation emission is observed within a definite time interval. So, for example, at the point in time t_2 in Fig. 13.7, suitable conditions are found in which the high intensity emission lines can be registered. In the set-up of Fig. 13.3, arranged in front of the detector, is an electro-optic shutter which opens for a particular length of time following the plasma ignition. The detector is exposed only during this opening time. It then registers a spectrum comparable with that at the instant of time t_2 in Fig. 13.7. A microchannel plate could, for example, be installed as a shutter, having opening times of between 5 ns and a few 100 ns, to which an impulse-type voltage would be applied, cf. Sect. 5.4.4.

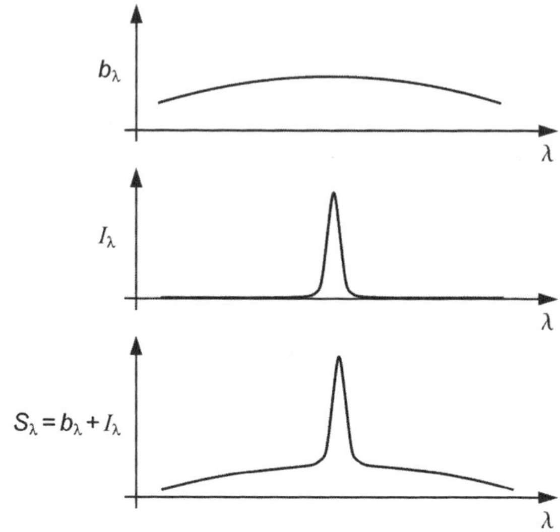

Fig. 13.8 Decomposition of the detected spectrum S_λ (*bottom*) into spectral line emission I_λ (*middle*) and background spectrum b_λ (*top*)

13.1.5 Data Evaluation

Element line intensities are determined from measured spectra, which generally exist in a digitized form in a computer, using software programs. The observed spectral emission is generally composed of various fractions. In the following, it will be assumed for simplicity, that the spectral intensity, S_λ at the wavelength λ is made up of only two fractions, namely the line radiation I_λ, and the background radiation b_λ, see Fig. 13.8:

$$S_\lambda(\lambda) = b_\lambda(\lambda) + I_\lambda(\lambda), \tag{13.11}$$

where S_λ is the spectral intensity having the dimension power per elementary area and wavelength interval, b_λ is the spectral intensity of the background radiation (in general caused by Bremsstrahlung and recombination radiation of the plasma and scattered light in the spectrometer) and I_λ is the spectral intensity of the line radiation.

In the evaluation, the so-called net line intensity will be calculated:

$$J = \int_{\lambda_1}^{\lambda_2} (S_\lambda(\lambda) - b_\lambda(\lambda))\,\mathrm{d}\lambda = \int_{\lambda_1}^{\lambda_2} I_\lambda(\lambda)\mathrm{d}\lambda, \tag{13.12}$$

with J net line intensity; λ_1, λ_2 limits of integration, within this interval $[\lambda_1, \lambda_2]$ the spectral line intensity I_λ assumes non-negligible values. The values of the function b_λ in the region of the emission line are interpolated from the progression of b_λ through the neighboring line-free zones.

In the evaluation, the observed emission lines are allocated to the various atom or ion species $k = 1,\ldots, m$ using tables or data bases [13–16]. For a selected number of lines, the net line intensity J_k is calculated using equation (13.12).

The intensities, J_k, each represent a measure for the number of particles, N_k, of the respective elements k. It will be assumed in the following, therefore, that the plasma is optically thin. In this case, the emitted radiation within the plasma is not, or only negligibly, absorbed. The line emission of the whole plasma volume is then directly proportional to the number of emitting particles, N_k, in the plasma, cf. equation (13.9).

Generally, the factors influencing J_k can be represented as follows:

$$J_k = F_k\left(I_0, W_{\mathrm{Laser}}, r_{\mathrm{b}}, t_{\mathrm{e}}, c_1, \ldots, c_m|_{i \neq k}\right) N_k. \tag{13.13}$$

The function F_k describes the dependency of the line intensity J_k on the laser intensity I_0, the laser pulse energy W_{Laser}, the beam radius r_{b}, the time of exposure of the spectrum t_{e}, as well as the concentrations c_i, of all the elements present in the sample except the analyte k. The latter arises because the vaporization process is not completely independent of the material composition and, furthermore, because all particle species including the electrons released from the heavy particles contribute to the excitation of the plasma radiation by way of their collisions. In analytical chemistry these effects are called matrix effects. The function F_k is generally unknown and must be found empirically through calibration measurements. In practice, line intensity ratios are frequently formed, such as $J_1/J_2 = N_1/N_2$, in order to compensate some part of these dependencies.

13.1.6 Measuring Range

Typical characteristic quantities and measuring ranges of laser material analysis are summarized in Table 13.1.

Since material vaporization in laser material analysis is purely optical, it is irrelevant to consider whether the sample under investigation is an insulator or a conductor of electricity.

The free distance between the optic and the workpiece is determined through the focal length of the focusing lens, L1, in Fig. 13.3. An enlargement of the focal length using the same lens aperture leads to a larger focus diameter, cf. Sect. 4.2.2. In this case, higher laser pulse energies are necessary in order that the vaporization irradiances are reached.

The lateral resolution is limited by the diameter of the laser beam at the surface of the workpiece and material properties. The beam waist diameter is again influenced by the focal length and aperture of the focusing lens, as well as the quality of the laser beam, cf. Sect. 4.2.2. The given values correspond to commercial Nd:YAG lasers.

Table 13.1 Characteristics and measurement ranges of laser material analysis based on laser-induced breakdown spectroscopy

Characteristics	Data	Remarks
Material	Metallic, nonmetallic	All states of aggregation
Free distance between optics and workpiece	< 300 cm	No principal limit, depends on laser pulse energy
Lateral resolution	1–300 μm	A consequence of the finite beam waist diameter and material properties
Detection sensitivity	1–1,000 μg/g	Element dependent
Measuring frequency	0.1–1,000 Hz*	Refers to evaluation of the spectral emission of each induced plasma

* higher frequencies up to 30 kHz possible, if the spectral emission of a series of plasmas is summed up within the exposure time of the detector

The detection sensitivities are element-dependent. In Table 13.1, typical values, determined using calibration curves gained with certified reference samples (CRM), are listed.

The measurement rate is determined by the repetition frequency of the Q-switched laser. For better measurement accuracy, several plasmas are ignited and their spectra are evaluated, in order to form average net line intensity values.

The reproducibility of laser material analysis measures how far the results of various measurements at a homogeneous workpiece deviate from one another. It depends on the stability of system parameters such as the laser pulse energy, focusing, and gas exchange rate in the interaction regions, as well as on the properties of the induced plasma over several measurements. By averaging a measurement over a series of spectra, the reproducibility is considerably increased. For example, with a pulse frequency of 20 Hz, 100 spectra can be recorded in 5 s. The net line intensities obtained deviate from those of another measurement at the same workpiece by only a few percent.

13.1.7 Examples of Applications

Laser material analysis is primarily used in those applications for which conventional analytical methods cannot be employed or have only a limited use. The requirements strive above all for a non-contact fast analysis under atmospheric conditions and, to a lesser extent, for a high level of accuracy or detection sensitivity. Focal points of applications of laser material analysis are material identification and inline control of material-dependent processes.

One of the first industrial applications established is the identification testing of high-alloy pipe fittings [19]. This application deals with the detection of material mix-ups in production lines of pipe fittings and tubes made of high-alloy steel grades. The laser material analysis system uses a flexible laser beam steering, where the laser beam is deflected by a *xy*-scanner to irradiate various inspection

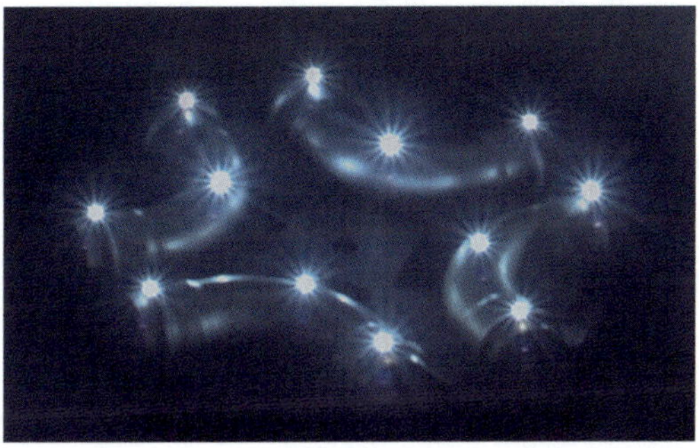

Fig. 13.9 Pipe fittings with laser-induced plasmas for mix-up detection

spots inside of a measuring volume of 400 mm × 400 mm × 100 mm [20]. Figure 13.9 shows the specimens with the laser-induced plasmas. Each specimen is measured at up to three positions to inspect the steel grades of the main body of the fitting and of the flanges.

In the production line the test objects are placed on carriers to assure a defined orientation of the fittings. These carriers are transferred to the inspection module. After closing of a protection gate the laser inspection is initiated. The scanner directs the laser beam sequentially to the pre-defined inspection positions which are transmitted automatically by the host computer where the coordinates of inspection positions for all types of pipe fittings are stored. Number and position of the inspection points can be chosen freely inside the measuring volume stated above. As soon as the scanner has reached the preset coordinates the measurement starts. For each measurement position the time needed amounts to about 1 s, within which up to 25 elements are detected simultaneously and the concentrations of the alloying elements are determined automatically. If these concentrations are within the ranges specified by the manufacturer for all test objects and measurement positions, then the inspected products on the carrier are cleared for subsequent processing steps and transported onwards. If a mix-up is detected the test object concerned is identified and discharged in the next step. All measuring results for each manufactured product are documented electronically.

The described inspection system fulfills the requirements for laser class 1 according to the IEC standard [21]. No further protection measures are needed for the user. Inspection equipment monitoring according to ISO 9001 and recalibrations are performed automatically. The described system is in use in three shift operation, 7 days a week. Since installation more than 20 million products were inspected automatically in routine industrial operation. The technical availability of the measuring system was tested according to established standards [22]. The availability is 99.94 % demonstrating the high degree of system performance achieved.

Figure 13.10 shows schematically a generic concept of applications of laser material analysis to identify and classify various materials [23–26]. A qualitative or semi-quantitative determination of the material composition frequently suffices in this case. The individual materials can, for example, travel past a laser measurement station on a conveyor belt, a laser pulse being triggered as the material passes over a photoelectric barrier. The radiation of the vaporized and excited material is evaluated, thereby identifying the material. The data is fed to a process control unit which separates the materials into individual classes. Since a large number of laser analyses can be carried out per second, this type of process is particularly suited to sorting out mixtures of materials for recycling purposes. For example, aluminum scrap metal is separated into different cast and wrought alloy fractions with the help of laser material analysis [26].

Aluminum is—after steel—one of the most important metals. Production of secondary aluminum saves up to 95 % of the amount of energy needed for primary aluminum production. However, secondary production of aluminum wrought alloys requires melting of high value primary aluminum or a pre-sorting of shredded aluminum pieces. The main processing route for non-sorted mixtures of wrought and cast alloys is the production of cast alloys. Float sink plants used to separate residues from old vehicles yield only a mixed aluminum shredder scrap consisting of a mixture of cast and wrought alloys. Up to now no large-scale, commercially available method exists, which would allow for an automatic separation of aluminum alloys.

Laser material analysis based on LIBS offers a new approach for the high-speed characterization and identification of scrap pieces. Figure 13.11 illustrates the set-up following the concept of Fig. 13.10 (the feeding vibrating chute is not shown in Fig. 13.11) [24]. Scrap pieces are singularized and accelerated on a belt conveyor up to velocities of 3 m/s passing two sensor systems. In a first step the position and size of the parts are detected by a 3D sensor, which is based on laser light section, cf. Chap. 10. With this data the beam of the LIBS module is directed onto the object to determine the composition. Depending on the measuring result the part is ejected by an array of pressurized valves to one of up to four sorting fractions, cf. module ejection in Fig. 13.11. Table 13.2 shows the technical data of the developed laser sorting system for aluminum alloys.

Figure 13.12 shows a laser sorting system with the laser module, which is mounted above a belt conveyor [24]. With a single laser pulse the spectral signals of all elements stated in Table 13.2 are detected simultaneously.

Figure 13.13 outlines the principle of an application for the quantitative inline analysis of a process, which depends—among other parameters—on the feeding of various starting materials.

For example, measurements can be carried out on liquid slag being a byproduct of steel production to determine the concentrations of the constituents and based on these the basicity of the slag is determined being decisive for the further use of this material [27]. The slags consist of mineral components in oxidic form as e.g. calcium, silicon, manganese and iron. The slags are liquid or partially solidified at the surface and have temperatures between 600 °C up to more than 1,300 °C. For the laser material analysis of these slags a series of laser pulses is focused at different points on the slag surface. The duration of the inline measurement of the basicity of slags amounts to less than 2 min. The correctness of the laser

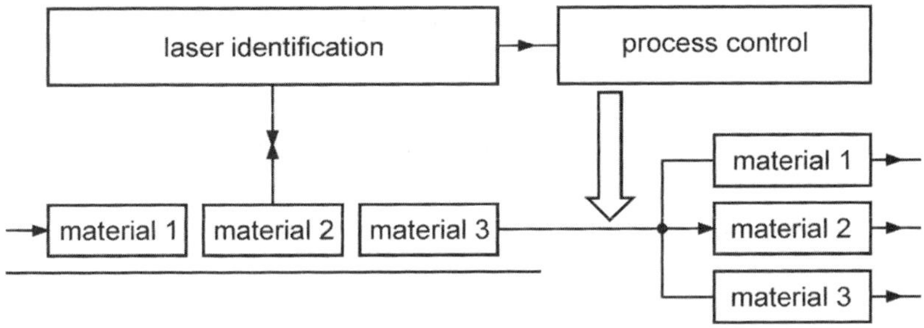

Fig. 13.10 Block diagram of laser material analysis for the identification and sorting of various materials

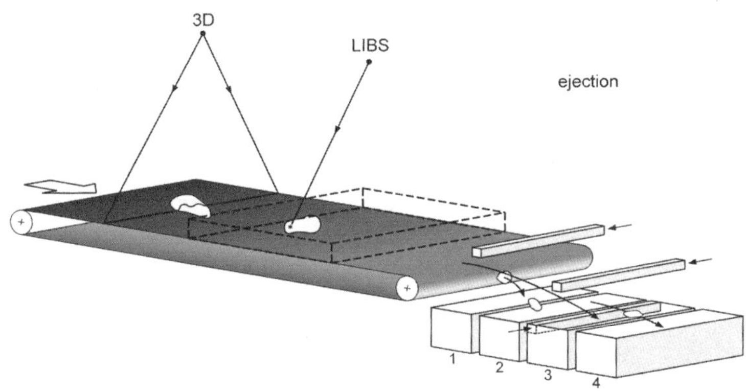

Fig. 13.11 Set-up of a laser sorting system with the three main modules. 3D = laser line section sensor to detect the position and shape of the pieces, LIBS = laser material analysis module based on LIBS with *xy*-scanner (cf. Fig. 13.12), *ejection* = array of pressurized valves for automatic ejection, 1–4 = boxes for the sorted fractions. The volume depicted by *dashed lines* is the measuring volume within which the laser beam of the LIBS module irradiates the pieces passing by

Table 13.2 Technical data of laser sorting system for aluminum alloy

Characteristics	Data
Laser	Nd:YAG
Simultaneously detected elements	Al, Si, Fe, Zn, Cu, Mg, Mn, C, Cd, Pb, Ti, Sn, Cr, Zr, Ni, Ca
Speed of belt conveyor	3 m/s
Sorting rate	40 pieces per second

measurement compared to a laboratory-based reference analysis lies within the specified range.

The data can then be fed to a process control system which varies the upstream process parameters in order to achieve a defined composition of the slag. Figure 13.14, left, shows a photograph of the installed laser measuring system to

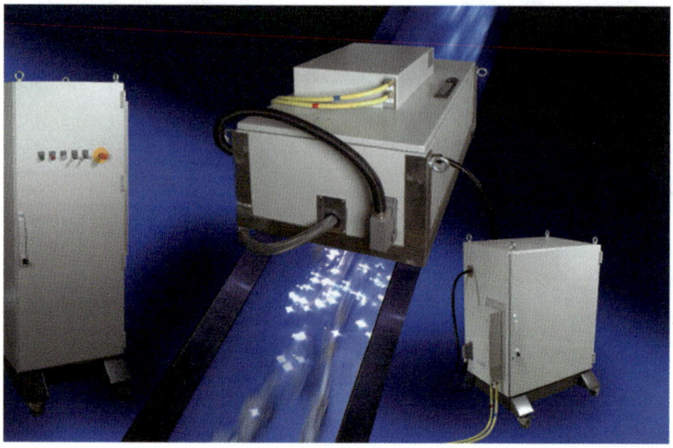

Fig. 13.12 System for the identification of scrap pieces transported on belt conveyors. *Left* electric control cabinet, *middle* laser module mounted above the belt conveyor with scrap pieces and laser-induced plasmas (time exposure comprising a set of measurements), *right* spectrometer module linked with the laser module via fiber optics

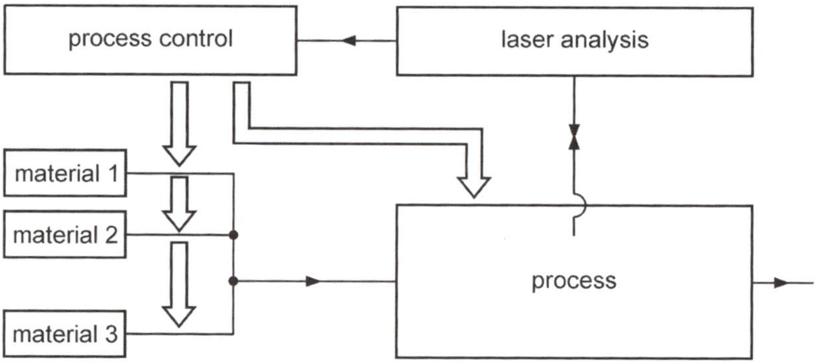

Fig. 13.13 Block diagram of laser material analysis for quantitative inline analysis of a process and upstream control loop

measure the basicity of liquid slag originating from a steel plant [28]. A truck with the slag bucket is stopping underneath the measuring system. Figure 13.14, right, shows an inside view of the cabin with the laser measuring system and the measuring lance inside which the laser beam propagates downwards in vertical direction to the slag melt surface to perform the analysis.

Figure 13.15 shows an application of laser material analysis to measure surface and depth concentration profiles [29]. The focused laser beam is guided along a line on the surface. At defined points, the material is vaporized, analyzed and a spatially resolved concentration profile can be measured. The lateral resolution is determined by the focus diameter, cf. Sect. 13.1.6. With the use of diode-pumped solid-state lasers (DPSSL) high-speed scanning microanalysis systems were

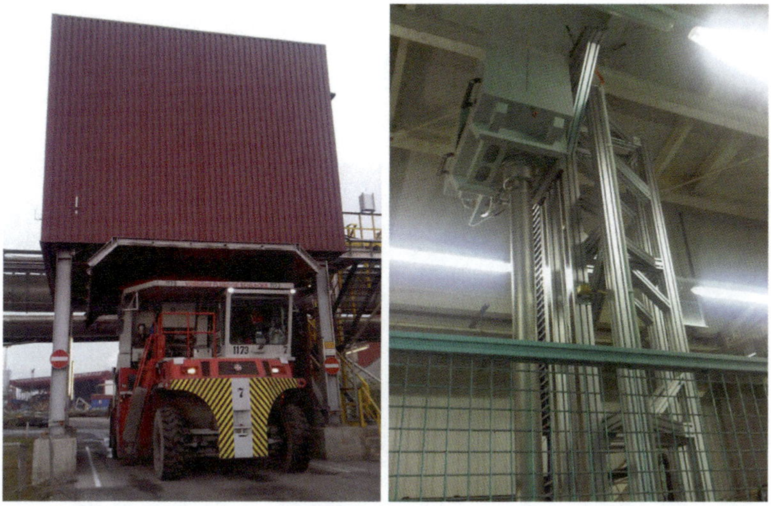

Fig. 13.14 *Left* Measuring system based on laser material analysis—mounted inside the *red cabin*—for the inline determination of the basicity of liquid slag originating from a steel plant and transported in a bucket with a truck stopping underneath the measuring system. *Right Inside view* of the cabin with the laser measuring system mounted at a scaffolding and coupled to a measuring lance (*cylindrical tube*) enclosing the laser beam being directed downwards to the slag surface in the bucket. Fotos by courtesy of voestalpine

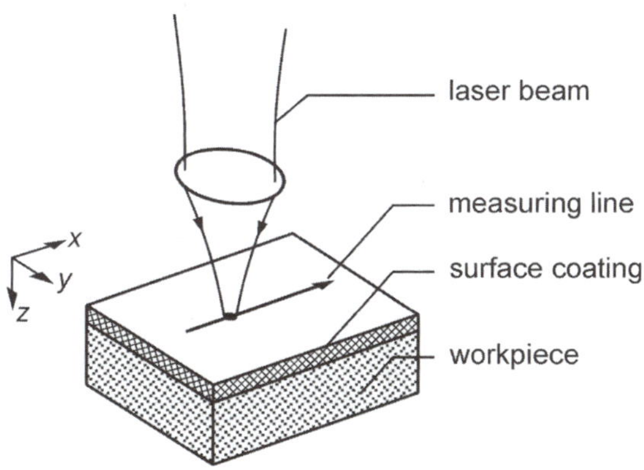

Fig. 13.15 Measurement of surface and depth concentration profiles using laser material analysis

developed to measure steel samples for the analysis of segregations, inclusions, decarburization zones and cementite laths [30, 31]. With laser pulse repetition rates of 1 kHz and more element mappings are generated showing the spatial distribution of analytes in scan fields of several square centimeters with step sizes of e.g. 20 μm [32]. In comparison with SEM-EDX (scanning electron microscopy energy dispersive X-ray fluorescence analysis) elaborate sample preparation and long measuring times are cut down significantly [33, 34].

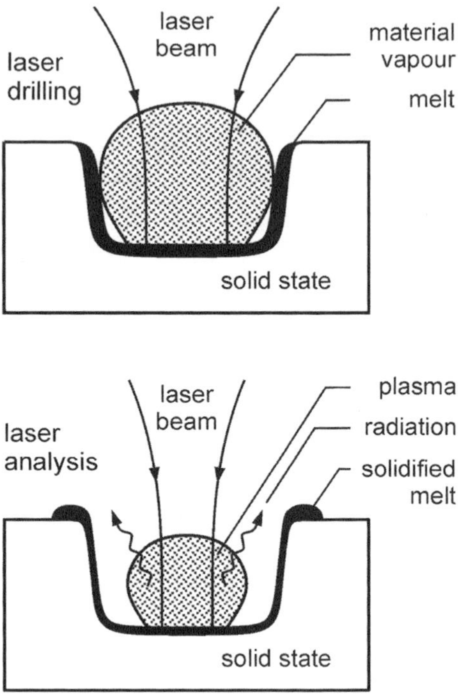

Fig. 13.16 Measurement of depth profiles by combination of laser ablation (top) and laser analysis (bottom)

If depth dependent material concentrations are found in the workpiece, i.e. along the *z*-direction in Fig. 13.15, then these can also be determined with laser material analysis. Figure 13.16 outlines the procedure [35, 36]. First, the laser drills into the workpiece, cf. Fig. 13.16, top, and when a certain depth is reached, the laser analysis begins, see Fig. 13.16, bottom. In this way, depth profiles are obtained.

The depth resolution, achievable with a metallic workpiece, is in the micrometer range. It is limited by the drill hole geometry and the melting of material which unavoidably occurs during the drilling. The original depth profile is altered in the region of the drill hole due to the material melting, which therefore should be limited to being as small as possible.

The thickness of thin metallic coatings on e.g. steel sheets is measured by penetrating the coating layer with a single laser burst [37, 38]. In such a way spectroscopic signals from the layer and from the substrate are measured at the same time. For a given laser burst energy and an approximately constant crater depth the ratio of spectroscopic signals originating from the coating to those coming from the substrate are a measure of the coating thickness. By this approach the thickness of zinc layers in the range of 2–12 μm on steel sheets can be measured with a root mean square error of 140 nm [38]. The measurement time to determine the thickness of the zinc coating can be as low as 100 ms enabling the inline monitoring of coating thicknesses in a process line.

Compact fiber lasers operating at pulse repetition rates of 30 kHz are the basis of mobile laser material analyzing systems with a handheld measuring probe [39]. Such devices can be used to identify a multitude of steel or aluminum grades for a material specific recycling of piece goods.

13.2 Light Detection and Ranging: LIDAR

LIDAR is a laser measuring process in which the concentrations of various substances in the atmosphere can be measured over large distances. The name LIDAR is an abbreviation for *light detection and ranging*. LIDAR systems are installed, for example, to monitor pollutant emissions and for the observation of the ozone layer.

13.2.1 Principle

The LIDAR principle is illustrated in Fig. 13.17. A pulsed laser beam is directed into the atmosphere via a telescope. The laser pulse irradiates scattering particles such as air molecules or aerosols in its path at a distance r. During the interaction of the laser pulse with these particles, a small fraction of the laser light is scattered back to the telescope. (At other values of r, the laser light would also be scattered back to the telescope. This will be discussed later). The amount of radiation scattered back is registered on a detector and amplified. A pulser delivers the synchronization signal which, on the one hand, triggers the pulse laser and, on the other hand, switches on the amplifier via an adjustable delay unit. Thus the signal of the backscattered light can be allocated to the distance r using the known time delay.

For the transit time of the light pulse from the LIDAR system to the scattering particles and back, the following is valid:

$$t = 2r/c, \tag{13.14}$$

with r distance between the LIDAR system and the scattering particles and c speed of light in the atmosphere.

The amount of power scattered back to the detector is calculated in the following. It must be considered, however, that the laser light is attenuated on its way from the telescope to the scattering particle. This decrease in laser light has two causes. Firstly, some part of the light is randomly scattered away from the incident direction as it propagates along its path. Secondly, a fraction is absorbed in the atmosphere. In the following, it will be assumed that the laser pulse profile can be represented by a Gaussian function. The laser power, as a function of distance and time, is then given by:

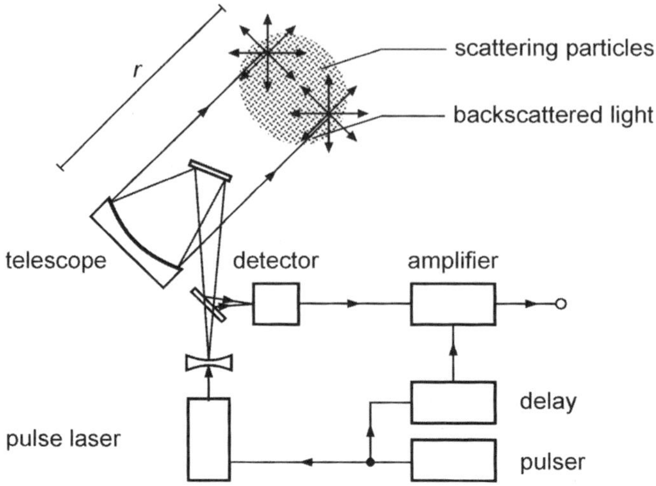

Fig. 13.17 Principle of LIDAR

$$P(r,t) = P_0 \mathrm{e}^{-\left(\frac{r-ct}{r_0}\right)^2} \mathrm{e}^{-\int\limits_0^r \mu(\lambda,R)\mathrm{d}R}, \tag{13.15}$$

where r is the coordinate along the propagation direction of the laser radiation, R is the integration variable, P_0 is the maximum laser power at $r = 0$, $t = 0$; r_0 is a measure for the spatial length extension of the laser pulse, the half width value of this spatial length amounts to $1.67r_0$ and $\mu(\lambda, R)$ is the position-dependent extinction coefficient with the unit m^{-1}.

The first exponential function in equation (13.15) describes the laser pulse during its progression into the space. At the position $r = ct$ this function takes on its maximum value. The decrease in the laser power is described by the second exponential function, corresponding to the Lambert–Beer law, see Sect. 3.3.1. In general, the extinction coefficient depends both on wavelength and location.

The laser pulse energy at $r = 0$ is given by:

$$\begin{aligned} W_\mathrm{L}(r=0) &= P_0 \int\limits_{-\infty}^{+\infty} \mathrm{e}^{-\frac{(ct)^2}{r_0^2}} \mathrm{d}t \\ &= P_0 \sqrt{\pi} \frac{r_0}{c} = P_0 \tau_\mathrm{L}, \end{aligned} \tag{13.16}$$

where W_L is the energy of the laser pulse and $\tau_\mathrm{L} = \sqrt{\pi}(r_0/c)$. The time τ_L is a measure for the duration of the laser pulse.

The power $P(r, t)$ is scattered at the location r and time t. The scattered power per solid angle amounts to:

$$\mathrm{d}P_{\mathrm{sc}}(r,t)/\mathrm{d}\Omega = P(r,t)\beta(\lambda,r)\ \mathrm{d}r, \tag{13.17}$$

where $\mathrm{d}P_{\mathrm{sc}}(r,t)/\mathrm{d}\Omega$ is the scattered power per solid angle at the location r and time t and $\beta(r,t)$ is the scattering coefficient per solid angle per path length with the unit $\mathrm{m}^{-1}\ \mathrm{sr}^{-1}$.

It is assumed in the following, that the laser radiation is scattered elastically, i.e. the scattered radiation and the laser radiation both have the same wavelength, cf. Sect. 3.4, as is the case with Mie or Rayleigh scattering. Also, for simplification purposes, it will be assumed that only single scattering occurs.

The solid angle of the receiving optic, with respect to the location of scattering, amounts to:

$$\mathrm{d}\Omega_{\mathrm{det}} = A/r^2, \tag{13.18}$$

with A aperture of the receiving lens or mirror.

At a time t, a signal is observed at the detector located at $r = 0$ which was backscattered from the location r at the time $t - r/c$. The radiation backscattered into the solid angle of the detector travels once more along the path r and is again attenuated, according to equation (13.15). For the power arriving at the detector, using equations (13.15), (13.17) and (13.18), the following is valid:

$$\begin{aligned}\mathrm{d}P_{\mathrm{det}}(t) &= \left(\frac{\mathrm{d}P_{\mathrm{sc}}(r,t-r/c)}{\mathrm{d}\Omega}\right)\mathrm{d}\Omega_{\mathrm{det}}\,\mathrm{e}^{-\int_0^r \mu(\lambda,R)\mathrm{d}R},\\ \mathrm{d}P_{\mathrm{det}}(t) &= P_0\mathrm{e}^{-[r-c(t-r/c)]^2/r_0^2}\beta(\lambda,r)\left(A/r^2\right)\mathrm{e}^{-2\int_0^r \mu(\lambda,R)\mathrm{d}R}\,\mathrm{d}r.\end{aligned} \tag{13.19}$$

This equation is integrated with respect to r, in order to obtain the total power incident to the detector. It is only in the region of $r = ct/2$ where the first exponential function takes on its maximum value, that the right hand side of equation (13.19) effectively contributes to the value of the integral. Therefore, the limits of the integration can be chosen so that they only cover the region surrounding $r = ct/2$, which corresponds to the extension of the laser pulse. Since the path length r is, as a rule, much larger than r_0, the other r-dependent terms are assumed to be approximately constant within the integration region and can be moved in front of the integral sign. The calculation yields:

$$P_{\mathrm{det}}(t) = P_0\xi(r)\beta(\lambda,r)\left(A/r^2\right)\mathrm{e}^{-2\int_0^r \mu(\lambda,R)\mathrm{d}R}\,(c\tau_{\mathrm{L}}/2). \tag{13.20}$$

Equation (13.20) is the so-called LIDAR-equation. (The relationship between the time t, on the left hand side of equation (13.20) and the location r on the right hand side is described by equation (13.14)). The overlap factor $\xi(r)$, introduced into equation (13.20), takes into consideration the fact that the observed scattering volume, detected by means of the receiver, does not completely represent the

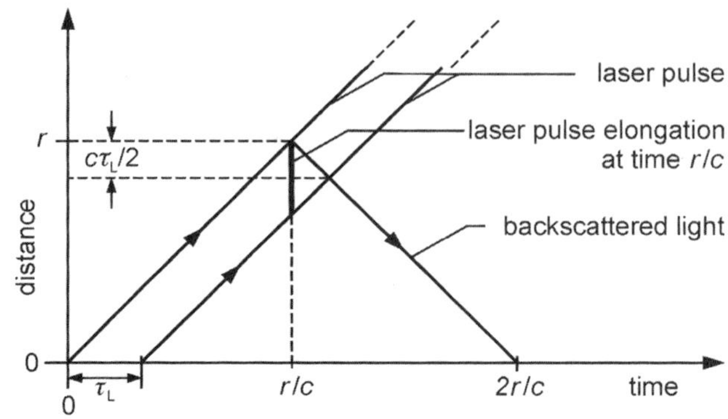

Fig. 13.18 Path-time diagram of laser beam and backscattered light in LIDAR

whole volume irradiated by the laser beam. The consequences of this will be more clearly explained, in relation to signal dynamics, in Sect. 13.2.3.

The spatial resolution depends both on the duration of the laser pulse τ_L, and on the time constant of the detector τ_D. These determine the uncertainty with which the arrival time of the backscattered signal is measured. This results in a limited spatial resolution. With equation (13.14), it follows that:

$$\Delta R = c(\tau_L + \tau_D)/2, \tag{13.21}$$

with ΔR resolution of range determination, τ_D time constant of the detector.

Figure 13.18 shows a distance-time graph of the emitted laser radiation and the backscattered light. First, the two lines, separated by a time interval of τ_L, describe the start and the end of the laser pulse. The black vertical bar marked on the diagram indicates the spatial position and extension of the laser pulse at the time $t = r/c$. At this point in time, the start of the laser pulse—upper end of the black bar in Fig. 13.18—would be scattered at the location r. The backscattered light reaches the detector position again at $t = 2r/c$. At this time, the light scattered at the location $r - c\tau_L/2$ from the end of the laser pulse also reaches the detector once again, as is shown in Fig. 13.18. The location, where the light is scattered, can therefore be determined with an uncertainty of only $c\tau_L/2$. The detector time constant influences the resolution in a similar manner.

The LIDAR-equation (13.20) forms the original basis for deducing the atmospheric optical quantities, such as the scattering coefficient $\beta(\lambda, r)$, and the extinction coefficient $\mu(\lambda, r)$ from the backscattered signal. These quantities again depend, among other things, on the density of a constituent part present in the atmosphere. In general, both $\beta(\lambda, r)$ and $\mu(\lambda, r)$ are unknown. Equation (13.20) is insufficient for their determination, if no relationship can be assumed between these quantities.

For further calculation, a dimensionless quantity, relevant to equation (13.20), will be introduced:

$$S(r) = \ln\left\{\left(P_{\text{det}} r^2\right)/\left(P_0 A\right)\right\}. \tag{13.22}$$

If r^* is a reference distance, then, using (13.20), the following is valid for the difference $S(r) - S(r^*)$:

$$\begin{aligned} S(r) - S(r^*) &= \ln\left\{\left(P_{\text{det}}(r)\cdot r^2\right)/\left(P_{\text{det}}(r^*)\cdot r^{*2}\right)\right\} \\ &= \ln\frac{\xi(r)\beta(\lambda, r)}{\xi(r^*)\beta(\lambda, r^*)} - 2\int_{r^*}^{r} \mu(\lambda, R)\mathrm{d}R. \end{aligned} \tag{13.23}$$

For purposes of simplification, it will be assumed for the overlap factor, that $\xi(r) \approx \xi(r^*)$ holds approximately. Differentiating (13.23) with respect to r then yields:

$$\mathrm{d}S(r)/\mathrm{d}r = \frac{1}{\beta(\lambda, r)}\frac{\mathrm{d}\beta(\lambda, r)}{\mathrm{d}r} - 2\mu(\lambda, r). \tag{13.24}$$

In the case of elastic scattering, which has been used as a basis up to now, the relationship between $\beta(\lambda, r)$ and $\mu(\lambda, r)$ can be represented in the following form:

$$\beta(\lambda, r) = \gamma\mu(\lambda, r)^{\kappa}, \tag{13.25}$$

where γ is a proportionality constant and κ an exponent, for which, corresponding to [40], the following range of values is valid: $0.67 \le \kappa \le 1$. Substituting equation (13.25) in equation (13.24) gives:

$$\mathrm{d}S(r)/\mathrm{d}r = \frac{\kappa}{\mu(\lambda, r)}\frac{\mathrm{d}\mu(\lambda, r)}{\mathrm{d}r} - 2\mu(\lambda, r). \tag{13.26}$$

Equation (13.26) is a differential equation of the Bernoulli type, the solution of which is:

$$\mu(\lambda, r) = \frac{\mathrm{e}^{-[S(r^*) - S(r)]\,/\,\kappa}}{\frac{1}{\mu(\lambda, r^*)} + \frac{2}{\kappa}\int_{r}^{r^*} \mathrm{e}^{-[S(r^*) - S(r'')]\,/\,\kappa}\mathrm{d}r''} \tag{13.27}$$

with $S(r)$ see equation (13.22), r^* reference distance, cf. equation (13.23).

This solution is valid for a clear, or only slightly foggy, atmosphere in which single laser light scattering processes are predominant. In an opaque atmosphere with thick cloud cover or thick fog, multiple scattering occurs. Therefore, the precondition accompanying the derivation of (13.27) is no longer fulfilled.

Using the extinction coefficient, determined from equation (13.27), as a function of the location r, the spatial density distribution can be determined for a gas of interest in the atmosphere, for example harmful gases. The extinction coefficient $\mu(\lambda, r)$ is described in the following way:

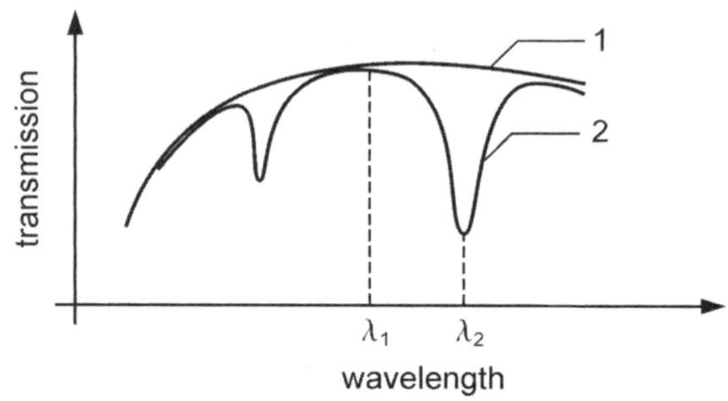

Fig. 13.19 Transmission of the atmosphere without (1) and with (2) the gas to be detected at two different wavelengths

$$\mu(\lambda, r) = \mu_{\mathrm{a}}(\lambda, r) + n_i(r)\sigma_i(\lambda), \tag{13.28}$$

where $\mu_{\mathrm{a}}(\lambda, r)$ is the extinction coefficient of the atmosphere without the gas of interest, $n_i(r)$ is the density of the gas i and $\sigma_i(\lambda)$ is the absorption cross-section of the gas i at the wavelength λ. For (13.28) it is assumed, that only one type of harmful gas is present in the atmosphere. The extinction coefficient of the atmosphere $\mu_{\mathrm{a}}(\lambda, r)$ is determined with the help of a reference measurement. With the known absorption cross-section $\sigma_i(\lambda)$, the density of the gas $n_i(r)$ follows from (13.28) as a function of the distance r.

13.2.2 Differential Absorption LIDAR

One common variant in the practice of the LIDAR method is the so-called differential absorption LIDAR, which operates with two different laser wavelengths. In order to determine the density of particular molecules in the atmosphere, the difference in absorption at these wavelengths is used. Two slightly different wavelengths are chosen, so that one is strongly absorbed by the gas to be determined, whereas for the other, the absorption is negligible. The absorption of the other gases in the atmosphere is virtually the same for both wavelengths. This situation is illustrated in Fig. 13.19.

From out of the LIDAR-equation (13.20) follows the ratio of the backscattered signals at the wavelengths λ_1 and λ_2:

$$P_{\mathrm{det}}(\lambda_1, r)/P_{\mathrm{det}}(\lambda_2, r) = \frac{\beta(\lambda_1, r)}{\beta(\lambda_2, r)} \cdot \mathrm{e}^{-2\int_0^r [\mu(\lambda_1, R) - \mu(\lambda_2, R)]\mathrm{d}R}, \tag{13.29}$$

where $P_{\mathrm{det}}(\lambda_1, r)$ corresponds to the quantity $P_{\mathrm{det}}(\lambda_2, r)$ in (13.20) at the wavelength λ_1 and $r = ct/2$ cf., (13.14). For (13.29) it is assumed, that the laser pulse duration τ_{L} as well as the laser power P_0 are the same for both wavelengths λ_1 and λ_2.

Converting equation (13.29) and using equation (13.28) gives:

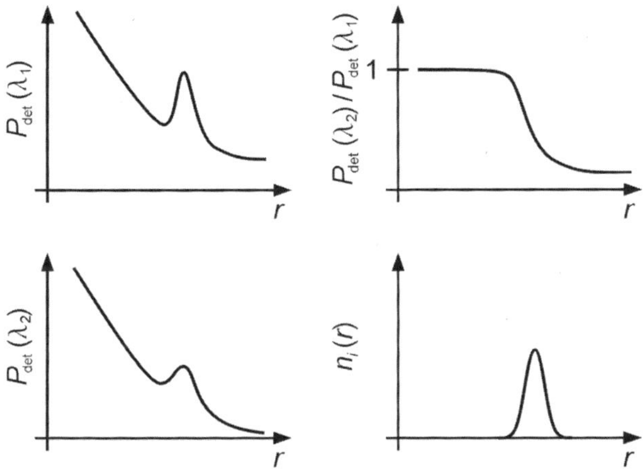

Fig. 13.20 Schematic signal sequence for differential absorption LIDAR

$$\frac{1}{2}\ln\left\{\frac{P_{\text{det}}(\lambda_2,r)\beta(\lambda_1,r)}{P_{\text{det}}(\lambda_1,r)\beta(\lambda_2,r)}\right\} = \int_0^r n_i(R)[\sigma_i(\lambda_1)-\sigma_i(\lambda_2)]\mathrm{d}R + \int_0^r [\mu_a(\lambda_1,R)-\mu_a(\lambda_2,R)]\,\mathrm{d}R \tag{13.30}$$

Differentiating equation (13.30) with respect to r delivers the sought after density $n_i(r)$:

$$n_i(r) = \frac{1}{2(\sigma_i(\lambda_1)-\sigma_i(\lambda_2))}\left\{\frac{\mathrm{d}}{\mathrm{d}r}\left(\ln\frac{P_{\text{det}}(\lambda_2,r)\beta(\lambda_1,r)}{P_{\text{det}}(\lambda_1,r)\beta(\lambda_2,r)}\right) - \mu_a(\lambda_1,r)+\mu_a(\lambda_2,r)\right\}. \tag{13.31}$$

Equation (13.31) allows still further simplification, if it can be assumed valid, that for the both closely spaced wavelengths λ_1 and λ_2 holds $\beta(\lambda_1,r)=\beta(\lambda_2,r)$ and $\mu_a(\lambda_1,r)=\mu_a(\lambda_2,r))$.

From equation (13.31), the density of the investigated gas is determined from the ratio of the measuring signal of the two different wavelengths and the known absorption cross-sections. The absolute density is determined from a relative measurement. In Fig. 13.20 the detector signals, the ratio of the detector signals and the density distribution are schematically shown.

In order to realize the differential absorption LIDAR, two different wavelengths are necessary, which are chosen, as shown in Fig. 13.19. Either two tunable lasers would be installed or one laser would be used, which possesses a series of emission wavelengths, from which two suitable wavelengths are selected.

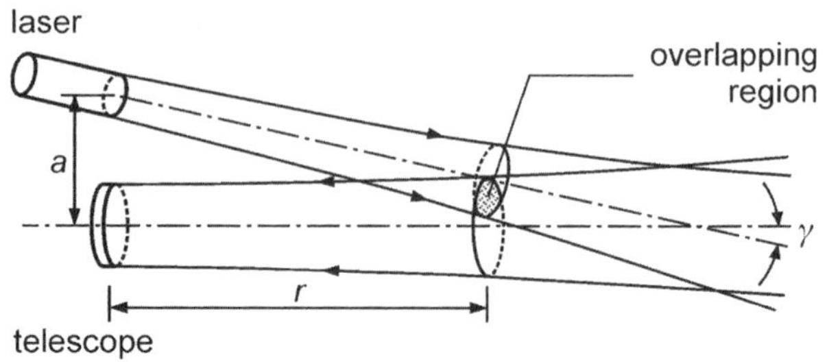

Fig. 13.21 Overlapping region between the observed scattering volume and the volume irradiated by the laser beam

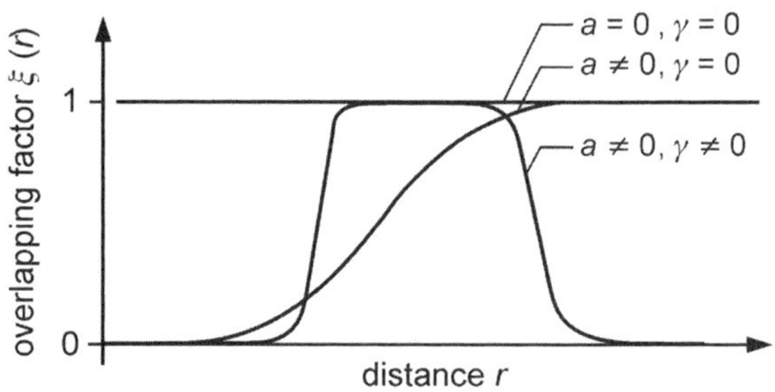

Fig. 13.22 Overlapping factor as a function of the distance for different parameters, cf. Fig. 13.21

13.2.3 Signal Processing

The LIDAR-equation (13.20) shows, that the backscattered signal decreases rapidly with an increase in distance r. Reasons for this are the $1/r^2$ factor and the exponential function, which describes the attenuation of the laser pulse along the path of the beam. A distance range from 100 m to 10 km therefore corresponds to a signal variation of more than 10^4: 1. In order to process such a signal with a high dynamic range different possibilities are on offer:

- variation of the signal amplification and the detector sensitivity,
- optical signal compression.

In the following, we limit ourselves to the latter possibility. Therefore, the overlapping factor $\xi(r)$ is considered as introduced in (13.20). Figure 13.21 illustrates the observed scattering volume, the volume irradiated by the laser beam and the overlapping region. The overlapping factor depends on the distance r, as well as on the distance of the optical axis a and its inclination angle γ, cf. Fig. 13.21. There are different arrangements possible for the laser and the telescope. With the co-axial arrangement, the axis of the telescope and the axis of the laser beam coincide. In this case, $a = 0$, $\gamma = 0$ are valid. If both axes are aligned parallel to one another, but not coinciding, we have a so-called bi-axial arrangement, i.e. $a \neq 0$, $\gamma = 0$.

The progression of the function $\xi(r)$ can be calculated by analyzing the beam geometry [40]. Figure 13.22 shows schematically the function $\xi(r)$ for different parameters a and γ. In the case of a co-axial orientation, the overlapping factor is

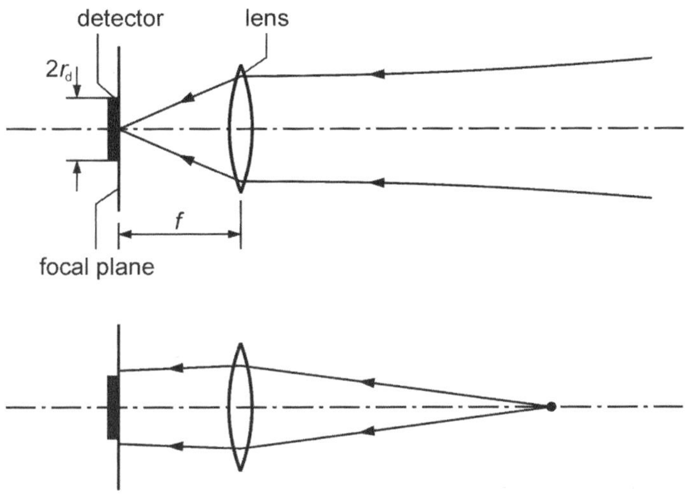

Fig. 13.23 Path of rays for the backscattered light for LIDAR

$\xi(r) = 1$, if the divergence angle of the laser beam is smaller than the opening angle of the telescope. For the bi-axial orientation with $\gamma = 0$, the overlapping factor increases with an increasing distance r. In the bi-axial orientation, with $\gamma \neq 0$, the overlapping factor first increases, then reaches a maximum value in the overlapping region and reduces for a further increase in distance r. At smaller distances both of the last two mentioned variants allow compensation for the strong distance dependency of the backscattered LIDAR signal, so that the otherwise necessary high signal dynamic can be restricted.

A further possibility for the optical signal compression is explained by Fig. 13.23. The receiving telescope of the LIDAR system is shown for simplification as a lens. The detector, with a diameter of $2r_d$ is arranged in the focal plane of the lens.

Backscattered light from a large distance is imaged within the aperture of the detector, see Fig. 13.23, above. At smaller distances defocusing appears, so that the back-scattered light is distributed over an area in the focal plane, which is larger than the detector aperture, i.e. only a part of the backscattered light would be detected, cf. Fig. 13.23, below. For smaller distances the defocusing is increased still further. The detected backscattered light would therefore be further reduced.

Figure 13.24 shows schematically the result of this effect on the power P_{det} registered by the detector as a function of the distance r for a coaxial LIDAR set-up. With regard to the LIDAR-equation (13.20) for an expected strong increase of the detector signal through small distances, this is considerably reduced by the influence of the detector aperture and the defocusing.

Through the choice of a suitable detector aperture, the dynamic range is allowed to be reduced, whereas the sensitivity for large measuring distances stays practically uninfluenced. Furthermore, a small detector aperture offers the

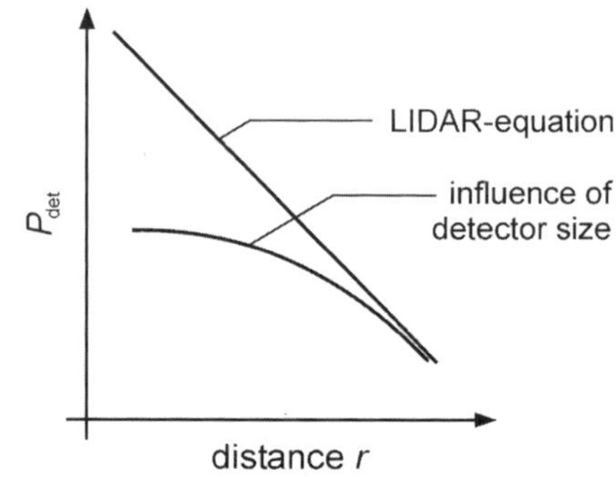

Fig. 13.24 Influence of finite detector aperture on the power of the detected signal as a function of the distance for the LIDAR method

Table 13.3 Measuring range of differential absorption LIDAR. $2f$ = frequency doubled

GAs	Laser	Wavelength [nm]	Range [km]	Detection limit [ng/g]
SO_2	$2f$-dye laser	299.38; 300.05	3	10
NO_2	Dye laser	448.1; 449.8	6	10
NO	$2f$-dye laser and Raman cell	227	0.9	–
HCl	DF-laser	3,636; 3,698	2	300
H_2O	CO_2 laser	10,250	6	–
O_3	Excimer laser and Raman cell	308; 353	40	10
Hg	$2f$-dye laser	254	–	–
Cl_2	$2f$-dye laser	303; 320	–	–

advantage, that the noise level can be kept lower. In addition, the lateral extension of the volume observed by the receiving telescope would be reduced, yielding a better spatial resolution.

13.2.4 Measuring Range

Since differential absorption LIDAR has proven itself to be the most significant method for applications, Table 13.3 shows a selection of gases, vapors and harmful substances, determined by this technique, the corresponding laser sources used as well as the accompanying measuring range [40, 41].

13.2.5 Examples of Applications

LIDAR-systems can be used for diverse applications in the field of environmental protection and atmosphere research [42]. An example is the measurement of the vertical ozone profile of the atmosphere [43]. The ozone in the stratosphere protects the surface of the earth from the ultraviolet rays of the sun. From model calculations, one expects that the destroying effect of the fluorochlorinated hydrocarbons on the ozone layer is particularly high, especially at the height of around 40 km. These heights are difficult to reach and normally can only be achieved with a balloon carried probe, yet through LIDAR technology, more accurate measurements in this region are made possible.

The German weather service operates, on the Hohenpeißenberg (988 m above sea level) in Bavaria, a differential absorption LIDAR to measure the vertical ozone distribution [44]. A XeCl excimer laser with a wavelength of 308 nm, a repetition rate of 20 Hz, a pulse duration of 30 ns and a pulse energy of 300 mJ is used. The absorption of the ozone at 308 nm is sufficient for the measurement although, on the other hand, it is not too strong, which allows it to measure the ozone at the height of between 20 and 50 km.

The reference wavelength, see Sect. 13.2.2, at 353 nm is generated by focusing the beam of the excimer laser into a hydrogen filled Raman cell. The reference wavelength is chosen significantly larger than 308 nm, since ozone has broad absorption bands at this wavelength. At 353 nm, the ozone absorption is lower than that for 308 nm by 3 orders of magnitude.

The backscattered laser light is received with a focusing mirror of 60 cm diameter. The backscattered light of both wavelengths is separated from one another through a dichroic mirror and registered with a photomultiplier. Figure 13.25 shows a measured ozone profile [43]. As a comparison, the results of a reference measurement are shown, performed e.g. with a balloon equipped with a probe based on a wet chemical method. Both measurements agree well for large height ranges. The remaining variations can be put down to the fact, that the measurements were not carried out at the same time.

In addition to stationary installed systems, LIDAR devices are operated also as mobile measuring stations in land vehicles or airplanes. Figure 13.26 shows a van, which contains a complete LIDAR system [45]. By using a mirror, which can be rotated about the vertical and horizontal axes, the measuring beam is directed in different directions in space, in order to measure horizontal and vertical concentration distributions. With such vehicles, SO_2 and N_2 distributions can be determined in large cities.

If aerosol particles are found in the atmosphere, the laser radiation is also scattered at these particles. This scattering process is called Mie scattering, see Sect. 3.4.2. It causes a local rising of the backscattered intensity, similar to the one shown in Fig. 13.20, so that the region in which an increased aerosol concentration occurs can be determined. With this process, planetary boundary layers, cloud formations and dust particles in the stratosphere are examined as well as the visibility at an airport [46, 47].

We have adopted up till now that, for the LIDAR method, the elastic scattering of the laser light is used, i.e. the scattered light has the same wavelength as the emitted laser beam. Besides elastic scattering, inelastic scattering processes can also be used, for instance fluorescence and Raman scattering, cf. Sects. 14.1 and 3.4.3.

For fluorescence LIDAR, the excited fluorescent radiation of the examined medium, whose wavelength is always greater than the laser wavelength, is

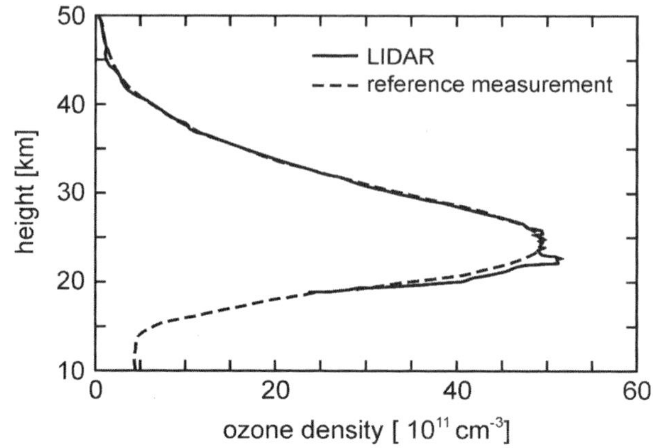

Fig. 13.25 Vertical ozone profile measured with LIDAR and reference measurement

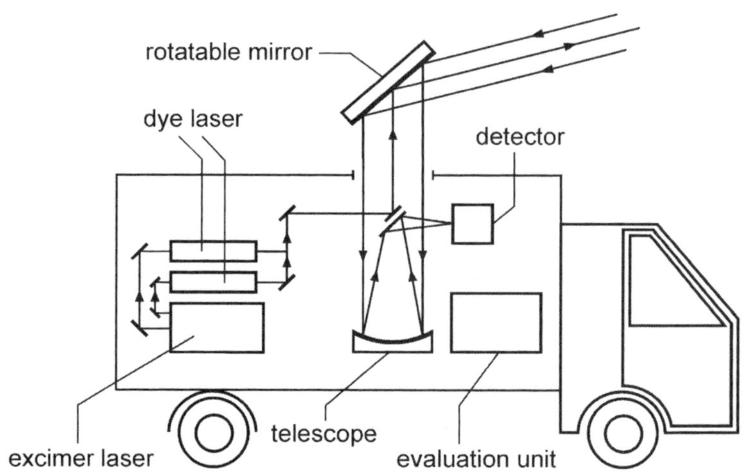

Fig. 13.26 Mobile LIDAR system installed in a light lorry

registered. With fluorescence LIDAR, it is possible to detect alkali metals in the atmosphere. Using this principle, atmospheric currents can be observed. Organic pollution in sea water or oil films on the water surface can be detected from the air with fluorescence LIDAR [40].

The range of applications cover the measurement of aerosol particles in the troposphere and stratosphere, visibility and cloud measurements, analysis of industrial emissions, water vapor and temperature profiling [48].

Space borne LIDAR systems allow to measure the location of the planetary boundary layer and clouds as e.g. stratospheric clouds [49]. Measuring results comprise cloud height, cloud thickness, cloud optical depth, cloud phase. The vertical resolution amounts to 120 m for heights up to 20 km and 360 m above 20 km. For the understanding of atmosphere dynamics a Doppler LIDAR is applied operating in the ultraviolet spectral region at 355 nm. From a height of 408 km winds are derived from the backscattered laser light, Doppler shifted by the movement of the scattering aerosols and molecules along the LIDAR line-of-sight. The primary mirror of the receiving telescope has a diameter of 1.5 m.

Complementary to photogrammetric techniques LIDAR allows to measure the three-dimensional shape of e.g. coastal settings where terrain is generally flat and subtle elevation changes have significant importance [50].

13.3 Coherent Anti-Stokes Raman Spectroscopy: CARS

The expression CARS is the abbreviated form for *Coherent Anti-Stokes Raman Spectroscopy*. CARS is a non-linear process of optical spectroscopy, with which the temperature and concentration of molecular gas components can be measured.

13.3.1 Principle

The atoms are polarized by the electrical field of a light wave in the propagating medium, see Sect. 3.5. A temporally variable electrical polarization acts as a source of electromagnetic waves. The electrical polarization $\vec{P}$ is in general a function of the electrical field $\vec{E}$ which affects the particles of the medium. This function can be described as a power series (the implementation of the powers of the vector $\vec{E}$ is explained in the following):

$$\vec{P} = \chi^{(1)}\vec{E} + \chi^{(2)}\vec{E}^2 + \chi^{(3)}\vec{E}^3 + \cdots, \tag{13.32}$$

where $\chi^{(1),(2),(3)}$ is the electrical susceptibility of first, second and third order.

Equation (13.32) is a generalization of the relationship (3.20a), given in Sect. 3. 5, where the vector character of the electrical field was not considered. The susceptibility χ is described through a tensor, which interrelates the components of the electrical field vector $\vec{E}$ to those of the polarization vector $\vec{P}$. The tensor determines the use of the powers with regard to the electrical field components in equation (13.32). For the x-component of the polarization $\vec{P}$ and the susceptibility of third order $\chi^{(3)}$, an example given is:

$$P_x = \chi^{(3)}_{xxxx} E_x E_x E_x + \chi^{(3)}_{xxxy} E_x E_x E_y + \chi^{(3)}_{xxxz} E_x E_x E_z + \chi^{(3)}_{xxyz} E_x E_y E_z + \cdots, \tag{13.33}$$

with $E_{x,y,z}$, $P_{x,y,z}$ vector components of the electrical field and the polarization, $\chi^{(3)}_{ijkl}$ components of the susceptibility tensor of third order.

In equation (13.33), not all the terms are listed. Due to the number of possible permutations of the field components E_x, E_y and E_z, there are $3^3 = 27$ terms which contribute to the polarization in the x-direction. Hence the tensor $\chi^{(3)}$ has in general cases a total of $3 \times 27 = 81$ components. Symmetry properties of the medium reduce the number of tensor elements. Therefore, for example, for isotropic media instead of 81 there are only 3 tensor elements independent from one another.

The first term in (13.32) depends linearly on the electrical field. If the electrical field $\vec{E}$ oscillates with an angular frequency ω, then this linear term causes an oscillation of the electrical polarization with the same angular frequency. Hence the susceptibility $\chi^{(1)}$ describes the interaction between electromagnetic waves and matter, in which no frequency change occurs, like for example refraction, absorption and Rayleigh scattering, see Sect. 3.4.

The second term in (13.32) describes a quadratic dependency on the electrical field, i.e. a change of the sign of the field strength does not change the sign of the polarization. However, this is not the case for isotropic materials without direction dependent properties, like for example gases or fluids. For these media, the susceptibility of the second order disappears.

The third term in (13.32) is the governing term for the CARS process. It is the first non-linear term, which does not disappear with isotropic media. In the following it is assumed that the electrical field $\vec{E}$ is made up of 3 different angular frequency components:

$$\vec{E} = \frac{1}{2} \sum_{n=1}^{3} \left\{ \vec{E}(\omega_n) \mathrm{e}^{\mathrm{i}\left(\omega_n t - \vec{k}_n \cdot \vec{r}\right)} + \text{c.c.} \right\}, \tag{13.34}$$

where $\vec{E}(\omega_n)$ is the amplitude of a wave with angular frequency ω_n; $\vec{k}_n$, ω_n angular wave vector and angular frequency of wave n, $\vec{r}$ is the position vector and t ist the time. The abbreviated term 'c.c.' stands for the conjugated complex number of the first term in the curved brackets, so that the electrical field $\vec{E}$ in the (13.34) is described as a real number. The real form is necessary for the further calculation of the non-linear term of the electric field. If only linear functions of $\vec{E}$ are examined, then the introduced complex notation in Sect. 6.1.1 is adequate.

Through the susceptibility of the third order a non-linear polarization is generated, which contains different frequency components. This becomes clear, if equation (13.34) is inserted into equation (13.32) and the terms are multiplied out.

For a clear depiction of the numerous terms, an abbreviated notation for the nonlinear polarization of the third order is used in the following:

$$\vec{P}^{(3)}(\omega_4) = \frac{1}{8}\chi^{(3)}\vec{E}_{g_1}(\omega_1)\vec{E}_{g_2}(\omega_2)\vec{E}_{g_3}(\omega_3)\,\mathrm{e}^{\mathrm{i}\left(\omega_4 t - \vec{k}_4\,\vec{r}\right)} + \mathrm{c.c.}, \tag{13.35}$$

with $g_n = \pm 1$, $\omega_4 = \sum_1^3 g_n\omega_n$, $\vec{k}_4 = \sum_{n=1}^{3} g_n\vec{k}_n$, $\vec{E}_1(\omega_n) = \vec{E}(\omega_n)$, $\vec{E}_{-1}(\omega_n) = \vec{E}^*(\omega_n)$.

$\vec{E}^*$ is the conjugated complex value of $\vec{E}$. The CARS signal is caused through the following frequency component of polarization ($g_1 = g_2 = 1$, $g_3 = -1$):

$$\vec{P}^{(3)}(\omega_4) = \frac{1}{8}\chi^{(3)}\vec{E}(\omega_1)\vec{E}(\omega_2)\vec{E}^*(\omega_3)\,\mathrm{e}^{\mathrm{i}\left(\omega_4 t - \vec{k}_4\,\vec{r}\right)} + \mathrm{c.c.}, \tag{13.36}$$

with $\omega_4 = \omega_1 + \omega_2 - \omega_3$ and $\vec{k}_4 = \vec{k}_1 + \vec{k}_2 - \vec{k}_3$.

The relationships for the frequency and the wavevector correspond to the conservation laws for energy and momentum in the photon model. Figure 13.27 illustrates this with an energy level scheme and a vector diagram. Starting from an energy level 1, a molecule absorbs a photon of energy $\hbar\omega_1$ and reaches an energy level $1'$. In general there is no energy level of the molecule available at this energy, hence this level is called a forbidden or virtual energy level. Therefore, in Fig. 13.27, left, this level is shown by a dotted line. Such energy levels can only be attained by a molecule for a short time. The molecule emits a photon of energy $\hbar\omega_3$ and reaches level 2. From there it absorbs a photon of energy $\hbar\omega_2$. A photon of energy $\hbar\omega_4$ is emitted in the transition from the virtual level $2'$ to the energy level 1.

The electromagnetic waves belonging to the angular frequencies ω_1 and ω_2 are described also as pump waves, with which the molecule can be 'pumped' into a higher energy level. The radiation with the frequency ω_3 and ω_4 respectively is called, like with spontaneous Raman scattering, the Stokes and anti-Stokes wave, see Sect. 3.4.3. The latter is the CARS signal.

The described event takes place with high probability, if the energy difference $\hbar\omega_1 - \hbar\omega_3$ is as large as the energy difference between the levels 1 and 2. The CARS signal with angular frequency ω_4 contains information about the population state of these levels. This then allows, as we will see later, temperature and concentration of the relevant molecular species to be determined. The deviation via the virtual levels offers amongst other things the preference, that the measured signal can be registered in the visible spectral range, although the energy difference between the levels 1 and 2 for typical rotational vibrational states of a molecule, lies in the region of 0.01–0.1 eV.

The wave vectors $\vec{k}_1$, $\vec{k}_2$, $\vec{k}_3$, and $\vec{k}_4$ determine the size and direction of the momentum of the photons involved in the CARS process. As with the momentum conservation in mechanics, the photon momentum must also remain conserved. In

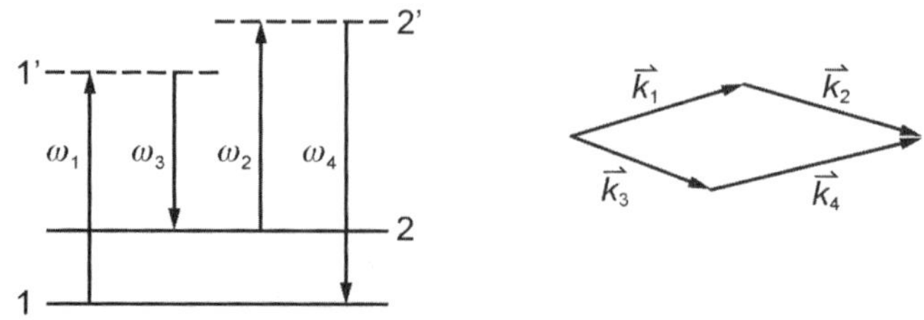

Fig. 13.27 Energy and momentum conservation for the CARS process

Fig. 13.27, right, is shown the vector diagram of the wave vectors in accordance with the given relationship for $\vec{k}_4$ in equation (13.36).

The CARS process can also be understood as a scattering process of an electromagnetic wave at coherent molecular vibrations. Coherent molecular vibrations means that a fixed phase relationship exists between the vibrations of different molecules. The stimulus of such vibrations can be described through the model of a harmonic oscillator, which is prompted into oscillation by an external electrical field. If a molecule possesses a dipole moment, an electrical field exerts a torque on the molecule. This torque is proportional to the cross product of the dipole moment and the electrical field: $\vec{p} \times \vec{E}$. Since the dipole moment and the polarization depend on the electrical field, see (13.32), the torque also depends on, amongst other things, terms proportional to E^2. If two waves with the angular frequencies ω_1 and ω_3 act on the molecule, a torque is generated, which oscillates with the difference angular frequency $\omega_1 - \omega_3$ (the similarly occuring sum frequency is not regarded, since this frequency is significantly greater than any vibrational mode of the molecule). If the difference angular frequency coincides with a vibrational mode of the molecule, this vibration is stimulated resonantly. This case gives:

$$\omega_m = \omega_1 - \omega_3, \tag{13.37}$$

with ω_m molecular vibration angular frequency.

Forced coherent molecular vibrations are produced through both the fields $E(\omega_1)$ and $E(\omega_3)$. If a further wave is irradiated into the interaction volume, this wave will be modulated by the stimulated molecular vibration. A wave is generated with a shifted frequency:

$$\omega_4 = \omega_2 + \omega_m. \tag{13.38}$$

Inserting equation (13.37) into (13.38) gives the relationship for the angular frequency ω_4, equal to the one shown in (13.36).

In the following, due to the simplification of the equation, we limit ourselves for the important case that the angular frequencies of the pump waves are equal:

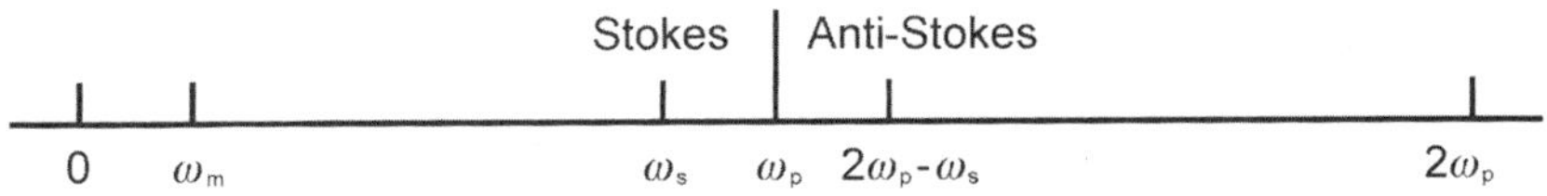

Fig. 13.28 Positions of the various frequencies for CARS

$\omega_1 = \omega_2 = \omega_p$. The angular frequency of the Stokes wave is denoted by $\omega_s = \omega_3$ and the CARS angular frequency with $\omega_{CARS} = \omega_4$, so follows:

$$\omega_{CARS} = 2\omega_p - \omega_s. \tag{13.39}$$

Figure 13.28 illustrates the positions of the different angular frequencies.

The intensity of the CARS signal is calculated using the Maxwell equations and the expression for the polarization in accordance with equation (13.36):

$$I_{CARS} = \left\{\frac{4\pi^2\omega_{CARS}^2}{c^2 n_{CARS}}\right\} I_p^2 I_s \left|3\chi^{(3)}\right|^2 L^2 \left\{\frac{\sin(\Delta k L/2)}{\Delta k L/2}\right\}^2, \tag{13.40}$$

with I_{CARS}, I_p, I_s intensity of the CARS signal, of the pump wave and of the Stokes wave, respectively; c speed of light, n_{CARS} refractive index of the medium at the angular frequency ω_{CARS}, $\chi^{(3)}$ susceptibility of the third order; L length, along which the pump and Stokes wave are interacting with the medium, $\Delta k = k_{CARS} - 2k_p + k_s$.

For the derivation of the equation (13.40) it is assumed, that the involved waves propagate in the same direction in space, so that only the absolute values of the wave vectors must be taken into account [51].

From (13.40), the intensity of the CARS signal I_{CARS} is proportional to $I_p^2 I_s$. This mirrors the multi-photon process exactly as is depicted in the diagram in Fig. 13.27, left. In order to produce a CARS photon, two pump photons and one Stokes photon are necessary.

The quantity Δk describes the deviation from the ideal momentum conservation. The valid fulfilment of the energy balance gives: $\omega_{CARS} = 2\omega_p - \omega_s$, but since $k = n\omega/c$ and the refractive index n is itself dependent on the frequency, we have in general $\Delta k \neq 0$. The CARS intensity attains a maximum according to (13.40), when $\Delta k = 0$ is fulfilled.

The measuring information is contained in the susceptibility of the third order. This depends, amongst other things, on the difference of the population numbers of the participating energy levels in the process. The population of the levels is a function of the temperature and particle concentration, so that these quantities of the medium can be determined from the CARS signal.

Figure 13.29 shows examplary typical CARS spectra for nitrogen gas at different temperatures. The intensity is shown as a function of the so-called Raman shift given by the difference of the wavenumbers $(1/\lambda_{CARS} - 1/\lambda_p)$ which is proportional to the angular frequency difference $(\omega_{CARS} - \omega_p)$. In Fig. 13.29, this

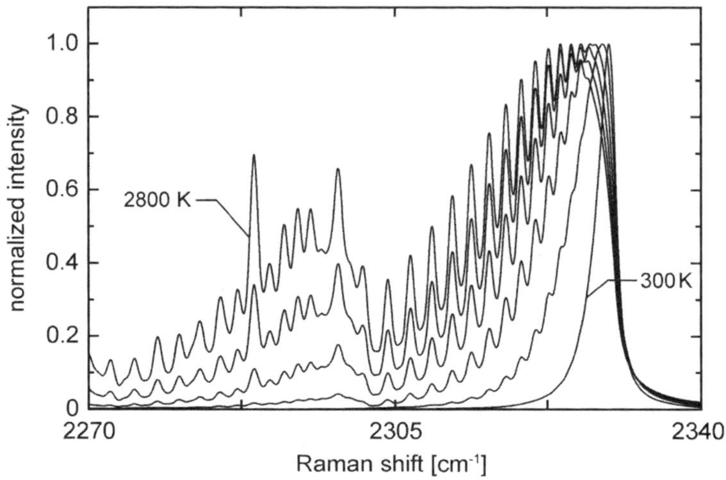

Fig. 13.29 CARS spectra of nitrogen gas at different temperatures. The spectra are normalized on the maximum value 1

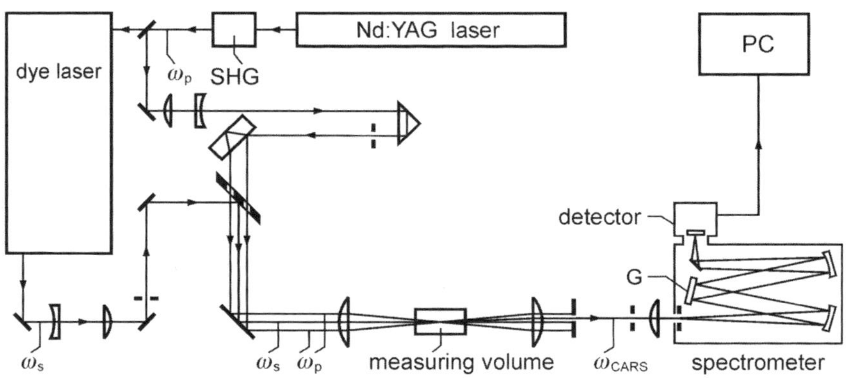

Fig. 13.30 Set-up for CARS. PC personal computer, SHG second harmonic generation, G grating

shift is given in the unit cm^{-1}, see Sect. 2.1. The lines in the spectra are caused through rotational vibrational transitions in the nitrogen molecule. The form of the spectra changes considerably in the displayed temperature range from $T =$ 300–2,800 K. With a computer fit process, the measured CARS spectra are compared with theoretical spectra, in order to determine the temperature.

Figure 13.30 shows the set-up for CARS spectroscopy [52]. The laser beam of a Q-switched Nd:YAG laser is frequency doubled and split into two beams. One of these beams pumps a dye laser, which generates the tunable Stokes frequency. Instead of a dye laser, there is also the possibility to use an optical parametric oscillator (OPO) as broadband Stokes source [53].

The dye laser is operated, so that not only a single angular frequency ω_s but a whole frequency band is emitted. In this way it is achieved that the condition (13.37) is fulfilled simultaneously for a series of different molecular vibrations.

The other split beam serves as pump angular frequency ω_p for the CARS process. Pump and Stokes beams are superimposed in a gas, in which the temperature is to be determined. The CARS signal emitted is separated from the pump and Stokes signal. Due to momentum conservation the CARS signal is emitted, cf. Fig. 13.27, right, in a defined direction and its frequency is easily distinguished from those of the pump and Stokes beams, allowing this to be coupled out simply using apertures and frequency filters. If a frequency band of Stokes frequencies is irradiated, a spectrum of CARS frequencies also occurs (13.39), which is spectrally resolved in a spectrometer. A computer programme compares the measured spectra with theoretical calculations and determines the temperature of the gas in the measuring volume.

The temperature information, can also be extracted from a CARS spectrum, evaluating line ratios in the spectrum. The molecular partial density is determined from the intensity of single lines.

In the energy level diagram for CARS, as is shown in Fig. 13.27, left, different energy levels of a molecule can be involved. Every level corresponds to a certain vibrational and rotational state of the molecule. The molecule can simultaneously vibrate and rotate, so that the vibrational levels are superimposed by rotational levels. If different vibration states of the molecule, but only one rotation state are involved in the CARS process, then this is called vibrational CARS. If only different rotational energy levels of the same vibration state are involved, then this case is called rotational CARS. Both processes are suitable for temperature and concentration measurements. Due to the low energy difference between the rotational levels, which are 1–2 orders of magnitude smaller than that of the vibrational levels, rotational CARS is especially suited for measurements at lower temperatures.

13.3.2 BOXCARS

The CARS methods, where the direction of the pump beam wavevector and the Stokes beams are different, as shown in Fig. 13.27, are known as BOXCARS. As the form of the enclosed area—see Fig. 13.27, right—is similar to a box, the variant is described as BOXCARS. With this arrangement, pump and Stokes beams cross over and interaction takes place with the medium in the overlapping region. In the set-up from Fig. 13.31 such a beam guiding is shown. The overlapping region of the beams defines the spatial resolution for temperature and concentration measurements.

By the so-called planar BOXCARS, the wave vectors lie all within one plane. In contrast to that, for folded-BOXCARS the wave vectors are orientated in different

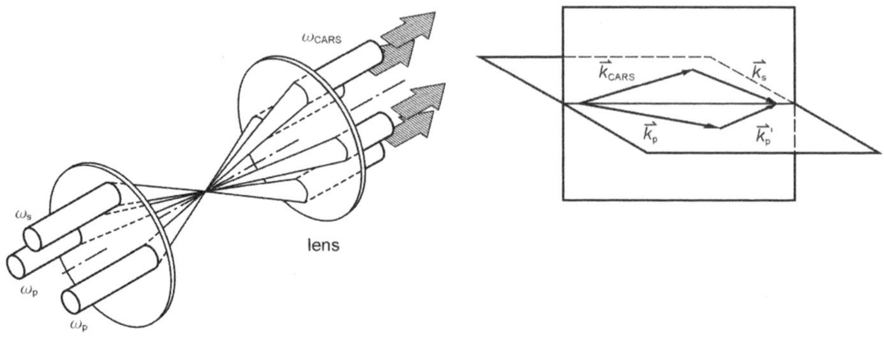

Fig. 13.31 *Left* beam guiding of pump beams and Stokes beam of the folded-BOXCARS method, *right* orientation of the wave vectors of the two pump beams, the Stokes and the CARS beam corresponding to the beam directions shown in the *left* image between the two lenses

spatial directions. Figure 13.31, left, shows the beam geometry for folded-BOXCARS, the wave vectors $\vec{k}_p$ and $\vec{k}_p'$ lie in one plane, and the wave vectors $\vec{k}_{CARS}$, $\vec{k}_s$ in a plane perpendicular to that, cf. Fig. 13.31, right. An advantage of this arrangement is that the CARS signal forms a large angle with the generating laser beams and can be separated from these simply through apertures.

13.3.3 Measuring Range

The spatial resolution that can be achieved with the CARS technique depends on the beam guiding and the focusing of the pump and Stokes beams. By the so-called collinear CARS, the wave vectors of the pump and Stokes beams are aligned in the measuring volume parallel to one another. Therefore the measuring signal originates along the whole length of the overlapping region, so that spatial resolution appears only in the perpendicular direction and not in the direction of propagation. The lateral resolution amounts, in this case, to about 1–20 mm.

With the BOXCARS method, the spatial resolution is determined by the crossover region of the focused pump and Stokes beams, cf. Fig. 13.31. It amounts to between 0.5–2 mm.

For temperature measurement, the evaluation of a single CARS spectrum is sufficient. This can be generated with a single laser pulse if a broad band dye laser is used, which supplies a spectrum of Stokes frequencies. In this case, the measuring time is given through the pulse duration of the lasers and amounts to about 10 ns. Depending on the repetition rate of the lasers, such measurements are carried out with repetition frequencies of up to 20 Hz.

In Table 13.4, typical measuring ranges of the CARS method are given for temperature and concentration determination [54]. The accuracy of the temperature measurements amounts to 2–3 %.

Table 13.4 Typical measuring ranges of the CARS method

Gas	Temperature (K)	Concentration (%)
N_2	300–3,500	> 2
H_2	300–2,500	> 1
O_2	–	> 0.5
CO_2	300–1,900	–
CO	–	> 0.1
H_2O	450–1,700	–
CH_4	–	> 0.2

A transfer standard for laser thermometry was developed based on the international temperature scale of 1990 (ITS-90) [55]. The standard provides optical access to a sample volume of a gas maintained at a stable temperature within the range of 300–1,850 K. The gas temperature is measured using built-in, calibrated thermometers including thermocouples, Accufiber™ optical fiber probes and an optical pyrometer for traceability to the ITS-90. Broadband CARS and scanning CARS experiments have been performed inside the furnace. The measuring uncertainty for broadband CARS amounts to less than 4 % between room temperature and 800 K. At temperatures between 800 and 1,850 K an uncertainty of less than 2 % has been achieved. For scanning CARS the uncertainty and the precision amounts to 2 % between 295 and 1,850 K. The pulse energy of the two pump beams at 532 nm is 40 mJ with a spectral bandwidth of 0.8 cm^{-1} (FWHM). The tunable dye laser yields the Stokes beam with a bandwidth of 150 cm^{-1} having a maximum at 606.4 nm and a pulse energy of 5 mJ. Temperatures are measured with N_2 CARS spectra.

13.3.4 Examples of Applications

The CARS method is used mainly for the examination of combustion processes and electrical gas discharges. Some examples are: ethyl-air-bunsen flame, methane-air flat flame burners, kerosene burner, turbulent gas diffusion flame, Diesel and Otto engines.

Figure 13.32 shows the combustion chamber of an Otto engine with the laser beam guiding for temperature measurement using the CARS method [56]. The engine is an air cooled one cylinder 4 stroke Otto engine. The laser beams are transmitted by quartz glass windows in and out of the burning space. The windows are designed so that they withstand a static pressure of 20 MPa. Figure 13.33 shows the determined progression of temperature as a function of the crankshaft angle. The motor speed was 2,400 min^{-1}, the used fuel is methanol at an air/fuel ratio of 1.

Within a short time after the ignition of the mixture, at a crankshaft angle of 149°, the temperature reaches the maximum value of about 2,500 K. During the

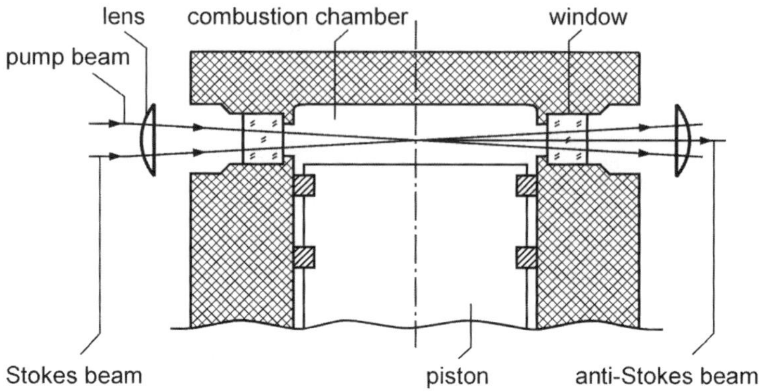

Fig. 13.32 Cross section of a cylinder head of an Otto engine and beam path for CARS measurements

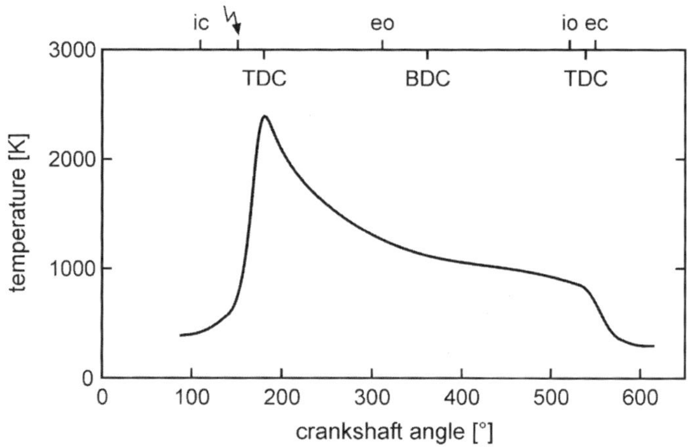

Fig. 13.33 Temperature measured with CARS as a function of the crankshaft angle. ic intake closing, eo exhaust opening, io intake opening, ec exhaust closing, TDC top dead center, BDC bottom dead center

working period the gas cools and after the opening of the suction valve an accelerated cooling is observed, due to the instream of cold fresh gas. From such measurements engine knocking effects are examined. The gained results contribute to a better understanding of the combustion processes and allow the possibility for specific further developments.

Figure 13.34 shows an example of temperature measurements with CARS in a graphite furnace, which is used for atomic absorption spectroscopy [57]. This furnace is constructed from a graphite pipe, which is electrically heated. The length of the pipe is about 30 mm, and has an inner diameter of 6 mm. With a graphite furnace a temperature of up to 3,000 K can be reached within a just few

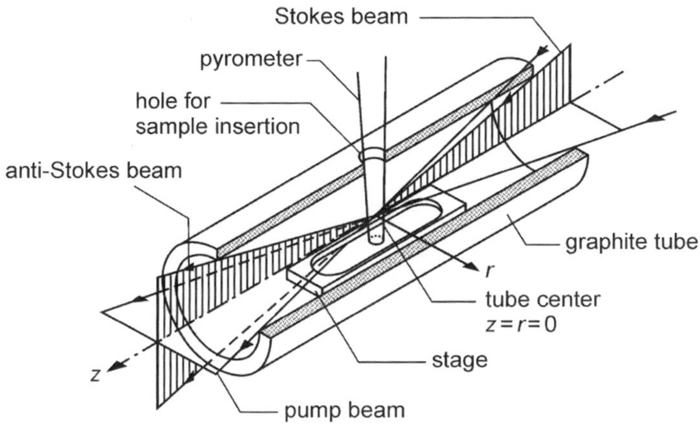

Fig. 13.34 CARS temperature measurement in a graphite oven

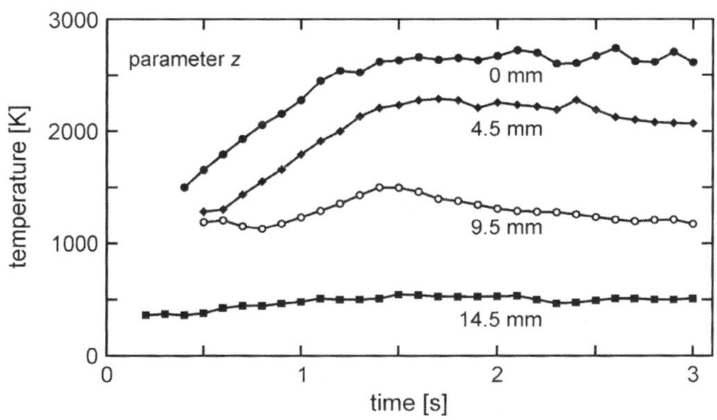

Fig. 13.35 Temperature profile in a graphite oven measured with CARS along the z-direction, cf. Fig. 13.34

seconds. They are used to evaporate substances dissolved in fluids in a short time, in order to examine the vapor with absorption spectroscopy.

Figure 13.35 shows the measured temperature profile as a function of time in the graphite furnace during the heat-up phase using CARS. The parameter is the z-position of the measuring point, cf. coordinate system in Fig. 13.34. The time and spatial temperature variation supplies important information for the examination of the furnace properties, like atomization rate, diffusion processes, gas phase reactions etc.

Figure 13.36 shows the set-up of a CARS temperature measurement in an electrical gas discharge, which is used to excite a CO_2 laser [58]. In a high frequency driven low pressure discharge, energy is coupled into a gas mixture of CO_2, N_2 and He in order to generate the required inversion for the CO_2 laser process. The properties of the gas discharge determine the efficiency and quality of

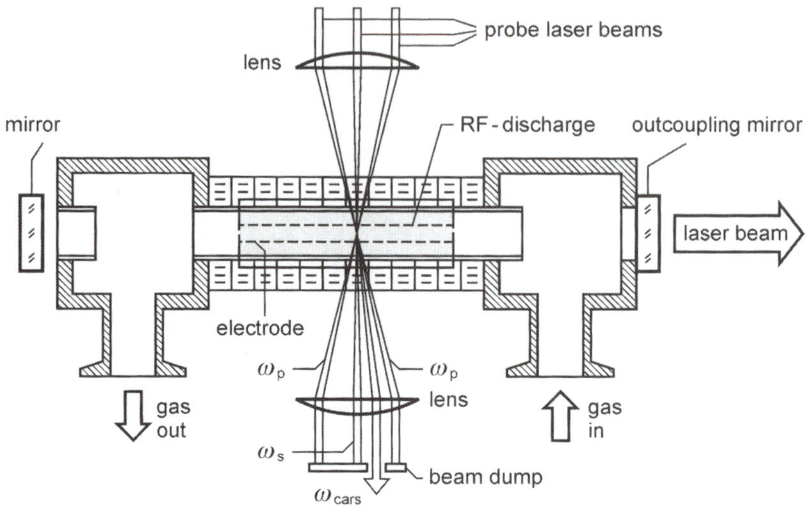

Fig. 13.36 CARS temperature measurement in a gas discharge of a CO_2 laser. RF radio frequency

Fig. 13.37 Photograph of the experimental set-up, cf. Fig. 13.36

the CO_2 laser beam. The discharge in the gas is examined with the help of CARS, in order to optimize the discharge. The CARS beams are aligned in a BOXCARS formation, perpendicular to the longitudinal axis of the discharge. The crossover region of the Stokes and pump beams defines the measurement region in the discharge volume. It is shifted relative to the discharging electrodes, in order to measure a spatial temperature profile. Figure 13.37 shows a photograph of the experimental set-up.

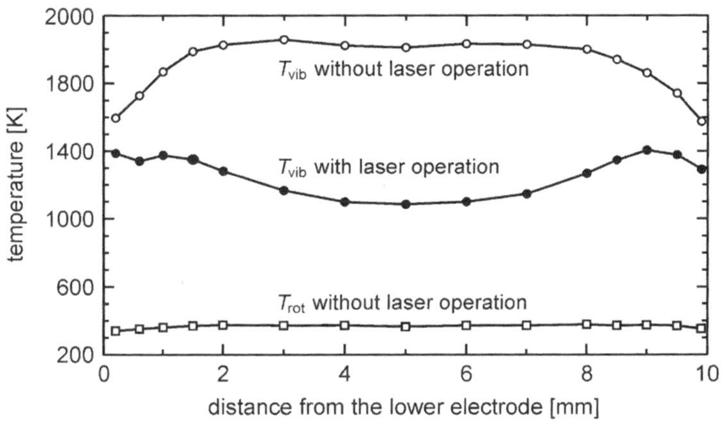

Fig. 13.38 Temperature profile in a CO_2 gas discharge measured with CARS

Figure 13.38 shows the N_2 temperature across the cross section of the discharge. The CARS method allows to distinguish between the vibrational temperature T_{vib} and the rotational temperature T_{rot} of the N_2 gas in the discharge. Due to the short energy exchange time between the rotational levels of nitrogen and the translation energy of the gas components, the rotational temperature corresponds to the gas temperature in the discharge.

The vibrational temperature reduces with the operation of the CO_2 laser. The vibration energy of the N_2 is transferred via collisions with the laser active CO_2 molecules and leads to a decrease in the vibrational temperature of the N_2.

CARS is also applicable to microscopy if objectives with high numerical aperture are used [59, 60]. In this case pump and Stokes beams need not to be irradiated separately, as shown e.g. in Fig. 13.31. These excitation beams are now guided collinearly and illuminate completely the aperture of the microscope objective with the result that in the focus volume, for each component of the wave vectors of the pump beams exists the corresponding Stokes beam wave vector component to fulfill the phase matching condition. The CARS signal is then observed in transmission of the microscopic object in forward direction (abbreviated as F-CARS). In this case holds (cf. (13.40)):

$$\Delta\vec{k} = \vec{k}_{CARS} - 2\vec{k}_p + \vec{k}_s \approx 0. \qquad (13.41)$$

In order to maximize the CARS signal, the sinc-function in (13.40) has to attain a maximum, which is the case for:

$$|\Delta k| L \ll \pi. \qquad (13.42)$$

For a very thin microscopic object with $L \approx 0$, the phase-matching condition is satisfied not only in the forward direction—given by relation (13.41)—but also in the backward direction [59]:

$$\vec{k}_{\mathrm{CARS}} = -2\vec{k}_{\mathrm{p}} + \vec{k}_{\mathrm{s}}. \tag{13.43}$$

For the absolute value of the phase mismatch we have in the case of the backscattered CARS signal:

$$\left|\Delta\vec{k}\right| \approx 2\left|\vec{k}_{\mathrm{CARS}}\right| = \frac{4\pi n_{\mathrm{CARS}}}{\lambda_{\mathrm{CARS}}}, \tag{13.44}$$

where n_{CARS} is the refractive index of the nonlinear medium. The backward propagating CARS signal is the basis of epidetected CARS, abbreviated E-CARS.

Via a spectral filter the forward CARS signal is separated from the pump and Stokes beams. By moving the object relative to the focus three-dimensional CARS images are generated. F-CARS allows to generate microscopic images of cellular structures of living cells as e.g. the membrane of the nucleus. This membrane contains fatty acid molecules which are made visible by vibrations causing stretching of the C–H distance with a resonance at 2,900 cm^{-1}. Radiant fluxes required for excitation of the CARS signal are in the range of a few 100 μW, so that bleaching is avoided.

Whereas for very thin objects the F-CARS and E-CARS signals are equal in intensity. With increasing L the forward signal increases $\propto L^2$, cf. (13.40). Therefore, the contribution of the solvent (large L), where the microscopic object is embedded, can overwhelm the contribution of the object (small L). In case of E-CARS the signal oscillates as a function of L: $I^{\mathrm{E}}_{\mathrm{CARS}} \propto \lambda^2_{\mathrm{CARS}} \sin^2 (2\pi n_{\mathrm{CARS}} L/\lambda_{\mathrm{CARS}})$. Therefore, the solvent contribution (large L) is significantly reduced compared to that of F-CARS and hence the signal-to-background ratio is increased.

The FWHM width of the images achieved with pump and Stokes beams at 800 and 917 nm and a polysterene bead with a diameter of 535 nm as microscopic object amount to 300 nm for F-CARS and 290 nm for E-CARS.

References

1. R. Scott, A. Strasheim, Time-resolved direct-reading spectrochemical analysis using a laser source with medium pulse-repetition rate. Spectrochim. Acta **26B**, 707–719 (1971)
2. J. Belliveau, L. Cadwell, K. Coleman, L. Hüwel, H. Griffin, Laser-induced breakdown spectroscopy of steels at atmospheric pressure and in air. Appl. Spectrosc. **39**, 727–729 (1985)
3. J. Millard, R. Dalling, L. Radziemski, Time-resolved laser-induced breakdown spectrometry for the rapid determination of beryllium in beryllium–copper alloys. Appl. Spectrosc. **40**, 491–494 (1986)

4. D. Cremers, The analysis of metals at a distance using laser-induced breakdown spectroscopy. Appl. Spectrosc. **41**, 572–579 (1987)
5. J. Henning, *Der Spektralapparat Kirchhoffs und Bunsens* (Deutsches Museum, Verlag für Geschichte der Naturwissenschaften und der Technik, Berlin, 2003)
6. R. Noll, *Laser-Induced Breakdown Spectroscopy—Fundamentals and Applications* (Springer, Berlin, 2012), ISBN 978-3-642-20667-2, 543 p
7. H. Carslaw, J. Jaeger, *Conduction of Heat in Solids*, 2nd edn. (Oxford University Press, Oxford, 1959) reprint 2000, ISSBN 0 19 853368 3
8. R. Klein, Bearbeitung von Polymerwerkstoffen mit infraroter Laserstrahlung, Dissertation, Aachen, 1990
9. *VDI-Wärmeatlas*, Springer-Verlag, Berlin, 11. Aufl., 2013, ISBN 978-3-642-19980-6, 1760 p.
10. R. Wester, Laserinduziertes Abdampfen als Basisprozess des Bohrens, Fräsens und Schneidens, Laser und Optoelektronik **23**, 60–63 (1991)
11. E. Beyer, Einfluss des laserinduzierten Plasmas beim Schweißen mit CO_2-Lasern. Schweißtechnische Forschungsberichte, Bd. 2, Düsseldorf, Deutscher Verlag für Schweißtechnik, 1985
12. N. Damany, J. Romand, B. Vodar (ed.), *Vacuum Ultraviolet Radiation Physics* (Pergamon Press, New York, 1974) ISBN 0-08-016984-8
13. A. Zaidel, V. Prokofev, S. Raiskii, V. Slavnyi, E. Shreider, *Tables of Spectral Lines* (IFI/Plenum, New York, 1970), 782 p.
14. W. Wiese, M. Smith, B. Glennon, *Atomic Transition Probabilities*, vol. I, II. National Standard Reference Data Series. (National Bureau of Standards 4, Washington, 1966)
15. NIST atomic spectra database, http://physics.nist.gov/PhysRefData/ASD/lines_form.html
16. P. Smith, C. Heise, J. Esmond, R. Kurucz, Atomic Spectral Line Database, ed. by R.L. Kurucz. CD-ROM 23, http://www.pmp.uni-hannover.de/cgi-bin/ssi/test/kurucz/sekur.html
17. R. Noll, R. Wester, Heuristic modeling of spectral plasma emission for laser-induced breakdown spectroscopy. J. Appl. Phys. **106**, 123302 (2009)
18. L. Radziemski, T. Loree, D. Cremers, H. Hoffmann, Time-resolved laser-induced breakdown spectrometry of aerosols. Anal. Chem. **55**, 1246–1252 (1983)
19. R. Noll, I. Mönch, O. Klein, A. Lamott, Concept and performance of inspection machines for industrial use based on LIBS. Spectrochim. Acta B **60**, 1070–1075 (2005)
20. C. Gehlen, J. Makowe, R. Noll, Automatisierte Verwechslungsprüfung von Edelstahlhalbzeugen in der Produktion. stahl und eisen **129**, S70–S72 (2009)
21. International Standard, *Safety of Laser Products—Part 1: Equipment Classification and Requirements*, 200 p. IEC 60825-1, Ed. 2.0, (2007)
22. German Standard, *Technical Availability of Machines and Production Lines, Terms, Definitions, Determination of Time Periods and Calculation.* VDI 3423, January 2002
23. H. Kunze, R. Noll, J. Hertzberg, R. Sattmann, Laser-Stoffanalytik, Proc. 10. Int. Kongresses Laser 91 Optoelektronik (Springer-Verlag, Berlin, 1992), pp. 181–185
24. R. Noll, C. Fricke-Begemann, P. Jander, T. Kuhlen, V. Sturm, P. Werheit, J. Makowe, Perspektiven der Lasertechnik zur Steigerung der Ressourceneffizienz, Hrsg. U. Teipel, *Rohstoffeffizienz und Rohstoffinnovation*, Fraunhofer Verlag, 2010, S. 287–298
25. R. Noll, V. Sturm, C. Fricke-Begemann, P. Werheit, J. Makowe, Laser-induced breakdown spectroscopy—new perspectives for in-line analysis of materials. Metall. Anal. **30**, 22–30 (2010)
26. P. Werheit, C. Fricke-Begemann, M. Gesing, R. Noll, Fast single piece identification with a 3D scanning LIBS for aluminium cast and wrought alloys recycling. J. Anal. At. Spectrom. **26**, 2166–2174 (2011)
27. K. Pilz, Online-Analytik zur Prozesskontrolle in der voestalpine Stahl GmbH. Berg- und Hüttenmännische Monatshefte (BHM) **157**, 250–257 (2012)
28. V. Sturm, R. Fleige, M. de Kanter, R. Leitner, K. Pilz, D. Fischer, G. Hubmer, R. Noll, Laser-induced breakdown spectroscopy for 24/7 automatic liquid slag analysis at a steel works. Anal. Chem. (2014). DOI: 10.1021/ac5022425

29. R. Noll, R. Sattmann, Lasergestützte Stoffanalyse für die online Prüfung von Oberflächenschichten, Proc. Surtec Berlin 1991, Ed. Deutsche Forschungsgemeinschaft für Oberflächenbehandlung, Düsseldorf, 1991, S. 367–374
30. H. Bette, R. Noll, High-speed laser-induced breakdown spectrometry for scanning microanalysis. J. Phys. D Appl. Phys. **37**, 1281–1288 (2004)
31. H. Bette, R. Noll, G. Müller, H.-W. Jansen, Ç. Nazikkol, H. Mittelstädt, High-speed scanning laser-induced breakdown spectroscopy at 1000 Hz with single pulse evaluation for the detection of inclusions in steel. J. Laser Appl. **17**, 183–190 (2005)
32. H. Bette, R. Noll, High-speed, high-resolution LIBS using diode-pumped solid-state lasers, *Laser-Induced Breakdown Spectroscopy*, Chap. 14, ed. by A. Miziolek, V. Palleschi, I. Schechter (Cambridge University Press, Cambridge, 2006), pp. 490–515
33. F. Boué-Bigne, Analysis of oxide inclusions in steel by fast laser-induced breakdown spectroscopy scanning: an approach to quantification. Appl. Spectrosc. **61**, 333–337 (2007)
34. F. Boué-Bigne, Laser-induced breakdown spectroscopy applications in the steel industry: rapid analysis of segregation and decarburization. Spectrochimica Acta Part B **63**, 1122–1129 (2008)
35. V. Sturm, J. Vrenegor, R. Noll, M. Hemmerlin, Bulk analysis of steel samples with surface scale layers by enhanced laser ablation and LIBS analysis of C, P, S, Al, Cr, Cu, Mn and Mo. J. Anal. At. Spectrom. **19**, 451–456 (2004)
36. H. Balzer, M. Hoehne, R. Noll, V. Sturm, New approach to monitoring the Al depth profile of hot-dip galvanised sheet steel online using laser-induced breakdown spectroscopy. Anal. Bioanal. Chem. **385**, 225–233 (2006)
37. H. Balzer, M. Hoehne, V. Sturm, R. Noll, Online coating thickness measurement and depth profiling of zinc coated sheet steel by laser-induced breakdown spectroscopy. Spectrochimica Acta B **60**, 1172–1178 (2005)
38. H. Balzer, S. Hölters, V. Sturm, R. Noll, Systematic line selection for online coating thickness measurements of galvanised sheet steel using LIBS. Anal. Bioanal. Chem. **385**, 234–239 (2006)
39. M. Scharun, C. Fricke-Begemann, R. Noll, Laser-induced breakdown spectroscopy with multi-kHz fibre laser for mobile metal analysis tasks—a comparison of different analysis methods and with a mobile spark-discharge optical emission spectroscopy apparatus. Spectrochimica Acta Part B **87**, 198–207 (2013)
40. R. Measures (ed.), *Laser Remote Chemical Analysis* (Wiley, New York, 1988)
41. C. Weitkamp, W. Lahmann, W. Staehr, Reichweite- und Empfindlichkeitsoptimierung beim DAS-LIDAR. Laser u. Optoelektronik **19**, 375–381 (1987)
42. Y. Carts, LIDAR proves useful in studies of the environment. Laser Focus World **11**, 53–66 (1991)
43. T. McGee, D. Whiteman, R. Ferrare, J. Butler, J. Burris, STORZ LITE: NASA Goddard stratospheric ozone LIDAR trailer experiment. Opt. Eng. **30**, 31–39 (1991)
44. H. Claude, Ozonmessung mittels LIDAR am Hohenpeißenberg, Proc. 10. Int. Kongr. Laser 89 Optoelektronik, Springer Verlag, Berlin, 1990, 408–411
45. H. Kölsch, P. Lambelet, H. Limberger, P. Rairoux, S. Recknagel, J. Wolf, L. Wöste, LIDAR-pollution monitoring of the atmosphere, Proc. 10. Int. Kongr. Laser ´89 Optoelektornik, Springer Verlag, Berlin, 1990, 412–415
46. R. Dubinsky, LIDAR moves toward the 21st century. Laser Optron. **4**, 92–106 (1988)
47. A. Beck, J. Fricke, LIDAR-Systeme erfassen atmosphärischen Aerosolgehalt. Physik in unserer Zeit **21**, 81–83 (1990)
48. C. Weitkamp (ed.), *LIDAR: Range-Resolved Optical Remote Sensing of the Atmosphere* (Springer, New York, 2005), 455 p
49. D. Lajas, P. Ingmann, T. Wehr, A. Ansmann, Aerosols and clouds: improved knowledge through space borne LIDAR measurements, in *3rd Symposium on LIDAR Atmospheric Applications*, P1.17, 4 p., 2007

50. J. Carter, K. Schmid, K. Waters, L. Betzhold, B. Hadley, R. Mataosky, J. Halleran, *Lidar 101: An Introduction to LIDAR Technology, Data, and Applications* (National Oceanic and Atmospheric Administration (NOAA) Coastal Services Center, Charleston, 2012), 76 p.
51. R. Hall, A. Eckbreth, Coherent anti-Stokes Raman spectroscopy (CARS): application to combustion diagnostics, in *Laser Applications, vol. V*, ed. by J. Ready, R. Erf (Academic Press, New York, 1984), pp. 213–309
52. J. Hertzberg, Einsatz der CARS-Spektroskopie zur Untersuchung einer CO_2-Laser-gasentladung, Dissertation, RWTH Aachen University, 1993
53. D. Brüggemann, J. Hertzberg, B. Wies, Y. Waschke, R. Noll, K. Knoche, G. Herziger, Test of an optical parametric oscillator (OPO) as a compact and fast tunable Stokes source in coherent anti-Stokes Raman spectroscopy (CARS). Appl. Phys. B **55**, 378–380 (1992)
54. I. Plath, CARS-Temperaturmessungen an laminaren und turbulenten Flammen - Untersuchung der Messgenauigkeit des Einzelpulsverfahres, Dissertation, University Stuttgart, 1991
55. C. Rieck, B. Bödefeld, R. Noll, G. Edwards, S. Boyes, M. Péalat, P. Bouchardy, N. Dorwal, J. Fischer, M. Stock, Development of a transfer standard for laser thermometry. SPIE Vol. **3107**, 74–85 (1997)
56. D. Brüggemann, Entwicklung der CARS-Spektroskopie zur Untersuchung der Verbrennung im Otto-Motor, Dissertation, RWTH Aachen University, 1989
57. B. Welz, M. Sperling, G. Schlemmer, N. Wenzel, G. Marowsky, Spatially and temporally resolved gas phase temperature measurements in a Massmann-type graphite tube furnace using coherent anti-Stokes Raman scattering. Spectrochim. Acta, Part B **43**, 1187–1207 (1988)
58. J. Hertzberg, R. Noll, P. Loosen, G. Herziger, Spatially resolved temperature measurements in a carbon-dioxide-laser discharge by folded BOXCARS, in *Proceedings of XI European CARS Workshop*, ed. by F. Castelucci, World Scientific Publishing, (1992), pp. 109–114
59. A. Zumbusch, A. Volkmer, Einblick in das Unsichtbare – Nichtlineare optische Phänomene ermöglichen die chemisch selektive Mikroskopie ohne Anfärbung. Physik Journal **4**, 31–37 (2005)
60. A. Volkmer, J. Cheng, X. Xie, Vibrational imaging with high sensitivity via epidetected coherent anti-Stokes Raman scattering microscopy. Phys. Rev. Lett. **87**, 23901–23904

Chapter 14
Laser-Induced Fluorescence

Abstract The basic principle of laser-induced fluorescence and fluorescence spectroscopy is presented to detect organic molecules with high sensitivity. We present various variants of this method: fluorescence correlation spectroscopy, fluorescence polarization spectroscopy, time-resolved fluorescence analysis. Examples of applications range from the inline measurement of the shape and deformation of metal sheets, measurement of organic residues on metal sheets, study of combustion processes, medical diagnostics of tumors to the detection of single labeled cells in microfluidic chips.

In the previous chapter we have seen, that laser radiation at high irradiances can generate a secondary radiation whether by transferring the measuring object to the plasma state, see laser material analysis in Sect. 13.1, or by a non-linear interaction as for CARS, presented in Sect. 13.3. In this chapter we describe laser measuring methods based on the excitation of fluorescence of the constituents with no change of the aggregation state of the object nor any use of non-linear interactions [1].

Fluorescence is a fast decaying emission of light from electronically excited states of atoms and molecules. The lifetimes of excited singlet states typically range from picoseconds to nanoseconds. If a reversal of the spin electronic wave function—triplet-singlet transition—is included, the phenomenon is called phosphorescence. Because such transitions are quantum mechanically prohibited, excited particles remain relatively long in a triplet state. Triplet state life times and the duration of phosphorescence are typically in the range of milliseconds to seconds.

Fluorescence excitation does not necessarily require laser light sources. However, due to the unique features of laser radiation—such as high spectral irradiance and focusability—lasers enable high-performance fluorescence methods. In this chapter laser-induced fluorescence (LIF) as well as fluorescence excited by incoherent light sources will be presented.

A. Donges and R. Noll, *Laser Measurement Technology*, Springer Series in Optical Sciences 188, DOI: 10.1007/978-3-662-43634-9_14,

14.1 Principle

Molecules can absorb energy not only in the form of electronic excitation, but also as vibrational energy of the nuclei. The various levels of vibrational excitation of an electronically excited molecule are called vibronic levels. Electronic transitions—responsible for optical absorption or emission—proceed on a much shorter time scale than the vibrations of the atomic nuclei around their equilibrium positions. The duration of an optical transition is typically in the order of 10^{-15} s, during this time the nuclei do almost not change their relative positions with respect to each other. In a term scheme, in which potential energy curves of different electronic states are drawn together with their vibrational levels—so-called Jablonski term scheme—optical emissions or excitations are therefore marked as vertical transitions (no change in the coordinates of the nuclei).

The wave functions of the nuclei in ground and excited state have a large overlap integral for transitions of high probability. With the help of the so-called Franck-Condon factors the transition probabilities between two vibronic states can be calculated.

At room temperature practically all molecules occupy their vibrational ground state. Therefore all optical excitations start from this initial state, see Fig. 14.1 showing a term scheme for the fluorescence of molecules. The larger the potential curves of primary and excited state are shifted against each other, the more likely the vertical transition leads to an excited electronic state that also has a vibrational excitation. The vibrational energy of molecules in condensed phases is distributed very quickly to the surrounding molecules via thermal relaxation. Fluorescence therefore usually is emitted from the vibrational ground state of the electronically excited state. This behaviour is referred to as Kasha's rule saying that in polyatomic molecules generally photon emission occurs in significant yield only from the lowest excited state of a given multiplicity [2]. For the emission of light, the Franck-Condon principle of vertical transitions must also be fulfilled, hence the emission process leads to a vibrationally excited level of the electronic ground state. This leads to the fact that the fluorescence experiences a red shift in relation to the excitation wavelength, which is called Stokes shift. In solutions the interaction between fluorescent molecules and the solvent molecules leads to a broadening of the vibronic transitions. Often the individual transitions cannot be differentiated any longer and only broad excitation and emission peaks can be detected. In gas phase experiments the individual vibronic levels can usually be separated spectrally.

Figure 14.2 illustrates the typical mirror symmetry of the absorption and fluoresecence spectrum of e.g. organic dye molecules [3]. The absorption spectrum is shown for anthracene in cyclohexane in terms of the molar extinction as a function of the wavelength. The fluorescence emission spectrum is drawn as dashed curve. The vertical scale of the emission spectrum, shown on the right y-axis, is adapted to clarify the symmetry properties.

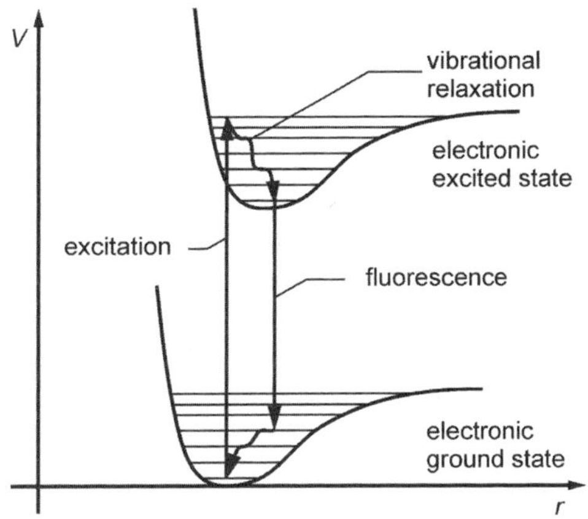

Fig. 14.1 Term scheme illustrating excitation and fluorescence transitions in molecules

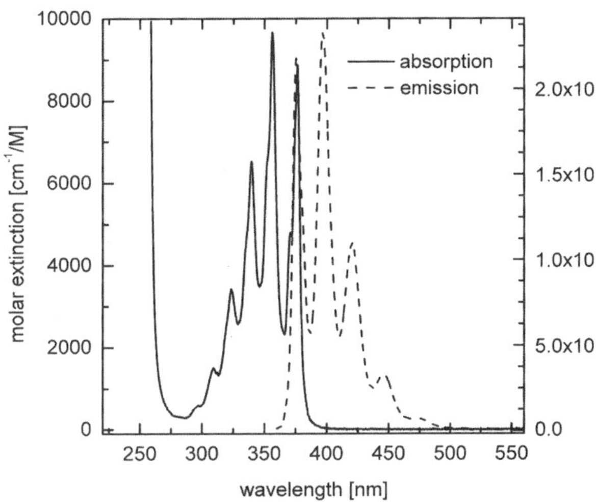

Fig. 14.2 Absorption and fluorescence emission spectrum of anthracene in cyclohexane [3]

Unlike molecules, atoms do not possess vibrational or rotational energy levels, only electronic excitations are possible, see Fig. 14.3 with an example of an atomic term scheme of thallium [4]. Their fluorescence spectra generally consist of individual, spectrally narrow separated lines. Only under increased pressure these transitions undergo a broadening caused by impacts between the particles, as is e.g. the case for a high pressure mercury lamp. The different electronic states have different spins, orbital angular momenta and total angular momenta, which are described by the term symbols shown at the four levels in Fig. 14.3. An introduction to the denotation of atomic terms is given in [5]. The dashed arrow linking two excited states illustrates a transition between the closely neighboring energy levels that is induced by the particle temperature or by impacts with neutral particles.

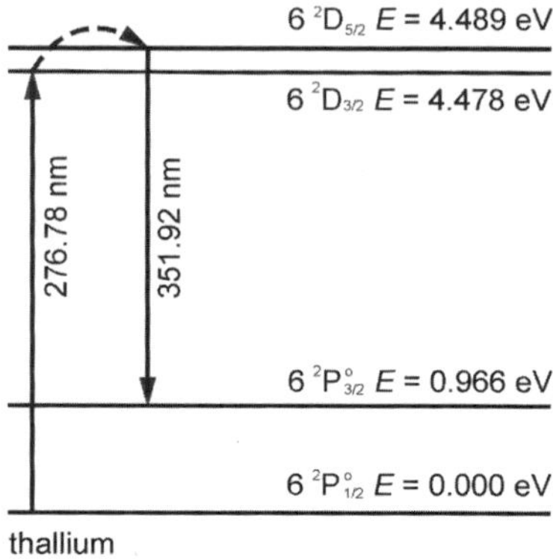

Fig. 14.3 Fluorescence in atoms. As example the levels for thallium are shown

14.2 Fluorescence Spectroscopy

Fluorescent particles can be excited in different ways. Atoms can be excited with X-rays or electron beams, the line emission of hot plasmas is based on the fluorescence of thermally excited atoms, cf. Sect. 13.1. Molecules are usually excited with ultraviolet or visible light.

In analytical chemistry fluorescence is mainly applied for a sensitive and quantitative detection of organic molecules [6]. Compared to absorption techniques, fluorescence has the inherent advantage that instead of a signal decline, i.e. a relative measurement, an absolute intensity is measured while the elastically scattered light of the excitation is efficiently suppressed by spectral filtering. For this reason the limits of detection are up to three orders of magnitude better than with absorption techniques. Typical detection limits lie in the range of 10^{-11} to 10^{-12} M, where the unit 1 M denotes a solution of concentration of 1 molar, i.e. 1 mol per liter [7].

Two principal approaches to measure the excited fluorescence are possible. For the recording of a fluorescence emission spectrum a sample is illuminated with monochromatic light of a fixed wavelength and the emitted fluorescence is detected and spectrally resolved. Provided that the resolution is sufficient, the vibrational modes of the electronic ground state can be resolved. For samples in solution usually only a wrapped contour can be measured.

An alternative way to measure fluorescence consists of a set-up with a tunable excitation source and fluorescence detection with a fixed spectral window with a filter or a monochromator to assure that no excitation radiation reaches the detector. While scanning the excitation source the red-shifted fluorescence is detected. The amount of detected fluorescence depends on the absorption of the excitation light, so a fluorescence excitation spectrum is recorded that can reveal the vibrational structure of the electronically excited state.

Figure 14.4 shows the principle of absorption spectroscopy. A tunable light source generates a beam of light, which is directed through a sample onto the detector. The light intensity is measured as a function of the wavelength. The

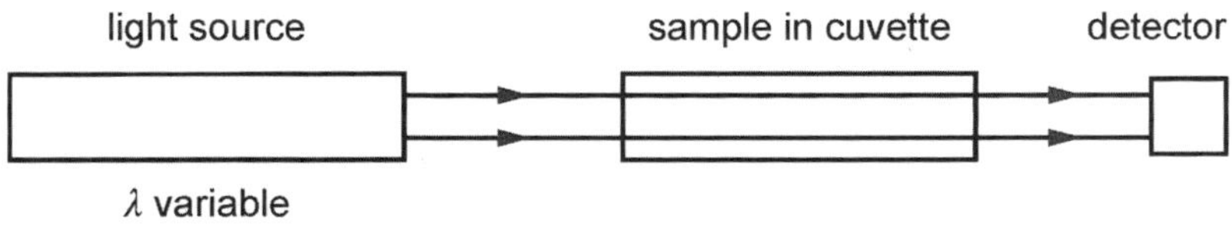

Fig. 14.4 Principle of absorption spectroscopy

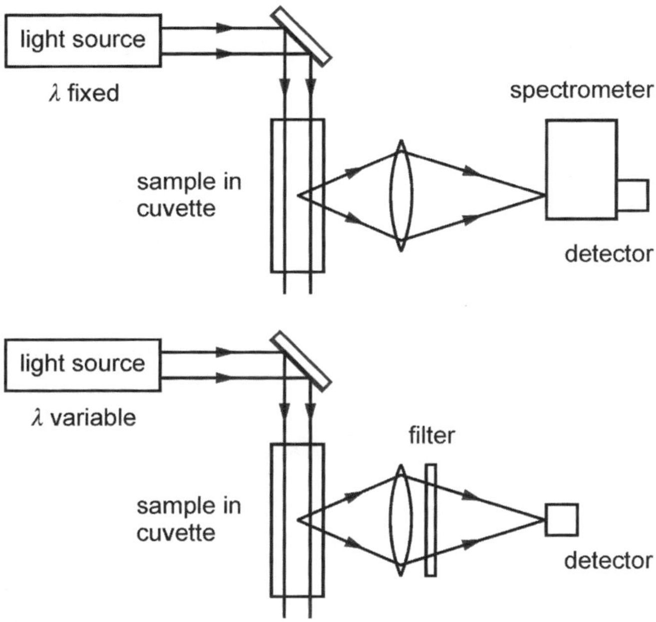

Fig. 14.5 Principle of fluorescence spectroscopy, *top* detection of fluorescence spectrum excitated by a fixed wavelength, *bottom* detection of a fluorescence excitation spectrum by variation of the irradiated wavelength

measuring result is the absorption spectrum of the sample. An alternative approach consists of a polychromatic light source in connection with a spectrally resolving detection, e.g. a CCD spectrometer [8].

In contrast to absorption spectroscopy the different methods of fluorescence spectroscopy—emission and excitation—are nearly background-free techniques. The limits of detection are significantly lower. In Fig. 14.5, top, the sample is illuminated with monochromatic light as generated by a laser light source, the fluorescence can be dispersed spectrally and a fluorescence emission spectrum results. Figure 14.5, bottom, shows the recording of a fluorescence excitation spectrum by scanning the excitation light source and simultaneously detecting the fluorescence in a defined spectral window using filters or a monochromator (observation usually at the maximum of the fluorescence emission) [6].

The usage of a laser for exciting fluorescence features several advantages compared to conventional light sources [9–11]. The laser delivers narrow bandwidth, monochromatic light that has excellent collimation and focusing qualities. All fluorescence applications demand a separation of excitation wavelength and fluorescence emission, because otherwise stray light might disturb the measurements. If narrow bandwidth laser light is used for excitation, the fluorescence emission can be separated efficiently from the laser wavelength with the aid of optical filters, cf. Sect. 4.1.1. If the electronic transitions of the sample have narrow bandwidths, laser radiation has additional advantages compared to broader light sources because of their high spectral radiance. Investigations of molecules in the gas phase or in so-called molecular beam experiments—with molecules moving in the same direction without interacting with each other—are usually carried out with tunable laser systems.

From high-resolution fluorescence spectra structural information like potential curves, vibrational frequencies, dissociation processes, the symmetries of electronic states or rotational constants can be derived. Additionally, transient species like radicals, photo fragments or intermediates can be investigated with pulsed laser radiation. Examples are the detection of concentration and distribution of nitrogen oxides, carbon monoxide, hydroxyl-radicals or hydrocarbons during a combustion process. These investigations are e.g. important for the optimization of motors, see Sect. 14.7.

14.3 Fluorescence Markers

Fluorescent dye molecules can be chemically coupled to biomolecules with specific binding capabilities, for example antibodies. With this principle, specific structures in cells or in tissue can be labelled with fluorescent molecules. This is the basic principle of a fluorescence marker. A selective fluorescent coloring can be visualized under a fluorescence microscope. The growing availability of various fluorescence techniques has led to a replacement of radioactive markers with fluorescence markers in many fields of application. Unlike with radioactive materials no special safety procedures are necessary and the disposal of used sample material is trouble free.

Fluorescent markers are used in bioanalytics and medical diagnostics in multiple ways. Flow-cytometry is based on the detection and sorting of single, dye-labelled cells with laser-induced fluorescence. In the field of high throughput analytics in microtiter plates and in modern DNA- or protein-biochips fluorescence markers are widely used. In microscopy fluorescence markers are necessary for the selective labelling of cells or cell-components. This enables a functional imaging [12]. In the field of medicine fluorescent markers can be used to selectively label tumors, cf. Figure 14.15.

14.4 Fluorescence Correlation Spectroscopy

The fluorescence correlation spectroscopy—abbreviated FCS—is a method for analyzing the diffusional behaviour of molecules. Since binding processes between molecules have an effect on the particle movement in solution, the interactions between molecules can be analyzed. Specific interactions between bio-molecules determine nearly all biochemical processes in living organisms. Examples for specific binding processes—so-called lock-key principle—are:

- Interactions between receptors and ligands
  Control and regulation mechanisms of the cell are based on these interactions. Their examination is important for the search for new active pharmaceutical ingredients.
- protein–protein interactions
  Enzymatic reactions or antibody–antigen interactions are examples.
- DNA–protein interactions
- interactions between nucleic acids.

The FCS technique is based on the analysis of diffusion. A laser beam is focused to a diffraction limited spot with a high numerical aperture microscope objective. Fluorescence from this extremely small focus volume is transferred with a confocal optics on a very sensitive detector, e.g. an avalanche photodiode, cf. Chap. 12. Optical filters are used to block stray light. If only a few fluorescent molecules are present in the focal volume, the Brownian motion of the molecules and the diffusion of the particles will lead to a strong fluctuation of the fluorescence intensity. The recorded fluorescence intensity is directly related to the movement of the molecules. The auto-correlation of the time resolved intensity delivers the average time it takes for a molecule to cross the focal volume (diffusion time) and the particle number in the focal volume (concentration).

As smaller molecules diffuse faster than larger ones they show shorter diffusion times. If a protein that is labeled with a fluorescence dye is added to a binding partner, the change in diffusion time reveals the chemical binding process and the reaction can be observed. This can be used to determine binding constants, see e.g. [13]. Figure 14.6 illustrates the principle of FCS. Graph (a) presents schematically the fluorescence intensity $I_F(t)$ as a function of time. Graph (b) shows qualitatively the auto-correlation function $G(\tau)$—cf. Sect. 11.3.2—corresponding to $I_F(t)$ if only the fluorescent species is present. After the characteristic diffusion time the auto-correlation function drops indicating that the corresponding fluctuations of the fluorescence signal are no longer correlated. If a binding partner is added, there are free and bound fluorescent particles in solution, that have different diffusion times. From the auto-correlation curve the relative fractions of free and bound molecules and the binding constant can be derived, see graph (c).

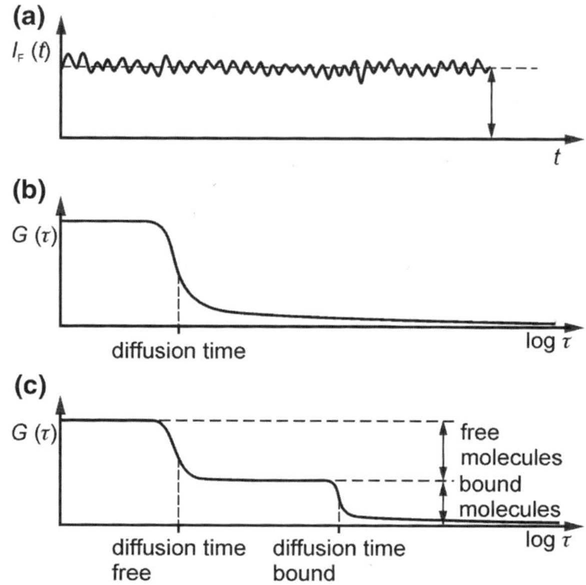

Fig. 14.6 Principle of fluorescence correlation spectroscopy. **a** fluctuations of the fluorescence intensity $I_F(t)$, **b** autocorrelation function $G(\tau)$ of $I_F(t)$ before addition of a binding partner, **c** $G(\tau)$ after addition of a binding partner. The indicated height difference is proportional to the respective fraction of free and bound molecules

14.5 Fluorescence Polarization Spectroscopy

Fluorescence polarization spectroscopy (FPS) makes use of the change of the polarization state of fluorescent light yielding as well information about binding processes between fluorescence-labeled analytes and larger molecules, as e.g. antibodies.

Figure 14.7 illustrates the principle. An analyte—e.g. an antigen—is labeled by a fluorescent molecule. The analyte is shown as a small rectangular box with a tiny notch. Linked to this analyte is the fluorescent marker, shown as small sphere. Graph (a) shows the incoming radiation being 100 % linearly polarized with a polarization direction presented by the arrow pointing to the right. The particles have an irregular orientation in the solution. Some of them have an orientation of their dipole moment which allows for a strong coupling with the irradiated polarization direction. In graph (a), left, these are the two particles shown on the left.

Let us assume that the characteristic fluorescence life time τ_f of these particles is greater than the rotation diffusion time τ_{rd}. This leads to the situation depicted in graph (a), right, the two particles with an initial strong coupling to the incident polarization state fluoresce, but—due to the small rotation diffusion time—have changed their orientation in space. Hence the fluorescence radiation traveling antiparallel to the incident radiation, has a different polarization state: it becomes partially depolarized, indicated by the dashed arrows in graph (a), right, pointing in two perpendicular directions.

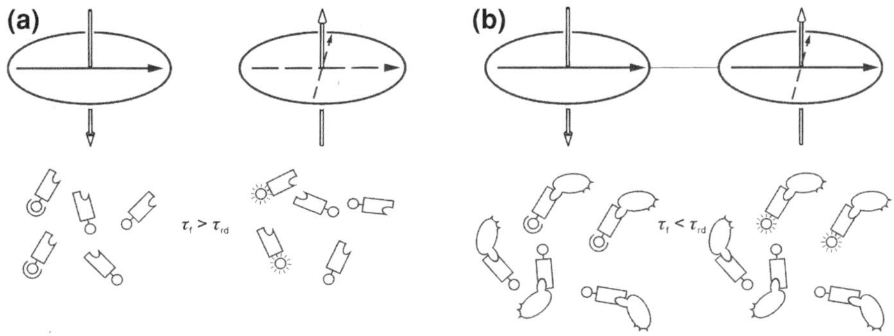

Fig. 14.7 Principle of fluorescence polarization spectroscopy (see text). τ_f fluorescence life time, τ_{rd} rotation diffusion time

If the analyte is coupled with an antibody, which fits to the above stated notch of the analyte, the situation changes, cf. Fig. 14.7, graph (b). Since the formed complex is now significantly larger, we have $\tau_f < \tau_{rd}$, i.e. the particle rotates more slowly. Thus the orientation of the complex is preserved for a longer time and as a consequence the emitted polarization state is mainly oriented parallel to the incident polarization and only a weaker part—illustrated by the dashed arrow in graph (b), right—points in a perpendicular direction.

The polarization state is described by a dimensionless quantity as follows:

$$P = \left| \frac{I_{f\parallel} - I_{f\perp}}{I_{f\parallel} + I_{f\perp}} \right|, \tag{14.1}$$

where $I_{f\parallel}$ denotes the fluorescence intensity in the direction parallel to that of the incident beam, $I_{f\perp}$ the fluorescence intensity perpendicular to $I_{f\parallel}$. Hence $P = 1$ corresponds to a completely polarized beam, and $P = 0$ is a completely depolarized radiation. With this method the fraction of bound antibodies can be determined by the anisotropy of the radiation. A measure of the degree of anisotropy is the difference of P-values:

$$\Delta P = |P_{ab} - P_{no\,ab}|, \tag{14.2}$$

with P_{ab} polarization degree with antibodies in the solution, $P_{no\,ab}$ polarization state without antibodies. For a quantitative measurement a so-called immunoassay is applied which contains labeled antigens and specific antibodies to these antigens. Prior to adding the sample to be determined, the anisotropy is measured with the help of (14.2). Due to the binding of the antibody to the labeled antigen the quantity ΔP will attain a high value. If the sample is now added to this solution additional antigens—those to be measured—start to partially replace the labeled antigens in a competitive process. In this way labeled antigens are set free again, and can rotate faster. As a result the anisotropy drops, i.e. ΔP decreases. Hence we expect that the function $\Delta P = \Delta P(c_a)$ has a negative slope for increasing analyte concentration c_a, cf. Fig. 14.16.

14.6 Time-Resolved Fluorescence Analytics

Besides the spectral color there is another feature that characterizes the fluorescence of a molecule, its life time. For measuring the fluorescence life time, a pulsed light source can be used that has shorter pulses than the analyzed life time. Lasers that emit pico- or femtosecond pulses can be used to analyze even short life times in the order of nanoseconds. The straightest way to record fluorescence life times is to excite a large number of molecules with a short laser pulse and record the fluorescence decay curve. For simple cases the fluorescence intensity decays with an exponential function, after a time τ the intensity has dropped to the fraction $1/e$ of the initial fluorescence intensity with τ being the fluorescence life time. When working with tightly focused laser beams and low sample concentrations, it is not possible to record a complete decay curve with only one excitation pulse, because the number of detected fluorescence photons is not sufficient.

One way of recording fluorescence decay curves in microscopy applications is the so-called *time correlated single photon counting* (TCSPC). With this technique the analysis of single molecules is possible. It is also frequently used in laser scanning microscopy for fluorescence lifetime imaging (FLIM). Instead of a fluorescence intensity the fluorescence lifetime gives the contrast in the images.

A laser with a repetition rate of several megahertz emits short pulses—in the picosecond to femtosecond range—and is focused with a microscope objective. The sample contains only a small concentration of fluorescent molecules. The excited fluorescence is collected with a confocal optical system and detected with a single photon counting detector. The laser power is so low, that only a few percent of all laser pulses lead to the detection of a fluorescence photon, so it is very unlikely that more than one photon per laser pulse will be detected.

Upon detection of a photon an electrical pulse is generated that is registered by a fast electronics. Additionally, a synchronization signal is emitted from the laser with every single laser pulse. The electronics precisely measures the time difference between the laser synchronization pulse and the detector pulse. This is realized with a time-to-amplitude converter. The laser synchronization pulse starts a voltage ramp that stops, when the detector pulse reaches the electronics. The ramp voltage is digitized with an analog digital converter (ADC). This voltage is a measure of the detection time of the fluorescence photon. Time resolutions below a picosecond can be achieved. Figure 14.8 shows the prinicple of TCSPC. The arrival times of many photons build a histogram that reveals the decay curve. The more photons are recorded, the more precisely the fluorescence lifetime can be estimated.

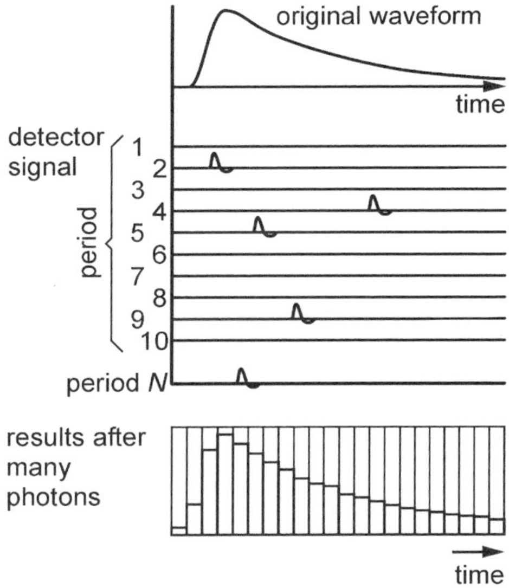

Fig. 14.8 Time-correlated single photon counting. Many single photons whose arrival times are recorded add up to a histogram, shown below, that reveals the fluorescence decay curve

14.7 Examples of Applications

At first industrial applications are described referring to technical measuring objects. A laser measuring system combining laser triangulation methods and induced fluorescence is presented to measure inline the deformation of metal sheets in the so-called bulge test [14]. Hydraulic bulging can be used to characterize sheet material under biaxial stress conditions. The test metal sheets are deformed frictionless under oil pressure. A measuring method applying laser light section for geometry analysis, cf. Sect. 10.3, and a camera device to observe UV-induced fluorescence for local strain determination enables a continuous inline calculation of flow curves. Effective stress and equivalent strain can be recorded until the failure of the material, thus providing extended information compared to the tensile test [15]. The testing procedure is predestined for simulation purposes and for an application in the quality control sector, allowing for a quick and profound material characterization.

Figure 14.9 shows a side-on view of the set-up of the laser measuring system. Two laser lines are projected on the measuring object being oriented perpendicularly to each other. These laser light sections are used to measure the shape of the deformed metal sheet, see Sect. 10.3. Prior to the bulge test, the metal sheet to be inspected is marked by a rectangular array of dots using a fluorescent dye. The ring light illuminates this grid of dots on the metal sheet with a wavelength of approx. 470 nm to excite the fluorescence. The fluorescing spots are observed by a camera to track their position during bulging of the metal sheet.

Figure 14.10 shows the top view of this set-up. In order to avoid cross talking between the two laser light sections two different laser wavelength are applied: 680 and 780 nm. These wavelengths are reflected by the dichroic mirror DM. On the

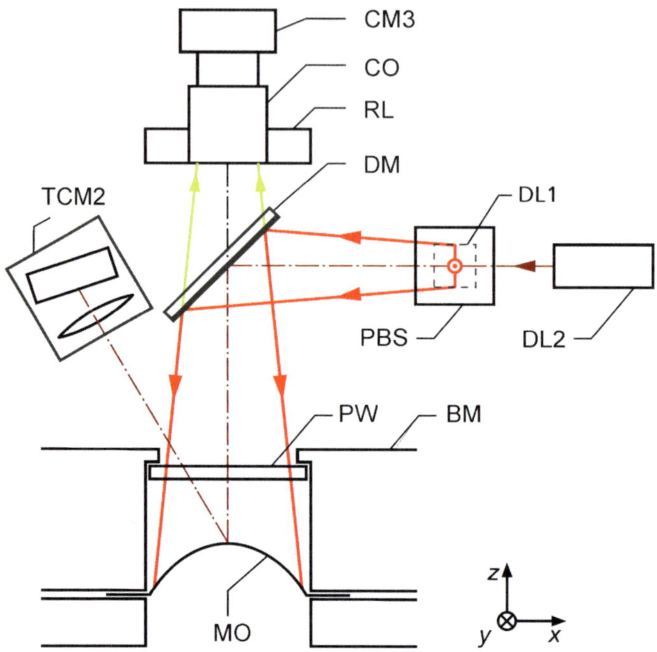

Fig. 14.9 Side-on view of the set-up to measure inline the shape and deformation of a metal sheet in a bulge test machine. BM bulge test machine, MO measuring object, PW protection window, DL1 diode laser module 1 with optics to form a laser line on the measuring object running parallel to the *x*-axis, DL2 diode laser module 2 with optics to form a laser line parallel to the *y*-axis, PBS polarizing beam splitter, DM dichroic mirror, TCM2 triangulation camera module 2 to measure the laser light section generated by DL2, RL ring light to excite the fluorescence, CO camera objective, CM3 camera module 3. *Red* rays of the diode laser modules, *yellow* fluorescence radiation

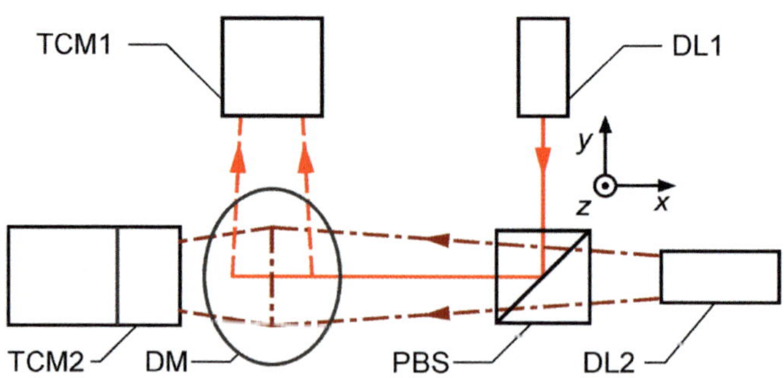

Fig. 14.10 *Top view* of the set-up of Fig. 14.9. TCM1 triangulation camera module 1 to measure the laser light section generated by DL1

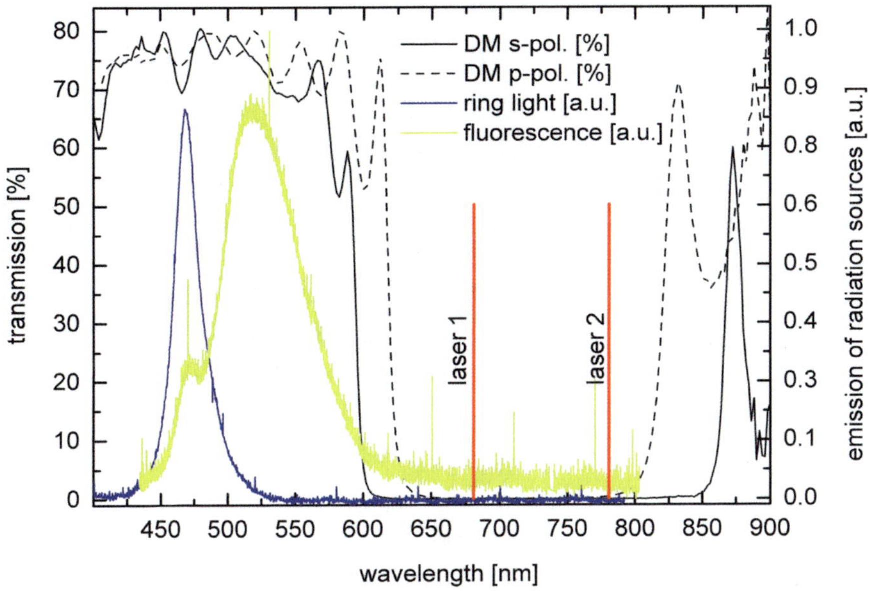

Fig. 14.11 Transmission of dichroic mirror, for s- and p-polarized light, emission ring light (*blue*), fluorescence signal (*yellow*) and position of the laser wavelengths (*red*)

other hand the excitation wavelength generated by the ring light should be passed (for simplicity, these excitation rays are not shown in Fig. 14.9). The center of gravity of the fluorescence is around 520 nm, which should as well be transmitted by the dichroic mirror.

Figure 14.11 shows the spectral characteristics of the dichroic mirror (shown for s- and p-polarization, cf. Sect. 3.2.2), the ring light emission, the induced fluorescence and the positions of the two laser wavelengths.

Figure 14.12 shows a photograph of the test object with the two crossing laser light sections and the fluorescing dot array. With this set-up the 3D geometry and the point displacements are detected simultaneously with a repetition rate of 5 Hz. The accuracy of determination of the cup height of the deformed sheet can be estimated to about 1 % of the measuring area which corresponds to about 30 μm at a cup height of 3 mm.

The inline capability allows for a direct control of the bulge test machine, e.g., to carry out constant strain rates during bulging. Measuring results gained are among others flow curves showing the dependence of the effective true stress (range 0–1,200 MPa) as a function of the equivalent strain (range 0–1). Deformation degrees can be increased by up to a factor of four compared to the conventional tensile test.

In the second application example laser-induced fluorescence is used to detect very thin residues of organic substances on metal sheets [16]. During production of sheet steel, steel strips are passing through several cleaning stages after the rolling

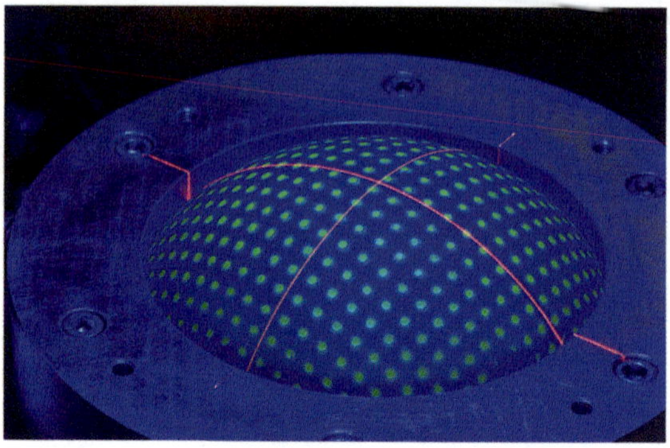

Fig. 14.12 Photograph of a bulged metal sheet illuminated with two laser light sections and the simultaneously excited fluorescence of a dot array

process and further treatment steps, in order to remove oil residues. The task of the fluorescence method is to monitor these cleaning stages and to detect deviations and irregularities inline. Thus interruptions of the production process are avoided and an uniform product quality can be maintained. LIF is applied to detect the oil residues which normally cannot be seen with the naked eye. The oil is excited with a pulsed ultraviolet laser beam operating at 266 nm, 20 Hz and a maximum pulse energy of 20 mJ.

This beam is attenuated and collimated onto the surface of the metal sheet. The diameter of the illuminated area is 5 mm. The resulting fluorescence radiation is observed perpendicular to the sheet surface and imaged to the input aperture of an optical fiber bundle guiding the light to a spectrometer. In front of the entrance aperture of the fiber filters are placed to block the excitation radiation. Expected fluorescence life times of oils are < 8 ns. The spectrometer is equipped with a gated MCP, cf. Sect. 5.4.4, to suppress ambient light and to synchronize the fluorescence light detection with the irradiated laser pulse. Figure 14.13 shows the principal set-up. The fluorescence spectra depend on the type of oil; typically these spectra cover a range from 320 to 450 nm. For calibration the spectrally integrated fluorescence signal is studied as a function of the oil residues area density using sets of artificial samples. Oil residues area densities vary between 0 and 300 mg/m^2. An approximately linear dependence is found between the integrated fluorescence signal and the oil residue area density. The estimated detection limits are in the range of 6 mg/m^2 which corresponds to an average oil film thickness of about 7 nm at the strip surface. This is the order of magnitude of the oil residues which can be present at the final cleaning stages.

Figure 14.14 shows images of the fuel distribution in a combustion chamber generated with laser-induced fluorescence where the laser beam is shaped as a plane, similar to the laser light section approach presented in Chap. 10 [17]. With this planar laser-induced fluorescence, abbreviated PLIF, the images are taken as a function of the crank angle (CA). The upper series of pictures shows a side view of the combustion chamber, the lower series of pictures show a top view. Fluorescence excitation was performed with a KrF excimer laser at 248 nm. The asymmetric distribution of fuel can be attributed to the influence of the open inlet valves.

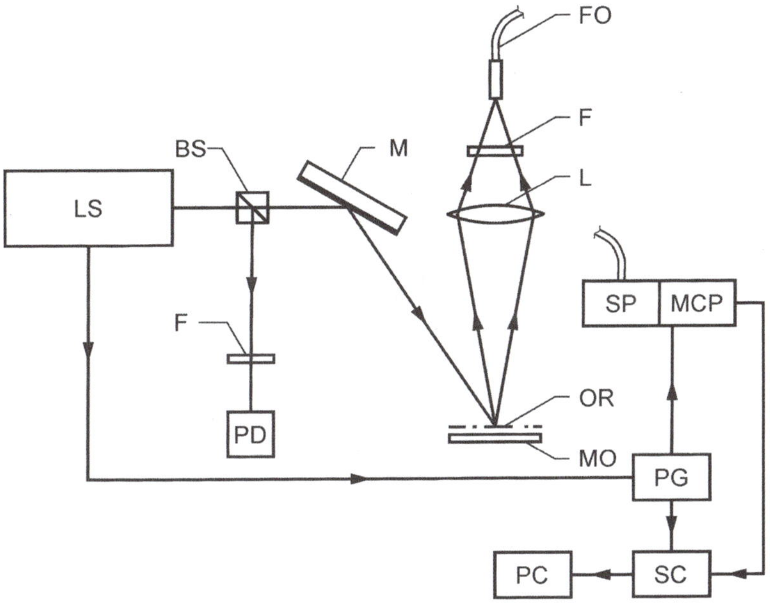

Fig. 14.13 Set-up for laser-induced fluorescence to measure oil residues on metal sheets. MO measuring object with oil residue, OR oil residue, LS laser source, BS beam splitter, M mirror, F filter, FO fiber optics, SP spectrometer, MCP microchannel plate detector, PG pulse generator, SC signal controller, PC computer

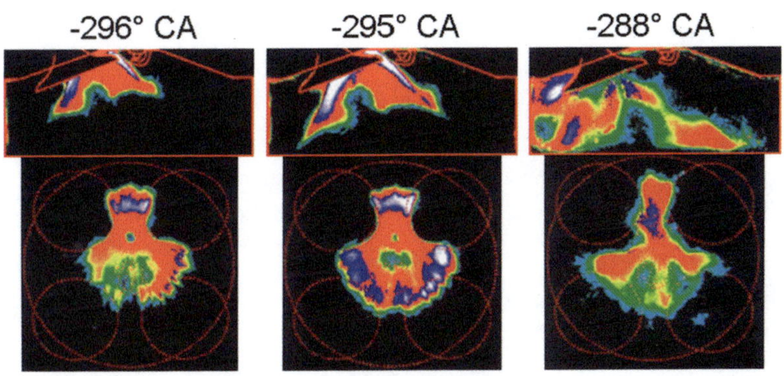

Fig. 14.14 Distribution of fuel in a combustion chamber visualized by planar laser-induced fluorescence (PLIF), *top row* side view of the combustion chamber, inlet valve and spark plug are shown in red contours, *bottom row* top view. CA crank shaft angle

Laser-induced fluorescence plays an important role as an analytical method for biological measuring objects. These include routine analytics, fluorescence microscopy, clinical diagnostics, the control of chemical or pharmaceutical reactions or basic research. Fluorescence has a role of special importance in the field of biochips and in vitro diagnostics (in vitro means "in glass", to be differentiated from "in vivo", which means in a living organism).

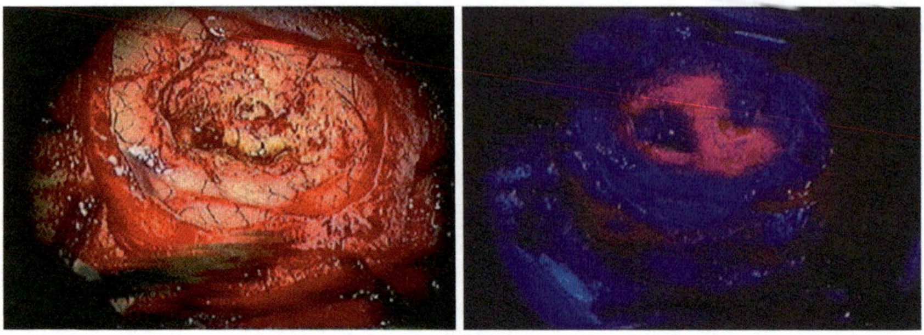

Fig. 14.15 Pictures of a brain tumor, *left* illuminated with white light, *right* illumination with blue light. The red fluorescence of the dye that specifically labels the tumor makes the tumor tissue clearly visible

Fig. 14.16 Degree of anisotropy ΔP as a function of the DON concentration measured with FPS

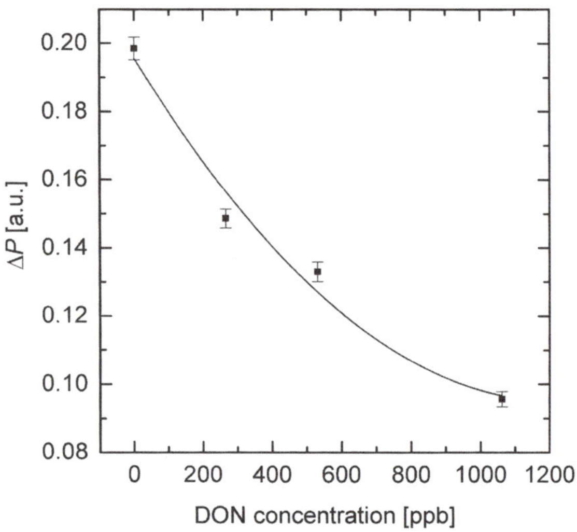

In the field of medicine fluorescent markers, cf. Sect. 14.3, can be used to selectively label tumors, see Fig. 14.15 [18].

An application of fluorescence polarization spectroscopy is the detection of mycotoxins in microtiter plates [19]. Mycotoxins are produced by molds that form in grains, food and feedstuffs and, thus, can cause health concerns when ingested. The occurrence of mycotoxins in foods is usually the result of mold contaminated grain at pre harvest or during storage. For a fast analysis of grain samples minute quantities of liquid must be handled and measured automatically. With an electronic pipette the assay components—sample extract, antibodies and fluorescence labeled toxines—are injected into the wells of a microtiter plate. The liquid volume where the measurement with FPS is carried out amounts to 200 µl. The duration of a single measurement is about 6 min. Figure 14.16 shows results for

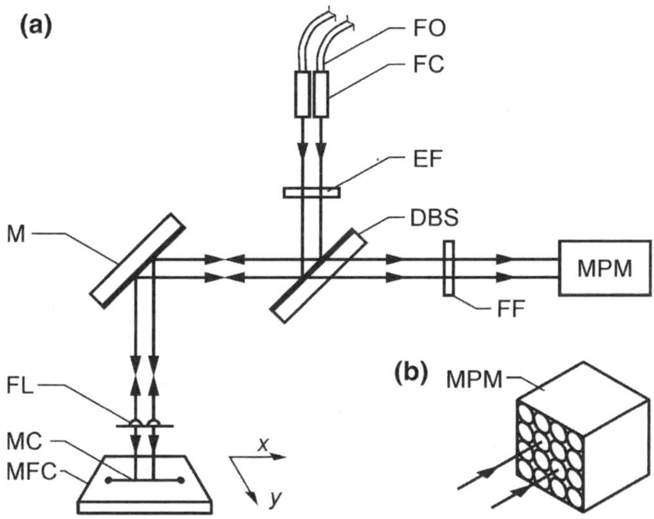

Fig. 14.17 Set-up of parallel fluorescence sensors to detect labeled objects passing through a microchannel of a microfluidic chip; **a** optical beam paths, **b** multianode photomultiplier. FO fiber optics, FC fiber collimators, EF excitation filter, DBS dichroic beam splitter, M mirror, FF bandpass filter for fluorescence radiation, MPM multianode photomultiplier, FL focusing lenses, MC microchannel, MFC microfluidic chip

the detection of the mycotoxin deoxynivalenol (DON). The degree of anisotropy ΔP is plotted versus the DON concentration, cf. Sect. 14.5. The detection limit for DON is less than 50 μg/kg, which is well below the maximum tolerable level of 500 μg/kg for bread including small bakery wares, pastries, biscuits, cereal snacks and breakfast cereals [20].

Laser-induced fluorescence is also applied to measure single cell properties in microfluidic devices [21]. Fluorescence emission has become a powerful tool because of the high sensitivity and versatility of available fluorescence staining methods, which enable microbiologists to stain microbial cells according to their individual biochemical, physiological or taxonomic properties. For high throughput applications, parallelization is a promising solution involving multiple probing locations. Figure 14.17 shows the principal set-up of parallel fluorescence sensors. Laser light at 488 nm is coupled to single mode fibers. Two collimators are used to collimate the excitation light which propagates via the dichroic beam splitter to the focusing lenses and finally to the microchannel. The collected fluorescence light passes the dichroic mirror and is detected by a multianode photomultiplier, see Fig. 14.17b. The width of the microchannel is 25 μm, the depth 150 μm. The excitation volume is adapted to this structure. Microbeads labeled with the fluorophore phycoerythrin (PE) are used to characterize the sensor. A suspension of microbeads with a diameter of about 8 μm are conjugated with different levels of PE. The conjugation level is quantified by the manufacturer and given as molecules of equivalent soluble fluorophore (MESF) [22].

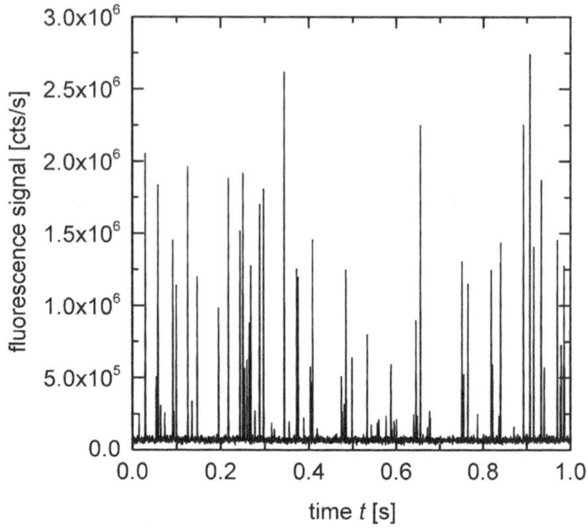

Fig. 14.18 Fluorescence signal as a function of time of labeled beads passing a microchannel at a flow rate of 10 μl/min

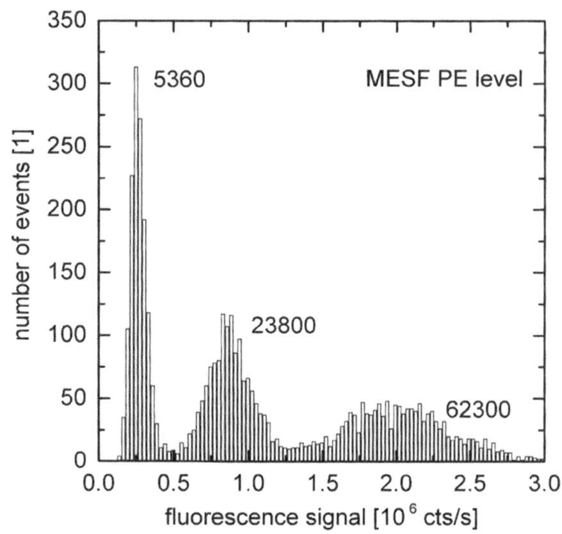

Fig. 14.19 Histogram of cumulated events for given fluorescence signal levels, parameter is the MESF PE level

Figure 14.18 shows the detected fluorescence signal of a solution of particles with different fluorophore conjugation levels taken with a flow rate of 10 μl/min and a dwell time of 0.4 ms. Particles passing the measurement volume cause signal peaks with different heights. The variations in height are due to different conjugation levels of the beads, the distribution of conjugation strength within one fraction and the nonuniform sensitivity within the measurement volume.

Figure 14.19 shows the heights of the peaks as a histogram revealing three populations of different conjugation levels. The mean value of each population is determined by fitting Gaussian curves to the respective distributions. The mean

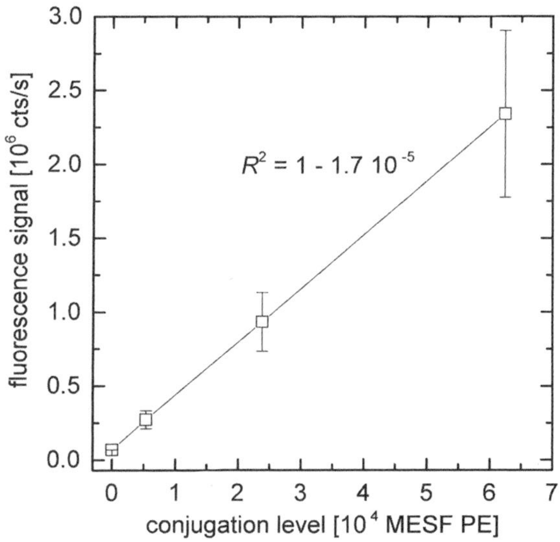

Fig. 14.20 Mean fluorescence intensities as a function of the conjugation level and a measurement with a blank sample, error bars correspond to 1 × *s*, R^2 coefficient of determination

signal levels and the related MESF values are given in Fig. 14.20 and show a linear relation with a slope of (36 ± 1) cts/(s MESF) with a coefficient of determination very close to 1. The estimated detection limit is about 2,900 MESF PE based on a type I error of 1/600000.

References

1. C. Janzen, R. Noll, Laser-induced fluorescence, in *Tailored Light 2, Laser Application Technology,* ed. R. Poprawe (Springer, Berlin, 2011) chapter 19.3, pp. 500–509. ISBN 978-3-642-01236-5
2. P. Klán, J. Witz, *Photochemistry of Organic Compounds: from Concepts to Practice* (Wiley, Chichester, 2009)
3. http://omlc.ogi.edu/spectra/PhotochemCAD/html/022.html, Feb 1, 2014
4. F. Hilbk-Kortenbruck, R. Noll, P. Wintjens, H. Falk, C. Becker, Analysis of heavy metals in soils using laser-induced breakdown spectrometry combined with laser-induced fluorescence. Spectrochim. Acta B **56**, 933–945 (2001)
5. H. Haken, H.C. Wolf, *Atom- und Quantenphysik—eine Einführung in die experimentellen und theoretischen Grundlagen* (Springer, Berlin, 1980)
6. D. Skoog, J. Leary, *Instrumentelle Analytik* (Springer, Berlin, 1996)
7. F. Settle (ed.) *Handbook of Instrumental Techniques for Analytical Chemistry*. (Prentice-Hall, New Jersey, 1997)
8. Römpp Chemie Lexikon, Online Version 2.0, Thieme Verlag. www.roempp.com
9. D. Andrews, *Applied Laser Spectroscopy* (VCH Publishers Inc., New York, 1992)
10. W. Demtröder, *Laserspektroskopie* (Springer, Berlin, 1997)
11. J. Lakowicz, *Principles of Fluorescence Spectroscopy*, 2nd edn. (Kluwer Academic, New York, 1999)
12. J. Berg, L. Stryer, J. Tymoczjo, Lehrbücher der Biochemie, Biochemie, (Spektrum Akademischer Verlag, Heidelberg, 2003)
13. R. Rigler, E. Elson, *Fluorescence Correlation Spectroscopy* (Springer, Berlin, 2011)

14. J. Vrenegor, R. Noll, Multi-Punkt-Lasertriangulation mit adaptiver Belichtungsregelung für die 3D-Formänderungsanalyse, cooperative project funded by the German Ministry of Research, Nr. 13 N 8111, final report, July 2005
15. W. Bleck, M. Blumbach, Laser-aided flow curve determination in hydraulic bulging. Steel Res. Int. **76**, 125–130 (2005)
16. V. Sturm, R. Noll, Rolling oil residue on steel strips: measurement using laser-induced fluorescence, Fraunhofer-Institute for Laser Technology ILT, Annual Report 2000, 73
17. H. Fuchs, Fuel distributions in engine is monitored with planar laser-induced fluorescence. Lambda Highlights - Lambda Physik **62**, 1–3 (2003)
18. Pictures by courtesy of Prof. W. Stummer, Laser Klinikum Universtiät München
19. C. Janzen, R. Noll, Biochemical analytics in microtiter plates, Fraunhofer-Institute for Laser Technology ILT, Annual Report 2012, 122
20. Commission Regulation (EC) No 1881/2006 of December 2006 setting levels for certain contaminants in foodstuffs, Sect. 2 Mycotoxins, p. L 364/17
21. G. Meineke, D. Flitsch, A. Lenenbach, R. Noll, Fluorescence sensors for parallel measurements in multichannel microfluidic devices covering the full channel cross sections. Proc. of SPIE **8615**, 86151C-1–8615C-7 (2013)
22. A. Schwartz, L. Wang, E. Early, A. Gaigalas, Y. Zhang, G. Marti, R. Vogt, Quantitating fluorescence intensity from fluorophore: the definition of MESF assignment. J. Res. Natl. Inst. Stan. **107**, 83–91 (2002)

Appendix A

A.1 Interference Visibility

In Fig. 2.19 we have illustrated the measurement of the longitudinal coherence length with a Michelson interferometer showing a curve of the intensity of the interfering waves versus the path difference $2\Delta L$. The degree of modulation of this curve decreases with increasing path difference. If the path difference is large compared to the coherence length L_c, then this modulation vanishes. In the following we present a more precise description of interference visibility related to the spectral width of the light source. This interference visibility or *contrast of interference* is defined as follows:

$$C = \left| \frac{I_{\max} - I_{\min}}{I_{\max} + I_{\min}} \right|, \tag{A.1.1}$$

where $I_{\max}$ and $I_{\min}$ are the maximum and minimum intensities of the superimposed waves. For $I_{\min} = 0$ the contrast achieves its maximum value $C = 1$. In general the interference of two waves leads to the resulting intensity described by (cf. 7.18):

$$I = I_1 + I_2 + 2\sqrt{I_1 I_2} \cos \delta, \tag{A.1.2}$$

with I_1, I_2 intensity of the single waves and δ the phase difference between the two waves at the point of interference. The phase difference is described by:

$$\delta = k(s_1 - s_2) + \delta_1 - \delta_2, \tag{A.1.3}$$

where k is the angular wave number for each wave, s_1 and s_2 the pathes travelled by wave 1 and wave 2; δ_1, δ_2 the initial phases of the waves at the chosen origin. In the case of monochromatic waves of the same angular wave number—as assumed so far—we obtain from (A.1.1) and (A.1.2):

$$C = \frac{2\sqrt{I_1 I_2}}{I_1 + I_2}. \tag{A.1.4}$$

A. Donges and R. Noll, *Laser Measurement Technology*, Springer Series in Optical Sciences 188, DOI: 10.1007/978-3-662-43634-9,

For simplicity, we consider the case, where the superimposed waves have the same intensity, i.e.:

$$I_0 = I_1 = I_2. \tag{A.1.5}$$

For two monochromatic waves follows then from (A.1.4) and (A.1.5), that the contrast of interference is $C = 1$. With (A.1.5) inserted in (A.1.2) we have:

$$I = 2I_0(1 + \cos\delta). \tag{A.1.6}$$

Let us now extend our considerations by assuming a light source which emits—instead of a single angular wave number, i.e. a pure monochromatic wave—a spectrum of waves. We describe the spectral intensity I_k of this light source by the following function:

$$I_k = \frac{\mathrm{d}I}{\mathrm{d}k} = f(k), \tag{A.1.7}$$

with k angular wave number. We can generalize (A.1.6) with the help of (A.1.7):

$$I = 2\int_{-\infty}^{\infty} f(k)(1 + \cos(\delta))\,\mathrm{d}k. \tag{A.1.8}$$

We assume that the function $f(k)$ describing the spectrum of the light source is symmetric with respect to an angular wave number k_0. Introducing new variables and definitions:

$$k' = k - k_0, \tag{A.1.9}$$

$$f(k) = f(k' + k_0) = f^*(k') \tag{A.1.10}$$

allows to rewrite (A.1.8) with (A.1.3):

$$I = 2\int_{-\infty}^{\infty} f^*(k')\,\mathrm{d}k' + 2\int_{-\infty}^{\infty} f^*(k')\cos[(k' + k_0)(s_1 - s_2) + \delta_1 - \delta_2]\,\mathrm{d}k'. \tag{A.1.11}$$

With the abbreviations $\Delta s = s_1 - s_2$ and $\Delta\delta = \delta_1 - \delta_2$ and Euler's formula the second integral in (A.1.11) is rewritten:

$$\int_{-\infty}^{\infty} f^*(k')\frac{\mathrm{e}^{\mathrm{i}[(k'+k_0)\Delta s+\Delta\delta]} + \mathrm{e}^{-\mathrm{i}[(k'+k_0)\Delta s+\Delta\delta]}}{2}\,\mathrm{d}k'$$

$$= \frac{1}{2}\int_{-\infty}^{\infty}\left[f^*(k')\mathrm{e}^{\mathrm{i}k'\Delta s \mathrm{i}(k_0\Delta s+\Delta\delta)} + f^*(k')\mathrm{e}^{-\mathrm{i}k'\Delta s}\mathrm{e}^{-\mathrm{i}(k_0\Delta s+\Delta\delta)}\right]\mathrm{d}k'. \tag{A.1.12}$$

With the substitution $\tilde{k} = -k'$ follows:

$$\int_{-\infty}^{\infty} f^*(k')\mathrm{e}^{-\mathrm{i}k'\Delta s}\mathrm{d}k' = \int_{\infty}^{-\infty} f^*(-\tilde{k})\mathrm{e}^{\mathrm{i}\tilde{k}\Delta s}(-\mathrm{d}\tilde{k}) = \int_{-\infty}^{\infty} f^*(\tilde{k})\mathrm{e}^{\mathrm{i}\tilde{k}\Delta s}\mathrm{d}\tilde{k}. \tag{A.1.13}$$

With (A.1.13) we have for (A.1.12):

$$\frac{1}{2}\int_{-\infty}^{\infty} f^*(k')\mathrm{e}^{\mathrm{i}k'\Delta s}2\cos(k_0\Delta s + \Delta\delta)\,\mathrm{d}k'. \tag{A.1.14}$$

Inserting (A.1.14) into relation (A.1.11) for the second integral then yields:

$$I = 2\left[\int_{-\infty}^{\infty} f^*(k')\,\mathrm{d}k' + \cos(k_0\Delta s + \Delta\delta)\int_{-\infty}^{\infty} f^*(k')\mathrm{e}^{\mathrm{i}k'\Delta s}\mathrm{d}k'\right]. \tag{A.1.15}$$

Finally, the interference contrast is with (A.1.15) and (A.1.1):

$$C(\Delta s) = \left|\frac{\int_{-\infty}^{\infty} f^*(k')\mathrm{e}^{\mathrm{i}k'\Delta s}\mathrm{d}k'}{\int_{-\infty}^{\infty} f^*(k')\mathrm{d}k'}\right|. \tag{A.1.16}$$

The result (A.1.16) shows that the interference contrast as a function of the path difference Δs is proportional to the Fourier transform of the spectrum of the emitted light of the source.

To demonstrate (A.1.16) we consider the following two spectra of a light source:

$$f_1(k) = \frac{I_0\mathrm{e}^{-\frac{(k-k_0)^2}{2\sigma^2}}}{\sqrt{2\pi}\sigma}, \tag{A.1.17}$$

$$f_2(k) = \frac{I_0}{2\sqrt{2\pi}\sigma}\left[\mathrm{e}^{-\frac{(k-k_1)^2}{2\sigma^2}} + \mathrm{e}^{-\frac{(k-k_2)^2}{2\sigma^2}}\right], \tag{A.1.18}$$

where σ determines the width of the Gaussian function, k_0 angular wave number at which the function (A.1.17) attains a maximum; k_1, k_2 angular wave numbers at which the function (A.1.18) attains local maxima, I_0 total intensity of both sources. The functions $f_1(k)$ and $f_2(k)$ are defined in such a way, that

$$\int_{-\infty}^{\infty} f_j(k)\mathrm{d}k = I_0, j = 1, 2 \tag{A.1.19}$$

holds. Spectrum (A.1.17) describes e.g. a single mode laser source, whereas (A.1.18) is a simple representation of a laser running at two axial modes. The full width at half maximum of the function $f_1(k)$ is given by:

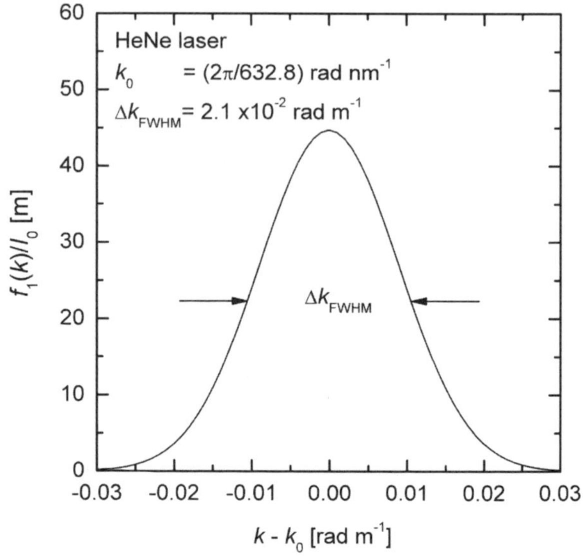

Fig. A.1 Spectrum of a HeNe laser as a function of the angular wave number difference $k - k_0$

$$\Delta k_{\mathrm{FWHM}} = 2\sqrt{2\ln 2}\sigma. \tag{A.1.20a}$$

Figure A.1 shows an example of the function $f_1(k)/I_0$.

(a) Contrast function (A.1.17):

Inserting (A.1.17) into (A.1.16) yields for the contrast function:

$$C_1(\Delta s) = \mathrm{e}^{-\Delta s^2 \sigma^2/2}. \tag{A.1.20b}$$

Figure A.2 shows this contrast function. The key point is that the quantity σ appears in the denominator of the exponent of the spectrum (A.1.17) whereas in the contrast function σ occurs in the numerator of the exponent. From this follows that a broader spectrum, i.e. greater values of σ, lead to a contrast function decreasing faster with increasing Δs. The behaviour of the two functions—spectrum and contrast—is inverted. The broader the first one becomes, the narrower is the second, and vice versa. We have discussed already a similar feature while introducing the temporal coherence, see Sect. 2.4.2, where the coherence time is increasing, if the spectrum of a light source becomes narrower.

Let us now take the relations (2.28) and (2.29) from Chap. 2 describing the dependence of the coherence length L_c and the angular frequency width of the light source $\Delta\omega$:

$$\Delta\omega \frac{L_c}{c} = \pi. \tag{A.1.21}$$

We assume that $\Delta\omega$ refers to the full width of the spectrum of the light source at half maximum. With $\omega = kc$ and the differential thereof, we can rewrite (A.1.21):

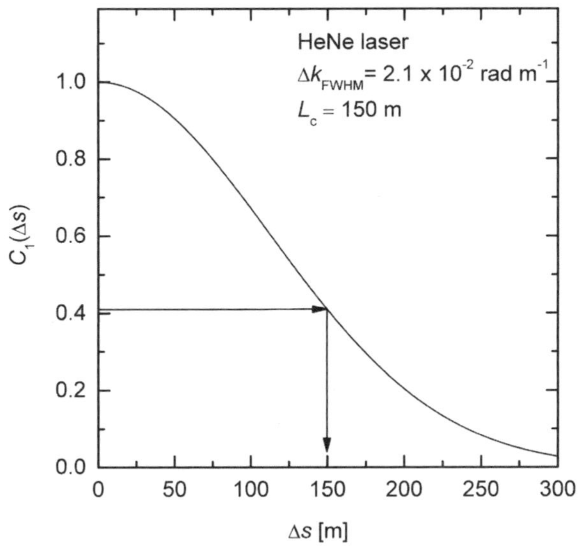

Fig. A.2 Contrast function belonging to the spectrum shown in Fig. A.1

$$\Delta k_{\mathrm{FWHM}} L_{\mathrm{c}} = \pi. \tag{A.1.22}$$

With (A.1.20a) and (A.1.20b) we can calculate the contrast if the path difference Δs is chosen to be just equal to the coherence length L_{c}:

$$C_1(\Delta s = L_{\mathrm{c}}) = \mathrm{e}^{-\frac{L_c^2\sigma^2}{2}} = \mathrm{e}^{-\frac{\pi^2}{16\ln 2}} \approx 0.41. \tag{A.1.23}$$

The coherence length defined with relation (A.1.22) corresponds to a decrease of the contrast to about 0.41. This is a way to measure the coherence length by successively increasing the path difference Δs caused by the two interferometer arms of different lengths until the contrast function reaches the limit given in (A.1.23), then we have $L_{\mathrm{c}} = \Delta s$. For the chosen example, Fig. A.2 indicates the resulting coherence length of about 150 m. For a laser with this coherence length the frequency full width at half maximum is calculated with (A.1.21) and $\Delta\omega = 2\pi\Delta f$: $\Delta f = 1$ MHz. This frequency width divided by the frequency at the maximum of the spectrum amounts to 2.1×10^{-9}. This illustrates the extremely high spectral resolution achievable with a Michelson interferometer.

Instead of taking the right-hand side of (A.1.21) equal to π another value could be chosen, so that e.g. the contrast function yields 0.5 for $\Delta s = L_{\mathrm{c}}$:

$$\Delta\omega \frac{L_{\mathrm{c}}}{c} = 4\ln 2. \tag{A.1.24}$$

This is just a question of definition. With the analogous calculation steps as above, we find with (A.1.20b) and (A.1.24):

$$C_1(\Delta s = L_{\mathrm{c}}) = \mathrm{e}^{-\frac{L_c^2\sigma^2}{2}} = \mathrm{e}^{-\frac{8(\ln 2)^2\sigma^2}{\Delta k_{\mathrm{FWHM}}^2}} = \mathrm{e}^{-\ln 2} = 0.5. \tag{A.1.25}$$

For the chosen example of Fig. A.1 and the right-hand side of (A.1.24) we obtain for the coherence length $L_c = 132$ m.

(b) Contrast function (A.1.18):

The contrast function (A.1.16) for the spectrum (A.1.18) is calculated with (A.1.9), (A.1.10) as follows:

$$C_2(\Delta s) = \frac{1}{2\sqrt{2\pi}\sigma}\left| \int_{-\infty}^{\infty} e^{-\frac{(k'+k_0-k_1)^2}{2\sigma^2}} e^{ik'\Delta s} dk' + \int_{-\infty}^{\infty} e^{-\frac{(k'+k_0-k_2)^2}{2\sigma^2}} e^{ik'\Delta s} dk' \right|. \quad (A.1.26)$$

The first integral can be solved with the help of the substitution $\tilde{k} = k' + k_0 - k_1$:

$$\int_{-\infty}^{\infty} e^{-\frac{(k'+k_0-k_1)^2}{2\sigma^2}} e^{ik'\Delta s} dk' = e^{-i(k_0-k_1)\Delta s} \int_{-\infty}^{\infty} e^{-\frac{\tilde{k}^2}{2\sigma^2}} e^{i\tilde{k}\Delta s} d\tilde{k} = e^{-i(k_0-k_1)\Delta s}\sqrt{2\pi}\sigma\, e^{-\frac{\Delta s^2\sigma^2}{2}}. \quad (A.1.27)$$

In an analogous way we have for the second integral of (A.1.26):

$$\int_{-\infty}^{\infty} e^{-\frac{(k'+k_0-k_2)^2}{2\sigma^2}} e^{ik'\Delta s} dk' = e^{-i(k_0-k_2)\Delta s}\sqrt{2\pi}\sigma\, e^{-\frac{\Delta s^2\sigma^2}{2}}. \quad (A.1.28)$$

Inserting (A.1.27) and (A.1.28) into (A.1.26) then yields:

$$\begin{aligned} C_2(\Delta s) &= \frac{1}{2} e^{-\frac{\Delta s^2\sigma^2}{2}} \left| e^{-i(k_0-k_1)\Delta s} + e^{-i(k_0-k_2)\Delta s} \right| \\ &= \frac{1}{2} e^{-\frac{\Delta s^2\sigma^2}{2}} \left| e^{-ik_0\Delta s} \right| \sqrt{(e^{ik_1\Delta s} + e^{ik_2\Delta s})(e^{-ik_1\Delta s} + e^{-ik_2\Delta s})} \ldots \\ &= \frac{1}{2} e^{-\frac{\Delta s^2\sigma^2}{2}} \sqrt{2(1+\cos((k_2-k_1)\Delta s))} = \frac{1}{2} e^{-\frac{\Delta s^2\sigma^2}{2}} \sqrt{2\cdot 2\cos^2\left(\frac{(k_2-k_1)}{2}\Delta s\right)}. \end{aligned} \quad (A.1.29)$$

So we have finally for the contrast function belonging to (A.1.18):

$$C_2(\Delta s) = e^{-\frac{\Delta s^2\sigma^2}{2}} \left| \cos\left(\frac{(k_2-k_1)}{2}\Delta s\right) \right|. \quad (A.1.30)$$

Let us consider a HeNe laser with a resonator length of L_r, then the frequency spacing between two neighboring longitudinal modes of this laser is given by:

$$\Delta\nu = \frac{c}{2L_r}, \quad (A.1.31)$$

with c speed of light and L_r distance between the two resonator mirrors, cf. Sect. 2.5. As numerical example we take $L_r = 30$ cm and with (A.1.31) we obtain

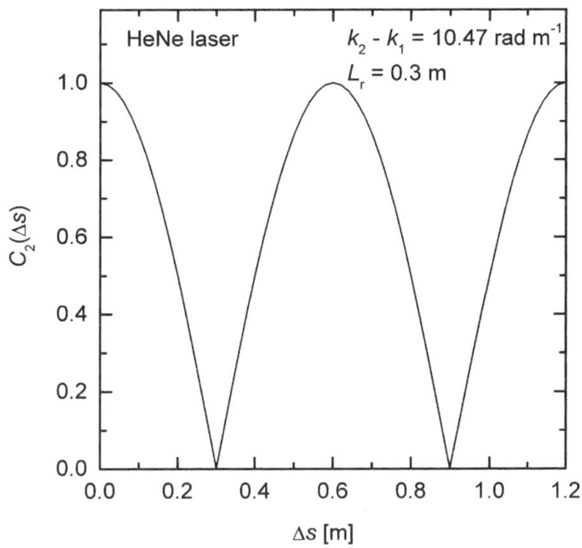

Fig. A.3 Contrast function belonging to the spectrum (A.1.18) of two neighboring axial modes of a HeNe laser

$\Delta\nu = 5 \cdot 10^8$Hz. The distance of the two longitudinal modes expressed with the angular wave numbers according to the definition chosen in (A.1.18) is then:

$$k_2 - k_1 = \frac{2\pi}{c}\Delta\nu = 10.47 \text{ rad m}^{-1}. \tag{A.1.32}$$

The resulting contrast function according to (A.1.30) is shown in Fig. A.3. The contrast varies periodically, a series of minima occur at odd integer multiples of the resonator length L_r. Starting with $\Delta s = 0$ and stepwise increasing Δs the first drop to $C_2(\Delta s) = 0.5$ appears for the following path difference:

$$\Delta s_{0.5} = \frac{2\pi}{3(k_2 - k_1)}. \tag{A.1.33}$$

A.2 Comparison of Spectral Radiance of the Emission of a Black Body and a Laser Source

The idea of this appendix is to provide the reader a numerical example of the significant differences of the spectral radiance emitted of a black body at thermal equilibrium and a laser source. The spectral radiance L_λ describes the emitted radiant flux divided by the unit area, the unit solid angle and the unit wavelength interval. Therefore, the unit of L_λ is W/(m^2 sr m). The spectral radiance of a black body is described by Planck's law [2]:

$$L_\lambda = \frac{2hc^2}{\lambda^5}\frac{1}{e^{hc/\lambda kT} - 1}, \tag{A.2.1}$$

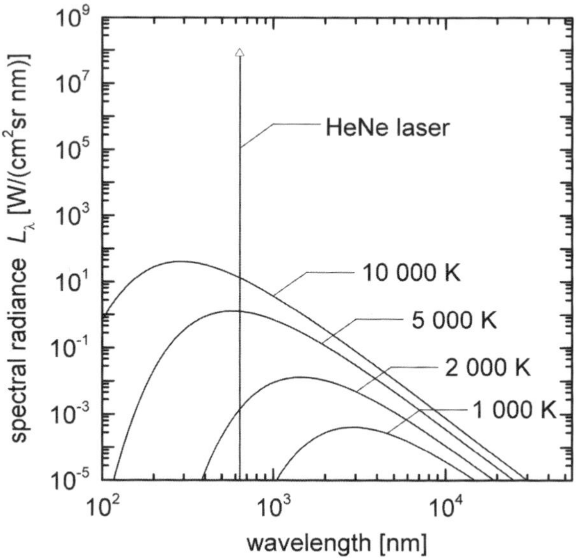

Fig. A.4 Spectral radiance of the radiation of a black body described by Planck's law for temperatures up to 10,000 K and spectral radiance of a HeNe laser with a flux of 1 mW, see text

with h Planck constant, c speed of light, λ wavelength, k Boltzmann constant and T temperature. Figure A.4 shows the spectral radiance of a black body for temperatures ranging from 1,000 to 10,000 K.

For comparison with a laser source we consider a HeNe laser emitting at $\lambda = 633$ nm taking the following exemplary data: flux $\phi = 1$ mW, full divergence angle $\Theta = 1$ mrad, beam diameter $d = 1$ mm, spectral line width $\Delta\nu = 1.5$ GHz. The latter can be transformed to a corresponding wavelength width with the help of the relation $c = \lambda\nu$:

$$|\Delta\lambda| = \frac{\lambda^2}{c}|\Delta\nu|. \tag{A.2.2}$$

Relation (A.2.2) yields then $\Delta\lambda = 2\,\text{pm}$. The solid angle of the laser beam in the far field for a given divergence angle Θ is described by:

$$\Delta\Omega = 2\pi(1 - \cos(\Theta/2)). \tag{A.2.3}$$

This relation follows from the integral:

$$\Delta\Omega = \int_0^{2\pi}\int_0^{\Theta/2} \sin\theta \, \mathrm{d}\theta \, \mathrm{d}\varphi, \tag{A.2.4}$$

where φ is the azimuth angle, θ the polar angle. We can check whether (A.2.3) is plausible by inserting $\Theta = 2\pi$ in (A.2.3) which yields $\Delta\Omega = 4\pi$ sr, i.e. the full solid angle of a sphere.

With the given divergence angle of the laser beam we obtain $\Delta\Omega = 7.85 \times 10^{-7}$sr. The spectral radiance of the laser source is given by:

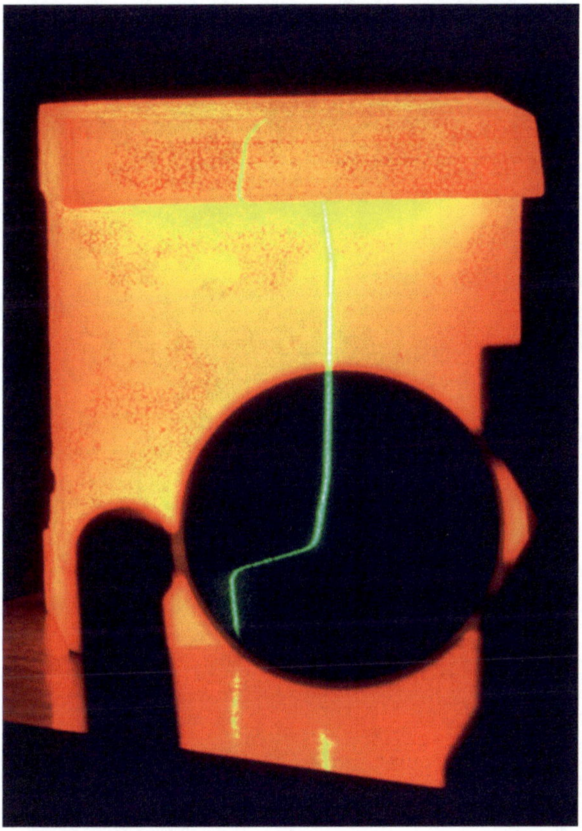

Fig. A.5 Photograph of a hot T-beam section illuminated by a laser light section. The dark black circle is a bandpass interference filter centered at the laser wavelength. The light passing the filter reveals a very high contrast which allows to follow the shape of the laser line even in the case of a strong luminous object

$$L_{\lambda}^{\text{laser}} = \frac{\phi}{\pi\left(\frac{d}{2}\right)^2 \Delta\Omega\Delta\lambda}. \tag{A.2.5}$$

Finally we obtain $L_{\lambda}^{\text{laser}} = 8.1 \times 10^7 \text{W}/(\text{cm}^2\text{sr}\,\text{nm})$, see data point shown as open triangle in Fig. A.4. This demonstrates that the spectral radiance of even a moderate laser with a flux of 1 mW only is by many orders of magnitude higher than the black body radiation calculated for temperatures up to 10,000 K.

Figure A.5 gives an experimental evidence of this significant difference in the spectral radiances. A laser light section, cf. Chap. 10, is projected onto a T-beam section. The scattered light is observed via a bandpass interference filter (black circular disk in Fig. A.5) with a width of about 10 nm centered at the laser wavelength. Behind this filter a very high contrast between the laser light and the black body radiation is observed which mirrors the above stated differences between the spectral radiances of a coherent and an incoherent light source. This unique feature is utilized to measure e.g. the flatness of hot heavy plates with laser triangulation, see Sect. 10.4, Figs. 10.24, 10.25 and 10.26.

A.3 Derivation of Equation (9.23)

We insert the spectrum of the light source (9.21) in (9.18) and integrate over the angular wave numbers to gain the quantity I, which we need to describe the signal for time-domain OCT, cf. (9.19):

$$I(z_r) = \int_0^\infty \left(1 + \sum_{j=1}^N R_{m,j}\right) \frac{I_0}{\sqrt{\pi}\Delta k} e^{-\left[\frac{k-k_0}{\Delta k}\right]^2} dk + \cdots$$

$$\cdots + \int_0^\infty 2\sum_{j=1}^N \sqrt{R_{m,j}} \frac{I_0}{\sqrt{\pi}\Delta k} e^{-\left[\frac{k-k_0}{\Delta k}\right]^2} \cos\left(2k\left(z_r - z_{m,j}\right)\right) dk. \quad \text{(A.3.1)}$$

The autocorrelation term is neglected, cf. third term with double sum in (9.18). The first integral in (A.3.1) is easily evaluated by use of (9.22):

$$A = \left(1 + \sum_{j=1}^N R_{m,j}\right) I_0. \quad \text{(A.3.2)}$$

As abbreviation for the second integral to be solved we use:

$$B = \int_0^\infty \frac{I_0}{\sqrt{\pi}\Delta k} e^{-\left[\frac{k-k_0}{\Delta k}\right]^2} \cos\left(2k\left(z_r - z_{m,j}\right)\right)\, dk. \quad \text{(A.3.3)}$$

With the substitution $k' = k - k_0$, and the abbreviation $\delta z_j = z_r - z_{m,j}$ we obtain:

$$B = \int_{-k_0}^\infty \frac{I_0}{\sqrt{\pi}\Delta k} e^{-\left[\frac{k'}{\Delta k}\right]^2} \cos\left((k' + k_0)2\delta z_j\right)\, dk'. \quad \text{(A.3.4)}$$

Assuming $k_0 > 0$ and $k_0 >> \Delta k$ the integrand of (A.3.4) has only significant positive values close to the origin of the k' axis. Since we assume that the width of this Gaussian function is small compared to k_0, we can approximate the integral of (A.3.4) by extending the lower integration limit from "$-k_0$" to minus infinity:

$$B \approx \int_{-\infty}^\infty \frac{I_0}{\sqrt{\pi}\Delta k} e^{-\left[\frac{k'}{\Delta k}\right]^2} \cos\left((k' + k_0)2\delta z_j\right)\, dk'. \quad \text{(A.3.5)}$$

With the trigonometric addition theorem $\cos(x + y) = \cos x \cos y - \sin x \sin y$, we have:

$$B \approx \frac{I_0}{\sqrt{\pi}\Delta k} \int_{-\infty}^\infty e^{-\left[\frac{k'}{\Delta k}\right]^2} \left[\cos\left(k'2\delta z_j\right) \cos\left(k_0 2\delta z_j\right) - \sin\left(k'2\delta z_j\right) \sin\left(k_0 2\delta z_j\right)\right]\, dk'.$$

(A.3.6)

The second summand is a point symmetric function with respect to k', whereas the Gaussian function is a symmetric function with respect to k' and the ordinate. Therefore, the integral of the product of the second summand and the Gaussian function becomes zero. So from (A.3.6) we come to:

$$B \approx \frac{I_0 \cos\left(k_0 2\delta z_j\right)}{\sqrt{\pi}\Delta k} \int\limits_{-\infty}^{\infty} \mathrm{e}^{-\left[\frac{k'}{\Delta k}\right]^2} \cos\left(k' 2\delta z_j\right)\, \mathrm{d}k'. \tag{A.3.7}$$

From [1] the following definite integral is known:

$$\int\limits_{-\infty}^{\infty} \mathrm{e}^{-a^2 x^2} \cos bx \;\, \mathrm{d}x = \frac{\sqrt{\pi}}{a} \mathrm{e}^{-\frac{b^2}{4a^2}}. \tag{A.3.8}$$

With $a = 1/\Delta k, b = 2\delta z_j$ and $x = k'$, we see that (A.3.8) is equivalent to (A.3.7) and finally get for (A.3.7):

$$B \approx \frac{I_0 \cos\left(k_0 2\delta z_j\right)}{\sqrt{\pi}\Delta k} \sqrt{\pi}\Delta k \mathrm{e}^{-\delta z_j^2 \Delta k^2} = I_0 \cos\left(k_0 2\delta z_j\right) \mathrm{e}^{-\delta z_j^2 \Delta k^2}. \tag{A.3.9}$$

Taking this result (A.3.9), we have now solved (A.3.1) and can write with resubstitution of the used abbreviations:

$$I(z_\mathrm{r}) = I_0 \left(1 + \sum_{j=1}^{N} R_{\mathrm{m},j}\right) + 2I_0 \sum_{j=1}^{N} \sqrt{R_{\mathrm{m},j}} \cos\left(2k_0\left(z_\mathrm{r} - z_{\mathrm{m},j}\right)\right) \mathrm{e}^{-\left(z_\mathrm{r} - z_{\mathrm{m},j}\right)^2 \Delta k^2}. \tag{A.3.10}$$

Relation (A.3.10) is identical to (9.23).

A.4 Derivation of Equation (9.26)

To solve the integral (9.26) we introduce a substitution $k' = k - k_0$:

$$\tilde{f} = \frac{\eta}{2\pi} \int\limits_{-\infty}^{+\infty} \frac{I_0}{\sqrt{\pi}\Delta k} \mathrm{e}^{-\left[\frac{(k-k_0)}{\Delta k}\right]^2} \mathrm{e}^{\mathrm{i}kz} \mathrm{d}k = \frac{\eta I_0 \mathrm{e}^{\mathrm{i}k_0 z}}{2\pi\sqrt{\pi}\Delta k} \int\limits_{-\infty}^{+\infty} \mathrm{e}^{-\left[\frac{k'}{\Delta k}\right]^2} \mathrm{e}^{\mathrm{i}k' z} \mathrm{d}k'. \tag{A.4.1}$$

Euler's formula says:

$$\mathrm{e}^{\mathrm{i}k' z} = \cos(k' z) + \mathrm{i}\, \sin(k' z). \tag{A.4.2}$$

With the help of (A.4.2) the definite integral in (A.4.1) can be decomposed as follows:

$$\int\limits_{-\infty}^{+\infty} \mathrm{e}^{-\left[\frac{k'}{\Delta k}\right]^2} \mathrm{e}^{\mathrm{i}k' z} \mathrm{d}k' = \int\limits_{-\infty}^{+\infty} \mathrm{e}^{-\left[\frac{k'}{\Delta k}\right]^2} \cos(k' z) \mathrm{d}k' + \mathrm{i} \int\limits_{-\infty}^{+\infty} \mathrm{e}^{-\left[\frac{k'}{\Delta k}\right]^2} \sin(k' z) \mathrm{d}k'. \tag{A.4.3}$$

The first term on the right handside of (A.4.3) is a symmetric function of z (symmetric about the y-axis), whereas the second term is a point symmetric function with respect to z (symmetric about the origin). Hence, the second term on the right handside of (A.4.3) is zero. The first term is a known definite integral [1]:

$$\int_{-\infty}^{+\infty} \mathrm{e}^{-\left[\frac{k'}{\Delta k}\right]^2} \cos(k'z)\mathrm{d}k' = 2\int_{0}^{+\infty} \mathrm{e}^{-\left[\frac{k'}{\Delta k}\right]^2} \cos(k'z)\mathrm{d}k' = \sqrt{\pi}\Delta k \mathrm{e}^{-\frac{z^2\Delta k^2}{4}} \tag{A.4.4}$$

for $\Delta k > 0$. Inserting (A.4.4) in (A.4.1) then yields:

$$\tilde{f}(z) = \frac{\eta I_0 \mathrm{e}^{\mathrm{i}k_0 z}}{2\pi\sqrt{\pi}\Delta k}\sqrt{\pi}\Delta k \mathrm{e}^{-\frac{z^2\Delta k^2}{4}} = \frac{\eta I_0 \mathrm{e}^{\mathrm{i}k_0 z}}{2\pi}\mathrm{e}^{-\frac{z^2\Delta k^2}{4}} \tag{A4.5}$$

Relation (A.4.5) is identical to (9.26).

A.5 Derivation of Equation (9.27)

We start with (9.25), insert (9.20) and for I_k (9.18):

$$\tilde{S}(z) = \frac{\eta}{2\pi}\left\{\int_{-\infty}^{\infty} I_{0,k}\left(1+\sum_{j=1}^{N} R_{\mathrm{m},j}\right)\mathrm{e}^{\mathrm{i}kz}\mathrm{d}k + \int_{-\infty}^{\infty} I_{0,k}2\sum_{j=1}^{N}\sqrt{R_{\mathrm{m},j}}\cos\left(k2\left(z_\mathrm{r}-z_{\mathrm{m},j}\right)\right)\mathrm{e}^{\mathrm{i}kz}\mathrm{d}k\right\}. \tag{A.5.1}$$

Relation (A.5.1) is decomposed in two summands. For the first summand we have with (9.21) and (9.26):

$$\tilde{S}_1(z) = \frac{\eta}{2\pi}\left(1+\sum_{j=1}^{N} R_{\mathrm{m},j}\right) I_0 \mathrm{e}^{\mathrm{i}k_0 z}\mathrm{e}^{-\frac{z^2\Delta k^2}{4}} = \left(1+\sum_{j=1}^{N} R_{\mathrm{m},j}\right)\tilde{f}(z). \tag{A.5.2}$$

For the second summand we obtain after insertion of (9.21) and the abbreviation $\Delta z_j = 2\left(z_\mathrm{r} - z_{\mathrm{m},j}\right)$:

$$\tilde{S}_2(z) = \frac{2\eta}{2\pi}\sum_{j=1}^{N}\sqrt{R_{\mathrm{m},j}}\frac{I_0}{\sqrt{\pi}\Delta k}\int_{-\infty}^{\infty} \mathrm{e}^{-\frac{(k-k_0)^2}{\Delta k^2}}\cos\left(k\Delta z_j\right)\mathrm{e}^{\mathrm{i}kz}\mathrm{d}k. \tag{A.5.3}$$

We abbreviate the definite integral with the letter B:

$$B = \int_{-\infty}^{\infty} \mathrm{e}^{-\frac{(k-k_0)^2}{\Delta k^2}}\cos\left(k\Delta z_j\right)\mathrm{e}^{\mathrm{i}kz}\mathrm{d}k. \tag{A.5.4}$$

With the substitution $k' = k - k_0$ follows from (A.5.4):

$$B = \int_{-\infty}^{\infty} \mathrm{e}^{-\frac{k'^2}{\Delta k^2}}\mathrm{e}^{\mathrm{i}k'z}\mathrm{e}^{\mathrm{i}k_0 z}\cos\left[(k'+k_0)\Delta z_j\right]\mathrm{d}k'. \tag{A.5.5}$$

With Euler's formula (A.4.2) this can be rewritten and decomposed in two summands:

$$B = \frac{e^{ik_0 z}}{2} \int_{-\infty}^{\infty} e^{-\frac{k'^2}{\Delta k^2}} e^{ik'z} \left[e^{i(k'+k_0)\Delta z_j} + e^{-i(k'+k_0)\Delta z_j} \right] dk', \tag{A.5.6}$$

$$B = \frac{e^{ik_0 z}}{2} (b_1 + b_2). \tag{A.5.7}$$

The summand b_1 can be converted as follows:

$$b_1 = e^{ik_0 \Delta z_j} \int_{-\infty}^{\infty} e^{-\frac{k'^2}{\Delta k^2}} e^{ik'(z+\Delta z_j)} dk'. \tag{A.5.8}$$

With Euler's formula (A.4.2) we see, that the integrand is again a sum of a symmetric and a point symmetric function, hence the integral over the second summand vanishes and we obtain:

$$b_1 = e^{ik_0 \Delta z_j} \int_{-\infty}^{\infty} e^{-\frac{k'^2}{\Delta k^2}} \cos\left(k'(z + \Delta z_j)\right) dk'. \tag{A.5.9}$$

This integral is known, cf. (A.3.8) and (A.4.4):

$$b_1 = e^{ik_0 \Delta z_j} \sqrt{\pi} \Delta k e^{-\frac{(z+\Delta z_j)^2 \Delta k^2}{4}}. \tag{A.5.10}$$

Analogously we have for the second summand:

$$b_2 = e^{-ik_0 \Delta z_j} \sqrt{\pi} \Delta k e^{-\frac{(z-\Delta z_j)^2 \Delta k^2}{4}}. \tag{A.5.11}$$

Now we insert these summands into relation (A.5.7):

$$B = \frac{e^{ik_0 z}}{2} \sqrt{\pi} \Delta k \left[e^{ik_0 \Delta z_j} e^{-\frac{(z+\Delta z_j)^2 \Delta k^2}{4}} + e^{-ik_0 \Delta z_j} e^{-\frac{(z-\Delta z_j)^2 \Delta k^2}{4}} \right]. \tag{A.5.12}$$

This can be rewritten with (A.4.5):

$$B = \frac{\sqrt{\pi} \Delta k}{2} \left[\frac{2\pi \tilde{f}(z + \Delta z_j)}{\eta I_0} + \frac{2\pi \tilde{f}(z - \Delta z_j)}{\eta I_0} \right]. \tag{A.5.13}$$

Going back to (A.5.3) and cancelling some terms we get:

$$\tilde{S}_2(z) = \sum_{j=1}^{N} \sqrt{R_{m,j}} \left(\tilde{f}(z + \Delta z_j) + \tilde{f}(z - \Delta z_j) \right). \tag{A.5.14}$$

Inserting (A.5.2) and (A.5.14) into (A.5.1) then finally yields:

$$\tilde{S}(z) = \tilde{f}(z)\left(1 + \sum_j R_{\mathrm{m},j}\right) + \left(\sum_j \sqrt{R_{\mathrm{m},j}}\left(\tilde{f}(z + \Delta z_j) + \tilde{f}(z - \Delta z_j)\right)\right). \quad \text{(A.5.15)}$$

The expression (A.5.15) is identical to (9.27).

References

1. I. Bronstein, K. Semendjajew, *Taschenbuch der Mathematik*, 19. Aufl. (Verlag Harri Deutsch, Thun Frankfurt, 1980), p. 118
2. A. Anders, *A Formulary for Plasma Physics* (Akademie-Verlag, Berlin, 1990)

Abbreviations

Abbreviation	Explanation	Chapter(s), Figure(s), Table(s)	Page numbers
A-, B-, C-scan	Depth scan, 2D scan, 3D scan	9.1, Fig. 9.2	228
ADC	Analog digital converter	14.6	382
BD	Blu-ray disc	12.3.1	315
BOXCARS	Variant of CARS, cf. Fig. 13.31	13.3.2	362
CD	Compact disc	12.3.1	314
DC term	Undiffracted reference wave	7.4.1	185
CARS	Coherent anti-Stokes Raman spectroscopy	13.3	355
CLSM	Confocal laser scanning microscope	12	305
CCD	Charge-coupled device	5.4.5, 10.2.4, Table 10.1, Fig. 5.38	135, 260, 261
CMOS	Complementary metal oxide semiconductor	10.2.4, Table 10.1	260, 261
CRM	Certified reference material	13.1.6	336
DON	Deoxynivalenol	14.7	388
DVD	Digital versatile disc	12.3.1	314
E-CARS	CARS in backward direction	13.3.4	368
ESPI	Electronic speckle interferometry	8.3.3, Fig. 8.11	213, 214
F-CARS	CARS in forward direction	13.3.4	367
FD-OCT	Fourier domain OCT	9.1.2, Fig. 9.19	235, 244
FLIM	Fluorescence lifetime imaging	14.6	382
FPS	Fluorescence polarization spectroscopy	14.5, Fig. 14.7	380, 381
FWHM	Full width at half maximum	12.1.2	307
IR	Infrared	5.4.3	131
LAM	Laser additive manufacturing	1	2
LDA	Laser Doppler anemometer	11.3, Fig. 11.3	292
LIDAR	Light detection and ranging	13.2, Fig. 13.17	343, 344
LIBS	Laser-induced breakdown spectroscopy	13.1, Fig. 13.1	324

(continued)

(continued)

Abbreviation	Explanation	Chapter(s), Figure(s), Table(s)	Page numbers
MEMS	Microelectromechanical systems	9.3	243
MESF	Molecules of equivalent soluble fluorophore	14.7	389
MOS	Metal-oxide-semiconductor	5.4.5, Fig. 5.37	134
MOSFET	Metal-oxide-semiconductor field effect transistor	5.4.5	134
NA	Numerical aperture	12.1.1, (12.1), 12.1.2, Table 12.1	306, 308, 315
OCT	Optical coherence tomography	9, Fig. 9.1	227
OPO	Optical parametric oscillator	13.3.1	360
PDA	Photodiode array	5.4.5	135
PET	Polyethylene terephthalate	9.3, Fig. 9.19	244
PIN	Positive-instrinsic-negative photodiode	5.3.2	121
PSF	Point-spread function	12.1.2	307
ROI	Region of interest	10.2.4	262
ptv	Peak-to-valley	7.5.2, Fig. 7.22	193
SEM	Scanning electron microscopy	12.3.1	315
TCSPC	Time correlated single photon counting	14.6, Fig. 14.8	382
TD-OCT	Time-domain optical coherence tomography	9.1.1, Fig. 9.3	229
UV	Ultraviolet	5.4.3, 9.3	131, 244
VCO	Voltage controlled oscillator	11.3.2, Fig. 11.16	296
VIS	Visible	5.4.3	131
2-D	Two dimensional	10.1, Fig. 10.2	249
3-D	Three dimensional	10.1, Fig. 10.2	249

Notations and list of symbols

The following notation rules are used in this book:

a) physical quantities are written with italic characters to visualize that these quantities have a unit, example: P, radiant flux; unit: Watt,
b) mathematical variables are also written with italic characters to clearly distinguish these from units or numbers, example: $j = 1, 2, \ldots$,
c) numbers including the imaginary unit $\mathrm{i} = \sqrt{-1}$, and the Euler number e are set with non-italic characters; mathematical functions are also set with non-italic characters, example J_0, zeroth order Bessel function,
d) units are set with non-italic characters to clearly distinguish these form the cases a), b); example: m, meter,
e) abbreviations for objects or denominations of points are set with non-italic characters, examples: r, stands for reference beam or P, denotes a point,
f) differentials are written with non-italic characters followed by a physical quantity written with italic character, example: $\mathrm{d}x$.

Symbol	Description	Unit	Chapter(s), Figure(s), Formula(s)
a	Typical dimension of diffracting aperture	m	2.3, (2.16)
$A(\tau)$	Autocorrelation function	$[s(t)]^2$	11.3.2, (11.38)
A	Absorptance	1	5.1, (5.1), 13.1.3, (13.8),
A	Resolution	1	6.2.4
A_{ji}	Einstein coefficient for spontaneous emission for the transition from the upper level j to the lower level i	s^{-1}	13.1.3, (13.8)
α	Absorption coefficient	m^{-1}	3.3.1, (3.12), (7.7)
α	Angle of incident laser beam	rad	2.2.3, (2.10), 10.2.1, (10.9)
α	Temperature coefficient	1/K	5.1.2, (5.18)
α_x	Angle of deflection in the x-direction	rad	10.2.5, (10.35)
(α)	Specific rotation	rad/m	4.1.2, (4.6)
b	Image distance	m	8.2.1, (8.3)
b_λ	spectral intensity of the background radiation	W/m	13.1.5, (13.12)
B	Size of the image	m	8.2.1, (8.4), 10.1.1
B_ω	Angular spectrum bandwidth	rad s^{-1}	11.2.2, (11.18)
β	Opening angle of the objective	rad	10.2.4, (10.31)
$\beta(r,t)$	Scattering coefficient per solid angle per path length	$m^{-1}sr^{-1}$	13.2.1, (13.17)
c	Speed of light	m/s	2.1, 3.2, (2.4), (3.4)
c	Specific heat capacity	J/(kg K)	13.1.2, (13.3)
C_Q	Heat capacity	J/K	5.1, (5.1)
$\chi^{(1)}$	Electrical susceptibility of first order	As/(Vm)	13.3.1
d_{fp}	Displacement in the film plane	m	8.2.1, (8.5)
d_{op}	Displacement in the object plane	m	8.2.1, (8.6)
d_{Plasma}	Plasma diameter	M	13.1.4, (13.10)
d_{sp}	Speckle size	m	8.2.1, (8.7)
d_x	Object translation within the object surface	m	8.3.2, Table 8.1
d_z	Object translation perpendicular to the object surface	m	8.3.2, Table 8.1
$\vec{d}$	Displacement vector	m	8.3.2, (8.26), (8.28)
$d\Omega_{det}$	Solid angle of receiving optic	sr	13.2.1, (13.18)
D	Thickness of hologram, diameter of aperture	m	7.1, (7.7), 8.1, (8.2), Fig. 8.6, 10.2.4, (10.11)
D	Optical density	1	5.4.1, (5.55)
D_c	Diameter of circle enclosing the interference pattern in the observation plane	m	8.2.1, Fig. 8.4
D_{ph}	Pinhole diameter	m	12.1.2, (12.8)

(continued)

(continued)

Symbol	Description	Unit	Chapter(s), Figure(s), Formula(s)
δ_h	Heat penetration depth	m	13.1.1
δ_{opt}	Optical penetration depth	m	13.1.1
$\delta(z_m - z_{m,j})$	Dirac function	m^{-1}	9.1.1, (9.11)
$\vartheta_{m,r}(x, y)$	Phase of scattered waves in the observation plane	rad	8.3, (8.21)
$\varDelta$	Path difference	m	2.3.1, (2.19)
$\Delta\phi$	Phase difference	rad	8.2.2, (8.17)
$\Delta\Omega_c$	Solid coherence angle	sr	2.4.3, (2.33)
ΔN_{sp}	Fluctuation of number of speckles per pixel	1	10.2.4
Δp	Width of pixel	m	10.2.4, Fig. 10.15
Δr_{min}	Lateral resolution	m	12.1.2, (12.5)
Δt	Temporal width	s	9.1.1, (9.1)
Δy	Height of pixel	m	10.2.4, Fig. 10.15
Δz	Uncertainty in the determination of the light spot on the z-axis	m	10.2.1, (10.9), Fig. 10.7, 10.2.4, (10.32)
$\Delta z'_{cg}$	Measuring uncertainty in the determination of the position of the light spot in the detector plane	m	10.2.4, (10.25), (10.29)
e	elementary charge	C	5.1.3, (5.50)
$\vec{e}_i$	Unit vector describing the direction of the incident wave	1	11.1
$\vec{e}_o$	Unit vector describing the direction of the scattered wave	1	11.1
$\vec{E}$	Electric field strength vector	V/m	2.1, (2.1a)
$\vec{E}_0$	Electric field strength amplitude	V/m	2.1, (2.1a)
$E_{m,k}, E_{r,k}$	Complex spectral field strength in the measuring arm, reference arm	V/rad	9.1.1, (9.10)
E_o	Complex amplitude of the electric field of the object wave	V/m	7, (7.1)
E_o^d	Complex amplitude of the electric field of the object wave of the displaced object	V/m	Fig. 7.5, (7.8)
E_r	Complex amplitude of the electric field of the reference wave	V/m	7, (7.1), 9.1.1, (9.5)
ε	Slope of the linear part of an amplitude transmission curve	$m^2/(Ws)$	7, (7.2)
ε_{AB}	Seebeck coefficient	V/K	5.1.1, (5.16)
ε_0	Permittivity of free space	$AsV^{-1}m^{-1}$	3.4.3, (3.20a)
$\varepsilon_{ji}^{\nu,d\Omega}(t)$	Line emission coefficient of the transition $j \rightarrow i$	$W/(m^3$ sr Hz)	13.1.4, (13.9)
η	Modulation index	1	11.2.2, (11.7)

(continued)

(continued)

Symbol	Description	Unit	Chapter(s), Figure(s), Formula(s)
η	Quantum efficiency	1	5.3.1, (5.43)
F	Area	m^2	5.1.3, (5.25)
f, f_F	Focal length	m	7.1, Fig. 7.3, 8.2.1, (8.3), 10.1.1, (10.2), 10.2.1, (10.6)
f	Frequency	1/s	2.1, (2.4), 5.1.3, (5.50)
φ_0	Phase angle	rad	2.1 (2.1)
ϕ	Phase of the object wave	rad	7.1, (7.8)
ϕ_i	Phase shifts	rad	7.3.3, (7.19)
ϕ_{sp}	Size of speckles	m	10.2.4, (10.11)
g	Object distance	m	8.2.1, (8.3)
G	Size of the object	m	8.2.1, (8.4), 10.1.1
$\Gamma(v)$	Line profile as a function of the frequency	1/Hz	13.1.4, (13.9)
h	Planck constant	Js	3.1, 13.1.4, (13.9)
H	Brightness on the monitor	a.u.	8.3.3, (8.50)
H	Radiant exposure	J/m^2	5.4.1, (5.56)
$\vec{H}$	Magnetic field strength vector	A/m	2.1, (2.1b)
$\vec{H}_0$	Magnetic field strength amplitude	A/m	2.1, (2.1b)
i_{cg}	Center of gravity of the counter variable i	1	10.2.4, (10.12)
I	Intensity, irradiance	W/m^2	2.1, (2.6), 6.1.1, (6.5b)
I	Current	A	5.1.3, (5.29)
I_j	Intensity of an element line of an analyte j	a.u.	13.1.1
I_{ph}	Photovoltaic current	A	5.3.2, (5.50)
I_r	Intensity of a reference line	a.u.	13.1.1
$I(r)$	Intensity profile	W/m^2	10.2.2
$I(x, y)$	Irradiance as function of the coordinates x, y	W/m^2	7, (7.1)
$I'(z')$	Intensity distribution of the imaged light spot in the plane of detection	W/m^2	10.2.2
I_k	Spectral irradiance	W/(m rad)	9.1.1, (9.16a)
J_0	Zeroth order Bessel function	1	8.3.4, (8.57)
J_k	Line intensity of species k	a.u.	13.1.5, (13.13)
k	Angular wave number	rad m^{-1}	2.1, (2.3)
$\vec{k}_1, \vec{k}_2; \vec{k}, \vec{k}'$	Angular wave vectors of incident and scattered light	rad m^{-1}	7.1, (7.16), 8.3.2, (8.26)
K	Number of beam quality	1	2.6.3, (2.39)
K	Kerr's constant	m/V^2	3.2.4, (3.10)
κ	Constant	1	10.2.4, (10.29)
κ	Thermal diffusivity	m^2/s	13.1.2
$\vec{l}$	Displacement vector	m	7.1, (7.16)
L	Distance between mirrors	m	2.5, (2.34)

(continued)

(continued)

Symbol	Description	Unit	Chapter(s), Figure(s), Formula(s)
L	Distance between scattering surface and observation screen	m	2.3, (2.16), 8.1, (8.1)
L_c	Coherence length	m	2.4 (2.28), 9.1.1, (9.1)
λ	Wavelength	m	2.1, (2.3), 8.1, (8.1)
Λ	Wavelength	m	4.1.1, (4.4)
m_L	Slope	1	10.1.1, (10.1), Fig. 10.4
M_φ	Angle enlargement	1	7.1
M_z	Enlargement in the z-direction	1	7.1
M_z	Measuring range in z-direction	m	10.2.1, (10.8)
M	Ray transfer matrix	N/A	4.2.1, (4.22)
M	Number of modes	1	4.3.1, (4.64)
M^2	Beam propagation ratio	1	2.6.3, (2.42)
$\mu(\lambda, R)$	Spatial dependent extinction coefficient	m^{-1}	13.2.1, (13.15)
n	Integer	1	7.1, (7.11)
n	Refractive index	1	3.2.2, 7.1, (7.7), 10.2.5, (10.35)
N	Resolving power	1	10.2.1, (10.10)
NA	Numerical aperture	1	12.1.2, (12.4)
N_a^z	Density of species a with charge z	m^{-3}	13.1.4, (13.9)
N_{sp}	Number of speckles per pixel	1	10.2.4
ν_{ji}	Frequency of the transition from $j \rightarrow i$	Hz	13.1.4, (13.9)
ω	Angular frequency of oscillation	rad s^{-1}	2.1, (2.1), 8.3.4, (8.53)
p	Momentum	kg m s^{-1}	3.1, (3.2)
p	Pyroelectric coefficient	C m^{-2} K^{-1}	5.1.3, (5.28)
$\vec{P}$	Electrical polarization	As/(m^2)	3.5, (3.20a), 13.3.1, (13.32)
P	Power	W	5.1, (5.1)
P	Radiant flux	W	5.3.2, (5.50)
q	Natural number	1	2.5, (2.34)
q	Complex beam Parameter	m	4.2.1, (4.18)
Q	Charge	C	5.2.1, (5.36)
Q_i	Radiant energy received by pixel element i	J	10.2.4, (10.12)
r	Electro-optical coefficient	m/V	3.2.4, (3.11)
r_m	Reflection coefficient of the measuring object	1	9.1.1, (9.6)
R	Distance between light source and diffracting aperture	m	2.3, (2.16)
R	Reflectivity, reflectance	1	3.2.2, (3.6), 4.1.1, (4.2) 13.1.2, (13.4)
R	Resistance	Ω	5.1.2, (5.18)

(continued)

(continued)

Symbol	Description	Unit	Chapter(s), Figure(s), Formula(s)
$R_{m,j}$	Reflectivity in the measurement path at location j	1	9.1.1, (9.17)
R_Q	Thermal resistance	$K/(Wm^2)$	5.1, (5.2a)
r_b	Beam radius	m	13.1.2, (13.2)
$r_{m,z_m}(z_m)$	Reflection coefficient in the measurement path per unit length	m^{-1}	9.1.1, (9.11)
ρ	Radius of curvature of the phase surfaces	m	2.6, (2.40)
ρ	Density	kg/m^3	13.1.2, (13.3)
$s(t)$	Local enlongation	m	8.3.4, (8.53)
s_{cg}	Standard deviation of i_{cg}	1	10.2.4, (10.19)
s_i	Distance between objective and image plane	m	10.2.4, (10.11)
s_{opt}	Optical path length	m	4.1.4, (4.15)
S	Spin	Js	3.1, (3.3)
S_k	Signal of Fourier-domain OCT depending on the angular wavenumber at constant position of the reference mirror	a.u.	9.1.1, (9.20), 9.1.2
S_λ	Spectral intensity	W/m	13.1.5, (13.11)
$S(z_r)$	Signal of time-domain OCT as a function of the position of the reference mirror	a.u.	9.1.1, (9.19)
$\tilde{S}(z)$	Inverse Fourier transform of S_k	a.u.	9.1.2, (9.25), (9.27)
$\vec{S}$	Poynting vector	W/m^2	6.1.1
$\sigma_i(\lambda)$	Absorption cross-section of the gas i at the wavelength λ	m^2	13.2.1, (13.28)
Σ	Sensitivity	K/W	5.1, (5.7)
t	Complex amplitude transmission	1	(7.7)
t	Time	s	
T	Transmittance	1	3.3.2, (3.15)
T	Integration time of the video camera	s	8.3.4, (8.56)
T	Temperature	K	5.1, (5.1), 13.1.2, (13.3)
T_0	Transmittance at low intensities	1	3.3.2, (3.15)
T_b	Beat period	s	2.2, (2.13), 11.2.1, (11.12)
T_c	Coherence time	s	2.4.1, (2.28), 9.1.1, (9.2a)
τ	Exposure time	s	5.4.1, (5.56), 7, (7.2)
τ_D	Time constant of the detector	s	13.2.1, (13.21)
τ_L	Duration of the laser pulse	s	13.2.1, (13.21)
θ_0	Opening angle of Gaussian beam	rad	2.6.1, (2.39)
θ_B	Brewster angle	rad	3.2.2, (3.7)
θ_t	Critical angle of total reflection	rad	3.2.2, (3.8)

(continued)

(continued)

Symbol	Description	Unit	Chapter(s), Figure(s), Formula(s)
u, v	Coordinates	m	10.1.1, (10.1)
$u_a^z(T)$	Partition function of species *a* with charge *z*	1	(13.1.4), (13.9)
U	Voltage of video signal	V	8.3.3, (8.48)
$U_{\lambda/2}$	Half-wave voltage	V	4.1.3, (4.10)
v_R	Velocity of the reference mirror	m/s	9.1.1, (9.1)
v_{th}	Average thermal speed of the atoms and ions in the plasma	m/s	13.1.4, (13.10)
$\vec{v}$	Velocity vector	m/s	11.1, (11.5)
V	Verdets's constant	rad/A	4.1.2, (4.7)
w, w_0	Waist radius of laser beam	m	2.6.1, (2.37a), 4.2.1, (4.19), 10.2.1
W	Photon energy	J	3.1, (3.1)
W_W	Work function	J	5.1.3, (5.33)
z_m	Position of measuring object	M	9.1.1, (9.6)
z_r	Position of reference mirror	m	9.1.1, (9.5)

(continued)

Index

A. Donges and R. Noll, *Laser Measurement Technology*, Springer Series in Optical Sciences 188, DOI: 10.1007/978-3-662-43634-9,

D

E

F

G

H

Printed by Printforce, the Netherlands